D0933594

Barriers and Bridges to the Renewal of Ecosystems and Institutions

Barriers and Bridges to the Renewal of Ecosystems and Institutions

Lance H. Gunderson,
C. S. Holling, and
Stephen S. Light
Editors

COLUMBIA UNIVERSITY PRESS
NEW YORK

Columbia University Press
New York   Chichester, West Sussex

Library of Congress Cataloging-in-Publication Data
Gunderson, Lance H.
Barriers and bridges to the renewal of ecosystems and institutions /
Lance H. Gunderson, C. S. Holling, and Stephen S. Light.
p. cm.
Includes bibliographical references and index.
ISBN 0-231-10102-3
1. Environmental policy. 2. Environmental policy—Case studies.
3. Social ecology. I. Holling, C. S. II. Light, Stephen S.
GE170.G86 1995
333.95′153—dc20   94-35061
CIP

∞

Casebound editions of Columbia Universtiy Press books are printed on
pernament and durable acid-free paper.

Printed in the United States of America
c 10 9 8 7 6 5 4 3 2 1
p 10 9 8 7 6 5 4 3 2 1

I am the love that initiates and the truth that passes away.
All that compels acceptance and all that brings renewal;
all that breaks apart and all that binds together;
power, experiment, progress, matter:
all this am I.

Hymn of the Universe
Teilhard de Chardin

Contents

Preface

This book explores ways for active adaptation and learning in dealing with uncertainty in the management of complex regional ecosystems. We present the book as twelve chapters grouped in four sections. The first section consists of the introductory chapter, where a common pattern of resource development is presented and diagnosed, and an emerging theory is suggested to explain that pattern. The heart of the book is in the next section, where six case studies of regional ecosystem management are presented in chapters 2–7. The goal was to test and expand the diagnosis of the first chapter and to refine and extend theory. This set of case studies is perhaps unique, because each case was explored by one or two individuals with considerable scholarship, direct experience in the policies and politics of the region, and the ability to communicate to a wide audience. The case studies are followed by a third section of four chapters written by social scientists who comment on the case studies and the emerging theories from a perspective of their own experience and expertise. These perspectives include political science, Third World development, learning theory and practice, and institutional psychology. We conclude with the fourth section, a chapter that presents a synthesis of social science theory and ecological theory to explain the observed patterns of frustrating resistance of institutions to change and the sudden lurches of learning that can occur.

How the Book Was Born

The book is the result of a three-year research project established to deal with two key questions: (1) Do institutions learn, and if so how? and (2) How do ecosystems respond to management actions? An international team of scholar/practitioners was organized to begin a series of comparisons of regional development experience. This was given focus by posing the series of postulates described in the introduction that seemed to represent common patterns observed over time in managed ecosystems. The postulates describe a progression of ecosystems that become less resilient, management agencies that become more efficient but more myopic, industries that become more dependent and static, and a public that loses trust. This can lead to a crisis triggered by unexpected external events, followed by a reformation of policy. The initial goal was to see if these patterns were observed in the history of regional ecosystems as interpreted by people firmly embedded in understanding the natural and human dimensions of the system. These case studies form the empirical foundation for the book.

Following a set of preliminary organizational meetings, the first workshop was held in October 1991 and was attended by most of the case study authors (Holling, Baskerville, Light, Gunderson, Costanza, Lee, Regier, and Francis) to discuss the applicability of patterns in their respective regions. A second workshop was held in April of 1992, when the case study authors met again to share findings and first-draft presentations with a broader group including the essayists: Steve Sanderson, Bill Clark, Frances Westley, and Don Michael. In addition, E. Walt Coward, Tim Clark, Greg Daneke, Gilberto Gallopin, Clyde Kiker, Steve Minta, and Harry Vredenberg participated in the April workshop and contributed to ideas that are presented in the text.

What We Learned in the Process

This book is the fourth in a set that deals with adaptive resource and environmental management. The first in that series, *Adaptive Environmental Assessment and Management* (Holling 1978), defined issues and approaches for dealing with the known, uncertain, and unknown dynamic facing and caused by management. The second, *Adaptive Management of Renewable Resources* (Walters 1986), is becoming something of a classic in its description of brilliantly innovative quantitative methods for analyzing, designing, and monitoring actively adaptive resource management systems. The third, *Compass and Gyroscope: Integrating Science and Politics for the Environment* (Lee 1993), uses the insights of a political scientist and the pen of a poet to describe the political context for adaptive approaches. It emphasizes the critical requirements for a democratic process in which the citizen must deal with the complexity and ambiguity of resource and environment issues.

This fourth book is one both of social and ecological theory and of empirical practice. Enough examples of regional experience have accumulated in the last fifteen years that we hoped the case studies would provide examples of pathologies and positive learning experiences, that is, of barriers and bridges to learning, especially in cases where adaptive techniques had been applied. The book deals equally with the way ecosystems are structured and behave and how institutions and the people associated with them are organized and behave.

One of the surprises of this analysis is that both ecological and social systems are inherently more dynamic and unpredictable than was first imagined. All the case studies exposed a profound but potentially transient pathology of resource development. This pathology generally results in a crisis, when the existing policies are recognized clearly as no longer being adequate, followed by a reformation and adoption of new policies. The effort to develop a coherent theory of structural and incremental change in ecosystems and institutions has led to a heuristic framework that seems to capture much of the dynamics of ecosystems and institutions as they continually co-evolve. Moreover, the heuristic suggests ways to identify critical needs during different phases of change, the ways to respond to those needs, and the ways not to respond. We have attempted to move beyond an analysis of stochastic events and patterns of behavior, and have begun to probe a logic of change itself. This and future studies will provide foundations for adaptive management of complexity—to learn to manage *by change* rather than simply to react to it.

We could summarize our specific findings in abstract scholarly terms, but these do not do justice to the reality. The reality is that individuals and small groups of individuals exert extraordinary influence by performing certain distinct roles within and outside institutions. It is this influence that provides a partial antidote to the people who perpetuate equilibrium-centered, command-and-control strategies that often permeate bureaucracies as they ally themselves, often unconsciously, with the power lobbyists to subvert the democratic process. We identified six critical roles, and each is associated with specific names of outstanding individuals who exemplify those functions:

- The creatively destructive role of public interest groups.
- The alerting role of loyal heretics within agencies.
- The importance of "gray eminences"—respected, wise individuals who synthesize, integrate, and communicate information.
- The redefining role of informal collegia of natural scientists, engineers, and social scientists operating outside formal institutions.
- The strategic design and research role of adaptive council in systemwide governance.
- The democratic political role of citizen science.

Acknowledgments

No project like this can succeed without fiscal and administrative support. A few individuals filled these crucial roles. Above all, E. Walter Coward, Jr., provided the encouragement and the bulk of the financing through the Rural Resources and Poverty Program of the Ford Foundation. By so doing he provided the independence and flexibility rarely possible when funding is constrained by government agencies. Timer Powers and Til Creel, executive directors of the South Florida Water Management District, were both brave and wise enough to look outside their boundaries to encourage the search for ways to learn to manage complex regional ecosystems. Consequently, the South Florida Water Management District provided partial funding for the work within Florida. Finally, the financial support of Eugene Stakhiv of the U.S. Army Corps of Engineers allowed us to pay the essayists. Without the collective foresight and trust of these people the project would have been impossible.

Financial support was also provided by the University of Florida Foundation, through the Arthur R. Marshall, Jr. Endowed Chair in Ecological Sciences. Administrative support was supplied by Candy Lane and Toni Carter at the University of Florida and Cherilyn Gerkovich at the South Florida Water Management District. Ilse Holling created the "Field of Dreams" atmosphere for the workshops and is rumored to have arranged for the lightning bolt at Ichetucknee River that reminded us continually to expect the unexpected. Other acknowledgments are found at the end of each chapter. We are grateful for the help of Ronald Harris for copyediting and all the people at Columbia University Press, especially Ed Lugenbeel, who saw the value in publishing this project. Of course, no acknowledgment would be too much for Ralf Yorque, University of McMurdo Sound, our mentor and source of inspiration.

Lance Gunderson, Buzz Holling, and Steve Light

Barriers and Bridges to the Renewal of Ecosystems and Institutions

Part 1
Introduction

1

What Barriers? What Bridges?

C. S. Holling

For the past few decades regional resource and environmental policy and management have been in and out of decision gridlocks in many regions of North America, Europe, and Australia. When issues are polarized, it is a time of deep frustration. Conflicts are extreme, mutual suspicions dominate, and cooperation seems the road to personal defeat. Identifying an enemy and utterly destroying him or her seem more important than finding win/win solutions. The result can be ecosystem deterioration, economic stagnation, and growing public mistrust. Alternatively, the result can be an abrupt reevaluation of the fundamental source of the problems, a redirection of policy toward restoration, and implementation of a process of planning and management that provides continually updated understanding as well as economic or social product.

The purpose of this book is to review a set of regional examples of resource and environmental policy and management where periods of crisis and polarization seem to be replaced either by a paralysis in decision making, exhausted apathy, or active adaptive learning. We want to discover if there are common features to all examples that identify critical barriers to and bridges for maintaining, renewing, or restoring the ecological attributes and institutional flexibility that underlie and provide services to the people and activities in a region.

The cases cover forest management in New Brunswick; water management for agriculture, cities, and ecosystems of the greater Everglades system in Florida; estuarine management in the Chesapeake Bay; salmon and power in the Columbia River; water quality, fisheries, and development in the Great Lakes; and the same issues in the Baltic Sea.

Such problems are not purely ecological, economic, or social. They are a combination of all three and require understanding of the interrelations between nature and people in different settings, performing different roles. Nevertheless, it will seem that much of this analysis is provided by recent advances in understanding the way ecological systems are structured and function—theories that have evolved out of examining and modeling natural processes. We sensed that some of these attributes of ecological systems are really attributes of any complex, evolving system, so that they might also structure the functioning of the economies and institutions that interact, often in hidden ways, with ecosystems. Thus the more fundamental aspect of the book was to evaluate those apparent similarities.

This is not, however, a formal effort to disprove alternative hypotheses in the traditions that led to the ecological theories. It is too soon for that. Rather, it is an effort to define a new set of interesting questions, the hypotheses that might well be testable, and the experiments in policy and management that might be part of those tests.

The need is generally evident. But we believe a new dimension is beginning to be added to local and regional problems that transforms them into a new class. At the same time that local and regional adaptive capabilities are eroding in some regions, intensifying global connections are becoming more evident. The resulting surprises seem to be almost archetypal unknowns. AIDS, the ozone hole, species extinction, and possible climatic change are occurring because of human transformations of local landscapes or of the atmosphere. These changes spread and become global.

The processes that make them problems are all fundamentally ecological, environmental and evolutionary. Although its origins are controversial, some evidence suggests that AIDS, for example, moved from simian to human populations. Initially, it was not virulent, but it evolved into a more deadly form as rates of transmission increased because of social disruptions arising from transformation of land, urbanization, and population increases (Morse 1993). That also has been the story of malaria in Africa (Desowitz 1991). But, unlike malaria, HIV requires no intermediate vector, so that the intensified movement of people around the planet turned a disease that was local and potentially self- extinguishing, into the present intensifying global pandemic. The consequences of these transformations reveal humanity as a planetary force, and perhaps one that is out of control.

So how does a regional politician react in these circumstances? How does the head of a regional resource management agency react? In the United States fewer administrators are now confidently proposing immediate solutions and practical actions to ecological and environmental problems than in the 1970s. The world is now too confusing. At the extreme, some are asking for more and more precision of data about more and more variables in order to be invulnerable in a courtroom! This is an attitude that does not see science as useful in diagnosing emerging problems or in providing a foundation for the integrated understanding needed for policy design. Rather, it views science's role as the provider of data needed for litigation.

But the issue should not be seen as a lack of certainty and precision of data or of predictions. Rather, there is a fundamental loss of certitude—loss in the belief that any of the ground rules work anymore. Any action seems to be full of costs and without benefit. The only comfort is a retreat to unsupported ideology and beliefs.

There are two responses. One is to seek a spurious certitude by increasing control on information and action. The U.S.S.R. learned the price of that strategy! The other is to seek understanding.

This is the motive for this book. For all the bad-news stories, there are signs of an alternative stream of experience. Certainly there are many local examples within developed countries where air has been made clearer and water cleaner. Areas of unsustainable agriculture have been successfully reforested. But how generic are the local successes? Is a new class of regional and global issues being dealt with in another complex phase of learning?

This chapter was written initially to provide guidance for the case studies that were chosen to answer this question. It was intended to provide a set of postulates to be tested and a consistent framework of analysis and synthesis, so that the project and resulting book could be much more than a simple compendium of independent studies.

The first section of this chapter diagnoses a fundamental pathology that has been identified in examples of ecosystem management, a diagnosis that leads to a set of postulates and a set of case studies to explore those postulates. The second section attempts to understand that pathology and its potential cure by first explaining why science and scientists seem so often to succeed in identifying potential problems but to fail in agreeing what to do about them. The third section lays the foundation for understanding the patterns of change in complex systems of nature and people as a possible framework to design creative responses to the inevitable surprises that nature and our actions generate. The final section uses that theoretical treatment to argue that sustained development is only possible if it is seen as a process of evolutionary change that rests on the capacity of nature and people for renewal.

A Diagnosis

My first sense that some of the present problems and responses fall into a new class came when I reviewed some twenty-three examples of managed ecosystems (Holling 1986). Those examples fell into four classes—forest insect, forest fire, savanna grazing, and aquatic harvesting. Two puzzling features were exposed by the initial comparison. One concerned the way ecosystems are organized. The other involved the way ecosystems are managed. Both have turned out to be the consequence of the natural workings of any complex, evolving system.

The first puzzle suggested that the great diversity of life in ecosystems is traceable to the function of a small set of variables, each operating at a qualitatively different speed from the others. The steps for solving that puzzle led to a grand journey collecting data and testing hypotheses that dealt with the morphology, geometry, and dynamics of ecosystems. It is presented in detail elsewhere (Holling 1992), and I shall review the results briefly toward the end of this chapter because they indirectly bear on the nature of policies that are adaptive and sustainable.

The second puzzle is the central focus for this book. It suggested that any attempt to manage ecological variables (e.g., fish, trees, water, cattle) inexorably led to less resilient ecosystems, more rigid management institutions, and more dependent societies. It was this puzzle of success leading to failure that, more than anything else, launched this book's effort to compare regional experiences. As a consequence, I shall dwell a bit more on the postulates that emerged as this puzzle was explored further. Those postulates guided the case study analyses.

All twenty-three examples were associated with management of a resource where the very success of management seemed to set the condition for collapse. In each case the goal was to control a target variable in order to achieve social objectives, typically maintaining or expanding employment and economic activity. In the case of management of eastern North American spruce/fir forests, for example—the target was an anticipated outbreak of a defoliating insect, the spruce budworm (Baskerville, chapter 2; Clark et al. 1979); for the forests of the Sierra Nevada Mountains the target was forest fires (Holling 1980); for the savannas of South Africa the target was the grazing of cattle (Walker et al. 1969); for the salmon of the Pacific Northwest coast the target was salmon populations (Walters 1986).

In each case the goal was to control the variability of the target—insects and fire at low levels, cattle grazing at intermediate stocking densities, and salmon at high populations. The level desired was different in each situation, but the common feature was to reduce variability of a target whose normal fluctuations imposed problems and periodic crises for pulp mill employment, recreation, farming incomes, or fishermen's catches.

The typical response to threats of fire or pestilence, flood or drought is to narrow the purpose, focus on it exclusively, and solve the problem. Modern engineering, technological, economic, and administrative experience can deal well with such narrowly defined problems. And these threats were countered: Insects were controlled with insecticide; fire frequency and extent were reduced with fire detection and suppression techniques; cattle grazing was managed with modern rangeland practice; and salmon populations were augmented with hatchery production.

At the same time, however, elements of the system were slowly changing as a consequence of the initial success of the policy. And because the problem was defined narrowly, such changes were not perceived. First, reducing the variability of the ecological target produced a slow change in the spatial heterogeneity of the ecosystem. Forest architecture became more contiguous over landscape scales, so that if defoliating insects or fire were released, the outbreaks could cover larger areas and have a greater impact than before management. Rangeland gradually lost drought-resistant grasses because of a shift in competition with more productive but more drought-sensitive grasses. If drought occurred, the consequences were more extensive, extreme, and persistent, so that grasslands turned irreversibly into shrub-dominated semideserts. Wild populations of salmon in the many streams along the coast gradually became extinct because fishing pressure increased in response to the increased populations resulting from enhancement. That left the fishing industry precariously dependent on a few hatcheries whose productivity declines with time.

In short, the success in controlling an ecological variable that normally fluctuated led to more spatially homogenized ecosystems over landscape scales. It led to systems more likely to flip into a persistent degraded state, triggered by disturbances that previously could be absorbed. This is the definition for loss of resilience (Holling 1973).

Those changes in the ecosystems could have been managed were it not for concomitant changes in two other elements of the interrelationships—in the management institution(s) and in the people who reaped the benefits or endured the costs. Because of the initial success, in each case the management agencies shifted from their original social and ecological objectives to the laudable objective of improving operational efficiency of the agency itself—spraying insects, fighting fires, producing beef and releasing hatchery fish with as much efficiency and as little cost as possible. Efforts to monitor the ecosystem for surprises rather than only for product therefore withered in competition with internal organizational needs, and research funds were shifted to more operational purposes. Why monitor or study a success? Thus the gradual reduction of resilience of the ecosystems went unnoticed by any but maverick and suspect academics whose research was driven simply by curiosity.

Success brought changes in the society as well. Dependencies developed and powerful political pressures were exerted for continuing the sustained flow of the food or fiber that no longer fluctuated as it once had. More investments therefore logically flowed to expanding pulp mills, recreational facilities, cattle ranches, and fishing technology. This is the development side of the equation, and its expansion can be rightly applauded. Improving efficiency of agencies should also be applauded. But if the ecosystem from which resources are garnered becomes less and less resilient, more and more sensitive to large-scale transformation, then the efficient but myopic agency and the productive but dependent industry simply become part of the source of crisis and decision gridlock.

So this is the puzzle: The very success in managing a target variable for sustained production of food or fiber apparently leads inevitably to an ultimate pathology of less resilient and more vulnerable ecosystems, more rigid and unresponsive management agencies, and more dependent societies. This seems to define the conditions for gridlock and irretrievable resource collapse. It seems to confirm one opinion that sustainable development is an oxymoron (Ludwig, Hilborn, and Walters 1993). Moreover, those pathologies occur not only in examples of renewable resource management but also in examples of rigid policies of regulation of toxic materials or in examples of narrow implementation of protection for endangered species.

It was this puzzle and its possible solution that set the postulates for the case studies; that is, crisis, conflict, and gridlock emerge whenever the problem and the response have the following characteristics:

- A single target and piecemeal policy.
- A single scale of focus, typically on the short term and the local.
- No realization that all policies are experimental.
- Rigid management with no priority to design interventions as ways to test hypotheses underlying policies.

The pathology continues and deepens when the reaction to conflict is to demand more data or more precision in data (e.g., for defense of lawsuits) and more certainty and more control of information and individuals.

The pathology is broken when the issue is seen as a strategic one of adaptive policy management, of science at the appropriate scales, and of understanding human behavior, not a procedural one of institutional control. This requires

- Integrated policies, not piecemeal ones.
- Flexible, adaptive policies, not rigid, locked-in ones.
- Management and planning for learning, not simply for economic or social product.
- Monitoring designed as a part of active interventions to achieve understanding and to identify remedial response, not monitoring for monitoring's sake.
- Investments in eclectic science, not just in controlled science.
- Citizen involvement and partnership to build "civic science" (Lee 1993), not public information programs to inform passively.

We decided to explore those postulates in a project that would engage an interdisciplinary team whose individuals together represented deep personal experience in specific cases of regional ecosystem analysis and management—people thoroughly grounded in a balance of theory, science, and practice. It was that group that provided the analyses of the case studies and authored the following six chapters. All have a broad range of experience outside their own field of specialization as well as experience with interdisciplinary and integrative modes of inquiry.

Part way through the project we invited another group of individuals active in the development of broad social science theory to join the original team and provide commentary from the perspective of their area of expertise—political science, institutional psychology, institutional management, social learning theory, economics, and Third World development. In addition to providing insight directly to case study authors, four provided commentary chapters for this volume ("Ten Theses on the Promise and Problems of Creative Ecosystem Management in Developing Countries" by Steven E. Sanderson [chapter 8]; "Governing Design: The Management of Social Systems and Ecosystem Management" by Frances Westley [chapter 9]; "Sustainable Development As Social Learning: Theoretical Perspectives and Practical Challenges for the Design of a Research Program" by Edward A. Parson and William C. Clark [chapter 10]; and "Barriers and Bridges to Learning in a Turbulent Human Ecology" by Donald N. Michael [chapter 11]). The full synthesis of this experience in theory and practice then became the focus for the editors, in consultation with the other authors, to prepare the last chapter and develop an expansion of general theory that would explain why the bridges identified succeeded in restoring the degraded renewal capacities of nature and people ("Barriers Broken and Bridges Built: A Synthesis" [chapter 12] by L. H. Gunderson, C. S. Holling, and S. S. Light).

The six cases are all regional-sized systems, and each centers around a recognized ecosystem (figure 1.1). The six systems include (1) maritime portions of the boreal forest and dependent lumbering industry (New Brunswick); (2) an internationally recognized wetland (the Everglades), which supplies fresh water to a burgeoning population, agriculture, and national park; (3) the largest estuarine system in the United States (Chesapeake Bay), with a seemingly sustainable consumption of marine resources; (4) one of the largest river basins in North America (that of the Columbia River), where people struggle to reconcile issues of producing electrical power and salmon; (5) the largest freshwater lake system in North America (the Great Lakes), where two

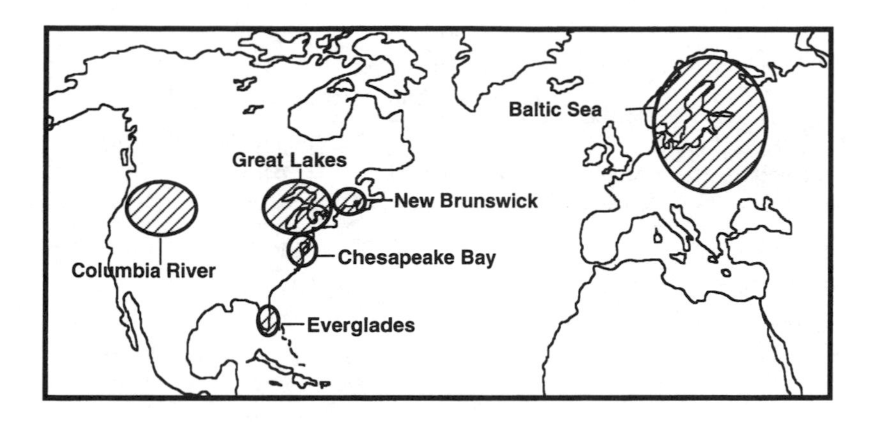

FIGURE 1.1
Location of the six study areas in the world. Ovals approximate the size of each ecosystem.

nations and ten states/provinces develop ways to manage water quality and quantity, fisheries, recreation, and economic opportunities; and (6) the largest brackish sea in the world (the Baltic Sea), surrounded by 15% of the world's industrial production, where nine nations, divided by political ideologies, develop responses to development and environmental deterioration. The case studies are presented in chapters 2–7 in a progression of increasing complexity, defined by the area of the system, population, and number of institutional or management units (table 1.1).

The other major criteria used for selecting the cases involve identifying authors who had a systemwide perspective, were expert in both the ecology and the management of the system, and most important, had lived in the system and participated in policy and management designs. Their individual and collective experience and understanding were tapped for this volume. That growing experience and knowledge have emerged from a pattern of exploitive development that has endured so long because ecosystems are remarkably resilient and because people do learn and adapt.

But the path of learning is not easy, partly because the new class of complex issues is sufficiently novel that the science is incomplete and the future is unpredictable. This is the topic to which I now turn in a search for a direction for understanding.

TABLE 1.1
Area, Population, Population Density and Political Units of Six Study Areas

Study Site	**Area*** (1000 km²)	**Population†** (million people)	**Population Density** (people/km²)	**Political Units**
New Brunswick	73	0.73	10	1 province
Everglades	23	5.19	224	1 state
Chesapeake	166	14.5	87	6 states
Columbia River	671	9	13	7 states, 2 provinces, 2 nations
Great Lakes	766	38	50	8 states, 2 provinces 2 nations
Baltic Sea	1641	75.2	46	9 nations

*Defined by catchment basin in all systems except New Brunswick.
†Dates of census vary from 1985 to 1990.

Seeking Understanding: Why Scientists Can Muddy the Waters

A critical minority of politicians and of the inquiring public is now not so much driven by fear of prophecies of doom as by the need for understanding. But if you seek understanding, to whom do you turn? Science is not helping, largely because there are not only conflicting voices but conflicting modes of inquiry and criteria for establishing the credibility of a line of argument.

In particular, the philosophies of two streams of science are often in conflict. The tension between them is now particularly evident in biology. One is brilliantly represented by the advances in molecular biology and genetic engineering. That stream of science promises to lead not only to health and economic benefits of biotechnology, but also to an uncertain sea of changing social values and consequences. It is a stream of biology that is essentially experimental, reductionist, and narrowly disciplinary.

The other stream is represented within biology by evolutionary biology and by systems approaches that include the analysis of populations, ecosystems, landscape structures and dynamics, and more recently, biotic and human interactions with planetary dynamics. The applied form of this stream has emerged regionally in new forms of resource and environmental management, where uncertainty and surprises become an integral part of an anticipated set of adaptive responses (Holling 1978; Walters 1986; Lee 1993). It is fundamentally interdisciplinary and combines historical, comparative, and experimental approaches at scales appropriate to the issues. This combination provides the necessary foundations for any kind of global science, if for no other reason than that we have but one globe to live on, for the present, at least, and cannot experimentally manipulate lost pasts. It is a stream of investigation that is fundamentally concerned with integrative modes of inquiry and multiple sources of evidence. This stream has the most natural connection to related ones in the social sciences that are historical, analytical, and integrative. It is also the stream that is most relevant for the needs of policy and politics.

The first stream is a science of parts (e.g., analysis of specific biophysical processes that affect survival, growth, and dispersal of target variables). It emerges from traditions of experimental science, where a narrow enough focus is chosen to pose hypotheses, collect data, and design critical tests for the rejection of invalid hypotheses. The goal is to narrow uncertainty to the point where acceptance of an argument among scientific peers is essentially unanimous. It is appropriately conservative and unambiguous, but it achieves this by being incomplete and fragmentary. It provides bricks for an edifice but not the architectural design.

The other stream is a science of the integration of parts. It uses the results and technologies of the first but identifies gaps, develops alternative hypotheses and multivariate models, and evaluates the integrated consequence of each alternative by using information from planned and unplanned interventions in the whole system that occur or are implemented in nature. Typically, the goal is to reveal the simple causation that often underlies the complexity of time and space behavior of complex systems. Often there is more concern that a useful hypothesis might be rejected than that a false one might be accepted. Since uncertainty is high, the analysis of uncertainty becomes a topic in itself.

The premise of this second stream is that knowledge of the system we deal with is always incomplete. Surprise is inevitable. Not only is the science incomplete, but the system itself is a moving target, evolving because of the impact of management and the progressive expansion of the scale of human influences on the planet.

In principle, therefore, evolving managed ecosystems and the societies with which they are linked involve unknowability and unpredictability. Therefore sustainable development is also inherently unknowable and unpredictable. Therein lies the issue that we address in this book. The essential point is that evolving systems require policies and actions that not only satisfy social objectives but also achieve continually modified understanding of the evolving conditions and provide flexibility for adapting to surprises.

This is the heart of active regional experimentation by management at the scale appropriate to the question—adaptive environmental and resource management (Holling 1978; Walters 1986; Lee 1993). Otherwise the pathologies of exploitive development are inevitable—increasingly brittle ecosystems, rigid management, and dependent societies leading to crises.

Faced with the partial understanding we have of the problems and with the conflicting views of science, it is no wonder that public concern and mistrust are great but public understanding disturbingly bad. Political responses have a weak foundation for confident action that will not make the cure worse than the disease. To whom can the public turn for insight? This is less a problem of trust in science than of trust in governance by all participants who, in the absence of firm foundations for understanding, are forced to shape their decisions by beliefs.

So much presently seems uncertain or unknown that many of the calls for action or inaction, however well supported by technical argument, are largely determined by such beliefs. Because each belief is partially relevant, impressive and convincing technical arguments can be mobilized for each, no matter how conflicting the resulting calls for action or inaction may be.

Four belief systems, and an emerging fifth, are driving present debate and public confusion. Each reflects different assumptions about stability and change, as I have suggested elsewhere (Holling 1987). Alternatively, they can be labeled (albeit unfairly) by a caricature of their causal assumptions, as I shall do here.

The first view (that of Nature Cornucopian) is one of smooth exponential growth where resources are never scarce because human ingenuity always invents substitutes. It was the basic view of Herman Kahn and is the foundation for Julian Simon's arguments (Simon and Kahn 1984). It assumes that humans have an infinite capacity to innovate and that nature changes gradually—fast enough to be detected yet slow enough to be managed.

The second view (that of Nature Anarchic) is hyperbolic, where increase is inevitably followed by decrease. It is a view of fundamental instability, where persistence is only possible in a decentralized system in which there are minimal demands on nature. It is the view of Schumacher (1973) and some extreme environmentalists. If the previous view assumes that infinitely ingenious humans do not need to learn anything different, this view assumes that humans are *incapable* of learning how to deal with the technology they unleash.

The third view (that of Nature Balanced) is one of logistic growth, where the issue is how to navigate a looming and turbulent transition—demographic, economic, social, and environmental—to a sustained plateau. This is the view of several institutions with a mandate for reforming global resource and environmental policy—of the Bruntland Commission, the World Resources Institute, the International Institute of Applied Systems Analysis, and the International Institute for Sustainable Development, for example. Many individuals are contributing skillful scholarship and policy innovation. They are among some of the most effective forces for change.

The fourth view (that of Nature Resilient) is one of nested cycles organized by fundamentally discontinuous events and processes. That is, there are periods of exponential change, of growing stasis and brittleness, of readjustment or collapse, and of reorganization for renewal. Instabilities organize the behaviors as much as stabilities. This was the view of Schumpeter's (1950) economics, and it has more recently been the focus of fruitful scholarship in a wide range of fields—ecological, social, economic, and technical. This has formed the body of my own ecological research for the past 20 years. I find striking similarities in Harvey Brook's view of technology (1986), Brian Arthur's and Kenneth Arrow's (1989) recent view of the economics of innovation and competition (Waldrop 1992), Mary Douglas's (1978) and Mike Thompson's (1983) view of cultures, Don Michael's view of human psychology (1984), and Barbara Tuchman's (1978) and William McNeill's (1979) view of history.

The emerging fifth view (that of Nature Evolving) is evolutionary and adaptive. It has been given recent impetus by the paradoxes that have emerged in successfully applying the previous, more limited views. *Complex systems behavior, discontinuous change, chaos and order, self-organization, nonlinear system behavior, and adaptive evolving systems* are all the present code words characterizing the more recent activities. This view is leading to integrative studies that combine insights and people from developmental biology and genetics, evolutionary biology, physics, economics, ecology, and computer science. The Santa Fe Institute is an interesting experiment (Waldrop 1992) in applying collaborative approaches to explore the insights and opportunities opened by such an evolutionary paradigm.

The point is not that one of these beliefs is correct and the others wrong. Each is true, but each is a partial truth. Because we are only now beginning to understand the changing reality, there is no limit to the ability of a good scientist to invent compelling lines of causal explanation that inexorably support his or her particular beliefs. How can even the best-intentioned politician possibly be expected to deal with that? How can even the most reflective citizen? With every issue having supporting evidence and contradictory counterevidence (all legitimate), the issues seem to involve no independent reality of nature, only moral issues that can be debated. Can we ever separate belief from fact?

Foundations for Integration: Science, Understanding, and Policy

The preceding argument explains my unease with calls for action that are dominated exclusively by prophesies of crisis. Certainly it is appropriate to cite clear examples of the critical new class of problems, particularly those that clarify the need for action (e.g., AIDS, the hole in the ozone layer, and carbon dioxide increase). These are so clear, so growing, so global, and so novel that action can be taken, which we would want to do in any case for other reasons of efficiency, health, and economic sustainability.

Perhaps I have been in the game too long to be sympathetic to "Chicken Little" stories of catastrophe. In 1969 *Time* magazine entitled an article "The New Jeremiahs" and featured six scientists who were prophesying doom—an environmental doom that may have been novel then but that is familiar now. I remember they included Paul Ehrlich, Barry Commoner, Ken Watt, and—me! Now, 25 years later, I find the articles, projects, and proposals that repeat the same litany of doom to be not necessarily wrong, but tiresome, unconvincing, and weak.

But what is really disturbing is that they ignore the remarkable advances, learning, and understanding that have occurred in the intervening years. They ignore the opportunities for conversation among and actions by previously polarized individuals that increase both understanding and the ability to develop and apply integrated and adaptive policies. The problems and topics revolve around five interrelated themes—regional resource management and development, ecosystem restoration, sustainable development, global change, and biodiversity. Population growth and technology drive them all.

The last 20 years have seen a stunning advance in understanding how the planet has evolved and functions in its physical aspects. The reconstruction of the composition of our atmosphere over the last 160,000 years (using bubbles trapped in the Vostok ice core from Antarctica) and its correlation with climate (using proxy biological and chemical signals) can be seen as an engrossing tour de force of international science. It is also useful for politicians. It tells them that the present concentration of carbon dioxide in our atmosphere is higher than it has been for the last 160,000 years.

The detection of the ozone hole in the Antarctic came as a complete surprise to existing "gradualist" theories of the atmosphere, and the demonstration of its reality and of the role of industrial emissions of chlorofluorocarbons on atmospheric chemistry has been an example of the passionate application of the best kind of cooperative, and at times combative, science in a complex new area. This has also been useful for politicians, as countries now move to ban CFCs as an act of international cooperation.

However narrow the mainstream of molecular biology might be, it has yielded techniques that now are transforming the evolutionary, ecological, and conservation sciences. Is it true that we can trace all human mitochondrial DNA back to an "Eve" in Africa (Vigiland et al. 1991)? Biologists now can certainly unravel affinities in related groups of species and individuals and can join the geophysicists in compelling reconstructions of the past that, at the least, put our present problems in a perspective—from the role of past extinctions to present declines in biodiversity.

The understanding needed for the changes we now experience or anticipate draws on this knowledge from geophysics, atmospheric science, and techniques of cellular and molecular biology. However, to understand such changes we must integrate ecosystem and community ecology with the more physically based earth sciences.

But we must recognize what this means and the challenge it presents. The relevant biophysical processes operate over an enormous scale, potentially from soil processes operating with time constants of hours or days in meter-square patches, to ecosystem successional processes of decades to centuries covering tens to thousands of square kilometers, to global biotic processes involved in the regulation and isolation of elements like carbon, which have time lags of millennia and a global impact. This is why satellite imagery, remote sensing, and geographic information systems now routinely available to analyze patterns are of such major consequence. Computer advances, both toward the portable but powerful and the large and parallel, have made it possible to visualize complexity in both space and time. It is a picture of discontinuous behavior, of multiple stable states, of the interaction between slow forces that accumulate environmental capital and fast processes that slowly exploit, suddenly release, and renew the capital. It is as far a cry from public perceptions of fragile, stable, and equilibrium nature as could be imagined. And that knowledge too is useful and used. It is the foundation for the regional experiments in adaptive policy design and management that are as much examples of institutional learning as they are of using science for public policy.

Moreover, emerging theories of hierarchical structure, of scale-independent geometry, and of nonlinear dynamics are providing the focus for posing the researchable questions about cross-scale interactions that are the first step toward usable understanding and useful policy. This is the topic to which I shall now turn.

On Theory

"Don't give me academic theory; give me practical advice and actions!" That's what I heard, appropriately, in the certainty of the 1970s. But at a time of confusion, such as the 1990s, promising and relevant theory is the only antidote to dated ideology or belief. And the intriguing paradoxes that have emerged by applying past incomplete theory of equilibrium, of gradual change, and of control have set a foundation for new theories of discontinuous change and evolution. Oddly, one of the most practical things we could recommend now is massive support for the expansion of new theory, but in combination with synthesizing theory from the reality of examples. An inductively based expansion of theory has the promise of yielding both integrated understanding and integrated actions.

The intensity and global nature of the changes now taking place are moving the planet and its occupants into totally new behavior. In this transformation some consequences can be predicted, others will be uncertain, and still others will be unpredictable. As a consequence, it is essential to be guided by theories of change that can contain short- and long-term changes, gradual and abrupt ones, and dynamic and structural ones.

These theories determine the questions we ask, the problems we perceive, the data we collect and analyze, and the policies and actions we initiate. Theories that do not match the problem can be at best delusions and at worst dangerous.

The discovery of the hole in the ozone layer is an example of the former. It was not detected initially by satellite imagery, because the smoothing algorithm applied to the data assumed that abrupt changes could only be caused by instrument glitches. The implicit theory presumed gradual, continuous change in atmospheric chemistry and chemical composition.

There are also many examples of theories that have had more disastrous consequences. The devastating events in the Sahel of Africa is one recent example, as Brian Walker and Tony Sinclair have described (1990). External changes in precipitation were partially responsible for the collapse in the region, but such changes have occurred and been absorbed before. The response was exaggerated by increased vulnerability of a culture and ecosystem caused in part by development aid that broke the patterns of nomadic movement and social adaptation that had evolved in these semiarid savannas. No adequate theory was utilized to relate the resilience of local ecosystems and the adaptive flexibility of people to mesoscale migrations of people and animals.

Regional changes of this nature and the anticipated global ones make the world we are entering one of surprises whose consequences threaten to overwhelm the adaptive capacities of individuals, business, and government. Investing in the development and testing of usable and useful theory is therefore not an academic luxury, but a practical necessity, particularly at times of such profound change.

The issue is not only one of change in general, but one of evolutionary change. The conceptual foundations therefore need to be drawn from the growing experience in understanding the operation of complex, nonlinear systems where discontinuous behavior and structural change are the norm. Scholars in an unusual variety of disciplines have contributed to the development of these theories—from thermodynamics (Nicolis and Prigogine 1977), oceanography (Broeker), climatology (Lorenz 1963), atmospheric chemistry (Crutzen and Arnold 1986), evolutionary and developmental biology (Kauffman 1992), and ecology (May 1977; Levin 1992; and my own work). All deal with the reality of abrupt changes organized by several equilibria, of the existence of multistable states, and of the interplay between order and disorder in evolving self-organizing systems. This is what our world is, and it is at the heart of feasible sustainable development.

Social scientists are also major contributors. Such historians as William McNeill (1979) have long argued in favor of a view of history that is a sequence of discontinuous events and of human responses to them. Wildavsky and Douglas (1982) argue for the inevitability and need for risk and surprise in any human development. More recently, Mary Douglas (1978) and Mike Thompson (1983) have used their background in cultural anthropology to characterize institutions as being driven by a similar interplay between stability and instability. And when someone of the stature of Kenneth Arrow suggests the need for transforming economics by nonlinear theory (as quoted in Waldrop 1992), a revolution in thought may be occurring in that field as well. In every instance these theories owe their force to the resolution of puzzles that appear when earlier incomplete or inadequate concepts encounter surprising reality.

The way key subcultures in the natural and social sciences view the world is converging on these theories of change. These theories rationalize the paradoxes of stability and instability, of order and disorder, and of stasis and evolutionary change. Since these are the same paradoxes inherent in the goal of sustainability and development, an avenue opens for directly relevant cooperation between critical parts of the social and natural sciences.

As a start, I shall describe this view of change as it applies to ecological systems. I do so not to force an inappropriate analogy on the way social and economic systems function, but to search for the common foundations for change that underlie the operation of any complex living system.

Ecosystem Function

Over the last decade, the literature on ecosystems has led to major revisions in a view of succession that was proposed by Clements early in this century (1916). That view was one of a highly ordered sequence of species assemblages moving toward a sustained climax whose characteristics are determined by climate and edaphic conditions. This revision comes from extensive comparative field studies (West et al. 1981), from critical experimental manipulations of watersheds (Bormann and Likens

1981; Vitousek and Matson 1984), from paleoecological reconstruction (Davis 1986; Delcourt et al. 1983), and from studies that link systems models and field research (West et al. 1981).

The revisions include four principal points. First, the species that invade after disturbance and during succession can be highly variable and determined by chance events. Second, both early and late successional species can be present continuously. Third, large and small disturbances triggered by such events as fire, wind, and herbivores are an inherent part of the internal dynamics and in many cases set the timing of successional cycles. Fourth, some disturbances can carry the ecosystem into quite different stability domains—mixed grass and tree savannas into shrub-dominated semideserts, for example (Walker 1981); thus there is more than one possible "climax" state.

In summary, therefore, the notion of a sustained climax is a useful but essentially static and incomplete equilibrium view. The combination of these advances in ecosystem understanding with studies of population systems has led to one version of a synthesis that emphasizes four primary stages in an ecosystem cycle (Holling 1986).

The traditional view of ecosystem succession has been usefully seen as being controlled by two functions: *exploitation,* in which rapid colonization of recently disturbed areas is emphasized, and *conservation,* in which slow accumulation and storage of energy and material are emphasized. For an economy, an economist might use such labels as *market* and *innovation* for the exploitation phase and *monopolist* or *hierarchy* for the conservation phase.

But the revisions in understanding indicate that two additional functions are needed (figure 1.2). One is that of *release,* or *creative destruction,* a term borrowed from the economist Schumpeter (as reviewed in Elliott 1980), in which the tightly bound accumulation of biomass and nutrients becomes increasingly fragile (overconnected, in systems terms) until it is suddenly released by agents such as forest fires, insect pests, or intense pulses of grazing. The second is one of *reorganization,* in which soil processes of mobilization and immobilization minimize nutrient loss and reorganize nutrients to become available for the next phase of exploitation. An economist might use such labels as *invention* and *reinvestment* for this stage.

During this cycle, biological time flows unevenly. The progression in the ecosystem cycle proceeds from the exploitation phase (box 1, figure 1.2), slowly to conservation (box 2), very rapidly to release (box 3), rapidly to reorganization (box 4), and rapidly back to exploitation. Dur-

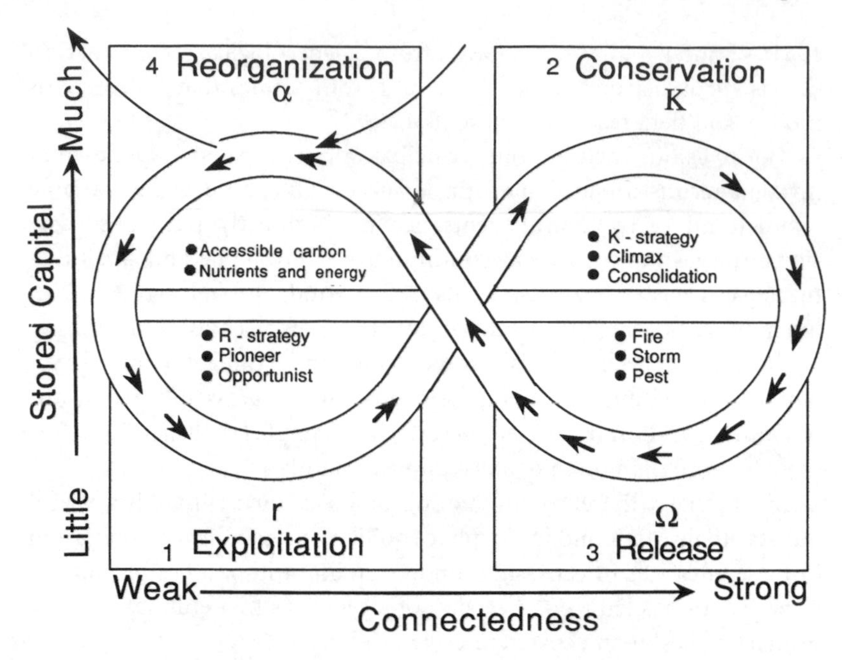

FIGURE 1.2
The four ecosystem functions and the flow of events between them. The arrows show the speed of that flow in the ecosystem cycle, where arrows close to each other indicate a rapidly changing situation and arrows far from each other indicate a slowly changing situation. The cycle reflects changes in two attributes; that is: (1) the *Y* axis—the amount of accumulated capital (nutrients, carbon) stored in variables that are the dominant keystone variables at the moment—and (2) the *X* axis—the degree of connectedness among variables. The exit from the cycle indicated at the left of the figure suggests the stage where a flip is most likely into a less or more productive and organized system (i.e., devolution or evolution as revolution!).

ing the slow sequence from exploitation to conservation, connectedness and stability increase and a "capital" of nutrients and biomass is slowly accumulated. That capital becomes more and more tightly bound, preventing other competitors from utilizing the accumulated capital until the system eventually becomes so overconnected that rapid change is triggered. The agents of disturbance might be wind, fire, disease, insect outbreak, or a combination of these. The stored capital is then suddenly released and the tight organization is lost to allow the released capital to be reorganized to initiate the cycle again.

This pattern is discontinuous and depends on changing multistable states to trigger and organize the release and reorganization functions. Instabilities and chaotic behavior trigger the release phase, which then proceeds in the reorganization phase, where stability begins to be reestablished. In short, chaos emerges from order, and order emerges from chaos! Resilience and recovery are determined by the fast release (or creative destruction) and reorganization sequence, whereas stability and productivity are determined by the slow exploitation and conservation sequence.

Moreover, there is a nested set of such cycles, each with its own range of scales. In the typical boreal forest, for example, fresh needles cycle yearly; the crown of foliage cycles with a decadal period; and trees, gaps, and stands cycle at a period of about a century or more. The result is a hierarchy in which each level has its own distinct spatial and temporal attributes (figure 1.3).

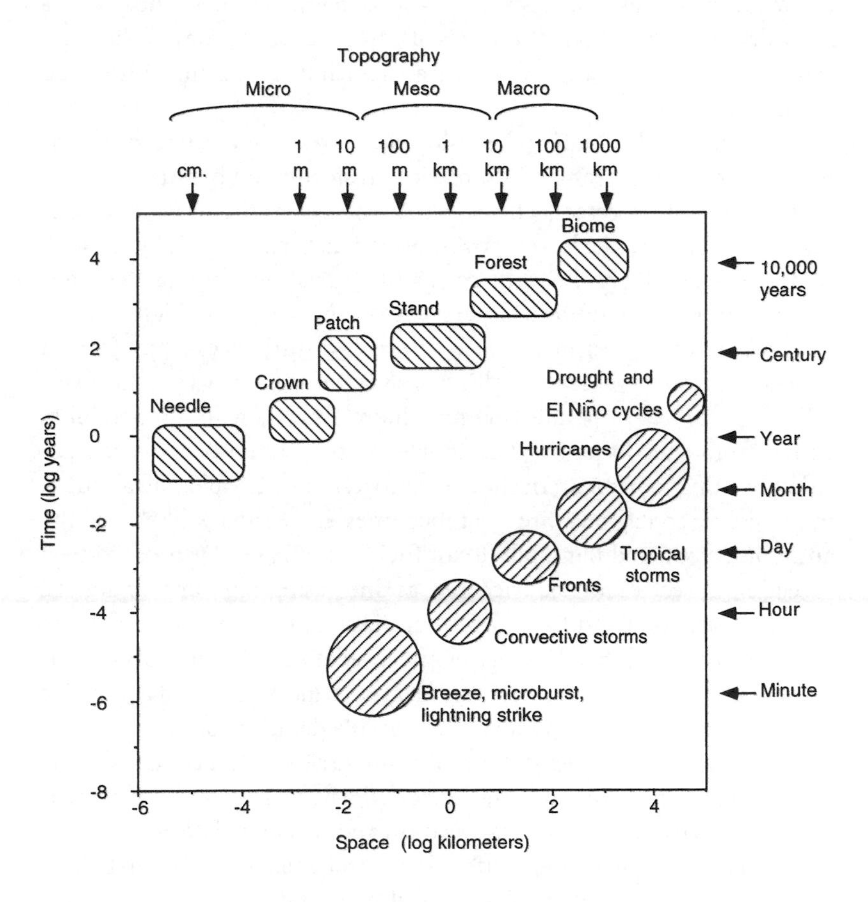

FIGURE 1.3
Space/time hierarchy of the boreal forest and of atmosphere.

Dynamics of Hierarchies

A critical feature of such hierarchies is the asymmetric interactions between levels (Allen and Starr 1982; O'Neill et al. 1986). In particular, the larger, slower levels maintain constraints within which faster levels operate. In that sense, therefore, slower levels control faster ones. If that is the only asymmetry, however, it would be impossible for organisms to exert control over slower environmental variables. This is the criticism that many geologists make of the Gaia theory (Lovelock 1988): How could slow geomorphic processes possibly be affected by fast biological ones? However, it is not broadly recognized that the birth, growth, death, and renewal cycle, shown in figure 1.2, transforms hierarchies from fixed static structures to dynamic entities whose levels are vulnerable to small disturbances at certain critical times in the cycle (Holling 1992). That represents a transient but important bottom-up asymmetry.

There are two key states where slower and larger levels in ecosystems become briefly vulnerable to dramatic transformation because of small events and fast processes. One is when the system becomes overconnected and brittle as it slowly moves toward maturity (box 2, figure 1.2). At these stages, there are tight competitive relations among the plant species. From an equilibrium perspective, the system is highly stable (i.e., fast return times in the face of small disturbances), but from a resilience perspective, *sensu* Holling (1987), the domain over which stabilizing forces can operate becomes increasingly small. Vulnerability comes from such loss of resilience. Hence the system becomes an accident waiting to happen. In the boreal forest, for example, the accident might be a contagious fire that becomes increasingly likely as the amount, extent, and flammability of fuel accumulates. Or it could be a spreading insect outbreak triggered as increasing amounts of foliage both increase food and habitat for defoliating insects and decrease the efficiency of search by their vertebrate predators (Holling 1988). It is also the phase where, in human organizations, the rebellion of aggressive interest groups can precipitate an inexorable demand for change.

Small and fast variables can also dominate slow and large ones at the stage of reorganization (box 4, figure 1.2). At this stage the system is underconnected, with weak organization and weak regulation. As a consequence, it is the stage most affected by probabilistic events that allow a diversity of entrained species, as well as exotic invaders, to become established. On the one hand, it is the stage most vulnerable to erosion and to the loss of accumulated capital. On the other hand, it is the stage from which jumps to unexpectedly different and more productive systems are possible. At this stage, instability comes because of loss of regulation rather than from the brittleness of reduced resilience. It is the phase in a system—ecological or human—where the individual or small groups of individuals can make the greatest structural change for the future.

It is this view of alternative phases in a cycle of birth, growth, death, and renewal that seems to underlie any complex adaptive system—ecological certainly, but human, institutional, and societal as well. This is one of the proposed foundations to the synthesis we explore in this book. Does such a view have generality? Does it suggest what to do, and equally important, what not to do? If so, then a possible foundation exists to turn sustainable development from an oxymoron to a plan of action.

The Paradox of Sustainable Development

Sustainable development is itself something of a paradox. The phrase implies that something must change but that something must also remain constant. The paradox appears in a number of forms, and its resolution can provide the direction to seek for investments that could sustain development.

In the introduction I described the two puzzles that appeared when I reviewed several examples of managed ecosystems. One concerned the organization of ecosystems. The other concerned the management of ecosystems. Both have turned out to be the consequence of the natural workings of any complex, evolving system. The resolution of those puzzles is central to the way ecosystems can be restored within a regional context of social and economic activities.

As described previously in this chapter, the first puzzle suggested that the great diversity of life in ecosystems is traceable to the function of a small set of variables, each operating at a qualitatively different speed from the others. The second puzzle suggested that any attempt to manage ecological variables inexorably led to less resilient ecosystems, more rigid management institutions, and more dependent societies. It was this puzzle, more than any, that launched this book's effort to compare regional experiences. I will now review each in turn, in order to show how the previous section on theories of change relates directly to the sustainability of development.

The Ecosystem Organization Puzzle

How could the great diversity within ecosystems possibly be traced to the function of a small number of variables? The models that were developed and tested for these examples certainly generated complex behavior in space and time. Moreover, those complexities could be traced to the actions and interactions of only three or four sets of variables and associated processes, each of which operated at distinctly different speeds. The speeds were therefore discontinuously distributed and differed from their neighbors, often by as much as an order of magnitude. A summary of the critical structuring variables and their speeds is presented in table 1.2. For the models at least, this structure organizes the time and space behavior of variables into a small number of cycles, presumably abstracted from a larger set, that continue at smaller and larger scales than the range selected.

But are those features simply the consequence of the way modelers make decisions, rather than the results of ecosystem organization? This uneasy feeling that such conclusions can be a contrivance of our minds rather than a reflection of the way ecosystems actually function led to a series of tests using field data to challenge the hypothesis that ecosystem dynamics are organized around the operation of a small number of nested cycles, each driven by a few dominant variables (Holling 1992).

The critical argument is that if there are, in fact, only a few structuring processes, their imprint should be expressed on most variables. That is, time series data for fires; seeding intensity; insect, mammal, and bird numbers; water flow (indeed, any variable for which there are long-term, yearly records) should show periodicities that cluster around a few dominant ones. In the case of the eastern maritime boreal forest of North America, for example, those periodicities were predicted to be 3–5 years, 10–15 years, 35–40 years, and more than 80 years. Similarly, there should be a few dominant spatial "footprint" sizes, each associated with one of the disturbance/renewal cycles in the nested set of such cycles. Finally, the animals living in specific landscapes should demonstrate the existence of this lumpy architecture by showing gaps in the distribution of their sizes and gaps in the scales at which decisions are made for location of region, foraging area, habitat, nests, protection, and food.

All the evidence we have so far confirms just those hypotheses—for boreal forests, boreal region prairies, pelagic ecosystems (Holling 1992),

TABLE 1.2
Key Variables and Speeds in Four Groups of Managed Ecosystems

The System	Fast Variable	Intermediate Variable	Slow Variable	Key Reference
Forest insect	Insect, needles	Foliage, crown	Trees	McNamee et al. (1981), Holling (1991)
Forest fire	Intensity	Fuel	Trees	Holling (1980)
Savanna	Annual grasses	Perennial grasses	Shrubs	Walker et al. (1969)
Aquatic	Phytoplankton	Zooplankton	Fish	Steele (1985)

and the Everglades of Florida (Gunderson 1992). A variety of alternative hypotheses based on developmental, historical, or trophic arguments was disproved in the fine traditions of Popperian science, leaving only the "world-is-lumpy" hypothesis as resisting disproof.

I conclude therefore that there is strong evidence for the following:

1. A small number of plant, animal, and abiotic processes structure biomes over scales from days and centimeters to millennia and thousands of kilometers. Individual plant and biogeochemical processes dominate at fine, fast scales; animal and abiotic processes of mesoscale disturbance dominate at intermediate scales; and geomorphological ones dominate at coarse, slow scales.
2. These structuring processes produce a landscape that has lumpy geometry and lumpy temporal frequencies or periodicities. That is, the physical architecture and the speed of variables are organized into distinct clusters, each of which is controlled by one small set of structuring processes. These processes organize behavior as a nested hierarchy of cycles of slow production and growth alternating with fast disturbance and renewal.
3. Each cluster is contained within a particular range of scales in space and time and has its own distinct architecture of object sizes, interobject distances, and fractal dimensions within that range.
4. All the many remaining variables, other than those involved in the structuring processes, become entrained by the critical

structuring variables, so that the great diversity of species in ecosystems can be traced to the function of a small set of variables and the niches they provide. The structuring processes both form structure and are affected by that structure. These structuring variables are therefore where the priority should be placed in investing to renew, maintain, or restore ecosystems.

5. The discontinuities that produce the lumpy structure of vegetated landscapes impose discontinuities on the behavior and morphology of animals. For example, there are gaps in body mass distributions of resident species of animals that correlate with scale-dependent discontinuities in the geometry of vegetated landscapes. Thus these gaps, and the body mass clumps they define, become a way to develop a rapid bioassay of ecosystem structures and of human impacts on that structure. It therefore opens the way to develop a comparative ecology across scales that might provide the same power for generalization that came when physiology became comparative rather than species specific.
6. Conversely, changes in landscape structure at defined scale ranges caused by land use practice or by climate change will have predictable impacts on animal community structure (e.g., animals of some body masses can disappear if an ecosystem structure at a predictable scale range is changed). Therefore predicted (from models or land use plans) or observed (from remote imagery) impacts of changing climate or land use on vegetation can also be used to infer the impacts on the diversity of animal communities.

The lessons for both sustainable development and biodiversity loss are clear: Focus should be on the structuring variables that control the lumpy geometry and lumpy time dynamics. They set the stage upon which other variables play out their own dramas. That is, it is the physical and temporal infrastructure of biomes *at all scales* that sustains the theater; given that, the actors will look after themselves!

The Ecosystem Management Puzzle

Earlier I identified a puzzle that launched the studies in this book. In many cases of renewable resource management, success in managing a target variable for sustained production of food or fiber apparently leads to an ultimate pathology of less resilient and more vulnerable ecosystems, more rigid and unresponsive management agencies, and more dependent societies. But something seems to be wrong with that conclusion, which implies that the only solution for humanity is a radical return to being "children of nature." The puzzle needs to be clarified in order to test its significance and generality.

The conclusion is based on two critical points. One is that reducing the variability of ecosystems inevitably leads to reduced resilience and increased vulnerability. The second is that there is no other way for agencies and people to manage and benefit from resource development.

Again some independent evidence is needed. Are there counterexamples? Oddly, nature itself provides counterexamples of tightly regulated yet sustainable systems in the many examples of physiological homeostasis. Consider temperature regulation of endotherms ("warm-blooded" animals), for example. That represents a system where internal body temperature is not only tightly regulated within a narrow band, but among present-day birds and mammals, at an average temperature perilously close to lethal. Moreover, that regulation requires ten times more energy for metabolism than is required by an ectotherm. This would seem to be a recipe for disaster, and a very inefficient one at that. Yet evolution somehow led to the extraordinary success of those with such an adaptation—the birds and mammals.

To test the generality of the variability loss/resilience loss hypothesis, I have been collecting data from the physiological literature on the viable temperature range of the internal body of organisms exposed to different classes of variability. I have organized the data into three groups, ranging from terrestrial ectotherms ("cold-blooded" animals) exposed to the greatest variability of temperature from unbuffered ambient conditions, to aquatic endotherms exposed to an intermediate level of variability because of the moderating attributes of water, to endotherms that regulate temperature within a narrow band. The viable range of internal body temperature decreases from about 40°C for the most variable group to about 30°C for the intermediate, to 20°C for the tightly regulated endotherms. Resilience (in this case the range of internal temperatures that separates life from death) clearly does contract as experience with variability is reduced, just as in the resource management cases. I conclude, therefore, that reduction of variability of living systems from organisms to ecosystems inevitably leads to loss of resilience in that part of the system being regulated.

But that seems to leave an even starker paradox of control inevitably leading to collapse. But, in fact, endothermy does persist. It therefore serves as a revealing metaphor for sustainable development. This metaphor contains two features that were not evident in my earlier descriptions of examples of resource management.

First, the kind of regulation is different. Five different mechanisms, from evaporative cooling to metabolic heat generation, control the temperature of endotherms. Each mechanism is not notably efficient by itself. Each operates over a somewhat different range of conditions and with different efficiencies of response. This overlapping "soft" redundancy seems to characterize biological regulation of all kinds. It is not notably efficient or elegant in the engineering sense. But it is robust and continually sensitive to changes in internal body temperature. This is quite unlike the examples of rigid regulation by management where goals of operational efficiency gradually isolated the regulating agency from what it was regulating.

Second, endothermy is a true innovation that explosively released opportunity for the organisms evolving the ability. Maintaining high body temperature, short of death, allows the greatest range of external activity for an animal. Speed and stamina increase and activity can be maintained at both high and low external temperatures. A range of habitats forbidden to an ectotherm is open to an endotherm. The evolutionary consequence of temperature regulation was to open opportunity suddenly for dramatic organizational change and the adaptive radiation of new life forms. Variability is therefore not eliminated. It is reduced in one place and transferred from the animal's internal environment to its external one as a consequence of allowing continual probes by the whole animal for opportunity and change. Hence the price of reducing internal resilience and maintaining high metabolic levels is more than offset by that creation of evolutionary opportunity.

That surely is at the heart of sustainable development—the release of human opportunity. It requires flexible, diverse, and redundant regulation, monitoring that leads to corrective responses, and experimental probing of the continually changing reality of the external world. Those are the features of adaptive environmental and resource management. Those are the features missing in the descriptions I presented of traditional, piecemeal, exploitive resource management and its ultimate pathology.

The case studies presented here have shown that the descriptions and postulates that launched this effort are seriously incomplete. For example, in New Brunswick, the intensifying gridlock in forest management, combined with slowly accumulated and communicated understanding, led to an abrupt transformation of policy whose attributes became much like those just described for homeostasis (Baskerville, chapter 2). It is a policy that functions for a whole region by transforming and monitoring the smaller-scale stand architecture of the landscape and by releasing the productive and innovative capacities of industry.

Even though the postulates are incomplete, they did provide the direction to reveal a number of new insights in the case studies. Informal collegia with contacts inside (rebel bureaucrats) and outside (maverick academics) the system are necessary to unlocking institutional gridlock, as is the case with New Brunswick, the Everglades (Light et al., chapter 3), and the Baltic (Jansson and Velner, chapter 7). The development of a sense of involvement, ownership, and belonging by the people at a regional scale was important to the generation of sustainable policies in the Chesapeake Bay (Costanza and Greer, chapter 4) and Great Lakes (Francis and Regier, chapter 6). An institution charged with regional strategic planning and supported by a research arm seems essential to provide that integrative and long-term view that is inexorably lost in agencies with a primary management or regulatory function (Lee, chapter 5). These and other conclusions are described in the concluding synthesis chapter.

These examples of regional resource management do suggest that institutions and societies achieve periodic advances in understanding and learning through the same four cycles of growth, production, release, and renewal that shape the spatial and temporal dynamics of ecosystems (figure 1.2). But each proceeds at its own pace and in its own space, and this creates extraordinary conflicts when there are extreme mismatches among the scales at which ecosystems, institutions, and societies function. If the scale of all three become more congruent, it is likely that the inevitable bursts of human learning can proceed with less conflict and more creativity.

This chapter has used metaphors and puzzles to provide some insight into what sustainable development is and how to harmonize relationships among people, nature, and enterprise. The ecosystem metaphor led to the conclusion that there is a cycle of slow growth and production that triggers fast disturbance and renewal. The slow growth and production phase accumulates natural capital. It is analogous to the processes of what we call development.

The fast disturbance and renewal phase releases bound and constrained capital and reorganizes it for a reestablishment of the ecosystem cycle. It is analogous to the conditions of what we call sustainability, and it is the phase where diversity is maintained. Therefore sustainability is measured by some attributes of disturbance and renewal, and development is measured by some attributes of growth and production.

The solution of a puzzle of ecosystem organization helped clarify what the specific attributes are that determine ecosystem sustainability. The puzzle was that a few simple processes seem to generate the great complexity and diversity within ecosystems. Ecosystems are hierarchically structured into a number of levels. Relatively few processes determine this structure, and each imposes distinct frequencies in space and time on the ecosystem over different scale ranges. They entrain all other variables.

Hence both sustainability and biodiversity are determined by the structuring variables of disturbance and renewal that control the lumpy geometry and lumpy time dynamics. To use another metaphor, they set the stage upon which other variables play out their own dramas. The health and viability of the physical and temporal infrastructure of biomes *at all scales* sustain the theater. Given that, the actors will look after themselves!

A second puzzle suggested that many existing examples of management of renewable resources inexorably led to more vulnerable ecosystems, more rigid management institutions, and more dependent societies. Its resolution came from another biological metaphor of regulation, that of homeostatic regulation of body temperature in endotherms. Indeed, successful control of variability there does reduce resilience within the system regulated. Unlike the pathology of management noted, however, the regulation is responsive to internal change and is functionally diverse and robust. It transfers internal variability externally to release opportunity for probing, creative opportunities.

This is at the heart of sustainable development—the release of human opportunity. It requires flexible, diverse, and redundant regulation; monitoring that leads to corrective responses; and experimental probing of the continually changing reality of the external world.

Finally, sustainable development is neither an ecological problem, a social problem, nor an economic problem. It is an integrated combination of all three. Effective investments in sustainable development therefore simultaneously retain and encourage the adaptive capabilities of people, business enterprises, and nature. The effectiveness of those adaptive capabilities can turn the same unexpected event (e.g., drought, price change, market shifts) into an opportunity for one system, or a crisis for another. These adaptive capacities depend on the processes that permit renewal in society, economies, and ecosystems. For nature it is biosphere structure; for businesses and people it is usable knowledge; and for society as a whole it is trust.

Citizen and politician are now frustrated because they are not hearing simple and consistent answers to the following key questions about present environmental and renewable resource issues:

- What is going to happen under what conditions?
- When will it happen?
- Where will it happen?
- Who will be affected ?
- How uncertain are we?

The answers are not simple or consistent because we have just begun to develop the concepts, technology, and methods that can deal with the generic nature of the problems. These generic features can be described in various ways, but here is my overly academic attempt:

- The problems are essentially systems problems where aspects of behavior are complex and unpredictable and where causes, although at times simple (when finally understood), are always multiple. *Therefore interdisciplinary and integrated modes of inquiry are needed for understanding. And understanding (not complete explanation) is needed to form policies.*
- The problems have a fundamentally nonlinear cause. They demonstrate multistable states and discontinuous behavior in both time and space. *Therefore the concepts that are useful come from nonlinear dynamics and theories of complex systems. Policies that rely exclusively on social or economic adaptation to smoothly changing and reversible conditions lead to reduced options, limited potential, and perpetual surprise.*
- The problems are increasingly caused by slow changes, reflecting decadal accumulations of human influences on air and oceans and decadal to centurial transformations of landscapes. These slow changes cause sudden changes in fast environmental variables that directly affect the health of people, productivity of renewable resources, and vitality of societies. *Therefore analysis should focus on the interactions between slow phenomena and fast ones, and monitoring should focus on long-*

term, slow changes in structural variables. The political window that drives quick fixes for quick solutions simply leads to more unforgiving conditions for decisions, more fragile natural systems, and more dependent and distrustful citizens.

- The spatial span of connections is widening, so that the problems are now fundamentally cross-scale in both space and time. National environmental problems can now more and more frequently have their source both at home and half a world away (witness greenhouse gas accumulations, the hole in the ozone layer, AIDS, and narrowing biodiversity). Natural planetary processes mediating these issues are coupling with the human, economic, and trade linkages that have evolved among nations since World War II. *Therefore the science needed is not only interdisciplinary but cross-scale. Yet the very best environmental and ecological research and models have achieved their success by being either scale independent or constrained to a narrow range of scales. Hierarchical theory, spatial dynamics, event models, satellite imagery, and parallel processing may open new ways to violate, successfully, the hard-won experience of the best ecosystem modelers (i.e., never include more than two orders of magnitude; otherwise the models will be smothered by detail).*
- Both the ecological and social components of these problems have an evolutionary character. That is why the phrase *sustainable development* is not an oxymoron. The problems are therefore not amenable to solutions based on knowledge of small parts of the whole or on assumptions of constancy or stability of fundamental relationships—ecological, economic, or social. Assumptions that such constancy is the rule might give a comfortable sense of certainty, but it is spurious. Such assumptions produce policies and science that contribute to a pathology of rigid and unseeing institutions, increasingly vulnerable natural systems, and public dependencies. *Therefore the focus best suited for the natural science components is evolutionary, that for economics and organizational theory is learning and innovation, and that for policies is actively adaptive designs that yield understanding as much as they do product.*

Part 2
Case Studies

2

The Forestry Problem: Adaptive Lurches of Renewal

Gordon L. Baskerville

New Brunswick, a province on the Atlantic coast of Canada (figure 2.1), is 86% forested, and its forests have been central to life in the province from the time of first settlement. New Brunswick always seems to have a forestry problem that is the basis of local debate. The nature of the problem evolves over time, and the heat of debate ebbs and flows but the forestry problem persists.

This chapter examines the evolution of the problem with reference to forest policy in New Brunswick from the late 1940s to the present. Beginning with a provincial decision for economic development, there followed an immediate conflict with an insect, followed in turn by conflict over the use of pesticides, recognition of a long-term timber supply problem, the necessity to change the rules of access to publicly owned forest, and redesign of forest policy. The process has been dynamic, often with discourse on these and other issues running on parallel tracks and then closing suddenly on one another.

The forestry problem always has technical aspects, and these have been earnestly debated by scientists and professionals on the assumption of public acceptance, but they have rarely been articulated in a form understandable to policy makers or to society at large. At the same time, the social aspects of the forestry problem, including the impact of tech-

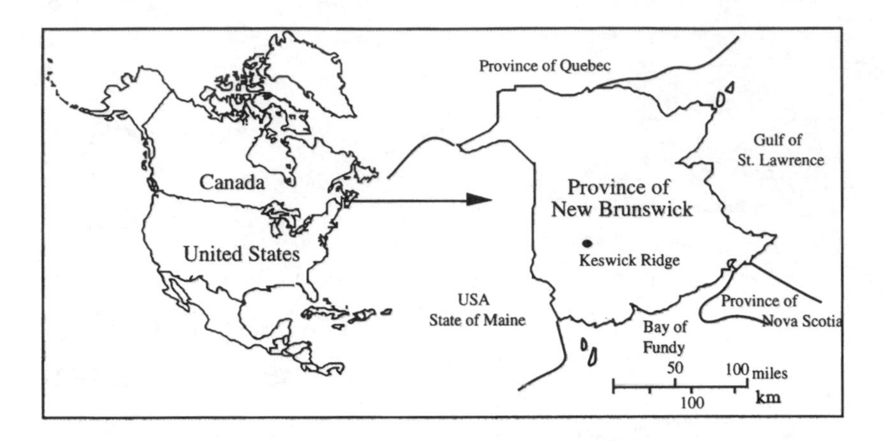

FIGURE 2.1
The location of New Brunswick in northeastern North America.

nologies, have been debated more or less continuously in the social arena, on the implicit assumption that there were no underlying technical and natural bases to the problems or that such bases were incredibly simple. Those responsible for the modulation of policy in government have tried to chart an "appropriate" course, given these seemingly inconsistent technical and social perspectives on the problem. The dichotomy of perspective, and government response, has fueled a continuing policy debate.

There is an enormous amount of written material about the forestry problem, but it is not uniformly available. The scientific elements are covered in the conventional scientific literature and professional journals, but the technical applications of this science are only partly described in working documents of the resource agencies. Academic policy literature has been essentially irrelevant to the unfolding real problem. The realistic policy literature exists solely in the form of government reports and internal documents for which there is no systematic access. A wide array of material on how the public views both technical and policy issues is found in media records and in reports released by a variety of agencies, citizen groups, and individuals. Although the latter material is not formally published, it is extensively circulated, and for much of the public it constitutes the real policy information. The true nature and dimensions of the forestry problem in New Brunswick may be found in the total of all these sources. The material referenced here is a small fraction of that total and can be found, except in a few cases, in conventional literature. The latter items may be found in the libraries of the New Brunswick Department of Natural Resources and Energy, or of Forestry Canada—Maritimes, both in Fredericton, New Brunswick.

What has been written in traditional forms rarely touches on the mechanics of policy issues as these actually operate. This is not surprising, for players have seldom put their policy views in comprehensive written form for public view. Thus most published inferences about policy development in New Brunswick are based on interpretation rather than on stated policy or on participation. At the same time, the players have been willing to state what someone else's policy was *believed to be.* Recognizing that distinction is crucial to understanding the controversy associated with the evolution of policy described here.

What follows is a précis of a complex set of events. The reader is warned that (1) the events described were orders of magnitude more complex than can be presented and (2) the description is being written with the benefit of hindsight. Given the nature of the information base, it is not possible (for anyone) to present an unbiased view of events. For the record, the author was, at various times in the story that follows, a student, a researcher and then a research manager in a federal lab, a professor of ecology, an assistant deputy minister in the provincial government, dean of a faculty of forestry, a policy consultant to the provincial government, a professor of forest policy, a policy consultant to industry, and chairman of a board that heard disputes with respect to Crown forest management in the province, as well as a citizen of the province of New Brunswick.

Forests and New Brunswick Society

Forests have always been important to the local society. Among the provinces of Canada, New Brunswick has the highest percentage of forest, all of which is easily accessible. The province has the highest proportion of the population with hunting licenses and fishing licenses. Access to the forests has been considered a birthright, whether or not there happened to be a law to say so, and each citizen claims the opportunity to enter forestland for hunting, fishing, bird watching, snowmobiling, or other recreational pursuit.

The province was first explored by the French and then settled by the British. It became home to Acadians at the time of the expulsion and home to United Empire Loyalists at the time of the American Revolution. The population is split almost equally between people of French and English linguistic backgrounds and is the only officially bilingual province. Both cultures have strong roots in the forests, and the forests are economically important to both.

The forests of New Brunswick are classified as Acadian except for a small area of boreal forest in the northwest (Rowe 1959; Loucks 1962). The typical softwood species are balsam fir; white, red, and black spruce; and white pine. The typical hardwood species are white and yellow birch, sugar maple, and beech.

Current ownership of the 6.1 million ha of productive forest is 48% public in the name of the province (referred to as Crown forest), 30% in some 30,000 small private ownerships known as woodlots, 21% in a half dozen large industrial ownerships, and 1% public in the name of the federal government (table 2.1). The Crown forest is mostly in the north and center of the province, usually back from settlements that line the river valleys and coasts. Woodlots tend to be concentrated along

TABLE 2.1
Some Current Statistics for New Brunswick

Population: ~730,000	
Largest city: ~80,000	
41 forestry-dependent communities	
Forest industry:	
12 pulp and paper mills	Total shipments: $2.7 billion
66 sawmills	Value of exports: $1.4 billion
10 shingle mills	Balance of trade: $1.3 billion
19 other forest related mills	Wages: $458 million
179 logging firms	Jobs: 18,000 direct/12,000 indirect
Total area:	7.3 million hectares
Forest area:	6.3 million hectares (86%)
Productive forest area	6.1 million hectares
Crown owned	2.9 million hectares (48%)
Woodlot ownerships	1.8 million hectares (30%)
Industry ownerships	1.3 million hectares (21%)
Federal ownership	0.1 million hectares (1)%
Pedominantly softwood forest, 46%	
Predominantly mixed softwood/hardwood forest, 27%	
Predominantly hardwood forest, 27%	

the edge of the settled areas, and the industrial ownerships lie between the woodlots and the Crown forest. The federal land is comprised of two national parks, various military reserves, Indian lands, and forest research reserves. The forest industry still dominates the provincial economy, as it has historically. It accounts for 25% of all goods produced in the province and for 50% of the exports. About 12% of all employment is in forestry, and 9% of the gross domestic production derives from the industry.

The provincial Department of Natural Resources and Energy has specific authority over Crown forests and general responsibility for forest resources management. This department and its various antecedents are referred to here as DNRE.

Some Early History

The forest policy history of what is now the New Brunswick Department of Natural Resources and Energy (DNRE) has been reviewed recently (Fellows 1988), and this section draws on that work. In 1810 the New Brunswick House of Assembly sought to regulate the use of Crown forests on the basis that they "would otherwise soon be rendered of no value." The forests referred to here were scattered stands of white pine, and individual white pine trees growing throughout a forest that was principally fir and spruce.

In 1816 a surveyor general was appointed with responsibility for the forest resource in the colony. There were few data on the resource at that time, but records of revenues from timber export show the government of the day was concerned that full value be obtained for timber harvested. By 1837 there was an agreement that the local assembly would have control of the Crown forests, including revenues from timber. In return, the assembly undertook to pay the salaries of local colonial bureaucrats.

Access to Crown forests was already a sensitive issue in 1837, when an act provided for the sale of harvesting licenses to the highest bidder, but objections that this would favor large companies forced immediate amendment. In 1841 an early example of environmental legislation forbid dumping sawmill slabs in the Miramichi River, as this practice was damaging the habitat of Atlantic salmon. At that time some 600 sawmills formed the economic backbone of the province, and the salmon waters of New Brunswick were becoming internationally famous among anglers.

New Brunswick entered the Canadian confederation in 1867, and in 1870 the provincial legislature passed an act entitling railway builders to claim 4000 ha of forest for each 1.6 km of rail line constructed. The land was awarded in blocks more or less parallel to the right-of-way, and some of these blocks survive today as the basis of a few large freehold ownerships. By 1901 there were only 315 sawmills in the province, and the average size had diminished dramatically from that of the mid-1800s. The first pulp mill was built in 1889, but the first records of round pulpwood harvest do not appear until 1907.

Throughout the 1800s, despite ups and downs of international markets, lumbering was the mainstay of the New Brunswick economy. The industry was based on white pine until the later decades, when spruce lumber increased in importance. For most of that century, reduced availability of large white pine was seen, and discussed, as a threat to the economy. Harvesting of the white pine was by selective cutting, not to be confused with the selection method (*méthode du controle*) of European forestry. In perhaps oversimple terms, selective cutting removes the best trees first (often described as "cut the best and leave the rest"), whereas true selection management cuts the best trees last. Top-grade pine trees were being felled at a rate faster than younger trees reached the age, and form, of the veterans (150–250 years). By the middle of the nineteenth century the remaining white pine trees in New Brunswick were of such size as to make production for the lumber markets of Europe difficult. While the lumber industry struggled on, the principal raw material became white spruce. Since this species grew primarily in stands (usually mixed with balsam fir), harvesting was somewhat more efficient than it had been for pine. At the same time, individual trees were not as big as white pine had been, nor was the wood as valuable for conversion to lumber.

The relative value of forest types was mirrored in the reaction to natural catastrophe. Records show the great Miramichi fire burned thousands of square kilometers in 1825 and was seen as a threat to habitation, but it was not much commented upon as a catastrophic loss of forest production. Equally, the Saxby gale of 1869 flattened extensive areas of forest in the south of the province but was noted more for the fact of its forecast than for its damage to the forest. There were outbreaks of the forest insect spruce budworm in 1770, 1806, and 1878 (Greenbank 1956), all of them geographically massive. Historical records make little mention of this damage by insect outbreaks, fires, and windstorm. Although the damage was of enormous proportions in terms of area covered, the valuable white pine had mostly been harvested from these forests, and fire, gale, and budworm disrupted a leftover forest consisting mostly of fir, which was considered a weed species.

An outbreak of spruce budworm that influenced the entire fir/spruce forest in the period 1913–1919 was recorded and commented upon with concern (Miller 1913; Tothill 1922a,b; Swain and Craighead 1924), as by that time the timber lost or damaged was important to the emerging pulp industry, and spruce had become the principal sawlog species. Surveys of the forest noted that damage by the budworm had been severe and extensive (Prince 1921; Flieger 1940). Nevertheless, the losses were accepted without much comment, as harvest levels for lumber and pulp combined in the 1930s approached the lowest levels in more than a century.

To strengthen industrial development and encourage management of the timber resource, harvest licenses with a 50-year term were introduced in 1927. Holders of licenses were expected to exercise stewardship over the forest on the license, and the tenure term almost equivalent to the time required to produce a crop was to ensure that the stewards would reap the benefits of their efforts. The term of tenure was reduced to 10 years by 1933.

In 1929, aerial photography was used as the province made its first attempt to obtain a comprehensive survey of the forest resource.

Markets recovered during World War II, but little new mill capacity was installed. By that time white pine was no longer a major species, because of historic selective high-grading and because of increasing impact of the introduced white pine blister rust and the native white pine weevil, which had spread through the white pine of the New Brunswick forest.

Although softwoods have been the dominant economic interest in the New Brunswick forests, the smaller hardwood resource was also utilized. Stands of sugar maple and yellow birch were selectively high-graded for lumber and veneer logs, particularly during and immediately after World War II. So intense was this high-grading that the hardwood forest was well on the way to weed status by the 1950s. The state of the hardwood resource has been a cause of lament, but never at a level that triggered a serious policy response.

New Brunswick did not run out of white pine trees; indeed, there may be more individual white pine stems in the forests today than there were in the early 1800s. The difference is that there are now very few trees 150–250 years old, and most of the maturing stems have been damaged by blister rust disease and weevil attack. Similarly, there are probably as many sugar maple in the forest now as there were historically, but the remaining trees are either young or old and of poor quality.

Each generation of New Brunswickers has strived to exact economic and social benefits from the publicly owned forest, and stewardship of these forests has been a continuing problem. Access to harvest rights in the Crown forests has gone through the full range from assignment of specific individual rights to sharing of a commons. None of these forms of access has worked; on the other hand, none has been left in place (without fiddling) long enough to learn much about its impact. One way or another, policy with respect to access to the Crown forests has changed two times or more during the life of every crop harvested in the last century (Fellows 1988). Since the dynamics of forest development operate on a time scale of centuries and political dynamics function on a time scale of weeks, achieving stewardship of a publicly owned forest is not straightforward, and New Brunswick has experienced the full range of problems.

An interesting review of Crown forest tenure from the early 1600s onward, and of its impact on stewardship of the resource, is found in Caverhill (1917). This work is both historical and prophetic. Caverhill noted, "Our past and present methods of handling Crown Lands are at fault," and his description of the forestry problem at that time is eerily similar to what follows here. He also foresaw the concept of sustainable development in his statement of policy intent:

> The object sought by the state in the management of the public Domains should be: The maximum permanent development of the state, or the greatest benefit which can be secured from this source for the greatest number of its inhabitants, consistent with future development and welfare of future generations, and any policy for land development which admits to other aims is bound sooner or later to meet with failure.

Although the forest played a dominant role in New Brunswick history, there is not much in that history to indicate an appreciation of the dynamic nature of the forest. Certainly, selective high-grading of white pine and hardwoods was carried on without much thought about the age of trees being harvested, about what would happen to the forest if only the biggest and the best trees were harvested, and about how long it would take nature to replace them with trees of similar size and quality. Yet selective high-grading was sufficiently intense to render local white pine and hardwoods noncompetitive as raw material for world markets.

Whatever the actual understanding of the pace and nature of forest development that may have existed among the woodsmen, the public, and the policy makers of the century and a half prior to World War II, there is little evidence that any such knowledge was used to control the impact of activities of man on the dynamics of change in the forests.

The Forestry Problem Post–World War II

Since 1945 the forestry problem has been variously described as the result of inadequate industrial capacity, the budworm, uncertain wood supply, raw material quality, stewardship, insecticides, environmental quality, economics, parks, markets, too many overmature stands, wildlife habitat, and so on, depending on who was talking, where, and when. The problem has had all these dimensions, and more, and whether or not they were acknowledged, they all have been present more or less continuously.

The Desire for Economic Development

The government of New Brunswick decided to reestablish the economic health of the province by renewing the forest-based industry following World War II (Brown 1950). The basis of development was to be the extensive area of mixed stands of balsam fir and white, red, and black spruce that made up half the forest area. The province, encouraged by a federal/provincial agency called the Atlantic Development Board, sought increased economic activity through industrial expansion of existing facilities, and new companies were enticed with prospects of plentiful fir and spruce raw material of high quality. Although there has been talk from time to time about revitalizing the hardwood forests, forest policy since World War II has focused on softwoods.

In the 1950s, the fir/spruce forest targeted for industrial development consisted almost entirely of two age classes. The older stands had originated around 1880 following a major spruce budworm outbreak, and these stands had been "thinned" during a less extensive budworm outbreak of 1913–1919. Surviving stands had open canopy conditions, which by the 1950s had resulted in the production of individual large-diameter trees. The second age class had originated about 1924 following a budworm outbreak in 1913–1919. These stands were so dense they acquired the name St. Michael (as pronounced in French), which is said to relate to a legend that the beard of St. Michael was thick enough to hide mice.

The historic dominance of budworm in the dynamics of natural development in the fir/spruce forest was such that on the order of 80% of the total area of the softwood forest was in one of the two age classes noted earlier. Harvesting for pulpwood in the fir/spruce forest from the early 1900s to 1945 had initiated new stands across an equivalent span of age classes. However, because markets limited harvest levels, these age classes were not extensive. In appearance, the fir/spruce forest of the 1950s was dominated by two classes of even-aged stands, both of budworm origin. The natural creation of these stands by budworm had occurred on a provincial scale where almost half the softwood forest area was "harvested" in the short space of an outbreak (typically about 8 years) with a period of about 30–60 years between outbreaks, leading

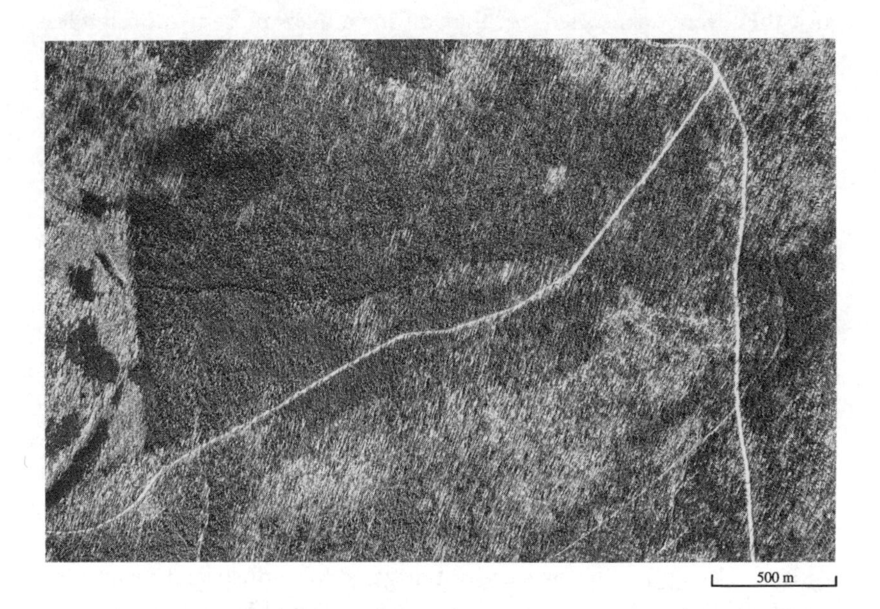

FIGURE 2.2
A portion of an aerial photograph taken in 1955 and showing fir/spruce stands of 1880 origin and 1924 origin. The uncut area consists predominantly of open stands of 1880 origin. The small patches in the clear-cut to the right of the photo are dense stands of 1924 origin. Similar patches may be seen scattered throughout the 1880 origin stands in the uncut block.

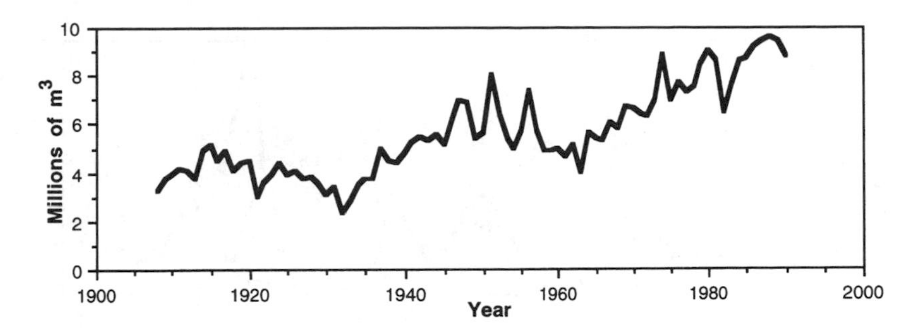

FIGURE 2.3
Total annual softwood harvest (mostly fir and spruce) in New Brunswick since 1907.

to a discontinuous even-aged structure, with average stand size on the order of hundreds of hectares. Although age classes were not recorded in the 1958 inventory (DNRE 1958), the two ages of fir stand were obvious in the forest, and they were distinct on aerial photographs (figure 2.2).

In 1957 a Forest Development Commission reported, "There is now plenty of wood in New Brunswick to support a great expansion of the forest industries without delay, but continued development 20 to 30 years from now will depend on early introduction of better forestry techniques aimed at building up the forest inventory" (Bates 1957). From the mid-1950s to 1980, installed mill capacity in the province doubled as the government marketed the fir/spruce forest (figure 2.3).

What About the Budworm?

Although the spruce budworm, an insect that feeds on the foliage of balsam fir and spruce trees, has been around for centuries, only since World War II has it become an economic threat. The insect periodically bursts into outbreak conditions across northeastern North America, defoliating large expanses of fir/spruce forest (figure 2.4). Usually, destruction of old stands results in immediate natural regeneration to new fir stands, eventually creating a forest with characteristics similar to those that were destroyed. The link between budworm and the host fir forest is evolutionary, and the insect is the driving force in natural development of the fir forest (Swain and Craighead 1924; Balch 1952;

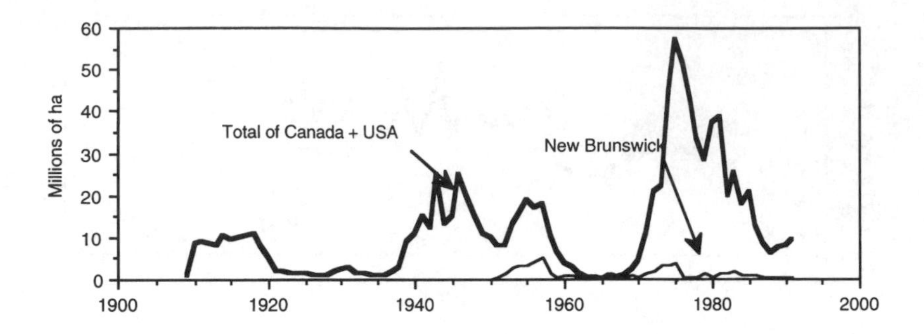

FIGURE 2.4
Area of moderate and severe defoliation caused by spruce budworm in eastern Canada and the United States from 1909 to 1991. The part of the most recent outbreak from 1949 onward in New Brunswick is also shown. (Data from Irving and Webb 1984a, DNRE, and *Forestry Canada.*)

Morris 1963). Since World War II the budworm has been at or near the center of the forestry problem.

Since the fir/spruce forest had not been historically important as an economic base, the budworm and its relationship to the dynamics of that forest were not widely recognized in the 1950s. As industrial development was being considered, some entomologists noted that the forest that was to be the basis of an expanded pulp and paper industry had been put there by a provincial-scale "harvest" by spruce budworm in the late 1870s. Further, the immature St. Michael stands were the product of a budworm outbreak in 1913–1919. It was argued that budworm outbreaks were triggered by a combination of climatic factors and the presence of large areas of mature fir stands, which in turn had been created by earlier budworm outbreaks (Balch 1946; Greenbank 1956, 1957; Morris 1963).

Since balsam fir stands rarely were referred to in provincial records prior to World War II, there was only a short human history of the fir/spruce forest and its relationship with spruce budworm. Nevertheless the naturalist/entomologists of the 1920s, 1930s, and 1940s saw a set of insect-driven forest dynamics that had dominated in the "unmerchantable fir/spruce forest" for centuries, and conditions were right in the 1940s for another spruce budworm outbreak (Swain and Craighead 1924; Balch 1946).

Forecasts of the entomologists proved correct. By 1950 there was widespread evidence of damage to balsam fir stands in northern New

Brunswick as a result of defoliation by spruce budworm (Balch 1952). At that time some could remember the devastation of the fir forests during the 1913–1919 budworm outbreak, and some had taken part in damage surveys following that outbreak. They foresaw that a budworm epidemic could result in near total loss of the mature stands of 1880 origin and in severe reductions in stocking among the stands of 1924 origin. This was forecast to happen over a period of less than a decade and was expected to result in collapse of the raw material supply for sawmills, major damage to the raw material supply for existing pulp and paper mills, and foreclosure of any option for industrial expansion based on the fir/spruce forest. Industry and government joined in an attempt to control damage by budworm to the fir/spruce forest. In 1952 aerial application of a pesticide was used operationally to prevent stand mortality caused by defoliation in northwestern New Brunswick (table 2.2) (Balch 1952; Morris 1963; Kettela 1975).

TABLE 2.2
Recent Budworm Infestation History in New Brunswick

Year	**Forecast Area***	**Treated Area†** (thousand hectares)	**After Treatment‡**
1950	n/a	102	
1951	na	563	
1952	n/a	75	~1536
1953	n/a	733	~2816
1954	n/a	462	~3328
1955	n/a	455	~3328
1956	n/a	799	~4096
1957	n/a	~2103	~5120
1958	n/a	105	~1220
1959	n/a	0	260
1960	n/a	107	~1024
1961	930	890	800
1962	570	550	No survey
1963	160	270	144
1964	690	840	640
1965	810	860	680
1966	~1580	800	960
1967	770	420	160
1968	610	~200	385
1969	~1860	~1260	969
1970	~2830	~1710	568
1971	~3480	~2430	~1406
1972	~4010	~1850	~1716

(continued)

TABLE 2.2
Continued

Year	Forecast Area*	Treated Area† (thousand hectares)	After Treatment‡
1973	~3030	~1840	~3310
1974	~5060	~2400	~3320
1975	~5740	~2700	~3567
1976	~5020	~3880	404
1977	~2540	~1680	481
1978	~2990	~1550	669
1979	~3200	~1600	~1320
1980	~3100	~1620	673
1981	~2600	~1900	~1221
1982	~4000	~1730	~1202
1983	~5300	~1740	~2028
1984	~4100	~1250	730
1985	~3570	730	~1070
1986	~3150	540	927
1987	~1710	580	430
1988	~1500	550	400
1989	~1650	610	396
1990	~1470	560	237
1991	990	320	266

*These are forecasts of the probable area of moderate to high infestation based on where the budworm was found the previous year. They were initiated in 1960.

†The actual area treated in a year is determined partly by the forecast and partly by the emerging defoliation pattern in the year of treatment. It can be greater or smaller than the forecast.

‡This is the area of moderate and severe defoliation in the forest after treatment. It is a result not only of untreated stands and dispersal of the insect but of inadequate protection and can be in the same or a different geographic location from the area treated.

How the Short Term Became the Long Term

In one form or another, the forest protection program begun in 1952 has continued to the present. From the outset, the use of insecticides to control defoliation of fir/spruce stands by budworm was designed to limit mortality when that became imminent in damaged stands. Since outbreaks had historically been of short duration, the strategy was to protect stands under immediate threat of mortality until the outbreak collapsed of its own accord. This protective action was initiated in stands only when it appeared there would be tree mortality the next year if the current year's foliage was not saved. This required an elaborate system of tracking each year both the insect and stand damage across the whole province. It was recognized that because of the population biology of the budworm and because New Brunswick represented a small portion of the total area afflicted with the highly mobile spruce budworm, local eradication of the insect was impractical (Balch 1952; Morris,1963; Kettela 1975; Kettela 1982) (table 2.2). As a feat of logistics, the program has been outstanding. To track the annual distribution of budworm populations and resultant damage over millions of hectares of forest with such precision and effectiveness of crop protection was exceptional (Irving and Webb 1984a).

The protection program targeted stands that were expected to die, but the insect dispersed each year, so that the spatial pattern of defoliation changed (table 2.2). As a result, the protection program "chased" the budworm around New Brunswick.

In the late 1940s an area of 54 square kilometers was set aside from budworm protection, and from harvesting, to allow long-term study of natural budworm/forest dynamics (the Kedgwick Check Area). By 1958, research surveys showed there were no stands of commercial-sized balsam fir surviving in the Kedgwick Check Area (Morris 1963), and immature stands in that area had suffered a severe thinning (Baskerville 1960). At the same time, it was clear that the regeneration pattern following destruction of the mature fir stands was toward another fir/spruce mixture (Baskerville 1955). These observations were consistent with those reported following the outbreak in the 1920s.

Population levels in the northern part of the province collapsed in 1959 as predicted, and in the Kedgwick Check Area they did not rebound (Morris 1963). Predictions of the entomologists for a short outbreak appeared correct. However, epidemic levels of spruce budworm continued to spread through the rest of the New Brunswick fir/spruce forest, which was being protected by aerial application of insecticides (table 2.2) (Kettela 1982; Webb and Irving 1983).

As the protection program began, the contemporary wisdom was that budworm outbreaks lasted 4–8 years and that protection would therefore be needed only for a few years until the outbreak subsided naturally. Meanwhile, periodic spraying of insecticides to reduce the number of feeding larvae would save foliage and allow threatened stands to survive and grow (Morris 1963). The biological mechanisms of population response to the maintenance of foliage in a stand remain the subject of scientific debate, but one thing is clear: Insecticides may kill almost the entire resident larval population, but either the remaining small fraction is sufficient to generate an immediate population rebound or the protected foliage is invaded during subsequent dispersal of adult moths from untreated stands harboring the insect. Whatever the mechanism, it is clear that budworm population dynamics are not independent of vegetative dynamics in the host forest.

Maintaining foliage levels by spraying insecticides was not expected to cause the insect population to collapse, but neither was it expected to permit insect populations to persist at outbreak-levels for periods far in excess of the normal outbreak period. A biological feature of the historic short period of outbreak level populations (4–8 years) appears to have been destruction of a substantial portion of the host forest and consequent loss of food supply. That is, the use of insecticides had not maintained the outbreak; rather, the outbreak was driven by the enormous area of untreated infestation outside New Brunswick (figure 2.4), and protective spraying maintained stands that were regularly reinvaded after protection.

Three conflicting realizations with respect to budworm dynamics and forest dynamics began to dawn on the New Brunswick forestry community in the 1960s.

1. It became clear that storing fir timber "on the stump" was not possible in the way spruce stands and white pine trees had been stored in the past. In the 1960s stands of the fir forest of 1880 origin were showing signs of breaking up because of overmaturity (Baskerville 1965). There was some argument as to whether these stands would remain stable long enough to permit their orderly harvest over a period of 15–20 years, even if they were successfully protected from budworm. Balsam fir stands had been severely damaged by budworm on such a regular basis that the natural tendency for stands of this species to break up at an early age (relative to spruce and pine) had not been noted until protection kept most fir stands alive.
2. It became clear that as long as there were significant areas of healthy mature fir forest, the budworm population might rise and fall, but the insect was unlikely to go away. That meant the protective spray program took on aspects of a forever thing if the fir forest was to be the basis of sustained industrial development.
3. Almost half the fir/spruce forest area was in the 40-year age class, and concern was growing that these immature stands

might also have to be protected until they were harvested—a period that would probably stretch from the 1980s into the 2020s.

The preceding realizations altered the timber supply and budworm perspectives on the forestry problem. It was said these were long-term problems, but despite annual lessons in natural dynamics being offered by the budworm throughout the province, there was little appreciation of the nature of budworm/forest dynamics. The budworm and the forest were discussed separately, mostly in separate forums.

A Dangerous Management Paradigm

In the 1950s and 1960s the dominant concern about the fir/spruce forest was what was referred to as "the inventory," which meant the total volume of all the trees in the forest (also known as growing stock). As the 1957 Forest Development Commission report had noted, there was plenty of wood, and the key was to build up the inventory, the all-purpose measure of the forest estate at that time. The 1958 forest survey had shown a considerable increase in growing stock over the 1937 estimate. Another survey in 1968 once again showed an increase in the total volume of fir/spruce growing stock (figure 2.5). These surveys were hailed as good news and evidence of successful forest management.

At this time, the larger the growing stock volume, the better the forest was considered to be. Related to that claim was the fact that the determination of sustainable harvest level amounted to dividing the volume

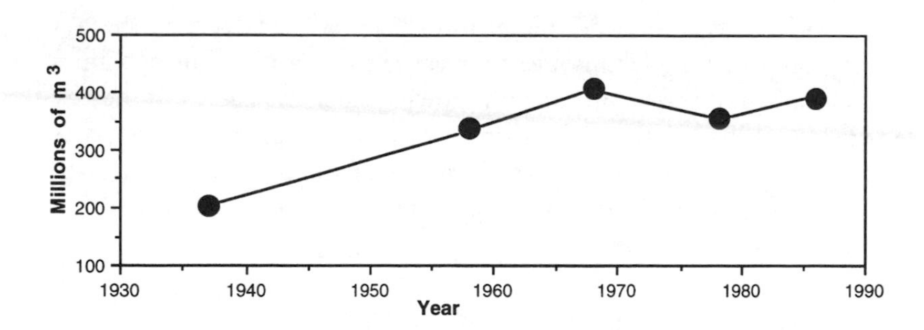

FIGURE 2.5
Total volume of fir/spruce growing stock in the New Brunswick forests as determined by surveys in 1937, 1958, 1968, 1978, and 1986.

of growing stock in the whole forest by the age of a single stand at the time of harvest. These formulas took the general form $H = GS/A$, where H is the amount (in cubic meters) of annual allowable harvest, GS is the total number (also in cubic meters) of growing stock of all fir/spruce stands in the forest, and A is the age for harvesting stands (Davis 1954). Thus the larger the volume of growing stock or the younger the stand age at harvest, the greater was the annual allowable harvest volume.

These simplistic annual allowable cut calculations were susceptible to manipulation. If a larger harvest was desired to permit expansion of a mill, it was only necessary to "discover" a lower cull factor that would lead to a larger volume of growing stock on the associated timber license (instantly) and thereby an immediate increase in allowable harvest. Equally, if a lower harvest level was needed in order to argue for retention of access to a large forest license, it was only necessary to "discover" a higher cull or higher age at harvest. Several reviews of allowable harvest levels for the province, and of individual timber licenses, were made in the 1960s, by DNRE and others. Although they were not published, it always seemed that with the appropriate adjustment in the growing stock for cull, proper merchantability rules, and a good rotation age, the level of harvest currently desired turned out to be just about right.

Although this view of forest production contained no biologically reasonable representation of forest dynamics, it was the textbook view of the day. Textbooks duly acknowledged the importance of forest dynamics, but since computers were not commonly available to perform the cumbersome calculations, growing stock was offered as an alternative basis for calculating sustainable harvest and as the measure of goodness of the forest estate. The problem lay in the failure of this approach to recognize the role of forest age-class structure in the dynamics of wood availability *over time,* especially relative to the dynamics of budworm. By concentrating on growing stock, managers of the day failed to see the risk that the existing age-class structure of the fir/spruce forest (created by budworm) might not maintain a balance of growth of the growing stock, with the chosen rate of harvest from that growing stock (by man) *over time.*

Two Problems, Two Debates

In the mid-1960s two issues began to consume the interests of the forest community, the use of insecticides, and forest dynamics.

Insecticides

Although the use of insecticides to protect fir/spruce stands threatened with budworm-caused mortality had been largely successful, it was often publicly portrayed as a failure because the budworm had not gone away. The public concern centered on the indefinite continuation of protective actions with chemical sprays.

Initially, debate in the media centered on environmental problems. Protecting hundreds of thousands of hectares of infested stands annually by using insecticides damaged nontarget populations. A wide array of studies examined the side effects of the protection program, constituting a scientific literature of their own. The use of this information in evaluating damage to populations of nontarget insects, trout, Atlantic salmon, and birds is summarized in Fowle (1988). In general, studies found varying degrees of local impact, sometimes severe, but little evidence of impact at the population level, except for bioaccumulation of DDT in some bird populations.

As DDT was phased out (1968) and new insecticides were phased into the protection operations, discussion of the environmental and health issues intensified. The newer chemicals were organophosphates (principally fenitrothion), which were more benign in their side effects. Although these chemicals did not bio-accumulate, they still affected nontarget populations, and these impacts were studied and reported extensively.

By the mid-1970s the emphasis in the pesticide discussions had shifted to the health risk to people exposed directly to the sprays or to the drift, which was sometimes extensive. Despite an absence of medically documented cases of human problems as a result of exposure to pesticides in the protection program (Spitzer 1982; Hatcher and White 1985), insecticides were consistently in the news. Although the demonstrated health effects in humans were anecdotal, the chemicals did pose a potential health threat, and fear of exposure was widespread among residents of the province (Varty 1975; Irland 1980; Thurston 1982; Irving and Webb 1984b; Fowle 1988). The public discussion tended to use scientific observations out of context. Given the nature of media hype, it was impossible to tell what proportion of the public actually was thinking about the issue as the media portrayed it.

Environmental and public health risks did exist, but the scientific dimensions of these problems were totally lost. (Some thought they were intentionally masked.) A succession of oversimple solutions appeared in the media that were said to be better (nonchemical) approaches that would bring budworm populations to endemic levels without the annual costs of crop protection, and without harm to the environment or human health. The main result of all this was to alarm the public.

The scientific community did not offer a usable basis for public discussion of health risks, so public discussion used science selectively. This in turn resulted in science spending an inordinate amount of effort to allay unfounded fears, which of course reduced the efforts aimed at legitimate concerns. As with other issues, this version of the forestry problem was discussed as if it was unrelated to the budworm, the forest, or the industrial aspects of the problem (Baskerville 1978; Baskerville and Brown 1985).

Information available to the public at this time encompassed incredible extremes. Sincere interest groups had read that wildlife would be killed and that losses of human life would occur if insecticides continued in use. These views were often grasped by those in the media looking for a chance to grind an axe and were offered to the public against a backdrop of a spray plane in action, dead insects floating on a stream, or the like. In addition, an appropriately scientific explanation was presented in a sandy dry format describing chemicals, dosages, and probabilities of impact on humans. The essence of this message was that insecticides carried some danger but that they were being used with caution and that no human impact should be expected. (However, there were no guarantees.) One had to conclude there was risk, but only a truly diligent citizen with a lot of time for reading could sort through the quagmire resulting from this mixture of information.

From 1973 on, the application of insecticides in the protection program was restricted in terms of buffer (or setback) areas, where the threatened stands were not treated. This began as a court injunction to prevent spraying near blueberry fields and the incidental kill of pollinators and developed into a set of regulations by the provincial Department of Environment. The latter included setbacks from habitations and water bodies, and the spray program came under tighter safety control (Irving and Webb 1984b). Because of the land ownership pattern in the province, setbacks had a disproportionate impact on woodlot forests. Private woodlots suffered grievous damage both to current growing stock and to the productive structure of these forests. The natural reaction of owners was to try to preempt budworm-caused losses in woodlots by harvesting, if necessary, at levels beyond those that were sustainable.

There was public discussion of decommissioning the World War II TBM aircraft that formed the basis of the protection fleet. This was based partly on the fact they were "warplanes" and partly on the minimum feasible target area for these aircraft. The planes flew in teams of three, treating blocks of about 4000 ha. The perception was that if smaller planes were used, the targets could also be smaller and thus there would be less risk from off-target drift. There was some validity to this point of view, for by that time a large part of the protection program was near populated areas, where the forest was broken into small stands. However, the planes were generally cost effective, and the mobile budworm did not recognize such geographic limitations as it continued to infest the susceptible forest wherever it occurred.

Throughout the preceding discussion and even after the implementation of setbacks, there was a general unwillingness (public, environmentalist, government, industry, professional) to acknowledge the timber management impact of the absence of protection within the forests of the setback area. The timber losses in these unprotected stands meant there must be a reduced future wood supply, but the health, environment, economic, timber supply, and forest management elements of the problem were conveniently kept separate in public, political, and usually technical debate. This made it possible not to protect part of the forest and yet to retain the annual harvest for the whole forest at a level that assumed there were no losses to budworm.

By the mid-1970s the infestation extended to some 50 million ha in eastern Canada and the United States (figure 2.4). A joint Forestry Canada/U.S. Forest Service budworm research project was mounted under the name CANUSA. This program resulted from the extensive damage to the fir/spruce forests of the eastern United States. This massive effort had no impact on the budworm, and as the infestation declined naturally, so did the program.

Perhaps the biggest change in the protection program was triggered by a sudden rise in oil prices in 1973–1974, which was reinforced by further increases in the early 1980s. The costs of insecticides and of their delivery rose so dramatically that the targeting and delivery systems were extensively restructured. Targeting stands for protection became much more discriminating, and technology to reduce the amount of material sprayed on a hectare was pursued actively in order to allow airplanes to cover more forest in each flight by using less volume of insecticide per hectare.

Some saw the possibility of avoiding the entire insecticide problem by using of a biological pesticide known as Bt (*Bacillus thuringiensis*). This had long been considered environmentally safe and was often referred to as a biological control. Actually, Bt is a biological insecticide used in a manner similar to chemical insecticides, and it has a similar temporary impact on insect population dynamics. The status of Bt has turned into a long-running controversy. It has been available and under active development for forest-level application for almost three decades. Yet it has not performed nearly as well on high populations and in varied weather conditions as it has in the relatively controlled situations of small experiments (Carter 1991).

Since the product does not consistently protect foliage at the desired levels of protection, and its cost per hectare treated is 50% more than that of chemicals, there is much argument as to the role of Bt in controlling defoliation by budworm. In general, Bt is efficacious in the trivial sense that treated stands suffer less defoliation than untreated ones. From a forest management point of view, however, efficacy relates to retaining enough foliage for a stand to recover its ability to produce wood volume, and Bt is not operationally efficacious in this context. In any event, Bt was (and is) widely proclaimed as an environmental advance, and apparently some believe it is worth using a publicly acceptable, if expensive and less effective, material if it reduces pressure from the environmental lobby. There has been no move to reexamine the sustainable harvest level in the context of the weaker protection achieved with Bt. Like so many elements of the forestry problem, the Bt discussion has run on a separate track, and the logic of its role in solving the forestry problem has never been thoughtfully addressed in the broader context.

The insecticide issue demonstrated the influence of special interest groups or nongovernmental organizations (NGOs) in forest policy. These groups choose an issue and tend to isolate it from the context in which the issue is embedded in reality. They then push "the answer" based on this out-of-context view. Government provides funding for many NGOs and regularly incorporates them into the policy process. The fact that a small group can create and maintain a dominant view of an issue in the media is seen by some as a danger to reasoned development in a technological society. At the same time, these groups have led in the recognition of problems, in some cases in advance of the appropriate bureaucratic structures. The continued existence of special interest groups and their support by government funding are seen by others as evidence that government lacks faith in the effectiveness of the regulatory agencies that are supposed to provide this same service.

In general, special interest groups stop things. The people involved usually are sincere and deeply convinced of the value of their position. Thus interest groups have been effective in getting issues before the public, largely by capturing TV time. Even when the original form of these issues turns out to be overly simple, or false, the issues become a proper part of public, scientific, and bureaucratic debate. At the same time, the media have been susceptible to manipulation by interest groups, as both are pleased (for different reasons) when an issue becomes high profile. The result, at least in the mind of one senior journalist, has not always been in the public interest (Malling 1991).

Forest Dynamics

The second debate, beginning in the mid-1960s, was among forestry professionals and dealt with the dynamics of development in the fir/spruce forest. There were two issues: the role of forest management in controlling budworm populations and recognition of forest dynamics in estimating the sustainable harvest level in a forest. With respect to budworm, there were two schools of thought. On the one hand, it was argued that when a managed forest was reached, with a balanced age-class structure having equal areas of stands in all age classes, such a "proper" forest would no longer be vulnerable to budworm. On the other hand, it was argued that the natural dynamics of the fir/spruce forest were driven by periodic destruction by spruce budworm and that the protection program was allowing the age-class structure to build to a balanced form that was neither natural nor biologically stable. The issue was whether the temporal and spatial dynamics in the forest created by management would cause the insect outbreak to go away or would cause it to remain forever, necessitating continuous protection.

A group of scientists presented a workshop in 1975 to offer a simple but scientifically based picture of budworm/forest dynamics and of the conflicts with forest management (see *Forestry Chronicle* 1975, 51:135–160). The natural dynamics of the fir/spruce forest were shown to be biologically inconsistent with the even-flow harvest paradigm of forest management. Natural forest dynamics, as driven by budworm, could never create a "normal" forest, in the technical sense of a balanced age-class structure, because the periodicity of extensive destruction by budworm would always result in the presence of only one or two (very large) age classes at any time. As one of these classes matured, it would become involved in an outbreak that would result in its destruction and in the creation, by regeneration, of a new age class in the forest. This was part of the basis of periodicity in budworm outbreaks. On the other hand, creation of a classic managed forest with all age classes continuously present would lead to a forest structure that the budworm could never produce (i.e., a forest that was continually suitable to support outbreak population levels [Baskerville 1975a, 1975b; Baskerville and MacFarlane 1975]). Success of classic forest management would make the budworm a perpetual problem for forest managers instead of a periodic problem. Many approaches to forest management were explored, and it appeared that any management that consistently maintained fir stands suitable for economic harvesting was likely to result in continuous attack by budworm.

The second issue was the explicit recognition of forest dynamics in estimating sustainable harvest levels. As computers became more available, the debate was between using static formulas and using simulation models that mimicked forest dynamics. Estimates averaged over time and over geographic area by the formulas (e.g., $H = GS/A$, as noted earlier) could be made with a desk calculator and a sheet of paper. With simulation models the continuous response of the forest over long time horizons to various levels of harvest, as well as various kinds and levels of silviculture treatment, could be explored, but this involved simulation on a computer.

The same assumptions had to be made both in the growing stock formulas and in the simulation models. The difference was that in the simulation model all the assumptions became explicit (inviting test and evaluation), whereas in the formulas they were implicit and hidden. For example, simulation models used stand yield curves to mimic future growth in each stand type, and these were clearly major assumptions. Some argued that no such assumptions were needed in the formulas. Actually, the formulas embodied a single implicit yield curve that could be extracted and that was shown to be biological nonsense.

The fundamental problem in the formulas-versus-simulation argument was the failure of the former to recognize dynamics at the level of the forest. The formulas assumed a yield curve and a nearly balanced age-class structure, neither of which existed in nature. From the simplest to the most complex, simulation models recognized the current real initial age-class structure, which consisted primarily of two age classes and showed the dynamics of age-class structure over time in response to harvesting by man and by budworm. In terms of simulated forest dynamics, harvest levels determined by formulas frequently were unsustainable. A direct attempt to dislodge the old formulas for calculating allowable harvests failed (Baskerville 1977a).

While this discussion was going on, it became apparent the budworm was not limited to feeding in old stands, as conventional wisdom had long held. Damage appeared in immature stands and even in regeneration. Clearly, if the budworm could sustain high population levels by feeding in immature stands, the traditional population management paradigm was meaningless. This "deviation" of the budworm became apparent during the broadest, most intense period of the outbreak and had enormous importance to the yield curves used (or assumed) in forecasting future wood supply.

The discussion of the preceding forest dynamics issues was not decisive. In general, entomologists did not offer much support, for they were unsure of the role of the forest in budworm dynamics. Their hypotheses about budworm outbreaks centered on a population dynamics paradigm in which the forest did not play a significant role. Moreover, the timber community did not offer much support. It appreciated the flexibility of choosing an annual allowable cut by manipulating cull factors, merchantability limits, and so on, and was uncomfortable with the notion of making forecasts based on age-class structures and stand yield curves, or with using a computer. In the end the conventional wisdom prevailed, and meeting after meeting resolved that heroic efforts to protect the forest from budworm should be continued. There was little acceptance of a need for change in forest management design and in the manner of estimating sustainable harvest levels.

Perhaps the most important issue from the point of view of public policy is that the two debates, about insecticides and about forest dynamics, were conducted independently. There was considerable overlapping among the people who took part in the two debates, but there was amazingly little carryover of the essence of the problems from one forum to the other.

Could There Possibly Be a Timber Supply Problem?

In the early 1970s it became clear the last of the fir stands of 1880 origin would be harvested by the end of the decade or early in the 1980s (Baskerville 1977b). It was equally clear that to sustain existing harvest levels, the next raw material source would have to be the 1924-origin stands. Simple simulation models of forest dynamics showed the transition, and it was evident in the forest that the jump from the 1880-origin stands to the 1924-origin stands would be a qualitative one of major proportions. The 1880 stands averaged about 800 trees/ha, and the trees tended to be large in diameter as a result of having grown in open conditions since the outbreak of 1913–1919. These stands offered a high proportion of sawlog material. The 1924-origin stands ran about 3000–5000 stems/ha, average tree diameter was proportionally smaller than in the older stands, and sawlogs would be a rarity unless size standards were relaxed.

The problem of tree size was a direct result of successful protection of the 1924-origin stands through the budworm outbreak from 1952 onward, particularly of the way protection altered development of those stands in comparison with what their development would have been without protection. In the normal course of events, the 1924-origin stands would have been severely thinned during the 1950s, and when the outbreak collapsed (as presumed in the absence of protection), residual stands would have developed in open-grown conditions. These stands would have been well on the way to a large average diameter in the 1970s. Instead protection had prevented a reduction in stand density as a result of budworm-caused mortality, and stand development during the immature stages was controlled by self-thinning. Since fir is a shade-tolerant species, stand structures developed that were extremely dense in comparison to the 1880-origin stands (Baskerville and MacLean 1979); that is, volume was relatively high, but the mean tree size (a crucial factor in harvest efficiency and in raw material value) was very small.

A regeneration problem also emerged partly as a result of protection. Balsam fir delivers a substantial supply of seed to the forest floor every second year from immaturity onward. The 1880-origin stands were characterized by an abundant advance growth (regeneration established in advance of harvest), ranging from 0.3 m to 2 m in height. Because of the open nature of these stands, there was an abundance of regeneration stems, with individuals large enough to survive logging and to provide instant regeneration. Clear-cutting, which was done mostly in winter, left the advance growth sufficiently intact to assure prompt return to the same species. The evolutionary mechanism for fir stand replacement worked well in the context of regenerating the open-grown 1880-origin stands following clear-cut harvesting.

The 1924-origin stands, protected from mortality caused by budworm, also had 100% stocking to advance growth, but because of the dense canopy that resulted from limited budworm-caused mortality, this advance growth tended to be less than 0.3 m tall. Also, stands in this age class had been protected from major stem loss due to budworm attack over a long period before harvest, and defoliation that occurred before the threshold for spraying had often resulted in budworm descending from tree canopies, where they destroyed any seedling they happened to land on. Over time, dense stand conditions, a switch from primarily winter logging with horses to year-round logging with machines, and chronic light defoliation resulted in a less-well-developed advance regeneration that was susceptible to postcut competition. When the heavy slash from these dense stands was deposited on top of the small seedlings following harvest, the result was uneven regeneration.

Instead of having an overlap of stand generations as had occurred in nature and with the harvest of the 1880-origin stands, small seedlings under the protected 1924-origin stands were often soon overtopped by shrubs following harvest. In these cases there was both a substantial delay in timing of regeneration and unevenness of stocking. Trials with shelterwood harvesting to enhance natural regeneration became common, as did use of herbicides to release seedlings in recent cutovers. Successful crop protection had converted a free guarantee of prompt regeneration to the same species into a substantial regeneration perspective for the forestry problem.

The combination of reduced sawlog availability and less consistent regeneration fueled debate about a future shortage of softwood raw material. It was clear that sawlog material of the former quality soon would be in short supply, but in terms of total quantity of wood (all forms) the dominant view in the 1970s was that the timber supply was secure. However, argument was emerging.

Models, a Tough Sell

Although models of one form or another have long been part of man's view of the forest and of budworm, the use of formally stated mathematical models emerged in the 1960s and 1970s. An elegant mathematical description of budworm population dynamics primarily based on observations in northwestern New Brunswick was published by a team of scientists as a multiauthor monograph in 1963 (Morris 1963). These models did not explicitly embody the forest; rather, they were models of budworm population dynamics, and forest characteristics were occasionally invoked as a factor triggering an outbreak.

This monograph and its models gained considerable respect in the scientific literature but did not contribute to the solution of the forestry problem, whether that was interpreted to be a matter of controlling budworm populations, preserving public health and environmental quality, or scheduling the harvest of fir/spruce stands in a sustainable manner. The models were neither intuitive nor easy to interpret. Furthermore, these analytical models had no spatial context and did not prove convenient for exploration of alternative approaches to dealing with budworm/forest dynamics at the temporal and spatial scales at which these dynamics were faced by resource managers and at which the forestry problem had to be solved.

In the early 1970s the federal forest research agency (now Forestry Canada) contracted a team of modelers from the University of British Columbia to work with local scientists to construct a budworm/forest simulation model. This was to mimic stand, forest, budworm, and industry dynamics, functionally linked to one another over geographic area so as to be usable in analyzing policies at the temporal and spatial scales of the systems involved. The project lasted about 3 years and involved about twenty local scientists and professionals from federal, provincial, and industrial agencies and from the University of New Brunswick as well as the UBC team. The result was a simulation model of mammoth proportions that for the first (and still only) time explicitly represented the dynamics of the forest, of industrial harvesting, and of insect populations in a single interlinked comprehensive format (Holling et al. 1976; Holling et al. 1977; Holling 1978). That is, in this model these forms of the forestry problem appeared in a dynamically linked manner.

About two-thirds of the New Brunswick forests were represented in the model as a set of rectangles, each of which was equivalent to a forest inventory map sheet (ca. 10 km x 19 km). The model grew the budworm-susceptible forest in each map cell by advancing an age-class structure for the classes of stands; by harvesting a portion of the stands in each cell according to a harvest rule designed to mimic local cutting patterns; by having the budworm population in each cell go through its annual life cycle, including damage to vulnerable stands; and reaction to protective actions; and by dispersing the population annually to mimic budworm migration into, across, and out of the province.

The model was used to examine response of the budworm/forest/industry system to an array of policies (Holling et al. 1976; Holling et al. 1977). At that time, the New Brunswick infestation covered about 3.5 million ha, and the infestation in northeastern North America involved about 51 million ha. Experience showed that no matter what protective actions were taken in the province, there was a high probability of reinvasion by moth flights the same or next year, demonstrating that the crop protection then being practiced was incapable of regulating the budworm population. This was inferred by the early work on dispersal by Greenbank (1957) and then shown definitively by work with radar (Greenbank et al. 1980). Since most jurisdictions other than New Brunswick had chosen not to protect their threatened forests or to protect only a small part of the full productive capacity of the vulnerable forest, the model was used to explore policies that might allow New Brunswick to buffer itself from the impacts of the much larger outside infestation.

The model representation of budworm/forest/industry dynamics performed reasonably at the geographic scale examined. When initialized for a budworm/forest/industry situation from the past, the model reproduced temporal and spatial responses that were similar to those that had been observed in the real budworm/forest/industry system in qualitative and quantitative form. Also, when initialized for a "present" budworm/forest/industry situation, the model generated reasonably possible futures in response to "implementation" of temporally and spatially complex policies operating on the budworm, on the forest, and on the industry (Holling et al. 1976). Because of its size—and more particularly because it embodied more than just a budworm view, a forest view, or an industrial view—the model was difficult to comprehend. A considerable number of people grasped some part of it, but only a handful (fewer than half a dozen) understood its nature and how it could be used realistically to explore policy options. To most people, including those central to the determination of provincial forest policy, the model was an extremely complicated and utterly mysterious black box. The fact that it was so big it could only be run at night on the university computer did not help this image.

Although the connection of this simulation model to the policy community was substantially better than that of the 1963 analytical model, its use was resisted. It was resisted by scientists (other than the team), professionals, policy makers, managers, environmentalists, and the public.

In retrospect, the ability of the model to permit examination of policies that had never been tried (i.e., for which there were no data) was at once its greatest strength and its fatal weakness. Some rather consistent features of the descriptions of reasonably possible futures generated by the model were the source of resistance.

The model forecasts embodied dynamics of the forest, budworm population dynamics, and industrial harvesting all on temporal and spatial dimensions of the real world. No scientist worked *across the full range of time and geographic area in any one of these subjects,* let alone in any two of these fields. The model thus challenged virtually all single-scientist and single-discipline views of the problem, which was not good news for scientists.

The model consistently forecast a reduction in volume of growing stock, even with perfect protection from budworm (or no budworm at all!). This was partly because of harvesting and partly because the 1924-origin stands began to break up in overmaturity before they were all harvested. The message was that current growing stock levels were as high as were likely in nature and that growing stock level was, in any event, an outcome of the dynamics of forest development rather than a driving force of these dynamics, which was not good news for proponents of maximizing growing stock as a forest management target.

The model consistently forecast a continuation of outbreak population levels as long as the fir/spruce forest was maintained by protection, and for a wide array of forest management interventions. This was not good news for proponents of a protection solution.

The model made it clear that any policy hoping to handle this forestry problem would have to be both complex and dynamic—at least involving the forest, the budworm, and industry—and would have to use a long time horizon. This was not good news for policy makers.

The model exposed a biological inconsistency between traditional views of forest management and ecological function in the forest. The natural cycle was boom and bust, whereas industrial man wanted continuous availability of stands suitable for harvest. The latter happened also to mean continuous availability of stands suitable for the support of budworm populations. This was definitely not good news for proponents of a forest management solution.

The only policies that brought budworm population levels over the large modeled area to endemic levels also involved enormous losses of merchantable volume, or periods of greatly expanded harvesting followed by periods with greatly reduced harvesting, neither of which was good news for industry.

If protective spraying was withheld in the model, the forecast was for widespread collapse of the fir/spruce forest, a loss of industrial capacity, and substantial loss of employment. This was not good news for advocates of abandoning the protection program.

The model descriptions of reasonably possible futures all showed a timber supply shortage around the years 2020–2040, no matter what action was taken to protect against damage by budworm in the interim. This problem had little to do with the insect and everything to do with the dynamics of a forest of two age classes created by past budworm outbreaks. Even if budworm was "turned off" in the model, the future supply problem persisted. The problem was related to harvest levels in the context of forest dynamics and not to budworm populations. This was definitely not good news for forest managers.

Policy players from federal and provincial governments, industry, woodlot owner associations, and NGOs were provided with outlines of the model and details of the policy analysis in a series of 1–2-day seminars. Some were frightened by the complexity of the problem, and a few grasped the potential and were impressed. However, there was no galvanization to initiate policy change.

Although time has shown the forecasts to have been essentially correct, the results were not what scientists, policy makers, resource managers, environmentalists, or the public wanted to hear in 1975! In one way or another, the model presented serious challenges to the reigning paradigms in each of those constituencies. Since there was no way to defeat the model on logic, the model, its unwanted messages, and the messengers were ignored.

Although much simpler in concept and operation, wood supply forecasting models being developed at that time were also resisted. The argument was made that the computer model was full of assumptions, whereas $H = GS/A$ (also a model but not acknowledged as one) was based on fact as reported in the forest inventory.

The dominant view of the time was that models were computer witchcraft (or just nonsense) and that the only safe way to deal with the future was to keep on looking back with more and more data.

The Forest Resources Study

About the same time as the budworm/forest/industry modeling exercise, a government Forest Resources Study was reviewing the status of forestry in the province. This study examined the full range of forest uses with the objective being "to recommend means by which New Brunswick can obtain the optimal mix of uses and benefits which its forests offer" (Tweeddale 1974).

With respect to timber the Forest Resources Study concluded, "There is now no available surplus of spruce, fir, or pine to be had for the taking; whatever more is needed must be cultivated, and once we start cultivating more wood, we should also cultivate better-quality wood." That is, there was no wood supply shortage, but neither was there room for industrial expansion. The study based its review of timber sustainability on a traditional growing stock formula approach. It was essentially silent on the subject of spruce budworm and its role in the dynamics of softwood timber supply. The main emphasis was on the need to continue protective action.

The Forest Resources Study represented a landmark in evolution of policy with respect to the New Brunswick forestry problem. Discussions went beyond timber issues in a manner that qualitatively altered the view of the forestry problem. Important features of the study were its recommendations for integrated forest management and for government management of the Crown forest. Its conclusions generated great interest in, and argument about, the role of forecasting in forest policy design. It marked the first time such a formal review had concluded there was no room to expand the forest industry. The study ushered in a qualitatively different process of challenge and change in provincial forest policy.

Working at Not Believing

The notion that the limit of timber supply had been reached was not readily accepted in the forestry community, and that community worked at not believing. The conclusion of the Forest Resources Study on wood supply was bleak, in that it assumed no losses to budworm, an assumption with little support even among those who did not recognize the importance of forest dynamics in forecasting wood supply. In all previous problems with timber supply it had only been necessary to alter a cull factor, or merchantability rule, or whatever, and shortages were made to disappear instantly. Many wanted to do that again and were unwilling to acknowledge that overharvesting might already be occurring, and no government or industry officials subscribed publicly to such a notion.

No one wanted to believe the bad news. This was partly unwillingness to admit that the problem had occurred under what were supposed to have been controlled forest management conditions. Rationalizing the situation was partly a matter of timing. A *real* wood supply shortage was when there was not enough wood *now*, whereas the report indicated only a limit on expansion, and even the dynamic models said the limitation would not occur for about 45 years. From a forest management perspective, the crucial issue that was rarely discussed was whether the limiting period was so close in time that there was no longer an opportunity to correct the supply problem with management.

Setting the Wedge for Models in Policy

Two events in the late 1970s resulted in increased credibility and increased use of models in forest policy design, the Budworm Task Force (New Brunswick Secretariat 1976) and the 1978 Provincial Forest Inventory (DNRE 1981).

The Budworm Task Force

In 1976 the area requiring protection from spruce budworm reached the highest level ever, 3.9 million ha, nearly the entire vulnerable forest in the province. There was intense concern about environmental and public health issues. At the same time, the suggestion that there was no more room for industrial expansion, even without losses to budworm, intensified concern with respect to wood supply. The government set up the Task Force to Examine Policy Alternatives with Respect to Budworm, which was charged with reviewing the status of the various approaches to controlling budworm populations and with exploring alternative strategies for maintaining industrial integrity in the face of the continuing outbreak. It was agreed that the examination of strategies would be built around an expanded version of the budworm/forest/ industry model.

The Task Force review of approaches available for regulating spruce budworm populations revealed that the then existing suite of fifteen tools was essentially the same as had existed in 1952, when the protection program began. Although there had been research programs in control methods over a period of 24 years, none of the alternatives appeared to have been brought significantly closer to operational use than they were at the start of the period. No new alternatives had been added. The only practical approach was insecticides, and although these had been developed in terms of application and efficacy, they still constituted only a protective action to limit the immediate local impact of the insect on stand losses. None of the tools under examination offered a realistic opportunity to get at the biological bases of the budworm/ forest/industry problem (N.B. Cabinet Secretariat 1976). Research appeared to be "learning about" pieces of the system, but there was little sign of progress with respect to acquiring comprehensive control of the system. Although the technology of application of pesticides has improved considerably since that report, there has been no significant improvement in the availability of alternatives, despite a further decade and a half of research.

Forecasts of the Task Force report, using an updated version of the budworm/forest/industry model that covered the entire province, again showed that as long as the fir/spruce forest was kept intact and moved toward the balanced age-class structure desired for even-flow sustainable timber production, spruce budworm would be a persistent problem. With regard to this and other fundamental issues of budworm/ forest management, the results were similar to those found by the group that developed the model in the early 1970s. However, this time use of the model had been agreed to by policy advisors in government in advance of the study.

The Task Force was unable to discover alternative protection strategies that met the two conflicting requirements of society: (1) that the use of insecticides be stopped and (2) that there be be no loss of jobs or of value added from the forest industry. Analyses did indicate that certain strategies with chemicals could be more effective than the traditional approach and would employ less pesticide.

Analyses of budworm/forest/industry dynamics indicated that if the protection program was stopped, the budworm population would flare for a few years, followed by collapse of the population as much of the host forest died. No tools or policies were available that could negotiate a soft economic landing for the softwood-using industries. No one was comfortable with these conclusions. Perhaps because of this an array of "solutions" persisted that were alleged to be capable of fixing the budworm/forest/industry problem at low cost if only "they" would listen. This silver bullet has yet to be identified in usable form.

Government was not pleased that the answer was not forthcoming from the Task Force. At the time it appeared that the efforts had not made much impact on government policy. In time it became apparent that the comprehensive nature of the model had helped restructure government (DNRE) discussions of budworm/forest/industry policy. The use of a comprehensive model had brought a broader grasp of the dynamics that so inextricably linked the budworm, the forest, and the industrial elements of the problem.

The 1978 Inventory

In the midst of the emotionally charged debate (public and professional) about the health risks of insecticides and herbicides, government was trying to reassure the public about the certainty of the timber supply. In 1978 the provincial forest inventory was updated. The approach was similar to that used in earlier inventories. Those who argued that growing stock was the key to forest management were anxious to demonstrate that growing stock was still increasing, thereby allaying fears of a wood shortage. Those who understood forest dynamics, as displayed in various simulation models emerging at the time, argued for an inventory that would provide a characterization of the biological structure of the forest. To the latter group, growing stock was not of much interest, whereas a description of the stand development types and of the areas at various stages of development would greatly facilitate improved forecasts based on forest dynamics (Hall 1978).

In the end the 1978 inventory was conducted in the old format. Surprisingly, to most in the forestry community it showed that the growing stock had declined since 1968—in the exact amount forecast by the budworm/forest/industry model in the early 1970s (figure 2.4). Since it was designed and implemented as a timber inventory rather than as an evaluation of forest structure, it was not apparent why growing stock had not continued to increase, and so for some the argument went on.

The use of models by the provincial budworm Task Force, the fact that growing stock had declined, and the fact that dynamic models had forecast the decline so closely opened the door to consideration of long-term forecasts of timber and budworm using dynamic models.

A Public Health Scare

Public concern about insecticides peaked as the 1970s came to an end. A claim that the usually fatal childhood disease known as Reye's syndrome was triggered by exposure to the insecticide used against budworm and/or its adjuvants was leaked from a research lab and publicized before it appeared in the scientific literature. When it appeared, the paper asserted that Reye's syndrome was more frequent in sprayed areas of New Brunswick than in nonsprayed areas. The province engaged a team of toxicologists and epidemiologists to review the scientific evidence and advise on the health risk. The team concluded that although spray mixtures were toxic to various degrees, their risks were low and were unlikely to be the cause of Reye's syndrome. Furthermore, they found no evidence for the claim of a difference in the incidence of the disease inside and outside the sprayed area (Spitzer 1982; Brookman et al. 1984).

By the time the special review was published there was considerable fear throughout the province. Medical science suspected the mechanism for triggering Reye's syndrome to be acetylsalicylic acid (ASA), and that is still considered most likely. The media, which had so aggressively promoted the false claim of risk to the population, gave scant attention to the refutation. Not surprisingly, there are still people who believe pesticides cause Reye's syndrome and who presumably are still not taking precautions with respect to the use of ASA for children with flu.

Science played a mixed role in this scary false alarm. The scare was initiated by a combination of loose scientific reporting and active (if nonunderstanding) promotion by a few scientists and special interest groups. Coupling a complicated technological risk with less than responsible scientific reporting, and less than responsible media representation, resulted in enormous public stress. Some said it was good to draw attention to the potential risk. Some said this was one more case of "Chicken Little crying that the sky is falling." Some called it an irresponsible act, equivalent to yelling "fire" in a dark theater without evidence that a fire exists. One legacy of the debacle was a general loss of credibility for science.

Meanwhile, Out in the Forest

Tree planting, which had been a part of forest management for some time, increased in the 1970s to major proportions, reaching a high of 62 million seedlings planted annually and constituting about 40% of the softwood cutover, with renewal of the remaining area by natural regeneration. Virtually all planting was with local species appropriate to the sites involved. Planting was widely hailed as "good forestry," particularly where black spruce, a species less vulnerable to budworm, was used. It became common for public evaluation of forest management to consist solely of a count of seedlings planted. That is, number of seedlings became the equivalent of growing stock as the single indicator of forest management performance. In terms of forest dynamics, the number of trees planted is even less meaningful than was total growing stock in the earlier discussions, and like the growing stock argument, it deflected serious analysis of the dynamics of wood supply.

As more and more seedlings were planted, it became apparent that some areas that were regenerating naturally were also being planted. This led to overstocking and to the necessity of thinning to bring stocking down to reasonable levels. At the extreme, it became necessary to decide whether planted trees or natural regeneration should be preferred during the thinning. This problem arose partly due to an unreasoned response to public pressure for planting and partly due to a change in natural stand dynamics. By the 1980s, the 1924-origin stands were beginning to open up as a result of persistent budworm damage (even with protection) and of advancing maturity. The result was better-developed advance regeneration and therefore less need for planting.

Planting currently runs at 18% of the annual cutover area and has now lost much of its virtue in the eyes of the environmental lobby because it is said to reduce biodiversity. Indeed, there is some pressure to reduce planting further, and if it is necessary, to use a mixture of the same species present before harvest and to eliminate chemical weeding.

In about half the plantations, shrub competition must be reduced to establish a stand promptly consistent with management goals, and this has been done primarily with herbicides. Public fear of chemicals extended to herbicides, and these too were restricted by setbacks from habitation and water. Here again the public discussion of planting and of herbicides as a potential health problem occurred separately, as if these topics were unconnected to each other or to other elements of the forestry problem. There was little, if any, recognition that if plantation growth was less than the levels used in the models to determine allowable harvest over time, then the forecast harvest level no longer was sustainable. Simulation models provided a ready means to estimate the needed reduction in harvest level, but there was little interest in discussing or making this calculation.

As a result of carrying on separate debates in different forums on the health risk of herbicides and on the management need for herbicides, there is little logic to the trade-off that emerged in the forest. Neither "side" likes the present situation, and the debate continues, still in different forums.

Registration procedures for insecticides and herbicides (controlled by the federal government) became sufficiently rigorous by the 1980s that chemical companies virtually ceased research on these products for forestry use. Forestry uses represented only 3% of total pesticide sales, and it seemed unlikely that they could ever pay back the developmental investment (Carrow 1985). There was some suggestion that chemical companies were willing to drop the small forestry market rather than risk publicity that could be attached to the vast majority of their products. Reduced availability of registered products was seen by some as a move in the right direction. Others were concerned that this froze pesticide technology without motivating development of alternative approaches and that it nearly invalidated growth and yield assumptions in forest forecasts.

Perhaps the most important message here is that planting and herbicides, like other tools of forest management, were evaluated by the public independently both of their function in the dynamics of resource development and of each other.

Responsible Management of the Commons

Stewardship of publicly owned forests has been a subject of continuing debate in Canada, where nationwide 90% of the total forest area is publicly owned (Pearse 1987; Baskerville 1988b). Although public ownership is only 48% in New Brunswick, it has been a subject of strong debate. The public, and therefore the government, desperately wants benefits of economic development as generated by the public forest. Management by government department has been tried in several formats and generally has been found to be ineffective. The tendency for inconsistent management budgets over time and the lack of motivation for managers resulted in "administration" of the use of the resources rather than "management" of the resources themselves.

Historically, with government responsibility for forest management there was no relationship of harvesting to technical management of the forest. In fact, management plans for Crown forests were conspicuously absent. Those that did exist were trivial and resided on a shelf, where they had no impact on what happened in the forest. Legislation has been weak with respect to resource management, and there has been little will on the part of Crown agencies (or, indeed, budget) to manage.

Industry has argued that if it was given tenure in a fixed area of forest for a reasonable length of time, and with reasonable conditions for renewal of the tenure, it would be in its interests to manage for the long term and it could manage the forest more effectively than government. The public, and therefore the government, is unwilling to allow industry to hold long-term tenure on Crown land. The public expectation that industry should manage Crown forests without having some assurance of the form of access to raw materials from that major resource has not been realistic. Nor is there public support for the auction of cutting rights, as this is believed to give an unfair advantage to large mills. In addition to the large industrial enterprises, there always are a variety of forest jobbers and contractors seeking random access to the forest to allow them to market wood. Characteristically, these are short-term ventures with no interest in long-term resources management.

Although many tenure forms have been tried, when a new law or regulation has not "fixed things" in 1 or 2 years, outcry from affected elements of the public and industry resulted in new reviews of tenure policy. Such instability, in the context of the pace of forest change and of the need for management planning horizons on the order of 50–100 years, was incredibly damaging to the progress of forest management. Each tenure process in succession was overturned by people who believed that a past change in law should have been instantly reflected in "a better forest," or in instant forest management performance (i.e., by people who did not grasp the pace of forest dynamics).

In classic tragedy of the commons form, the public, as owner of the Crown forests, wanted to be able to use these forests to capture the biggest possible share of economic return *now.* As there have been few controls on the amount and quality of material harvested from Crown forests over the years, the quality of those publicly owned forests has been run down. The dynamics here are very similar to the pathological process described by Hardin (1968). Loss of quality raw material from white pine and hardwoods is an extreme example.

Although political tinkering with tenure and management responsibilities at the request of various elements of the population was incessant, there is no evidence of technical forest management of the New Brunswick Crown forests prior to the 1980s. The government *administered* the public forest resource, but it did not *manage* that resource. Some very good silviculture programs were implemented, but there was no forest-level planning to give context to that stand-level effort. Paradoxically, in the sense of looking to the future of the *resource*, protection of the forests from damage by budworm, the source of so much public outcry, stands out as the principal evidence of public stewardship at the scale of the forest.

The Forest Resources Study had spoken to the stewardship issue and had recommended a trial whereby a Crown corporation would conduct all harvesting in the Crown forest and sell raw material to industry. In this model there would be only one harvester on the commons, who would mete out supply through a controlled market structure. In the mid-1970s, government canceled the 400,000-ha Crown timber license of one pulp and paper company, and a Crown corporation known as the New Brunswick Forest Authority was created to carry out harvesting in that forest and to distribute the raw materials. The DNRE retained responsibility for forest management. Thus responsibility for harvesting and resource management was split in the forest between a government line department and a Crown corporation. Each had different views of its role in the venture, and there was little cooperation.

The belief was that the Crown corporation would operate with the financial efficiency of a company and with the public responsibility (not responsiveness) of government. If the N.B. Forest Authority was successful, the approach was to be extended to all Crown forests. The venture was short-lived.

There was general resistance to the N.B. Forest Authority from the forestry community at large. Many saw it as a new form of government boondoggle, with no motivation for better forest management. The principal focus of the Forest Authority was on road building and harvesting, which were short-term economic imperatives. Management at the forest level never got started. The project quickly got into debt and was closed after about 5 years; the assets, including a high-tech sawmill, were sold (Watson 1984). Rightly or wrongly, this pilot project became just one more test of long-term forest planning that was concluded to have failed in less than 5 years.

Other than a gain for credibility of dynamic models in policy analysis and forest management design, the 1970s were characterized by argument and confusion. Participants usually seemed more intent on discrediting the "other side" than on achieving a common understanding that could lead to the overall betterment of New Brunswick. There was little evidence of gain with respect to managing the forest resources. A decade of technical and social disarray ended in technical and social disarray (Baskerville 1979).

The Breaking Point

Late in the 1970s there was general discontent in the forestry community about incessant changes in tenure, uncertain wood supply, problems associated with the protection program, and ineffective forest management. Government undertook a review of the context of Crown forest management. Two task forces were struck to provide advice on allocation of shares of the allowable harvest from the Crown forest and on management of the Crown forest, including the form of tenure. The Task forces were large and had difficulty in coming to agreement. The Allocation Task Force was unsure of the level of harvest that was sustainably available for allocation, as there was little agreement on the long-term availability of timber. The Management Task Force was stuck in an outdated view of forest management procedures. There was much activity but little progress.

The breaking point for forest management came suddenly in December 1979 in a large government/industry meeting that was called to discuss the progress of the task forces, both of which were present. Some industry members expressed exasperation at the general resistance to recognizing what they considered to be the fact of an emerging timber supply problem and of the need for aggressive forest management while there was still time to correct the situation. Points were made bluntly about the need to act decisively. After a number of confrontations between those who wanted to continue talking about the *potential* problems and those who wanted to begin fixing the *existing* problems, those who wanted to act took charge. From that point onward, the argument in New Brunswick has been about the timing and nature of wood supply problems and how to design management to deal with those problems, and about how to make management a reality in the Crown forest—not about whether a wood supply problem existed or whether existing management was adequate.

The old formula calculations of allowable harvest level have not since been used in the province. They have been replaced by wood supply analyses that explicitly recognize forest dynamics and that use an 80-year time horizon for forecasts. The approach to forest management in Crown forests has become the most aggressive and technically complete in Canada.

Similarly, a variety of frustrations with protection of the forests from budworm came to a head in the late 1970s and early 1980s. Government, which was the dominant shareholder, appointed active players to the board of directors of Forest Protection Ltd. and for the first time played an active role in managing the company. This action was triggered not by poor protection of the forest, but by the continuing controversy over pesticides. It seemed there was always a new health scare, with pressure to stop using a particular chemical. At the same time, medical researchers pushed for reasoned change, in a manner that permitted incremental gains in knowledge, rather than continually starting with a new spray mixture. One adjuvant used in the spray mixture, and challenged through the media, also happened to be a common ingredient in the undercoating of most automobiles, in which context it had met the full spectrum of health standards. The manufacturer tried to withdraw the product quietly, out of fear of being identified as a health risk by association with the N.B. spray operations. Liability concerns were high, and the argument ran along the extreme lines of "no risk" versus "deadly risk."

At a long and rancorous meeting of industry and government (including the premier of the province) there was recognition/acceptance of the problem and a decision was made to act. The spray program was converted from a private company to an operation of government. A toxicologist was engaged to advise on medical research and testing needs with respect to human health, and the province supported a research program dealing with health risks associated with a variety of chemical materials.

A Period of Rapid Change

The minister responsible for forests at this time was newly elected and was decisive. He created a small, informal group that became known as the super task force, made up of ten industry CEOs and senior advisors from other constituencies. The original Task Forces lingered on but were overrun by events, and their final reports were presented after the action was over. The super task force agreed that there was a long-term supply problem, that its best definition lay in the output of models that represented forest dynamics explicitly, and that a need existed for a tightly structured approach whereby forest management was made to happen in the Crown forest. Perhaps most important, the approach was to define the forest management problem in terms of forest dynamics and to analyze these dynamics to discover an appropriate solution, whereas previous approaches all began with a solution and tried to make the forestry problem conform to it.

The planned change was extensive. It was agreed that a new law would cancel all existing forms of licenses and leases to Crown timber and would mandate a new form of renewable tenure based on area. The law would mandate industry responsibility for forest management, and the principal condition of tenure would be satisfactory forest management performance by the industrial licensee. Crown forest productivity would be estimated recognizing forest dynamics, and Crown forest productivity would be shared proportionally based on estimates of long-term production capability of the forest as determined by the then existing timber supply model. It was recognized that this could amount to "sharing a shortfall." It was decided that all this was to be done by April 1982.

A survey of actual production and of the installed capacity for all processing plants in the province was carried out, revealing that the capacity of all softwood mills in the province, taken together, exceeded the long-term even-flow sustainable timber supply from all ownerships of softwood forest in the province. Forecasts indicated the forest could continuously support about 80% of installed capacity! The supply picture could not be made to change significantly for any credible alteration of forest dynamics as depicted in the wood supply models used to make the forecasts.

Three things prevented that finding from being totally disastrous. First, because of weather, strikes, markets, and so on, no mill actually operates continuously at full capacity. With this flexibility there was a potential fit of the mills to the forest in terms of quantity, but not in terms of quality, of raw material. That is, the availability of fiber might be sufficient to sustain the pulp and paper mills, but the sawlog portion of the resource had been sufficiently run down (high-graded) that there would be a period when the supply of logs of current quality was not sufficient for all the sawmills. Second, the problem of supply was not a forever thing. The limiting period was from about 2020 to about 2040, after which the restructured forest could sustainably deliver an annual harvest equal to the 1980 installed capacity with some room for expansion (Baskerville 1982). Third, in the absence of proper yield curves for wood supply models, the first approximations had been created on the conservative side. In 1980 the potential problem of forest structure so earnestly debated as a possibility in the 1960s and 1970s was seen as the reality to be dealt with decisively.

The timber supply forecast assumed a level of protection from budworm damage equivalent to what was being achieved at that time. In the midst of change in forest management structure, the crisis with respect to use of pesticides burgeoned, and government acted to take control. Some elegant research on spray delivery, targeting, and drift led to substantial improvements in spray application. There have been few challenges with respect to health risks in recent years; however, the protection program has become smaller as budworm populations waned and as the infestation moved north in the province and away from areas of population.

Although the comprehensive budworm/forest/industry model did not play a direct role at this stage, industrial wood supply and budworm elements of the problem were now effectively linked. Wood supply forecasts since 1980 have dealt with budworm damage as an explicit factor. Since 1980, discussions of wood supply issues have featured analyses based on forecasts using dynamic models (Hall 1978; Hall 1981; Clifford 1981; Baskerville and Kleinschmidt 1981; MacLean and Erdle 1984; Erdle and Baskerville 1986; Wang et al. 1987; Erdle 1990; Edmundston A.T.G. 1991). One of the most reassuring early signs was that several models emerged. The most widely used and long-lasting, in its variant forms, was produced by a forest company. There has been considerable debate about dynamic models in terms of such things as the initial conditions to use, how best to mimic stand response to silviculture, how to account for budworm, how to get the best harvest schedule, and so on, and all this debate has been constructive.

Because the 1978 inventory had been aimed at estimating the growing stock instead of the growth characteristics of the forest, timber supply forecasts were being built with rather sketchy data. It was clear in 1981 that better estimates of the dynamic structure of the existing forest were essential for defining the forest management problem and for designing a solution. Despite the closeness to the last inventory, the problem was seen as sufficiently important that a new inventory, called a Forest Development Survey, was designed to meet the needs of forecasting forest dynamics (DNRE 1987b). A decision was taken in 1980 to rephotograph the forests of the province in full color, at a scale of 1: 12,500. New data could not be available for the first 5-year planning period under the new management system beginning in 1982, but the Forest Development Survey was complete by 1985, and information was available to initialize simulation models for planning the second 5-year period under the new forest management system in 1987.

Planning for timber, wildlife habitat and protection from budworm became strongly linked and required geographic specificity. Whereas previous "planning" had been on average for the whole forest (i.e., nonspatial, and therefore, strictly speaking, not implementable in the forest in the same form as planned), the emerging approach allocated treatments and showed responses at the stand level. In this context, the dynamic structure of a forest-level problem was disaggregated into stand-level actions and responses that in total reflected dynamics of the forest-level problem. To facilitate this cross-scale connection, a geographic information system was purchased to map the entire province at a scale of 1:12,500 to a stand resolution of the order of 10 ha. Candidate geographic information systems were tested, and one that could demonstrably handle the problem was purchased, installed, and loaded with Forest Development Survey information as it became available.

There has been interactive geographic coverage for the forests of the entire province (all ownerships) at the stand level since 1985. Activities in the Crown forest are reported in both numerical and geographic form. Although geographic information system coverage extends to the entire forest, its use so far in management design is primarily on the Crown forest. Two companies with large freehold forests own and operate the same system and share information with the provincial system. Availability of full geographic coverage has introduced the possibility of geographic design and brought timber planning, protection planning, and wildlife habitat planning all to entirely new planes. Availability of the geographic information handling made the direct link between local silviculture, local crop protection, and forest-level management feasible. The integration of stand silviculture and protection and of wildlife habitat into forest-level planning has begun (MacLean and Erdle 1984; Erdle 1990). All this has not come easily, and geographic system access is still developing. Here, as elsewhere, there was a tendency to overestimate the ease of introduction and the power of new technology. A key feature of this overestimation is failure to recognize that a technology cannot think.

There was a qualitative change in thinking about all aspects of managing Crown forests in the period 1979–1982 that cannot be overemphasized. There had really never been any degree of forest management in New Brunswick in the sense of directed control of development in a whole forest system over time. The change was from doing some "good" things in more or less anecdotal fashion (silviculture, crop protection, and so on) to an adaptive approach involving (1) forecasting forest-level performance under various tentative plans for actions distributed across time and across geographic area; (2) choosing one forest plan to implement; (3) defining local silvicultural, protection, and habitat actions in context of the long-term dynamics associated with that forest plan; (4) implementing that plan in the forest while recording deviations and assessing the impacts (Erdle and Frame 1991). In designing the system, government and industry alike showed openness to innovative approaches and to borrowing useful tools wherever they could be found. Government formally adopted a long-term perspective on forest management, especially as it related to planning (Baskerville 1983; Baskerville 1986).

Facilitating the preceding technological change was a rapid evolution of forest policy led by the minister of natural resources and the super task force. A Crown Lands and Forest Act was drafted and put through the legislature in July 1980 with unanimous consent. It took effect on April 1, 1982. The law canceled all existing forms of access to the Crown forest and put a cap on industrial development based on the Crown forest until that forest was managed to support increases. The law specified an area-based tenure system. That is, a group of mills would be assigned to a fixed land base, and the harvest for those mills was limited to what could be produced sustainably with even flow (the same volume annually) from the designated forest.

One tenant on the license was designated a licensee, and this company entered into a Forest Management Agreement (FMA) with government. The FMA is a contract to manage the whole license, including arranging the supply of raw materials to other companies, known as sublicensees. Before the act, Crown forest management had been fragmented with several hundred pieces and a variety of tenure arrangements. Most of these had been held by mills that sought raw material in a single form. There was a tendency to high-grade "their forest area" for the material they used and either to leave the rest in the forest or to sell it to whomever was interested. After the Act took effect, there were only ten licenses with one tenure form, and all spruce and fir raw material available for harvest from a license was allocated to one of the mills on the license. In the geographic design of a license, combinations of mills were sought relative to the productive capability of a specific forest. The intent was that a mixture of mills with different raw material requirements could efficiently use all the raw material from a specific forest.

The quality and quantity of wood available to each licensee and the sublicensees would depend on their own forest management capabilities. Since all available wood supply from all ownerships was accounted for in the design of the ten Crown licenses, the licensees would be motivated to manage well. Any increase in sustainable production from the Crown forest that could be achieved by dint of superior forest management would remain with the license.

In April 1982 the licensees took responsibility for design and implementation of management on ten Crown forest licenses. The main task of the DNRE became to set goals for timber, wildlife, and recreation and to evaluate licensee performance once every 5 years. The principle was that the licensee took the property in a 25-year contract with an obligation to manage the resources on an 85-year time horizon. The licensee was expected to bring the forest to a specified state at the end of the first 5-year period and to serve the supply needs of the sublicensees. At that time, licensee performance would be evaluated *in terms of the forest state,* and if found satisfactory, the FMA contract would be extended by 5 years, again to 25 years, and so on. The guiding rules were structured to provide adaptive management and comprehensive forest models.

In this manner management of wood supply was in the hands of the users, and allocation was removed from the day-to-day political fiddling that featured the previous century and a half. Timber management was in the hands of those with the most to gain from good management (a chance to expand), and there was a law that required the licensee to provide a 25-year management plan (80-year time horizon) revised every 5 years, with geographic detail for management actions. One advantage of the approach was that since the forest management plans resulted in identification of stands scheduled for harvest over the next 25 years (by individual stand for the first 5 years and by stand class from years 6–25), it provided a rational basis for the design of protection from budworm and for the design of wildlife habitat management.

The policy flurry from 1979 to 1982 left no part of forest legislation or tradition of forest operations in the province untouched. It was unusual to have the mix of people, problems, tools, and stress that resulted in such extensive change. Not surprisingly, there were many glitches in the product of such sudden and extensive change. Despite the grief, a structure was in place to lead to management of timber, wildlife habitat, and recreation in the Crown forests (Baskerville 1990; DNRE 1991a, 1991b)

Recovering/Reacting After Rapid Change

On April 1, 1982, the first period of the management plans for the Crown forest began, and so did the real problems. Because of the fast start and the change of responsibilities for design and implementation of forest management, it was necessary for DNRE and licensees to share certain tasks in the first 5-year period, and there was naturally a considerable amount of consultation as appropriate decisions were sought. Also, the new forest planning data were not available until about halfway through the first planning period; therefore, to the extent they existed for 1982–1987, the management plans came from the DNRE analyses of the licenses in 1979–1982.

Since DNRE had always been "the manager" and industry had always been "the harvester," the 1982 change in roles was a challenge for players in both industry and government. Not surprisingly, some individuals never met the challenge. Some in DNRE believed industry was incapable of managing and that it was their personal role to prevent it from becoming the manager. Rather than allow industry to learn from mistakes (that it surely would make), these people inserted themselves into the day-to-day management process, where they did not belong. In so doing, they tended to freeze the system in the old form of anecdotal stand treatment, preventing the evolution of forest-level management, but in general, DNRE made a solid effort to install the new system as a form of working forest management.

For its part industry was hesitant. Some seemed to think the acquisition of harvest rights was all that had happened and did not realize the responsibility they had taken on forest-level management. Some did not believe the new tenure arrangement had been made by government sincerely (or perhaps even knowingly), and they believed government would still play politics with timber allocation. Most in industry wanted to capture the opportunity to demonstrate what a free-enterprise approach could achieve with respect to long-term forest management.

However, on the whole, industry badly misjudged the skills and effort required to design and actually implement technically sound management of large forests in the form required under the new policy.

Since industry had never borne responsibility for managing the forest in the sense of temporal and spatial design, the forest management authority lines that existed in industry were weak and usually reported within the harvesting operations division of the companies. Industrial forest management designed and implemented from the vantage point of a logging operations department lacked technical finesse at a time when technical change was pervasive.

The cap on harvesting that the act had put in place caused industry to worry about loss of market share. Clearly, if world markets continued to expand while their mill output was frozen, they would lose market share. Industry/government discussions began on how aggressive management might allow more than just holding even on raw material supply through the transition period (until 2020), beyond which there is some room to expand.

Although there were local problems, the parties generally operated in good faith, and most believed progress was being made in terms of actually implementing management control over the temporal and spatial evolution of pattern in the forest. Given that forest-level management was a new venture and given the scale of the endeavor, the preceding might be considered normal start-up difficulties in introducing a high-tech approach to Crown forest management.

One Step Backward, Two Sideways, One Forward

After the act passed but before the first evaluation of industrial performance with respect to forest management (a period of about 6 years) there were five different ministers in the DNRE portfolio. Just as the first 5-year review of licensee performance came up in 1987, a provincial election resulted in the first change of the political party in power in 17 years. This did not seem to be a problem for forest management because the winning party had supported the legislation and appeared to have a solid resource management ethic. Also, although the senior bureaucrats from 1979 to 1982 had either retired or moved on, members of the bureaucracy who had built the working parts of the management process were still in the system. Despite the good signs, the central intent with respect to achieving management *in the forests* was seriously damaged as the second management period began.

Given the array of start-up problems, notably joint responsibility for some matters, and the fact that DNRE had prepared the first management plans, evaluation of licensee performance at the end of the first 5-year period (in 1987) against the final intent of the act was not realistic. The attempted evaluation was a disaster in all respects. Industry believed it was being faulted for not being good at filling out forms, not being on time with reports, and so on, rather than for not doing its job in the forest. Although there were substantive issues that could have been raised, there was no mention of the state of the forest in the evaluation of industrial forest management performance by government (DNRE 1987a, 1987c). The problem was that DNRE evaluators also were learning forest-level management for the first time. Instead of the first period being a learning experience for both industry and government, it ended with many in industry believing DNRE really did not want the forests managed in the manner the act intended and with many in DNRE believing industry did not have a serious commitment to manage the Crown forest (Baskerville 1987).

A strict evaluation of *forest management performance* might well have failed all ten licensees. This would not have been surprising, given the learning process, and should have motivated better management effort, for the licenses were continued with a 20-year rollover. Instead, on weak bases, given the intent to use an adaptive approach to management and given the start-up problems and the long-term nature of a forest management venture, three of the ten licensees were said to have defaulted on management performance, but on vague grounds none was penalized by government. With that the most elegant attempt in any province to bring an entire Crown forest under technically solid management began to falter. As the second 5-year period began, the trust and understanding that had been built in the first 5 years deteriorated when the government attempted to change license areas and management roles in a manner that industry perceived as a significant government withdrawal from the management contract agreed to in 1982. From 1987 through 1991 there was more or less continuous discussion between industry and government about the basis of the forest management contracts. Relations between the Crown forest management partners, government and industry, were strained in a manner that did not bring out the best in either side.

With the 1992 evaluation of licensee management performance, changes in government facilitated a renewal of the original forest management intentions, based largely on a desire to introduce management of wildlife habitat as part of a program to achieve population targets.

Where Is the Forestry Problem Now?

Since the forestry problem evolves continuously, the question is not so much what it is as where it is. Great progress has been made in bringing the New Brunswick Crown forests under management. The problem is more defined, in all its dimensions, than it was 45 years ago. For the most part the elements of the problem are now technically tractable. The principal difficulty is achieving the stable level of understanding among the technical players, and within society, required to make choices at the scale of the whole forest and to implement technically appropriate solutions over appropriately long time horizons.

Living the forestry problem has brought the technology of resources management completely up-to-date. Not surprisingly, tools such as the geographic information system and formal models have had a double-edged impact. Formal and geographically explicit management provides a bridge, eliminating a large number of traditional arguments that tended to focus on anecdotal information. The same information access is a barrier in the sense that it creates a whole set of arguments that were unknown before the advent of this technology.

Has New Brunswick achieved integrated resource management? Not yet. At best, the flow of timber is being managed so as to be sustainable in explicit amounts and quality. At worst, integrated resources management is an illusion in that the broad (integrative) goals of society are not articulated—indeed, are unknown—thereby making *any* general claim contestable. Establishing a real intersection of what is biologically possible with what is socially acceptable in terms of goals remains a major problem.

In New Brunswick integrated resources management has progressed far enough that it is now possible to see how complex such management really is as it takes form in the forest. Learning has not been uniform among or within agencies, organizations, and elements of society. Although the general level of understanding has increased, the problem-solving structures that have emerged do not foster learning that has the uniformity and breadth to support provincial policy. Although discussion within the context of timber has acquired substantive logic, there has been relatively little progress with respect to other forest values. The problem is exacerbated by the fact that criteria of social acceptability shift more or less continually without any means of systematic tracking. Such shifting is inherent in change, but in the absence of a set of measures that allow tracking, adaptive management is impossible or provides very incomplete feedback. There has been great value in the public debate of the elements of the problem, and such debate is necessary. Adaptive management requires an environment in which all the players can take part in the *same* debate.

There is considerable disagreement with respect to where New Brunswick forestry is now and where it should go next. Economic goals, specified by legislators on behalf of the public, have not been formally reviewed in a public sense, and some challenge the assumption of economic primacy in goals. The public became somewhat sanguine about the forestry problem once the budworm, and the protection program, moved away from populated areas. Some believe the public has given up proprietorship of the forestry problem to government agencies, and these people believe that this will result in the continuation of traditional policy approaches inconsistent with "best" management of the forest resources for the province as whole. The fundamental problem here is that the multiple visions and objectives of the variety of constituencies have not been articulated in a common form. There are many spokespersons for the proverbial "we," but it is unclear what each of these groups is trying to do with the forest system.

Although the various debates of the past four decades have facilitated some important learning, it has been difficult to capture in usable form. Feedback control in government policy tends to be strong and weak at the same time. It is readily responsive to vague general messages in terms of equally vague statements of response, but it is weakly responsive in terms of change in the decision structure. That is, there is weak feedback inside government with respect to technical control and strong feedback in the political context. A difference between policy as it is stated and policy as it is executed still persists. There are still difficulties in determining the truth of general information about the forestry problem. Bureaucracies have great difficulty with adaptive management, for it requires admitting, "I don't know; I must learn," and this results in removing the standard reference point in a bureaucracy.

With a careful choice of examples some naysayers hold that technical forest-level management has not worked. There is some public sentiment of antitechnical management and an associated preference for local stand-level decision by consensus. In this format local actions would be subject to public review out of context of forest-level considerations. Some members of the public, and indeed some professionals, have referred to modelers in DNRE as the "Nintendo kids." These people know what should be done in "their" part of the forest to achieve the value they seek. None of these local solutions is based on analysis of resource dynamics, nor do they deal with values beyond those defined in the local solution.

Industry is leery of the intent of government and suspicious that there is insufficient government and public commitment to achieve long-term management of the resource. At present, the perceived commitment of government is sufficiently low that industry shows little willingness to invest money in managing publicly owned forests. A reasonably conducted performance evaluation in 1992 could go far toward rectifying the situation.

A blue-ribbon Premier's Roundtable on the Environment initiated a series of sectoral discussions that dealt extensively with public involvement. In its conclusions the roundtable suggested involving the public in goal setting and in broad trade-offs of forest values, rather than in the mechanics of management design and implementation (Premier's Roundtable 1992). Nevertheless, polls and media reports suggest the public would like to get the "theoretical stuff" out of the way and get down to designing and implementing management by popular consent (town hall meetings). It sounds ominously like a return to the tragedy of the commons.

It is not at all clear if the public really wants sustainable resource management or just the application of actions that currently enjoy the public label of "good," and then only as long as it does not interfere with their access to the forest. *The problem is one of establishing public expectations that are consistent with public willingness to allow/permit/enforce long-term efforts to make management work in the forest.*

The people of New Brunswick are basically impatient and do not recognize the long time lags in the dynamics of forest systems. One interpretation of this is that, despite protestations in the media, people really do not care about *the forest* or about *the wildlife*, except as these directly affect their personal daily life. In this context the widely claimed conservation ethic of society may be a thin veneer covering individual self-interest. The Crown Lands and Forests Act was designed to achieve reasoned long-term management in the Crown forests. Yet many, including some woodlot marketing boards and environmental groups, apparently believed the Crown Lands and Forests Act was all-purpose legislation to fix all forest-related ills on all ownerships quickly, and since there is little evidence of such an instant impact, they are campaigning for major change in the act. There is no call for an analogous review, and repair, of forest resources and environment law applicable to private forests.

Analyses of a problem as big as managing the Crown forests of a province always leave some things out. For example, it is impossible to get all the issues into the space allowed for writing, one person does not have all the necessary information and it is impossible to get the relevant people together, and so on. Actually, the problem is more fundamental: The authority of agencies is not defined at the system level or in a system context, nor do the public owners think in a systems context.

Where Is the Forestry Problem Going?

It would be very easy for evolution of the forestry problem to take yet another unexpected turn, and in that case successive crops will grow through several contradictory policy changes, as have all the crops before. It remains to be seen if an adaptive approach to forest resource management will be tested in New Brunswick over a time frame appropriate to the natural system dynamics involved. Although changes in the management process are necessary, the key will be the persistence, or lack thereof, of the basic policy and of maintaining a sufficiently long time perspective. Failure to achieve consistent implementation of the process will result in unrealistically high expectations with respect to timber and wildlife habitat. That would be disastrous, as the foregoing history suggests it would be unusually insightful for government and/or society to acknowledge such a situation and act to constrain resources use.

The 1982 policy still appears to be a sound basis for controlling Crown forest development, for keeping industrial development in tune with the forest, and for introducing management of wildlife and recreation. The question becomes how thoughtfully the policy will be implemented. In effect, the question has become what level of confidence can be established among DNRE, its industrial partners, and the public with respect to focusing on the forest resources. On a positive note, the 1982 policy direction with respect to integrating management on Crown forests has resulted in a major effort to introduce technically sound wildlife management objectives, with both temporal and spatial dimensions, into the 1992 plans.

The budworm problem still persists. Insect populations still rise and fall (they are low right now) and remain a crucial driver in the dynamics of development in the fir/spruce forest. There is considerable risk that collapse of budworm populations would result in diminished attentiveness to the well-being of the forest.

The current recession appears likely to force a major rationalization of the forest-related industry in the province. On the pulp and paper side, timber supply is secure, but the delivered cost of that raw material is comparatively high, and some mills are weak competitors because of age and size. The sawmill industry faces raw material of poorer and poorer quality until at least 2020, after which quality should begin to recover slowly. The pressure cooker of economic rationalization will test commitment to long-term policy.

Attention to structures that facilitate all party learning is urgent. There is a need to learn how to embrace error and to break from the "priesthood" approach where only a single group or agency holds wisdom. Clearly, the forest will continue to change, regardless of the policy in place, and industry will change in response to the forest, to international market pressures, and to public policy. The key issue will be the degree to which these policy and forest management changes are reasoned (adaptive) approaches to the uncertain future, as opposed to nostrum solutions that capture contemporary public support, even though they do not manage resource dynamics.

Messages

Messages from this complex case are themselves complex.

Policy Is a Process

Policy literature has recognized that policy is a process rather than a finite thing. For the public in the present case, that recognition has not yet occurred. There is still a perception that policy is created de novo, with an expectation that by virtue of policy creation, problems go away—instantly. There is little recognition that *policy effects cannot appear any faster than the operation of dynamics in the system being managed.* There is little comprehension that policy must be consistent with natural processes as these operate in natural systems and especially as these relate to spatial pattern. For the most part, scientists and academics are indistinguishable from the public in the latter respects.

Over time, there are periods of rapid policy change and periods of slow policy change. These possess characteristics similar to the continuous cycle of exploitation/conservation/creative destruction/mobilization that Holling (1992) has postulated for natural systems. His diagram (figure 1.2) was intentionally structured to suggest that the cycle continues infinitely. The present review suggests that the actions and reactions of human society to these natural changes follow the same cycle. The cycle of Crown forests policy described earlier began with mobilization and exploitation by society following World War II. Forest policy changed slowly in a conservation mode until the creative destruction of 1979–1982. The latter was followed by rapid mobilization, and exploitation of the policy vacuum. By the end of the first management review in 1988, policy change was back in a long, slow conservation mode. It appears that elements of societies exhibit the characteristics of *r* and *K* strategies, just as what we call natural populations do.

The important point here is that management of a natural system such as a forest is an attempt to prevent the cycle of exploitation/conservation/creative destruction/mobilization from operating normally. Indeed, if the cycle did not operate, there would be no need for society to manage those systems. In general, management is invoked to prevent the creative destruction step in order to allow orderly capture of economic and social value that has accumulated in a natural system during the period of conservation. In the broadest sense, this value is captured by "harvesting," whether for timber or for aesthetic values, from the resources over time.

Crop protection that prevented creative destruction in the natural forest system in this case was the basis for the introduction of management of the fir/spruce forest system. Coincidentally, crop protection set in motion a complete new suite of public policy concerns. For example, the frequently cited notion that pesticides create a treadmill with no escape is inaccurate. New Brunswick can choose at any time not to protect the fir/spruce forests from budworm, but then the natural mode of creative destruction will, sooner or later, operate in the forest system. The latter has two important side effects: (1) It will no longer be possible to manage the fir/spruce forests with respect to the temporal and spatial availability of habitat for pulp mills and sawmills or of habitat for deer and pine marten, and (2) it will no longer be possible to sustain consistent industrial capacity or consistent wildlife population levels. If protection is a treadmill, it is because society is unwilling to allow the creative destruction of nature and thereby give up the benefits of forest and wildlife habitat management, not because it is impossible to stop spraying.

It will be difficult to achieve the level of public understanding of a policy that permits more than a temporary escape from the Holling infinity cycle. The tendency of society to want immediate results leads to a policy focus on the tools used now, rather than on the process and the results management is to achieve over time. Thus tree planting is seen as an immediate impact and is characterized as good policy (or bad, depending on the viewer) independently of how planting alters long-term forest resource dynamics. In fact, the focus on tools relieves society of any obligation to measure whether the policy goals are being achieved over time; that is, there is a belief that as long as good tools are employed, good forest-level management will be the outcome. This rationalization represents the conservation mode of society, and an event of creative destruction (preferably of our choice) is necessary to motivate society to react, and change policy, if it is to avoid the creative destruction of nature.

Goals and Trade-offs

The forestry problem in New Brunswick illustrates an almost complete lack of power to deal credibly with economic and social goals or with trade-offs among these goals across time and across geographic area, at the level of the whole forest and of the whole society.

Goals

It is clear that as a society we do not have a way to decide what we want over large geographic areas and over long time horizons. Various analyses of policies relating to the budworm/forest/industry system have been done (e.g., Marshall 1975; N.B. Cabinet Secretariat 1976; Runyon 1984). These starkly reveal that the only values amenable to classic goal setting and measurement were related to economic conversion, and these mostly by estimation. Despite their common use, it appears that standard economic indicators, such as value added and total wages, do not enjoy wide public understanding or acceptance. Clearly, nonmonetary values, such as wildlife habitat and aesthetics, are important to society, and goals for these values are not credibly represented in dollar form. The only economic study of the budworm/forest/industry system that gained any consistent acceptance in the forest policy community was that associated with the 1976 task force. It presented the economics in the context of a companion set of forecasts of performance in the biological system.

Whether in economic or noneconomic terms, there is confusion in establishing value because of the predominance of cross-scale interests and/or concerns. For example, the individual does not experience value (of a deer, of a job) in the same context as does society. Even the value of a specific balsam fir stand of known location, size, age, volume, and so on, is ambiguous. Ambiguity can be reduced by such conventions as calculation of conventional stumpage value as the remainder of all the costs of converting the stand volume subtracted from the sale value of the converted product. Such a process says that all the value possessed by the stand is "now" and that all the value is captured in a return from manufacturing into lumber. Few people, including those who make these calculations, believe that to be realistic, if for no other reason than that it fails to account for the need for a continuous flow of stands to sustain the value-producing sawmill.

Determination of *value* with respect to the many reasons for considering a particular fir stand is complicated by the absence of agreed-upon units of measure. Specifically, the problem is *not* the absence of an economic measure, but rather the lack of a measure that reflects value *in the units in which the value is held.* For example, the estimation of the dollar value of a deer commonly results in discussion of expenditures on rifles, shells, guides, and so on, none of which has anything at all to do with the *value of a deer* in a particular forest at a particular time.

Economists argue that dollar values, if necessary in shadow units, are appropriate to allow equal consideration. Although that may be true for direct conversion of a stand into lumber or paper, the complete absence of such putative dollar values in the real discussions of the forestry problem suggests that few, if any, of the players believe these dollar values represent a reasonable basis upon which to make choices in integrated resources management.

A fundamental dichotomy of goals underlies much of the argument about the forestry problem, in whatever guise that takes. The broadly accepted goal is to attain the best forest management control toward the present public desires (as expressed by legislators) with respect to a sustainable forest industry, subject to sustaining wildlife habitats. That goal is conveniently assumed in most discussions, whereas a portion of society wishes to explore alternative goals for the forest resources, notably ones that are not linked to pulp and paper or to sawmilling. The question, "Should New Brunswick continue to build its economic future on forestry?" is not formally asked, or answered. The assumption that the goal is economic development is a continuing (often unstated) source of contention in the forestry problem. The goals of New Brunswick forestry have not been debated in the broadest sense.

Trade-offs

There are two elements to the problem of intertemporal and interspatial trade-offs among goals: First is the determination of the *value* involved, and second is the basis for *exchanging* that value across geographic area and/or time. As noted earlier, the determination of values has not been formally addressed. Therefore any trade-off of an economic production goal for timber as against a goal for aesthetics (or wildlife, or whatever) is perceived as being one-sided in favor of economic value, and since the exchange is made without formal description of the values involved, the perception persists in society.

In the present case, the absence of a consistent determination of value (whether for trees, deer, or whatever) tended to drive trade-offs into arguments about the measures used, to the point of excluding consideration of actual trade-offs. The continuing debate about the value of chemical pesticides in forest management versus the health risks to society is the classic example. There is little acceptance in society of the notion of trading a decrement of economic development for a decrement of threat to public health resulting from pesticide use. Experience in this case suggests that representation of nonmonetary values in dollar form can be strongly counterproductive in the design of natural resource management and equally counterproductive in the evaluation of public health risks. An economic approach may give the illusion of consistent treatment (in terms of economic efficiency), but it is seldom (perhaps never) accepted by society as equitable with respect to the issues at hand. Public resistance to economic dominance in goals and in trade-offs seems to be more related to the indicators used to *measure* goal value by the sides than to the mechanism of making the trade-off itself.

With respect to trading value across space and time, it is clear enough that values are not constant as they are exchanged, but there is no agreement on how the values are transformed by the exchange, at least for nonmonetary values.

Trading from place to place at any one time involves only the problem of the *value* being seen differently in the two places. For example, altering access to the forest, and therefore raw material availability for two mills, can involve trading a loss of jobs at one location for a greater number of jobs at another location, which results in a net benefit to the province. To the villages involved, however, the trade-off can be social disaster at one location, compared to economic windfall at the other. Although the long-term net impact of the trade-off for the province is positive, the community that loses jobs does not *see* the benefit at the provincial level. In the end, such trade-offs are strongly resisted by the public and typically result in political intervention to retain (by subsidy) the jobs that would be lost at the first location while increasing the jobs at the second location. That solution defies the original socioeconomic problem and almost always makes that problem worse.

The difficulty here is not a lack of efficiency in economic decisions (after all, they do follow the proper economic protocols), but a lack of acceptance of these decisions as socially equitable. To generalize: (1) A decision is announced engendering an expectation of good results, in total and particularly at some locations; (2) it is noted that some bad results, perhaps at other locations, are associated with implementing the decision (i.e., there are side effects); (3) the latter are resisted by those who must suffer them, to the point of artificially preventing (masking) their occurrence; and (4) because of (3) the good results do not appear as forecast, and the policy is seen to have failed. As with all decisions, if economic trade-offs cannot be implemented, they do not have the forecast impacts; indeed, their impacts become part of a new, even less-well-understood policy problem.

Trading values across time is a major problem in forest resource decisions, because there can be two or three human generations between the cost and the benefit. Conventional approaches, such as net present value, assume the present generation (the one making the trade-off) and the future generation that will live during the trade-off value the things involved in the same way. For example, an economic evaluation carried out in the 1930s would have decided not to protect fir stands from fire or insect attack, since these stands had little value to the generation making the decision. As time has shown, the generation of the 1950s valued the same fir stands highly (perhaps because of newly available technology) and invested heavily in crop protection, forest management, and industrial development. Violation of the assumption of equivalent value forms in a 1920 economic analysis could not have been foreseen, but in retrospect it would have invalidated such an early analysis in the eyes of the 1950s generation.

Failure or inability to forecast when coupled with discounting of the future can be a pernicious form of option foreclosure. Norgaard and Howarth (1991) argue that it is necessary to recognize separately (1) the efficient use of resources available to the present generation (a situation where discounting is appropriate) and (2) the equitable redistribution of resource opportunities open to this and future generations (a situation where discounting is inappropriate). In all attempts to apply economic analysis in the present case study, these two perspectives were merged into a single classic economic evaluation, and the analyses had no visible impact on the public or on the policy makers. Although decisions have turned out to be inefficient in economic terms, there is no measure of the gains achieved in equitability as a result of bearing this cost. It does seem clear that equitability is a real issue in the eyes of the public.

It is rare in decisions about a distributed resource, such as a forest, that there is agreement among players on what constitutes an equitable *distribution* of either the costs or the benefits. For example, most net present value approaches evaluate expenditures at specific locations against future returns achieved at the same location (e.g., at the level of stand silviculture). This misses the whole point of resource management. An elegant calculation of the probable value of timber on 1 ha of plantation in 50 years against the present cost of creating that hectare of plantation may be academically interesting, but it totally misses the policy issue that part of the value of the plantation lies in the context of managing the whole forest, and particularly in the way the existence of the plantation alters forest-level dynamics and thereby the evolving value of the forest as a whole (Regier and Baskerville 1986; Baskerville 1988a; Reed and Baskerville 1991).

There has been considerable public resistance to formalization of goals and trade-offs in the New Brunswick forestry problem. Whether or not it is because of the low credibility with respect to the values assigned, the policy makers, public, and media largely ignored attempts to formalize goals in measurable form or to state trade-offs of values formally. All sides have refrained from stating measurable goals in written format and have reacted even to the use of the term *trade-off.* It appears that most people preferred to keep the argument about values loose and open, rather than to give credence to any one proposed value set by accepting an analysis.

The important point here is that conventional methods of estimating value and of exchanging value time and across geographic area were not accepted as reasonable, either by policy makers or by the public. In the absence of credible analysis to the contrary, the policy reaction was to "keep our options open until the dynamics of the system cause a crisis, or until events outside the system cause a crisis that forces intervention" (i.e., the conservation mode).

Stability of Policy Intent

It will be difficult for society to achieve consistent stewardship with respect to resources having developmental cycles on the order of several decades as long as it continues to base policies on tools that happen to enjoy current public support rather than directly on the goals to be achieved. The underlying issue here is that social mechanisms have difficulty absorbing ecological (scientific) understanding and that compromise as a result of weak understanding of system dynamics inevitably causes instability of policy intent.

The classic example is analogous to the earlier problem of economic adjustment between communities. A natural system goal is agreed to by legislators on behalf of the public. A policy is stated that relates goals to the general types of action needed to achieve the goals. To a greater or lesser extent, this statement is based on a technically rational view of the policy issue. As that policy is being implemented, the public sees some of the actions as unacceptable (e.g., use of clear-cutting to implement even-aged management or plant closures to implement rationalization of industry). When sufficient public pressure has been built up, the politicians rule out (or overrule) the offending tools but leave the original goals in place. This "better" situation is hailed by all as progress. In fact, for ecological systems such compromises of the action set (as opposed to trade-offs at the level of goals) frequently invalidate the entire policy. That is, with the revised tools as announced, it is no longer possible to achieve the original policy goals, but the latter are rarely, if ever, acknowledged.

In the present case, there has been great progress in the management of the forest resources, but for some who resisted the use of insecticides the real issue was not so much the chemicals as the fact that they wanted a different forest management goal. Since these people believed they had been left out of the policy decision that chose economic development based on the the forests, they took whatever means were at hand to thwart the policy they did not support. The policy process for handling goals and trade-offs did not function smoothly, preventing the stability of forest policy that will be necessary to manage (for any of the forest resources) over time horizons of 50–100 years.

This case study illustrates that a barrier in one policy context is often a bridge in another policy context. This emphasizes the need to change, artificially if necessary, the "boundaries" of policy discussions in order to move away from defining policy in terms of tools used and to move toward characterizing the system state that is to be achieved.

Finally, this case study illustrates the importance of transition policies. That is, an emphasis on *new* policy inevitably triggers unrealistic expectations in terms of timing and amount of response that can be achieved in the natural system. There is a need for bridging policies, which are different from existing ones and lead to the new one but which *do not* create undue expectations and *do* create a need for the new policy.

Roles in Policy

There are many roles to be played in forest resources policy, and recognition of the players is essential to the success of policy design and implementation. However, the critical role in policy evolution is learning—learning what society wants, how natural systems function, and how to match these often disparate embodiments of learning. Neat descriptions of this three-sided problem are illusory. Politicians, bureaucrats, and organizations (public and private) must *all* learn, in a representative democracy of citizens, and they must all remember to permit the evolution of sustained policy, which is necessary to sustainable achievement of *any* goal.

This case study illustrates that learning is uneven in the policy structure and that the repositories of learning usually are *not* the formal structures. Learning seems to accumulate in what have been termed subcultures of the various agencies and organizations. General access to this learning is limited to those willing to invest the necessary time to access the relevant subcultures. There is virtually no written record of this "policy learning" with respect to the forestry problem in New Brunswick.

The role of legislators is obvious. As the elected representatives of the public, they are expected to make policies and legislation consistent with the desires of the public that elected them. If they fail to reflect public desires, they are not reelected. Such a pure view of the situation ignores the prevalence of special interest groups of all stripes (environmental, industrial, wildlife, labor, fishing, and so on). The goal of these groups is to move government policy toward a position more acceptable to their contemporary interests, even if that means running contrary to the general interests of society in the broader long-term context. In the present case, such groups, again of all stripes, had a powerful influence on the direction of government policy with respect to resources management. Periodic elections provide partial protection against this problem. Although that feedback is slow in operation, the legislators must ultimately reflect the people.

The role of bureaucrats is also obvious. The bureaucracy was invented to provide stability (inertia) as elections changed governments and to provide consistent "technical" input to government policy as it evolves. In general, rewards in a bureaucracy are greater for finding better ways to do what we already do than they are for finding different (better) things to do. At the same time, there is a very fine line between improved technical application and change of policy intent. Since the bureaucracy is not directly responsible to the populace, there is great resentment when the bureaucracy is perceived to be making, or assuming, policy by its actions. Because the technical expertise of government resides in the bureaucracy, there is frequent public suspicion that bureaucrats, not elected representatives, are making policy. In the present case, this feeling reached sufficient proportions that many proposals have been made to bypass both elected government and the bureaucracy and to decide resource management issues by the equivalent of a local referendum.

The role of science was ambiguously and inefficiently performed in the present case. Although an enormous research effort was maintained over the period of time reviewed (there were more than 20 full-time forest scientists in the province continuously over the period), traditional science failed to answer key policy questions in usable form. Scientists frequently assert that the answers are available when what they mean is that some work on a related topic has been published in a scientific paper. Almost by definition, science that is acceptable to a journal will not be in a form usable in addressing a policy issue at the scale of a whole resource or of a whole society. The scientific community is still arguing about budworm reactions in artificial lab environments, and the competing views only constitute noise in the evolution of policy that must recognize natural dynamics at the enormous time and geographic scales at which these operate in nature. It is argued that when the scientific detail has been established, science will deliver "the answer." This involves an assumption (so far unrealistic in the present case) that the problem disaggregation used as the starting point for research is dynamically associated with the policy problem. When the scientific rhetoric is stripped away, the search for a budworm population control that recognizes and acts in concert with population dynamics at the geographic and temporal scales at which these function in nature has not yet begun.

There were some outstanding contributions of research with respect to the forestry problem, and these related mostly to operational applications, such as spray equipment design, droplet patterns, geographic information systems, stand yield forecasting, damage sampling, and so on. Very little of this useful work appears in traditional research literature, but it enjoys wide respect among people who live the problem, both in New Brunswick and elsewhere, *because it was made available in usable form.* To be effective in policy issues, research needs to seek approbation of the proposed users, not just of other researchers. The case study suggests that altering/improving a policy is best accomplished by direct infiltration and subversion of the system by the bearer of scientific (or other) understanding, rather than by academic commentary from the relative comfort and safety of an epistemic community.

The role of the media may be the single most powerful element in contemporary policy evolution. Contemporary social communications structures are not suited for carefully measuring steps that, *over time,* get us from where we are to our goal. In the present case, journalists professed to inform the public on a variety of issues. In fact, an issue was addressed only if it had potential for headlines, and background preparation for presentation of an issue seemed to dwell more on how upset someone was than on informing the public of the dimensions of the problem. Perhaps the most insidious impact is that inaccurate media information remains in the public domain and continues to be "consumed" long after it has been shown to be false. It is hard to find an illustration of false information delivered via the media being effectively retracted/refuted by the media in the present case. Acknowledgment and rectification of error are not strengths of contemporary mass communications. It has been argued that the media are not at fault, because they only give people what they want to hear, and the message for solving the forestry problem, and other problems of a similar comprehensive nature, is bleak indeed.

Finally, and most important, the role of people in overcoming the tragedy of the commons syndrome is crucial. The public does not trust industry with long-term tenure in Crown forests. Industry is not motivated to manage the resource unless it sees some opportunity to capture a competitive advantage. Management by Crown agencies has proven too susceptible to local pressure and to political (public) whim to be a realistic alternative for the long term. Breaking this three-way impasse is essential to progress in renewable resources management.

It is unfortunate that in carrying out their roles, failure of the few in a constituency to perform their roles credibly can overshadow the good efforts of the many from the same constituency. The forestry problem illustrates that a failure tends to be attributed to everyone (i.e., everyone associated with a group is painted with the color of the most "delinquent" member of the group). It also shows that the work of the diligent, thoughtful players can provide a consistently solid basis for communication to ease the evolution of the problem.

Natural Resource Policy Must . . .

To be effective, natural resource policy must enjoy the support of the public for as long as it takes to implement the policy and for the natural system to respond to the policy actions. Policy must be technically sound relative to the natural system dynamics involved. The two imperatives are in strong contradiction today. Therefore policy must encourage the players to embrace error as the basis for learning. Policy must be made specifically to operate in transition. Policy must be left in place long enough to evaluate its impact accurately on the dynamics of the system involved. Policy must address systems at the time and in the spatial scales at which system dynamics operate. Policy must be built in, and promulgated from, an enduring structure.

3

The Everglades: Evolution of Management in a Turbulent Ecosystem

Stephen S. Light, Lance H. Gunderson, and C. S. Holling

The Everglades . . . the name evokes many images. The flat, grassy marsh has been described in many ways, depending upon the values and perspective of the descriptors. The Native Americans called the area "Pa-hay-okee," which loosely translates as "grassy lake." British cartographers (Vignoles 1823) coined the term *everglade* to describe the seemingly endless clearing or glade. Marjory Stoneman Douglas (1947) coined the phrase "river of grass" in her landmark book that describes and pleads for preservation of the natural values. Ecologists (Gunderson and Loftus 1993) use terms such as subtropical wetland ecosystem comprised of graminoid and halophytic hardwood plant associations. Some describe it as a god-forsaken, mosquito-infested marsh that one has to drive through in transit from one coast to the other of southern Florida. Farmers see the deep histosols resulting from thousands of years of accumulating plant debris as black gold, giving them cause for "sweet" dreams. To water managers it is a source of water for millions of east coast residents, the national park, and agriculture. To environmentalists it is the scene of conservation battles—some won, some lost, and some rained out.

The Everglades is truly many things to many people, but can it be all things to all people? The answer is uncertain, because of the unknown ability of current institutions to adapt to inherently unpredictable fluctuations in the environment, notwithstanding future shifts in social preferences. Environmental surprises and the human responses account for much of the water management history. Human impacts on the system were relatively minor until the latter part of the nineteenth century. The past century has been a period of intense development underlain by a quest for control of the region's hydrology. The complex, tightly coupled human/ecological system of today is characterized as a highly partitioned pattern, where competing demands for water resources and myopic, conflicting management objectives have led to a never-ending series of water wars and bureaucratic gridlock. In an attempt to make sense of the enduring management predicament, this chapter reviews how the current situation came about and reveals fundamental patterns of consistency and change (i.e., strategies) that might be general. The past also continues to structure and influence prevailing and future management options. As much as this chapter is about the past, we are really looking forward, because as William Faulkner said: "History is not what was but what is."

The goal of this paper is simple: to determine how to advance the aim of restoring the natural system called the Everglades. By *restoration* we do not mean returning the system to the way it used to be, but rather renewing its vitality by reuniting the systems' functions. It is in this vein that we attempt to understand the dynamics of both the natural system and human institutions in order to learn how man might coevolve with this complex resource.

The renewal of regional ecosystems, from an ecological perspective, is an attempt to rehabilitate the variability and expand the resilience of the natural system that has become fragile and degraded through societal or management actions. One assumption of this book is that renewal can be achieved by the proper matching of changes in the natural environment with people's understanding and with those inherent in management institutions. Prevailing myths of change are linked to the time horizon over which one evaluates the change. Within some time periods, systems move toward stages of climax or persistent equilibrium. Over longer periods, patterns of catastrophic change and evolution occur (Holling 1986; Mintzberg and Waters 1982). Movement toward the goal of ecosystem renewal can be initiated by comparing models of ecosystem dynamics with observed changes in human systems.

This chapter chronicles the evolution of water management in this unique wetland ecosystem. At the onset, we present a set of postulates that characterize a persistent pattern of change observed in the history of the system. The recurring sequence appears to be that ecological changes are perceived as crises and are followed by periods of alteration that provide opportunities for innovation, creativity, and renaissance. Next, the spatial and temporal scales of the environmental and human systems are identified. The bulk of the chapter is devoted to historical accounts of the four eras of water management, where crises and responses have guided the system to its current configuration. The next section outlines the barriers and bridges to system renewal observed during the development of the system. The paper ends with a summary and recommendations that outline alternative futures for ecosystem renewal.

Postulates: Coupling Ecological and Institutional Change

Postulate 1: Crises inevitably occur in resource management systems. They are precipitated by surprises that take at least two forms. One form is unexpected variation in the natural system due to discrete events or changes in longer-term trends. Another source of surprise occurs when human reaction to previous events result in changes to slower variables that result in unpredictable shifts in the system response (e.g., eutrophication causing a massive conversion of sawgrass to cattails, the loss of sea grass beds in Florida Bay). As systems mature, technology, institutions, and values tend to become outmoded and prevailing strategies fail.

Postulate 2: Crises precipitate a relatively quick sociopolitical reframing or restructuring that results in the emergence of new consensus and strategy. The reconfiguration either can be a dramatic alteration of the system, involving new components or can encompass a recombination of previous components. The scope of a reconfiguration depends upon the degree of fundamental learning that has taken root in the shadow networks of existing institutions. These networks are comprised of informal collaboration of sagacious practitioners, consortia of scientists, and citizens of concern. Since the new understanding usually emerges from the periphery, it often appears as heresy to the prevailing myths and dogma.

Postulate 3. Once new understanding has emerged and achieved or attracted a certain level of support, a catalyst triggers political action based on the new understanding, leading to a new system configuration that may involve new components (i.e., rule sets or operating premises at the institutional or operational level). Groups involved in management (politicians, agencies, scientists, nongovernment groups, respected individuals) collaborate as a result of the catalyst. Political action occurs with a new vision for system management, in relation to existing technologies, myths and institutions. The resulting action is implemented, thereby reconfiguring the system until the next crisis occurs. Each new configuration may result in a narrowing of options or in a broadened sense of system responsibility and authority. Because of the ever-present nature of ambiguity in managing ecosystems, at some level residual competency traps or policy deadlocks are likely to persist.

Spatial and Temporal Patterns in the Everglades

One of the principal assumptions of this chapter is that in order to coevolve with nature, human systems must adapt at rates similar to the rates of change (or variation) in the natural system. Humans, or any other organism for that matter, have two ways of survival in a dynamic biophysical environment: Either control the variability or adapt to it. The history of development in the Everglades has been one of attempting to control the fluctuations in the environmental processes of the region. Although this approach works for short periods in human systems, the long-term result can be an increase in brittleness or loss of resiliency in both the human and natural systems. To lay a foundation of understanding for restoration, the space and time domains of critical ecosystem processes in the Everglades are described.

Everglades Ecosystem

The Everglades ecosystem is relatively new, dating back only 5000 years. Within a few hundred years of this date, the sea level rose approximately 5 m (Scholl and Stuiver 1984; Robbin 1984), resulting in a change from an arid, sclerophyllous forest to the freshwater wetland complex now present (Gleason et al. 1984). The configuration of land forms seen today was established at this time: large inland freshwater marsh, large freshwater lake, and an upland, coastal ridge. Peat accumulations were initiated in the freshwater marshes. Since aboriginal Indians were already settled on the much wider peninsula, humans witnessed this extraordinary transformation (Carr and Beriault 1984; Holling et al. 1994).

The geographical setting of southern Florida determines much of the variation in the physical environment of the Everglades. Both the hydrologic and temperature regimes can be attributed to the location of the system at the southern end of a peninsula on the cusp of the tropics. The climate of southern Florida is characteristic of a transitional region between the tropics and temperate areas, with hot summers and mild winters (Hela 1952; Dohrenwend 1977).

The watershed of southeastern Florida has three major basins: the Kissimmee River, Lake Okeechobee, and the Everglades (figure 3.1). Although most water management activities have occurred in these regions, other basins to the west, primarily the Caloosahatchee River and the Big Cypress Area, are components of the south Florida hydrologic system. The main watershed, also referred to as the KOE system, extends for some 500 km in a north/south direction and 100 km in an east/west axis and covers 23,300 km^2 (Parker 1984). The freshwater marshes of the Everglades covered approximately 10,000 km^2 prior to development (Davis 1943; Gunderson and Loftus 1993). Lake Okeechobee covers approximately 1900 km^2 but is a shallow lake, with maximum depths on the order of 10 m. The Kissimmee River basin covers 11,400 km^2. The freshwater system historically provided runoff to the estuaries of Florida Bay.

Rainfall is the primary hydrologic input to the system, providing the renewable resource for all sectors of consumption. The annual mean rainfall is about 130 cm, with a measured range over the past 40 years of 95–270 cm (MacVicar and Lin 1984). These measures provide a sense of the bounds of the variation, but little more than a start at understanding for management. The temporal pattern of rainfall is characterized by variation at different periods, ranging from seasonal to longer-term cycles. These periodicities are generated by the characteristics of sets of processes acting at different scales.

Annually, there are two seasons in southern Florida: wet and dry. Approximately 80% of the annual rainfall occurs from mid-May through October (Thomas 1970). During the summer a differential heating between the land mass of the peninsula and the surrounding

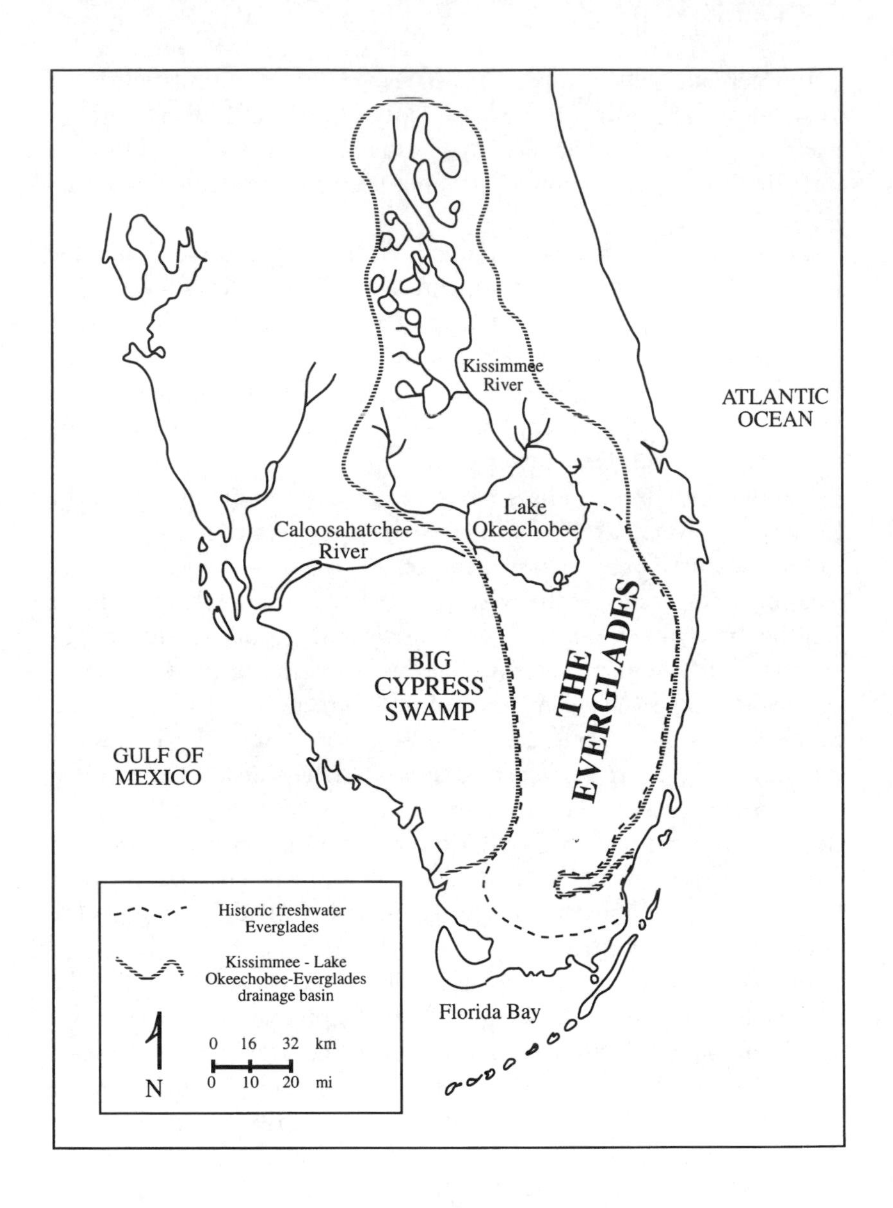

FIGURE 3.1
Boundaries of the Kissimmee, Okeechobee, and Everglades hydrologic system and the historic Everglades.

waters drives the sea-breeze circulation pattern. This mesoscale circulation drives the development of convective thunderstorms that account for the majority of the rainfall. Broader-scale storms, such as tropical depressions and hurricanes, can dramatically add to wet season rainfall amounts. Rainfall during the winter is usually associated with the passage of cold fronts. The dry season occurs during the spring, when cold fronts no longer reach southern Florida and before the summer convective activity has started.

The interannual pattern of rainfall can be attributed to a variety of factors. Changes in the mesoscale circulation patterns resulting from changes in the land use that affect albedo are thought to change the amount of rain in the convective thunderstorms (Gannon 1978). Such global events as the El Niño Southern Oscillation not only decrease the frequency of hurricanes in the Atlantic but also increase rainfall in the southeastern United States (Rasmusson 1985; Ropelewski and Halpert 1987). Periodicities of high- and low-rainfall years occur at intervals of 3–7 years (Thomas 1970).

The temporal variation in the rainfall and the flat topography of southern Florida create a wide range of surface water conditions throughout the system. The waters of the upper system, primarily a chain of lakes south of Orlando, drain into the Kissimmee River, which formed a meandering channel and drained into Lake Okeechobee. The lake had a number of input rivers, mostly along the northern rim. The southern rim of the lake was characterized by marshes and swamps with a few creeks (Davis 1943) and overflowed to the south at stages of more than 6.1 m (Johnson 1974). The coastal ridge, formed by ancient limestone reefs in the south and deep sands in the north, provided an eastern dam that impounded water in the interior marshes. A few rivers penetrated the eastern rim, allowing some outflow to the Atlantic Ocean.

The biota of the system, including humans, operate in a system that varies through space and time. To simplify and understand, systems of concern are bound by domains in these dimensions. The important processes and structures that define and influence the Everglades system can be shown in a plot with axes of space and time scale. Structures (leaves, trees, storms) originating from either human or nonhuman sources, have domains in space and time. That is, they can be defined at a minimum resolution, cover a specific area, and exist for a certain period of time. Categories of important structures in the natural areas of the Everglades, such as leaves, individual plants, plant associations (tree islands), and landscape units, have characteristic domains (figure 3.2). A number of discrete hydrologic events—ranging from showers to convective storms, polar fronts, and global cycles—form hierarchies of input variation to the system (figure 3.2) (Clark 1985; Holling et al. 1994).

Current Components of the System

Currently, land in the historic Everglades (excluding the Kissimmee River Basin and Lake Okeechobee) is divided into four sectors of use: urban area, agriculture, water control, and protected park. The upland areas of the coastal ridge and some of the eastern wetlands and transverse glades have been developed for urban and residential purposes. Much of the original Everglades' freshwater marshes has been drained and developed. Urban uses have consumed about 12% of the original ecosystem, and various forms of agriculture have taken another 27%. Of the original Everglades, less than half remains in some semblance of a natural state (Gunderson and Loftus 1993; Davis et al. 1994). The central third of the historic system marshes has been designated as Water Conservation Areas. About 21% of the historic system is preserved in Everglades National Park or Preserve (figure 3.3).

The Natural System

The conservation movement has been active in politics and land use since the turn of the century. Initial efforts by the National Audubon Society were aimed at protecting the wading bird populations from illegal plume harvest. Early natural historians recognized the unique tropical features of southern Florida (Heilpren 1887; Harshberger 1914; Simpson 1920; Harper 1927; Small 1929; Davis 1943). The first area set aside for conservation was Royal Palm State Park (Blake 1980). The Everglades National Park movement started in the 1920s, guided by Ernest Coe. The preservation came to fruition in 1935, when the U.S. Congress authorized the park, to preserve and protect the resources within the freshwater wetlands, tropical uplands, mangrove forests, and shallow marine system of Florida Bay. The park was dedicated in 1947, and in the 1980s was designated as an international biosphere reserve and world heritage site. As the water control system developed, one of the water conservation areas was given additional protection as the Lox-

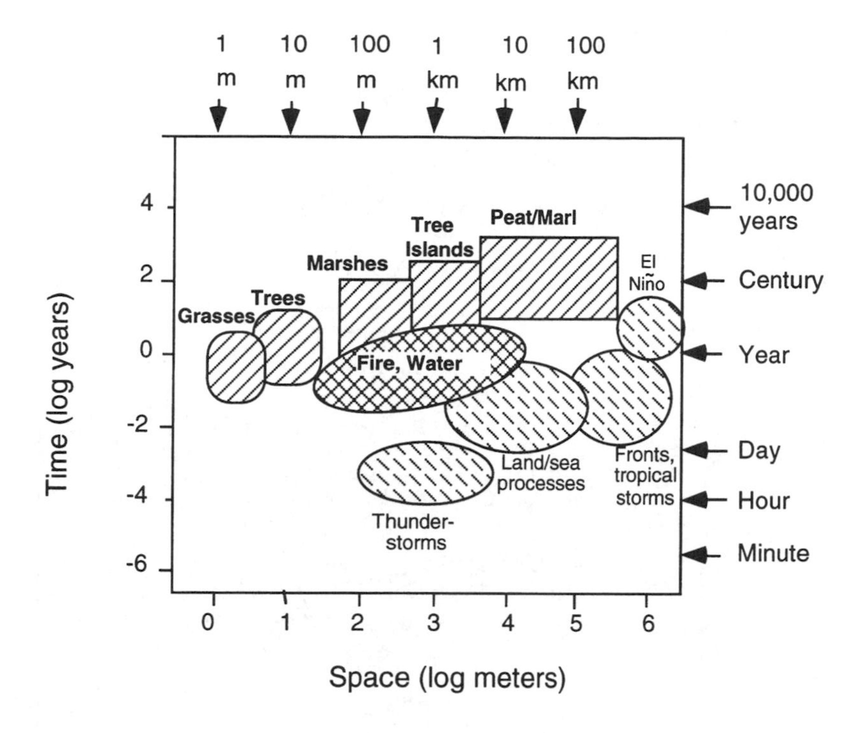

FIGURE 3.2
Spatial and temporal domains of vegetation/landform structures (squares), atmospheric variation (circles), and key processes (cross-hatched area) in the Everglades ecosystem.

ahatchee National Wildlife Refuge (NWR) and was later renamed the Arthur R. Marshall Loxahatchee NWR. The Big Cypress National Preserve was established in 1976 to protect the drainage basin for portions of Everglades National Park.

The vegetation of the Everglades region is comprised of flora that originated in four areas. About half of the approximately 1600 plant species are from tropical regions, primarily the Antillean Region of the Caribbean. Other species have temperate affinities, from the southeastern coastal plain of the United States. Less than 5% of the flora are endemic; the remaining exotic taxa were introduced by humans.

Current vegetation issues pertain to the vegetation as an indicator of broad environmental factors. The native wetland flora, especially sawgrass, has been displaced by cattails in areas of phosphorus enrichment

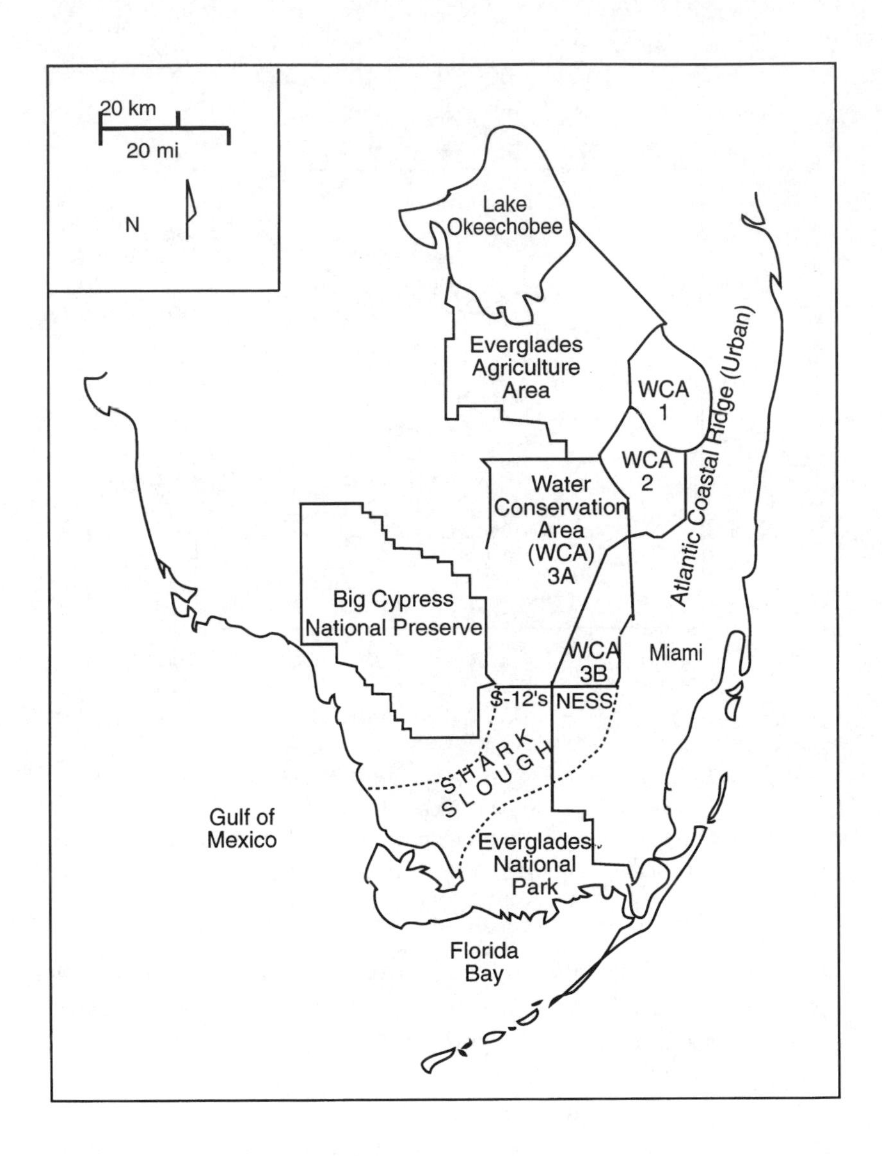

FIGURE 3.3
Current land uses in the historic Everglades Ecosystem, showing agricultural area, water conservation areas, urban area, and Everglades National Park.

because of agricultural runoff (Davis 1989, 1990, 1991). Other shifts include a systemwide decrease in wet prairie, the association within which the bulk of energy transfer to higher trophic organisms occurs (Davis et al. 1994; Gunderson and Loftus 1993). The populations of exotic plant species, especially melaleuca (*Melaleuca quinquenervia*) and Australian pine (*Casuarina* spp.), have dramatically increased since introduction in the mid-1920s. This increase is due to the adaptations of these species to drained marshes with increased fire frequencies (Myers 1983).

Whereas the flora of the area is dominated by tropical taxa, the fauna is mostly temperate in origin. About 160 species of vertebrates are native to the Everglades. Groups with large numbers of species include the breeding birds (70 species) and the herpetofauna (44 species of reptiles and amphibians). The wading birds and reptiles are the characteristic species in the Everglades habitats.

The critical issue with respect to animals in the Everglades is the dramatic decline over the past 50 years in abundance and numbers of individuals that are characteristic of Everglades habitats. The nesting success of many wading bird species in the Everglades has declined at least 95% since the mid-1930s (Ogden 1978). Populations of wood storks, white ibis, tri-colored herons, and snowy egrets have declined at least one order of magnitude (Frederick and Collopy 1989; Bancroft 1989). These declines are largely attributed to alterations in the system because of water management (Ogden 1978; Frederick and Collopy 1989; Bancroft 1989; Walters et al. 1992).

Agriculture

The deep organic soils of the northern Everglades have attracted agriculture since the late 1800s. Various crops are grown in the area. Sugarcane is cultivated on about two-thirds of the 280,000 ha of the Everglades Agricultural Area (EAA); a number of vegetables and landscape grass are grown on the remaining third (Izuno et al. 1992). The farming practices, including tilling and water management, have resulted in subsidence of the organic soils. At current rates of loss, it is estimated that farming can continue for another 20 years (Stephens 1984). Runoff from the farm fields has been pumped into both Lake Okeechobee and the freshwater marshes of the Everglades, causing eutrophication problems in both areas.

Despite very modest beginnings (c.1900) on test plots southwest of Lake Okeechobee, sugarcane is now the dominant crop in the Everglades Agricultural Area. J. O. Wright (1909) estimated that the cultivation of sugarcane in the Everglades could yield a crop worth ($180 million), equal to all the assessed property in the state at that time. This vision of the Everglades as one of the greatest wealth-producing regions of the nation was predicated on massive and intensive water management schemes, which Wright, among others, had championed. With the aid of a federal flood control project, a Cuban sugar embargo, and U.S. government subsidies, sugarcane production in the Everglades has flourished during this century.

Developed Sector

Development of the Everglades is scarcely 100 years old. Population growth in the region has been dramatic, exploding from less than 30,000 at the turn of the century to just over 5 million in 1990 (table 3.1). The development of the area was predicated on a number of factors that enabled humans to control the undesirable aspects of surviving summers in southern Florida. Among these factors were the development of air conditioning, control of insect populations (mostly mosquitoes), and control of the natural flooding of many areas, including eastern marshlands of the Everglades.

Much of the economic development is linked to the natural environment. Retirees have been attracted by the cheap land, low taxes, and mild weather. Tourism is the primary industry (along with agriculture), because of the natural features of the area, mild temperatures, beaches, fishing, and other recreational activities. South Florida is becoming an international banking center for the Caribbean and South America. Water from the Everglades supplies important recharge to the coastal aquifers used by the residents of the coastal ridge.

Management Institutions

The federal operations or participation in water management in the Everglades has been historically entrusted to the U.S. Army Corps of Engineers (COE), which has jurisdiction (U.S. House of Representatives Document 643) for the design and construction of the water manage-

TABLE 3.1
Some General Statistics on the Everglades Hydrologic System

Statistic	Measure	Units	Percent
Dimensions			
North/south axis	210	km	—
East/west axis	77	km	—
Area	23,220	km^2	—
Population			
Total population	5.19	million people	—
Population density	224	people/km^2	—
Dominant towns			—
Miami			—
Orlando			—
Gross Regional Products			
Finance	15.61	billion dollars	20
Services	14.94	billion dollars	19
Retail and Wholesale Trade	14.89	billion dollars	19
Utilities	7.97	billion dollars	10
Government	7.92	billion dollars	10
Manufacturing	7.88	billion dollars	10
Construction	5.32	billion dollars	7%
Agriculture	2.12	billion dollars	3%
Total	76.89	billion dollars	
Gross Regional Product/Capita	14,815	$/person	
Land Use (Everglades only)	10,520	km^2	
Water Conservation Area	3,484	km^2	33
Agriculture	2,830	km^2	27
National park, Preserve	2,193	km^2	21
Urban, Developed	1,262	km^2	12
Other	750	km^2	7%

*Most data were compiled for the entire hydrologic system (defined as the South Florida Water Management District). Land use types represent only the historic Everglades region.

ment infrastructure. Everglades National Park and Big Cypress National Preserve are managed by the U.S. National Park Service (NPS). The Arthur Marshall National Wildlife Refuge is managed by the U.S. Fish and Wildlife Service (FWS); other branches of this agency are responsible for endangered species enforcement. The U.S. Environmental Protection Agency has maintained a relatively low profile in Everglades matters, notwithstanding its review of COE permitting of wetlands development.

The state of Florida has a number of agencies that have management responsibilities of various portions of the system. The South Florida Water Management District (SFWMD) operates and maintains the water management system to meet the goals of flood protection, protection of the water supply, and fulfillment of environmental mandates. The Department of Environmental Regulation (DER) retains supervisory authority over the SFWMD and has exercised that control most frequently on water quality matters. The Florida Game and Freshwater Fish Commission manages for wildlife habitat as well as the harvest of game and fish.

A number of nongovernmental organizations are involved in the politics and management of the Everglades. Agro-business concerns are represented by the Florida Sugar Cane League and the Florida Fruit and Vegetable Association. The Everglades Coalition, Friends of the Everglades, and the Florida Defenders of the Environment, among others, are all structured around conservation interests.

Crises: Events that Shaped the System

In the past century, the Everglades ecosystem has been transformed from a vast subtropical wetland into a highly managed, multiple-use system as a result of one of the largest public works projects in the world. This transformation was not a linear process, but it was characterized by turbulence and punctuated by profound change. For the most part, the interplay between humans and their increasing control over the system was driven by a series of events that were perceived as crises that threatened exploitation of the resources. Each crisis precipitated actions (or a series of actions) that resulted in a reconfiguration and the emergence of a new system. Crises appear to have followed two pathways: those created by external environmental events and those created by human activities.

Surprises and crises caused by environmental events are mostly exogenous to the region and arise from larger-scale processes. These events are perceived as local surprises (Brooks 1986). The environmental events occur dramatically (hence the local surprise) but reflect a bounded variation in a broader system. One example is the occurrence of tropical cyclones (hurricanes). These storms occur at irregularly spaced, rather long intervals, as perceived by a human observer at a point in south Florida. In the Atlantic each year, tens of hurricanes occur. For the most part these events are described statistically; the probability that a hurricane will strike along an 80-km segment of south Florida is about 15% (Gentry 1984). The environmental crises in the Everglades take the form of either too much (flood events) or too little rainfall (droughts) over the system.

The other type of crises occurs over a longer period of time. These events tend to be endogenous to the system and reflect a chronic problem defined by a slower variable. These phenomena are described mathematically as cusps or catastrophes (Casti 1979) or as changes in slow keystone processes in ecosystems that result in a nonlinearity in a response variable (Holling et al. 1986). Walters (1986) describes these as pathologic "surfing." Examples of these changes include water-quality degradation that leads to dramatic shifts in dominant taxa. This class of crisis appears to originate as a result of human involvement and development of natural resources. Another way of casting the problem is to say that humans have accelerated the rate of natural processes that were much slower and that might have occurred over a much longer time span. In the Everglades these types of crises have been associated with agricultural activities, primarily in the form of soil loss and water pollution. Other slow variables that have created crises as they change include human myths, institutions and technologies that fail, and poorly conceived operating criteria that have altered hydrologic cycles in Everglades National Park.

The crises perceived in the Everglades, both environmental and human induced, have resulted in at least four major eras of water management. Flooding (natural and man induced) has been a recurrent and dominant problem that has shaped most of the water management infrastructure in the Everglades during the first part of the century (figure 3.4). The first two eras were a result of flooding from high-rainfall events. The third era was related to drought events, and the fourth era resulted from attempts to rectify latent or previously unattended problems.

The earliest settlers were intent on reclaiming land "lost" to natural flooding, in order to farm the rich muck soils. Early attempts at drainage were able to control water levels during average water conditions. Periods of dry years allowed for expansion, until the next wet year. The approach was to dig canals and drain the land as fast as possible; a strategy labeled "Cut 'n Try" was coined by Gov. Bonaparte Broward. However, these attempts at drainage were unable to cope with the full

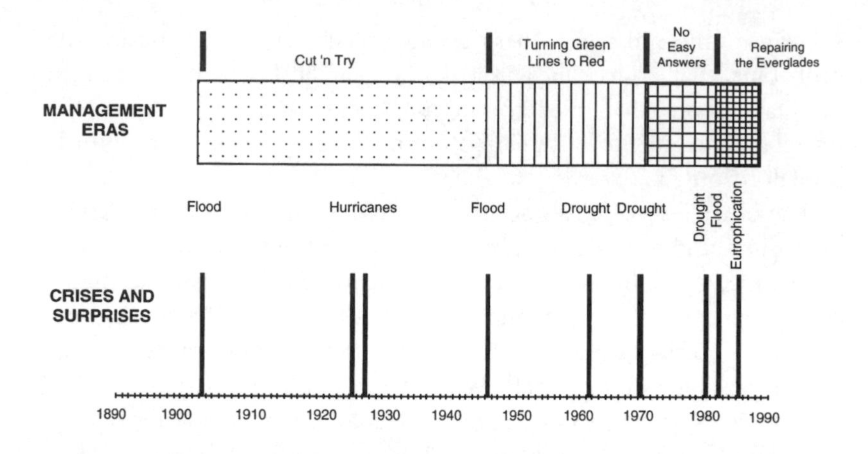

FIGURE 3.4
Time course of four management eras (top tier) and the crises and surprises (bottom tier) that precipitated transformation of eras from 1890 to 1990.

variation in climatic regimes that would inevitably occur. Flood crises occurred in 1903 as a result of high rains and in 1926 and 1928 as a result of severe hurricanes.

The "Cut 'n Try" era persisted until 1947, when more than twice the normal amount of rainfall fell on south Florida, severely impacting the coastal communities. The acute flooding resulted in the implementation of a widespread plan to avoid this type of flooding in the future. This massive control plan, developed by the U.S. Army Corps of Engineers, called for the creation of specific land use areas (agriculture, water conservation, and national park) spanning more than 46,800 km^2 (18,000 mi^2), and the water management infrastructure (2240 km of canals and levees, pumping stations with 3.8 billion liters/day capacity, and requisite water regulation schedules) needed to regulate flood waters. Proposed structures on the planning maps were colored green, and once construction was completed, the lines were colored red. This era of water management, dubbed "Turning Green Lines to Red," lasted from 1947 through the early 1970s.

The crisis that precipitated the next era of water management was a drought that occurred in 1971. Record low rainfall occurred between June 1970 and May 1971, when less than 49 cm (32 in.) fell over the system, and was recorded as the worst drought in 40 years (Blake 1980). By the early 1970s the population of southern Florida topped the 2 million mark and sugarcane production in the EAA more than tripled following the communist takeover in Cuba. The low rainfall coupled with increased urban and agricultural water demands prompted concern for an adequate supply of water to meet these needs. A number of serious problems arose, in correcting observed deficiencies in a water management system that had been operational for about a decade. Another problem arose in trying to retool a system that was designed for flood control to meet water supply concerns. The situation was exacerbated by the difficulty inherent in decisions involving trade-offs among water use sectors. This era of water management is referred to as "No Easy Answers," a paraphrase from Jack Maloy, then the executive director of SFWMD. Maloy advised his board that all the easy answers had been used up and that only difficult choices (many requiring changes in social behavior) were ahead in trying to solve competing objectives of water management.

The current era of water management (Repairing the Everglades), which began in 1983, is attempting to restore and revitalize the natural system. The growth and change that had diminished the remaining slack in all the sectors of water use was dramatically revealed during the record drought in 1981. This dry cycle was followed immediately by wet years in 1982–1983. These events fostered a number of initiatives seeking to reverse the downward trend in the natural system, to mitigate interactions among land uses, and to attain goals of sustainability for the entire system, although much of the inherent pathologies still exist.

Evolution of Ecosystem Management: The Management Eras

The Early Years (1800–1903)

South Florida during the nineteenth century changed from an unknown entity to a vast area ripe for colonization. The extent and nature of the Everglades remained largely undocumented until the wars between the United States and Seminole Indians (1835–1842). Florida became a state in 1845. Shortly thereafter, a report by Buckingham Smith (1848), commissioned by the U.S. Congress, claimed that draining the Everglades by 1.3–1.6 m would produce a "tropical breadbasket of no trifling advantage to the whole nation" (Dovell 1947). The Swamplands Act of 1850 deeded 8 million ha to the state. This transfer allowed for internal improvements, such as railroads and drainage districts, that could attract growth. An Internal Improvement Trust Fund was established in 1855 by the state legislature to manage these newly acquired lands. For the most part, the fund was controlled by the Bourbons, who dominated state politics until the turn of the century (Blake 1980). Lavish land grants were given to the railroads. By 1898 Henry Flagler had invested $30 million and constructed the Florida East Coast Railroad from Jacksonville to Miami (Carter 1974).

Hamilton Disston, a Philadelphia millionaire, made the first major attempt at internal improvements in the interior of the southern half of the peninsula. In 1881 he purchased 1.6 million ha, and by his death in 1893 he had succeeded in opening a channel route that linked the Kissimmee's upper chain of lakes near Orlando with the Caloosahatchee River and Ft. Myers on the southwest coast. Disston experimented with the production of a wide range of agricultural goods, such as sugar, rice, and peaches. Although his dream was consumed in the bank panic of 1893, Disston had at least partially verified Smith's report to the Congress.

The history of economic expansion in the region prior to 1900 was dominated by wealthy and politically powerful individuals from Northern states. Development was decidedly haphazard, uncoordinated, and heavily dependent on the whims of a few commanding figures who wielded extraordinary influence over the trustees of the fund (Izuno et al. 1992). Immediately after the turn of the century, the system was about to experience the first of many natural crises.

Cut 'n Try (1903–1947)

Crises and Surprises

The major crises that launched this management era came in the form of floods and hurricanes. Although significant drainage experiments had been attempted through the use of private funds, all had failed to tame the natural forces. The flood of 1903 destroyed 70% to 90% of the entire crop within the region and ostensibly caused $500,000 in damages to farms and cropland in south Florida. Hurricanes in 1926 and 1928 claimed 400 and 2000 lives, respectively, and caused damages of well over $76 million (Blake 1980). Dry years exposed the excesses of drainage and development. Wellfields in Miami were threatened by saltwater intrusion in 1925, and muckfires in the Glades south of Lake Okeechobee generated plumes of smoke so thick that roads became impassable.

Cut 'n Try

Governor W. S. Jennings viewed the flood of 1903 as an opportunity to wrest control of the Internal Improvements Trust Fund (IITF) from the Bourbons and the railroads. The flood provided justification to expend the necessary money to reclaim the Everglades. The flood also prompted an appeal for assistance from President Roosevelt, who awarded more land to the state (1 million ha) but no money. The gubernatorial election of 1905 altered the future of the Everglades. Napoleon Bonaparte Broward ran on a populist platform and accused the railroads and the Bourbon-dominated IITF of "draining the people instead of the Everglades." Broward won the 1905 election and interpreted the outcome as a referendum to reclaim the Everglades (Dovell 1947).

Broward assumed personal responsibility for the Everglades project. He was his own planning engineer, construction superintendent, and promoter of the project (Johnson 1974). His plan was simple—cut 'n try. Broward viewed the duty of the IITF "to dig canals the same as the water mains in a city, owned by the city, laid by the city and tapped by private individuals" (Dovell 1947). By 1915 four major canals—the Miami, North New, Hillsboro, and West Palm Beach canals—dissected the Everglades from Lake Okeechobee to the Atlantic Ocean. Additional outlets on Lake Okeechobee were added by the mid-1920s, and a muck dike was affixed to the lake's southern perimeter.

To finance this ambitious undertaking, Broward overcame considerable legislative opposition and lawsuits to establish the Everglades Drainage District (EDD). The district had the authority to tax and issue bonds, but soon came up short on public confidence and commitment to the project. Governor Broward and his successors relied on a series of reports (Wright 1909; U.S. Congress 1911; Marston et al. 1927) as rubber stamps to the prevailing approach and its progress. The U.S. Department of Agriculture (USDA) added support to the reclamation in order to "determine the practicality of draining and otherwise improving the hidden resources of the wetlands of Florida" (Dovell 1947).

Over the years, the USDA provided valuable research that has enhanced agricultural production in the EAA. An experimental station was opened at Canal Point in 1920 with the expressed purpose of developing sugarcane varieties suited for the region. Although the USDA contributed to the production of new cultivars, some of its most important breakthroughs involved issues of fertility and subsidence. During the 1920s soils were found to be low in major plant nutrients, primarily phosphorus and potassium, and micronutrients of copper and manganese (Izuno et al. 1992), which limited crop production. The rate of subsidence has been well documented since the 1920s and shown to be directly correlated with the water table depth (Stephens and Johnson 1951).

In the aftermath of the tumultuous hurricanes of 1926 and 1928, the Hoover dike was constructed around Lake Okeechobee. This marked the first presence of the federal government in water management of the Everglades. Prior to that time federal management-related efforts had been limited to surveys of the Kissimmee, Okeechobee, and Caloosahatchee water systems. The federal agency charged with the flood control mission was the U.S. Army Corps of Engineers, under the Mississippi and Tributaries Act of 1928 and the Flood Control Act of 1936. The flood protection advanced reclamation south of the lake, increasing public confidence and land values in the mucklands. Ultimately, large agro-business, not the small family farmer, harvested the fruits of the populist movement to reclaim the Everglades.

Other Voices

The voices of discontent resisting the reclamation of the Everglades arose from persons who valued its natural heritage. Willoughby (1898) recounted the natural beauties and hardships from traveling across the southern Everglades. Early naturalists (Simpson 1920; Small 1929) chronicled in disbelief the "senseless vandalism" of this watery wilderness. Arthur Morgan testified before Congress in 1912 that the "haphazard reclamation of the watershed would finally result in unpredictable confusion in the balance of life in the Everglades" (Dovell 1942, 1947). J. K. Small (1929) prophesied that "this reckless and even wanton devastation has now gained such headway that the future of North America's most prolific paradise seems to spell DESERT." Other advocates, U.S. Congressman Mark Wilcox and Ernest Coe, a landscape architect, championed preservation and conservation in the Everglades. They proposed protection of the the submarginal lands of the southern Everglades and the Gulf Coast. In conjunction with R. V. Allison, Wilcox campaigned for a stronger role for the USDA in conserving Everglades land and water resources. At the turn of the century, a coalition of women's clubs and Audubon Societies helped ban plume hunting, and later urged the state and national governments to establish a tropical national park. The park was authorized by the U.S. Congress in 1935 but did not materialize until later (Blake 1980). Marjory Stoneman Douglas's book *River of Grass* (1947) created the compelling image of the system as a "river of grass" and increased public awareness of the need to preserve the system. The Florida legislature eventually appropriated $2 million in 1947, and the Everglades National Park was dedicated in December of that year. Unfortunately, the original size of the park (1934 congressional authorization) was reduced by one-third to exclude privately held lands north of Tamiami Trail.

In 1921 the Soil and Crop Science Society (SCSS) was established as a forum for the exchange of information among soil and water experts in the Everglades. This society was the first consortium or shadow network in the system that developed deeper understanding and alternative views of the system. The SCSS was instrumental in documenting the problems related to overdrainage. A number of critical issues were addressed by the society related to the conservation of soil, wildlife, and vegetation; saltwater intrusion; natural balance between land and water; rewatering process; data-gathering needs; and institutional problems (DeGrove 1958, 1973, 1978). The culmination of the society's work was the "Rewatering Plan" of 1939, spearheaded by Dr. R. V. Allison. The plan addressed the problems of overdrainage, advocated that some areas revert to wetlands, and proposed reorganization at the state level that would consolidate the functions of the EDD, IITF, and the Okeechobee Flood Control District (Blake 1980).

Interpretive Analysis

The flood of 1903 triggered a major shift in sociopolitical relationships within Florida. Governors Jennings and Broward used the flood to launch a political campaign that pitted the populist movement (anti-establishment and anti–big business) against the landed aristocracy (Bourbons) and the railroads that had received large land grants from the state. Instead of using Florida's capital (land) to subsidize the development of coastal resorts for the wealthy, Broward sought to open up the southern frontier (bread basket) of the state to small family farmers.

Following his election, when Governor Broward was informed that his idea to drain the Everglades would take 15 years to determine its practicality. He replied, "I will be dead by that time" (Dovell 1947). Without the data needed on which to base a plan, Broward proceeded with his "cut 'n try" philosophy. As a result, the project suffered from gross underestimation of the complexity of the task or the water control problems involved. Subsequent reports that reviewed the projects were motivated more by the need to reaffirm commitment to existing policy than to provide feedback and offer mid-course corrections.

One historian of the Everglades (Dau 1934) observed that "no part of Florida has been so much written of and so little understood as the Everglades." Despite the work of the Soil Science Society and Allison's Rewatering Plan, which served as the basis for the federal flood control project to come, no one captured the essential character of the Everglades as did Ms. Douglas. She called it a river of grass. In three words Douglas portrayed the Everglades as a unique hydrologic and ecological system and not just a big swamp. This radical departure from conventional wisdom has formed the conceptual well from which new understandings of how the system functions still ascend. Unfortunately, Douglas's provocative image of the Everglades as a natural system came too late. The fundamental integrity of the Everglades, physically and politically, had been significantly altered by 1947, when her book was published.

Although the natural historians that wrote during this period were greatly dismayed by the irreversible damage they saw, the tide of reclamation continued to gain momentum. Nonetheless, the state through the eyes of their more conservation-minded leaders (e.g., Messrs. E. Coe, J. Pennekamp, Rep. M. Wilcox, and Gov. S. Holland) mustered the political will to preserve a significant portion of the Everglades south of Tamiami Trail.

Strategy and Relevant Myths

This management era was marked by a global change in state policy in favor of drainage as the primary function of the Internal Improvement Fund Trustees. Armed with the vision of Florida's interior as the nation's winter "bread basket" occupied by a corps of small family farms, Broward took bold and decisive steps that shaped the destiny of the region for the better part of the century. The myth that the reclamation of the Everglades would be a haven for small family farmers faded as they were displaced by large-scale agro-business once the federal government began construction on the Hoover Dike. "Taming of the Everglades" to conquer nature had been replaced by Broward's more contemporary view that supported the advance of urbanization that likened the drainage effort to laying out the water mains for a burgeoning city. Also prevalent during this era was the premise that a "love of land" bonded farmers to the "rigorous observance of rules" that would guard the earth against man's destructive tendencies (Dovell 1947).

South Florida's capital was bound up in the exploitation of its rich muck soils and subtropical climate as a bread basket and as an attractive winter haven for the nation. Massive drainage and advances in soil science coupled with railroads and seaports opened up the region for agricultural production and urban development. In summary, many of those who witnessed this quantum shift in the land use during this era gave special meaning to *B.C.* and *A.D.*—"Before Canals" and "After Drainage" (Brown 1948).

Turning Green Lines into Red (1947–1970)

Crises and Surprise

Following the most severe drought on record in 1944–1945, heavy rains occurred in south Florida during 1947–1948. Spring and summer of 1947 were unusually wet and were followed by major hurricanes in September and October. A total of 274 cm (108 in.) of rain fell that year (MacVicar and Lin 1984). About 1.2 million ha were under water for 6 months. Estimates of damage ranged from $60 to $500 million. In the fall of 1948 the region was struck by another severe hurricane. The state published a book demonstrating the severity of the flooding conditions, also known as the "weeping cow" book because of the depiction of a cow standing belly deep in water. The book became a national symbol for federal flood control (Blake 1980).

The Federal Project: Turning Green Lines into Red

The Corps of Engineers physical plan for the Everglades was called the Central and Southern Florida (C&SF) Project for Flood Control and Other Purposes. The project was composed of four principal technologies: levees, water storage, channel improvements, and large-scale use of pumps to supplement gravity drainage. The initial objective was to stop water in the Everglades from moving toward the coast. To that end the eastern perimeter levee and borrow canal stretching from Palm Beach to Dade County was completed in 1952. It featured a 3–6-m-high continuous wall running for over 160 km roughly parallel to the coastal ridge. The next objective was to partition the Everglades into areas designated for agriculture and water conservation. The next steps constructed the levees around the Everglades Agricultural Area (1955–

1959) with 3.8-billion-liter/day pumping stations on every major canal where they intersected the water conservation areas (1957–1958). By 1962 the water conservation areas and control structures regulating discharges were completed, giving man control over water flow to the park's Shark Slough. Work on the infamous canal that channelized the Kissimmee River ended in 1964. The last portions of the project finished in this period (1969) were the structures that paralleled the eastern boundary of the park and emptied into Barnes Sound.

The problem of confused and divided responsibilities among the Everglades Drainage District, the Internal Improvement Fund, and the Okeechobee Flood Control District was resolved by the state legislature. At the behest of the federal government and based on the recommendation of a twenty-one-member gubernatorial appointed task force, the legislature consolidated state water management and coordinating functions into one entity—the Central and Southern Flood Control District (FCD). Its primary function was to make the Corps master plan come alive. The FCD was governed by a five-member appointed board with ad valorem taxing authority and the power of eminent domain for project purposes. The sharpest debate over the creation of the FCD involved how the costs of the project would be distributed. The options were beneficiary pay versus an ad valorem tax based on assessed value. The confab pitted urbanizing counties against rural and agriculturally dominated ones. Eventually, urban legislators agreed to pay 95% of the state's share of the project (Blake 1980). To ensure that fish and wildlife concerns were tended to, the FCD turned supervision of Water Conservation Area (WCA) 1 over to the U.S. Fish and Wildlife Service in 1961, and the Florida Game and Fish Commission was asked to manage WCAs 2–3, which contained the largest vestiges of the natural vegetation.

No Water Deliveries to the Park

No sooner had the Water Conservation Area and water delivery structures just north of the park been completed than a dispute over water supply erupted. The Flood Control Act of 1948 authorizing the C&SF Project declared the project and the park "complimentary features of Federal activity necessary to restore and preserve the unique Everglades region" (House Document 643). The park called for a cooperative agreement (specifying magnitude, spatial, and temporal parameters) to be worked out, but the corps and the FCD declined. The corps claimed that the matter needed to be settled between the National Park Service and the FCD. In effect, between 1963 and 1965 the structures in Shark Slough were closed and sheet flow was interrupted while the newly completed WCAs were filled (Wagner and Rosendahl 1987). Moreover, the levee and structure design across the northern boundary of the park completely severed the sheet flow connection between the WCA-3B and Northeast Shark Slough—an area that had historically received over half of an estimated 3.2 million cubic meters per year (260,000 acre-ft/yr) across Tamiami Trail. Rookeries of ibis and egrets in East River and Cuthbert Lake failed to form for 3 successive years (Farb 1965). The plight of the Everglades made national television for the first time. The drought situation produced widespread public awareness that competition for water might become severe. Heavy rains during 1966 led to excessive pumping into the Water Conservation Areas, which in turn resulted in the death of hundreds of deer and raccoons (Carter 1974).

In response to these crises, the corps developed a report in 1968 that recommended (based on NPS and U.S. Geological Survey calculations) a 3.9 x 10^6 m^3/yr (315,000 acre-ft/yr) minimum-flow schedule for the park, increased storage in Lake Okeechobee, and backpumping from the coast to the WCAs. Water quality was not discussed in the report. The park was skeptical of any overtures that fell short of a binding agreement (Carter 1974). The Florida Freshwater Game and Fish Commission was highly critical of proposals to raise the regulation schedule of Lake Okeechobee and backpump water into the WCAs. Threats to water quality (inflows of nutrients, pesticides, etc.) were the primary reason for objection. After 2 years of bureaucratic wrangling and political opposition, Sens. Gaylord Nelson (D-Wis.) and Edmund Muskie (D-Me.), supported by the national environmental community, overrode opposition from Sen. Holland (D-Fla.) and succeeded in having a congressional act passed in 1970 that established minimum flows to the park (Carter 1974).

Other Voices

Little opposition arose to the concept of an enormous federal flood control project. Edwin Menninger, publisher of the *Stuart News,* was a notable exception. He wrote Senator Holland, "what we need is hard-shelled conservationists ... we are not interested in getting rid of any more water" (Blake 1980).

Stronger objections surfaced at the national level over the apparent insensitivity by the state to the needs of the Park and the WCAs for water. The indignation of the national environmental community descended on Governor Kirk first, for drying up the park and then for drowning the WCAs. Articles in the *New York Times* chided the state for choosing people over alligators in attending to the interests of the nation's only subtropical preserve. It was clearly a case of "mindless technology triumphing over life" (Blake 1980).

In the 1960s Florida was a study in contradictions. Riding the crest of the space age boom, the state was portrayed as a cornucopia of seemingly inexhaustible economic, social, and environmental wealth. As Thielen (1963) affirmed, riches even flowed from infertile soil (e.g., the practice of rock plowing enabled the production of winter vegetables). On the other hand, the specter of vanishing places for nature's threatened and endangered species raised concern about the dark side of explosive growth. The perception was mounting that the ecological fate of the Everglades would be determined by the turn of the century. As Archie Carr, one of the premier conservation biologists of the twentieth century, mused in 1967, "Will man always make provision for natures's wild creatures in the world of tomorrow? Or will multiplying man and his machines crowd more and more animals off the earth forever?" (Carr 1967).

The Jetport Controversy

The first major successful challenge to the unrestrained growth scenario in south Florida came in the form of organized resistance to the proposed Dade and Collier counties' citing of an international jetport (November 1967) on the northwest boundary of the Everglades National Park in the Big Cypress Swamp. A national coalition of environmental groups (National Audubon Society, Sierra Club, National Parks and Conservation Association) rallied to oppose the project (Blake 1980). The Department of Interior was persuaded to undertake an environmental impact study of the jetport proposal. Luna Leopold chaired the interdisciplinary team effort, and Art Marshall was instrumental in drafting the final document. Marshall, a former FWS employee intimately familiar with the region, decided to leave government employment to mount a campaign to protect the remaining wilderness in the Everglades. The final ecological impact statement, the first of its kind in Florida, stated emphatically that the impact of the jetport on the Big

Cypress Swamp would inexorably destroy the south Florida ecosystem and ultimately the park (Leopold 1969). Political support for the jetport began to erode quickly.

Foreign Policy and Local Agriculture

A change in foreign policy coupled with the C&SF project resulted in sugarcane becoming the dominant crop grown in the Everglades Agricultural Area during this era. Even though significant advances in research and management increased sugarcane yield by solving nutrient and pest problems and developing new cultivars of sugarcane, acreage in sugarcane production remained less than 20,000 ha prior to 1960. The Sugar Act of 1934 stabilized prices and guaranteed farmers in the Glades a percentage of market share (Clarke 1977). However, sugar production could not overcome the effects of a national depression and financial wows that plagued the EDD. The enhancement of flood control and water supply capacity through authorization of the C&SF project (1949) followed by the accession of Castro to power in Cuba (1959) changed farmers' minds. The embargo on Cuban sugar and the temporary lifting of restrictions on domestic production dramatically increased the acreage in sugarcane (from 20,000 to 81,000 ha) during the 1960s. Between 1970 and 1975 acreage under production soared again, putting another 40,000 ha in sugarcane production—a sixfold increase in 15 years (Clarke 1977). During the 1980s, the increase continued, so that by 1988, 170,000 ha were in sugarcane in the EAA.

Interpretive Analysis

Despite declarations to the contrary, the C&SF Project was not designed as a complementary feature to the park and a federal activity necessary to restore and preserve the unique Everglades region. Principally, the project became an engine for economic growth. The tussle over water deliveries to the park during the 1960s revealed the fundamental contradiction in values and objectives that the two federal ventures represented. Carter (1974) contended that only vague assurances were given in the 1948 project report and that "there were no plans to meet the park's minimum needs ... the FCD was not prepared to release water to the park except in wet years." The underlying conflict in values between the park and the C&SF project would continue to manifest itself in subsequent decades. The Florida Freshwater Game and Fish Commission anticipated the adverse water quality impacts of several of the measures proposed by the corps' 1968 water supply plan.

The polemics over the ultimate character of the region found expression in the disposition of the international jetport proposed for the Dade/Collier County border in 1967. Coupled with the announced shutdown of construction of the Cross-Florida Barge Canal, the defeat of the jetport proposal served notice on proponents of "unrestrained growth" that limits to environmental degradation within the region did exist.

If the era of "Cut 'n Try" severed the integrity of the Everglades ecosystem, this era of the "Turning Green into Red Lines" balkanized the Kissimmee/Okeechobee/Everglades system. The excessive compartmentalization of the KOE structure for the purpose of comprehensive control effectively put the natural system at war with itself. The master water control plan absent in the previous management era but envisioned during the 1930s and 1940s had been realized—the C&SF Project. However, in attempting to solve the problems of excess drainage, the new project began to disclose a fresh set of more fundamental resource problems and political questions.

The acts of man during this water management era further departed from the image of the Everglades as a "river of grass." Lost to extinction were three of the seven major physiographic landscapes (i.e., eastern cypress band, peripheral wet prairies, and the custard apple swamp forest) indigenous to the Everglades (Davis et al. 1994). The configuration of WCAs north of the park resulted in

Loss of the transverse glades that provided early-season feeding habitat for wading birds;
Modification of flow pattern (attenuated to pulsed), which reduced hydroperiods;
Unnatural pooling and overdrainage as a result of diking;
Accelerated reversal of muck building to rapid oxidation; and
The abandonment of nesting areas in the park because of loss of total volume to Florida Bay (Walters et al. 1992; Davis and Ogden 1994).

From the corps project's inception ignorance of (or worse, lack of interest in or arrogant disregard for) predrainage conditions was profound. By his own account Lamar Johnson (1958), engineer for the

Everglades Drainage District, attempted to inform the park in 1950 that the intent of the project would deprive the southern Everglades of 50% of the water inflows it had received in primeval times. In fact, two early studies of predrainage flows to Shark Slough were conducted. The studies by the Flood Control District (1950) and Parker and colleagues (1955) concluded that average annual predrainage flows to Shark Slough were over 2 million acre-ft. These results were ignored (Wagner and Rosendahl 1987).

Eventually, increasing tension over park water needs during the 1950s led the park to undertake two studies to determine a minimum delivery schedule. A preliminary study by NPS researcher Allyson Van Dunn was conducted using the discharge records from a 6-year period (1947–1952). James Hartwell of the U.S. Geological Survey was subsequently engaged to conduct an independent analysis of stage duration curves based on records from October 1953 to September 1962 (Hartwell et al. 1964). Later, the works of Dunn and Hartwell were combined by the Department of Interior to form the National Park Service recommendation of 3.9×10^6 m^3/yr as the park minimum delivery schedule from the project (Wagner and Rosendahl 1987). The volume of water that passed through Shark Slough is still being debated. Some subsequent studies suggest that the project delivered more water to the park, because of impoundments and the eastern levee (Kushlan 1987). However, others suggest the minimum delivery amount grossly underestimated the historic volume of water flowing into the southern Everglades (Smith et al. 1989; Walters et al. 1992).

The complexity of managing water and maintaining the natural ecosystem of the Everglades system was becoming apparent in the mid-1960s. The Everglades National Park developed a research plan in 1966 that was notable for several reasons. First, it recognized the wide natural fluctuations of rain, fire, flood, drought, hurricanes, and salinity that influenced organisms living on the extremes of their geographic range. Second, the plan identified the basic requirement of acquiring sufficient information to formulate solutions but realized that much of the information obtained may not be immediately usable by management. Third, the plan acknowledged that corrective steps must be taken immediately in spite of the lack of exhaustive studies that would produce the wealth of understanding needed to ensure sound management practices. George Sprugel, chief scientist of the National Park Service in 1966, expressed the need for hard evidence of biological effects in order to "defend its appeals for a portion of the surface water available elsewhere in south Florida" (Robertson et al. 1966). The lack of Park Service research capacity severely restricted the Everglades National Park's ability to communicate its needs effectively to the FCD and the Corps of Engineers. According to DeGrove (1958), during the period 1949–1958 the park developed no ordered liaison with the FCD.

Strategy and Relevant Myths

The C&SF project was a further elaboration of the drainage and flood control strategy of the previous era. The project was intended to correct the problems of overdrainage and strike a balance between man and nature. The natural system was assumed to be benign, but as the project unfolded, nature took on a more ephemeral character. Ignorance of the system's fundamental nature (i.e., river of grass) was replaced with the belief that the fundamental nature of the system had been miscalculated. The alternating cycles of dry and wet periods of the 1960s exposed the fundamental ineptitude of control strategies attempting more sophisticated management of the Everglades. Opposition to the jetport and the barge canal resulted in victories for the preservationists, reflecting an erosion of the metastrategy of comprehensive control and the need to develop viable alternatives.

In summary, the seemingly inexhaustible bounty of the region propelled south Florida even faster toward the direction of unrestrained growth despite the victory by conservationists on the jetport. The C&SF project provided a sense of security in a wildly fluctuating environment that would encourage unlimited growth and economic development. Shifts in foreign policy launched sugarcane production in the EAA to heights that had only been dreamed of before.

No Easy Answers (1970–1980)

Crises and Surprise

The drought years (natural and man-made) during the 1960s had made the nation painfully aware of Florida's water management deficiencies. The return of drought in 1970–1971 made the problem situation and the current way of doing business even more untenable. In twelve consecutive months (June 1970–May 1971) rainfall at the Miami Airport was below normal. Only 80.8 cm (31.8 in.) of rain, a new record, fell during that year. The Everglades were so dry that 300,000 ha burned.

This extraordinary dry period extended, though it was less intense, through 1975.

The Governor's Conference

The record low rainfall of 1970–1971 exacerbated the water shortage and caused more political anguish than the state could tolerate. Governor Reubin Askew took action in September of 1971. He convened 150 professional resource managers, policy makers, and representatives from many interest groups to discuss the matter. Dr. John DeGrove was asked to conduct the meeting and arrive at a consensus on action that needed to be taken. The final report to the governor was compelling and far-reaching. The following is the introduction to the report:

> There is a water crisis in South Florida today. This crisis has long-range and short-range aspects. Every major water area in the South Florida basin, the Park, WCAs, Lake Okeechobee and Kissimmee Valley is steadily deteriorating in quality from a variety of polluting sources.... The quantity of water, though potentially adequate for today's demand, cannot now be managed effectively over wet/dry cycles to assure a minimum adequate water supply in extended drought periods [DeGrove 1978].

A new institutional capacity at the state level was viewed as essential. Although the FCD had authority to manage the project works in conjunction with the corps, it had no mandate to manage consumptive water use, surface water installations, water quality, and wetlands comprehensively for the region. The governor's conference urged that a managing agency for the region (based on hydrologic boundaries) be established with the power necessary to develop and implement a comprehensive land and water use plan under state direction. The group responded to the governor's challenge to stop "viewing our environment through prisms of profit, politics, and geography or local or personal pride" (Blake 1980) and displayed a remarkably broad-based understanding of water quantity and quality problems. In the minds of those assembled, the region needed to stop draining wetlands, control growth, and give prime consideration to the survival of ecosystems in establishing water priorities.

The work of the governor's conference was extended by a fifteen-member task force charged with the responsibility of crafting pieces of legislation for the 1972 session of the Florida legislature. The result was the passage of four landmark pieces of legislation (including the two following acts) that had a profound effect on regional management.

The Water Resources Act. This act established regional water districts. Although the initial commission pertained to control of surface water management and water allocation, Chapter 373 of the Florida Statutes gave it jurisdiction over water quality and environmental protection as well. Moreover, the districts were armed with the Reasonable-Beneficial Use Doctrine (Maloney et al. 1972) as the basis for allocating water.

The Florida Comprehensive Planning Act. This act established the requirement for every state agency to develop aspects of a statewide set of goals and policies (e.g., land use, transportation, and water use) to be coordinated by the governor. It also initiated some local planning requirements.

The Water Use Supply and Development Plan

The state enactments were intended to transform the Flood Control District into a comprehensive regional water management entity. The first major test of how the district embraced this expanded role was embodied in the water use plan it was required to submit to the state in 1978. In 1973 the district engineer, Bill Storch, published a framework for the water use plan that recognized that allocation of more water was not necessarily better and that technology should not be left to determine how demand should be satisfied. Storch's prescription for water planning assumed that social controls could be used to alter growth patterns and per capita consumption trends. This water planning strategy was a radical departure from the traditional thinking of an agency preoccupied with constructing and maintaining a massive "plumbing system" for the region.

Unfortunately, Storch died before the plan was completed. Without his leadership the blueprint of the plan seemed to change. The plan relied more on traditional approaches to meeting future demands. It assumed a "value-neutral" stance, suggesting that there was as much water as residents were willing to pay for in cash or environmental damage. The district held firmly to the position that water was not the limiting factor in growth. "Control and planning of population growth and development are the direct responsibility of local government through its land use plan" (SFWMD 1978). The principal options incorporated in the plan were backpumping, wellfield development, desalinization, increased Lake Okeechobee storage, water conservation, and deep aquifer storage.

For months subsequent to the publishing of the draft plan the district conducted an extensive public participation campaign. Ten workshops attracted well over 1000 attendees. The results of the workshops were surprising. Analysis of the comments showed emphatically that the public did not want water supply improvements at the expense of water quality. Second, by overwhelming margins (94% to 96%) the most popular alternatives from the public's perspective were water conservation, regulation, and wastewater reuse (SFWMD 1978).

The *Miami Herald* took exception to the draft report: "As we see it, the report leans too heavily on increased management of water resources and not enough on technologies which ... imitate natural systems to the fullest extent possible" (Blake 1980). The *Herald's* editorial was consistent with the conclusions of the H. T. Odum (1977), who depicted the net effect of development in the region as having reduced overall storage capacity of the natural system yielding greater fragmentation; increased the loss of fresh water to tide; and intensified reliance on energy-dependent technology.

The Water Quality Threat to Lake Okeechobee

Art Marshall appeared before the Florida governor and cabinet on two occasions (April 1971 and December 1972) "with all the fervor of an Old Testament prophet," warning of the need for "environmental repentance" (Blake 1980). In both instances his remarks emphasized that water supply and water quality were degraded in the Kissimmee/Lake Okeechobee watershed because of human actions. The channelization of the Kissimmee River, an intolerable ecological insult in its own right, was "sluicing sewage from Orlando directly to Lake Okeechobee" (McCaffery 1981). In his second appearance he presented a report by an interdisciplinary team of university professors and consultants that raised a series of technical and scientific questions and propositions regarding the interrelationship of hydrology, marsh ecology, and lake eutrophication processes (Marshall 1972).

The Marshall report prompted Governor Askew and the legislature to address the problems in the basin. In response, the legislature appropriated money for a state study entitled "Special Project to Prevent the Eutrophication of Lake Okeechobee." The state Division of Planning was given primary responsibility, with support from DER and the FCD, and took an advocacy position, which made the project controversial. The final report recommended the control of nutrient in-flow from the Taylor Creek Nubbin Slough and the lower Kissimmee Valley. Partial reflooding of the Kissimmee was recommended. Furthermore, the report recommended the elimination or maximum feasible reduction of backpumping into Lake Okeechobee from the Everglades Agricultural Area. To offset the need for backpumping, Everglades Agricultural Area (EAA) water should be recycled through the Holey Land Tract. This action was later blocked by hunting groups (Florida and National Wildlife Federation as well as Florida Freshwater Game and Fish Commission).

The special report triggered two immediate agency actions. First, the state legislature passed a law supporting the restoration of the Kissimmee River and requesting the Corps of Engineers to restudy the C-38 project and consider alternatives for dechannelization (McCaffery 1981). Second, the secretary of the Department of Environmental Regulation (DER), Victoria Tschinkel, ordered the SFWMD to apply for a temporary permit for EAA backpumping. By November of 1978 the DER and SFWMD had worked out an agreement calling for backpumping to be phased out within 30 months. The agreement and changes in the disposition of agricultural runoff set the stage for crises in the 1980s (Blake 1980).

Other Voices

In Florida during the 1970s no voice was more visionary or prophetic than that of Art Marshall. Marshall used professional experience as an ecologist, U.S. Fish and Wildlife Service researcher and administrator, professor at the University of Miami, and board member of the FCD to create and execute his quest for Everglades restoration. At every opportunity he attempted to inject an ecological systems perspective into the agenda of the region's political machinery (Cornwell et al. 1970). Marshall and his cohorts viewed restoration of the ecological system as essential. They also recognized a fundamental link between the human and natural system. Marshall particularly understood the failings of "mission-oriented agencies."

Marshall recommended taking an inventory of all sources of pollution and removing them, reflooding lake and river marshes throughout the system, restoring sheet flow to the system, acquiring the Big Cypress Swamp, reversing the pattern of overdrainage of wetlands, and buying the east Everglades and turning that land over to the Park Service (McCluney 1971). Throughout the decade Marshall continued to elaborate on these ideas and actively seek supporters. By 1981 his simple yet profound plan of action was formally published as "Repairing the Everglades" by Friends of the Everglades and endorsed by every environmental group in Florida (a.k.a. the "Marshall Plan").

In the 1970s debates raged between corps and park staffers over whether Lake Okeechobee was hydrologically connected to the Everglades, while the FCD administrator, Ed Dail, continued to define his mandate as simply "removing floodwater and, in times of deficiency, carrying water where it's needed" (Ward 1967). Meanwhile, the attitudes of average Floridians were changing dramatically. As Patrick Caldwell, a political pollster, reported, "A new issue appears to be emerging with quality of life replacing quantity of life with environmental concerns clashing with economic concerns" (Blake 1980). Caldwell was stunned by the overwhelming saliency of the environmental issue. Moreover, as one person noted, "Florida is everyone's second state, and what happens to the Everglades is of national significance" (Ward 1967). Issues like the fate of the international jetport received national treatment from the likes of the *New York Times, Life, Time, Look, National Geographic,* and NBC's "Today Show" (Carter 1974).

Marjory Stoneman Douglas summed up the problems in south Florida during the 1970s this way: The "future of south Florida depends on intelligent management of water.... [W]ithout that the region will be a desert.... Overdrainage, more canals, land loss, water pollution, salt intrusion, fires, bad land management must be stopped. Perhaps there is time" (Ward 1967).

Interpretive Analysis

Water management in the Everglades during the 1970s was turbulent. The C&SF project had bolstered unparalleled population and economic growth. Operation of the project revealed man's shortsightedness and inadequacy to manage for wet and dry cycles in a manner that ameliorated rather than exacerbated fluctuating hydrologic conditions (i.e., man-made droughts and floods). The south Florida experience seemed to confirm Barry Commoner's Third Law of Ecology: "Whatever man touches he makes worse."

The transformation of the agency from the Flood Control District to Water Management District created a number of problems. The 1978 water supply plan focused on traditional structural solutions and underestimated shifts in social preference toward water conservation and nonstructural solutions. The district did not view growth control as part of its mission and embraced more of an instrumental role of providing water when needed and removing water if too much occurred. One scientist of note complained about this strategy and warned that the region was running out of options that did not impose extreme economic and/or ecological hardships (Tabb et al. 1972). The plan showed little deference to natural system fluctuations.

During the 1970s the district appeared hesitant to tackle the emerging water quality problems of Lake Okeechobee and the lower Kissimmee Valley. It took external pressure in the form of Marshall's report on the Kissimmee/Okeechobee Basin, the state's special report, and subsequent state regulatory measures to get the FCD/WMD's attention and action on these matters. By 1981 the district had folded water quality into its mission and adopted strategy for Lake Okeechobee (SFWMD 1981). The reluctance may in part be attributed to ambiguity over which state agency was responsible for water quality—the Department of Environmental Regulation or the district. Another reason for the hesitancy could be attributed to a carryover from the old FCD mission, wherein water quality management was simply not a primary purpose of the C&SF project. To the district's credit, Bill Storch, the district engineer, recognized in the 1960s that quantity and quality could not be separated, and toward the latter part of that decade monitoring programs for pesticides and nutrients in Lake Okeechobee were initiated within the SFWMD. Furthermore, in response to the nationwide water quality planning program launched by the U.S. Environmental Protection Agency, six local planning efforts were initiated in 1978 throughout the district's sixteen counties. During this period the district also modified its surface water management permitting requirements to provide for water quality treatment before stormwater entered district canals (Huser 1989).

In this era of "No Easy Answers," Florida and its political leadership turned to institutional reform to cope with the immediate and long-term crises in water management. The state legislatively grafted new mandates in water quality—environmental protection in the form of water use and surface water management—that indirectly imposed land use controls on development. But even as Florida was beginning to view itself as a model for the nation in how to respond to the "environmental crisis," the voices of those that understood the Everglades best were profoundly skeptical. Frank Craighead, for instance, doubted if the life systems in the park could survive if development continued increasingly to isolate and fragment the natural system (McCluney 1971).

Strategy and Relevant Myths

The decision to transform the existing FCD into a comprehensive regional water management agency was based upon the presumption that institutions could be perfected and that resource problems could be solved. Regionally, the agency pursued traditional failsafe, technological-fix strategies underpinned by the myth of comprehensive water control (Light 1983). During this time frame an alternative strategy emerged that placed greater reliance on self-restraint and regulation and that took a more ecocentric view of the world that championed a "steady-state" economy—an approach to development that was intended to be more synchronous with natural rhythms. Regionally, the steady-state concept received only passing consideration; however, state growth management initiatives did take hold and begin to influence land use decision making in the region. Although the SFWMD appeared to be neutral on growth management at the regional level, its fledgling regulatory programs in surface water and water use management were de facto influencing development at the local level. This period revealed another myth that had permeated water management—decision making should be left to the technocratic elite. The results of the public participation campaign for water supply planning showed emphatically that government must not take public sentiment for granted.

In summary, the consequences of excessive development became apparent during this management era. The leadership in Florida turned to governmental, top-down solutions. The environmental degradation on a grand scale could now be tamed, it was thought, by vesting government institutions (principally the SFWMD) with the power and authority to resolve the imposing resource problems in the region.

Repairing the Everglades (1980–Present)

Crises and Surprises

The crises of the 1980s involved a severe drought, large rainfall during the dry season, and the reappearance of chronic eutrophication problems associated with agricultural runoff. Little rain fell in the 16 months from February 1980 to June 1981. Lake Okeechobee dropped an unprecedented 2.1 m, and on July 29, 1981, it reached an all-time post-development low (2.95 m MSL). The drought was centered in the interior of the peninsula, and the return frequency was calculated to range from 1 in 125 years to 1 in 200 years. Tropical storm Dennis in August 1981 helped to relieve the drought by dropping 63 cm in 2 days. An extreme El Niño Southern Oscillation event produced unusually heavy rains during the dry season district-wide (1982–1983). This excessive rainfall triggered emergency measures to protect the threatened deer population, which eventually became a national media event. In August 1986 a massive algae bloom covering one-fifth of the lakes' 1924-km^2 focused attention on nutrient conditions. Nutrient-laden discharges of agricultural runoff were documented to be dramatic shifts in the vegetation in Water Conservation Areas. These crises created four major institutional responses: modified water deliveries to the park, the governor's "Save Our Everglades" program, and the state's Surface Water Improvement (SWIM) Act and subsequent federal lawsuit.

El Niño Event and Crisis in the Park

After a decade of experience, the park was not getting the desired results from the congressionally mandated minimum-delivery schedule. The pattern of delivery bore no resemblance to a natural hydrologic regime and was thought to be seriously disrupting wildlife nesting and feeding tendencies. At the same time, the SFWMD was following a state agreed-upon plan that called for movement of nutrient-laden water out of the EAA into the WCAs and eventually the park instead of into Lake Okeechobee. Unusually heavy rains associated with El Niño persisted through January and February of 1983, requiring heavy, undesirable, and unseasonable regulatory releases to the park. On March 9, 1983, Gary Hendrix, research director for the park, appeared before the SFWMD board and declared that an emergency existed in the Everglades National Park. He presented a plan that called for (1) structural and operational changes that would disperse flows over a broader and more natural frontier, (2) field tests of a new delivery schedule, and (3) a more rigorous water quality monitoring program.

The SFWMD and the corps acted swiftly in response to all the measures requested by the park. The U.S. Congress, in the Supplemental Appropriations Act of 1984, authorized the corps to conduct an experimental water delivery program to the park. MacVicar (1987) developed a statistical formula that was based on historic (preproject) rainfall/flow relationships. The flows predicted by this formula were to be spread across a wider cross section of the southern Everglades, into the east Everglades. The agencies and private individuals agreed to test this formula in an iterative process and thereby adopted the first adaptive strategies in the system (Light et al. 1989). In 1989 the Everglades National Park Protection and Expansion Act authorized the U.S. Department of Interior to add 42,800 ha of the Shark Slough and East Everglades to the park.

The corps plan and corollary impact statement that would make significant structural and operational modifications to water deliveries to the park reached an impasse. Although the corps, SFWMD, and the park agreed on a rain-driven operating strategy, the FWS objected on the grounds that this approach "would jeopardize the continued existence of the snail kite and adversely modify its critical habitat" (U.S. Corps of Engineers 1990). The park and FWS disagreement reflects conflicting agency missions. The FWS would give priority to single-species management alternatives, whereas park officials view ecosystem restoration as being preeminent (Graham 1990). As of this writing the tension continues.

The Governor's Response to the Drought (1980–1981) and Flood (1982–1983) Surprises

Prior to the 1981 drought, $76 million was spent to hold more water in Lake Okeechobee in order to "drought-proof" the region through 1985 and handle most needs through the year 2000. The 1981 drought, however, exceeded all expectations; half the counties in south Florida were under mandatory restrictions, and water was rationed in the Everglades Agricultural Area.

The drought brought more clearly into focus the inherent contradictions in water management. Arguments centered on the use of Lake Okeechobee as the source of water for the region. Early in the drought, the SFWMD wanted to increase flexibility and thought that needs could be met by structural modifications. By the end of the drought, the executive director of the SFWMD, Jack Maloy, had come to the conclusion that

> People aren't going to stand for any of these fancy schemes to augment water supply. . . . People are beginning to say, "This is a problem that technology can't solve." . . . When you look at the

> costs . . . and consider the amount of water being wasted, you can't avoid the conclusion that adding more and more equipment is perpetuating the problem rather than solving it. If everyone would learn to tighten their belts, we could put all the crazy ideas to rest. . . . I'm talking about the need for a major social change and it's not going to be easy, but there isn't any choice [Hutchinson 1981].

In response, over the past decade the SFWMD has embraced the need for both demand and supply-side conservation as a fundamental component of the management strategy. A regional water shortage plan impacting all sectors of the water economy was adopted. Public education programs coupled with the most innovative demand management measures were established and have received national recognition. For example, a program was adopted that curbed water consumption by planting natural or drought-tolerant vegetation in urban landscape ("xeriscaping"). The district now requires water conservation planning of municipalities requesting new or renewed water use permits. Agricultural users are also required to convert to water-conserving methods of irrigation.

The sense of crisis in the Everglades following the drought of 1981 and the flooding of 1983 moved Governor Graham to action. In August 1983 he announced the "Save Our Everglades" (SOE) program. Graham's vision of the Everglades was that by the year 2000 the system should look and function more like it did at the turn of the twentieth century. Much of his program paralleled the Marshall Plan and capitalized on initiatives that were already under way: (1) restoration of the Kissimmee River; (2) restoration of the Holey Land/Rotenberger tracts; (3) restoration of the natural flows to the park; (4) environmental improvements to the conversion of Alligator Alley to I-75; and (5) protection for the Florida panther and deer herds. A significant departure from the Marshall Plan was in area of water quality. The Marshall Plan's first recommendation was to resolve pollution problems in those areas through the process of "purifying and restoring" sheet flow to the KOE system. However, the initial "Save Our Everglades" program made no reference to the nutrient problems in Lake Okeechobee, WCAs, and EAA.

To demonstrate his commitment to the Everglades further, the governor established a state resource planning and management committee in February 1984, to analyze land and water issues in the park and east

Everglades. Soon thereafter he met with leaders of the environmental community to organize the Everglades Coalition, which held its first meeting in January 1986. As an additional show of support for the program, Governor Graham instituted a Save Our Everglades report card that served as a way to document progress as well as to fortify his environmental image.

Following 7 years of study at the federal level, pressure mounted on the SFWMD to act in Kissimmee River Valley. The corps released a feasibility study recommending no action under current federal guidelines. As a result, in 1985 the SFWMD was compelled to strike out on its own. It began a large-scale experiment to restore the Kissimmee River. Engineers began placing sheet metal weirs on a short stretch of the channel to divert canal flow into marshes and old oxbows. The success of this effort led to a large-scale physical and numerical modeling effort. In 1988 the University of California at Berkeley began a 3-year study that led to the development of an overall restoration plan that was approved by the SFWMD governing board. The plan was subsequently referred to the corps by Congress for review and approval before submission for full authorization. The Kissimmee restoration was buoyed by the 1986 and 1992 Water Resources Development Acts (WRDA), which authorized federal funds to the corps to support environmental restoration projects. Section 1135 of the 1986 act authorized modifications to existing corps projects for environmental purposes. The 1990 act directed the corps to complete a feasibility study on Kissimmee River restoration. In addition, the 1990 WRDA unequivocally conferred a restoration mission on the corps. These congressional actions not only cleared away the federal obstacles previously encountered by the state but paved the way for a new generation of environmental rehabilitation projects.

In addition to breaking new ground from a policy-making perspective, the Kissimmee River restoration cut a new path scientifically by establishing ecological integrity as the restoration goal. "The ecological integrity goal shifted the focus of restoration planning from independent objectives involving discrete taxonomic components or ecological functions to the organizational determinants and self-sustaining properties or river/floodplain ecosystems" (Toth and Aumen 1993). This goal was operationalized through a set of criteria called "determinants of ecological integrity." These determinants consisted of five mutually reinforcing and interlocking hydrologic criteria thought to reestablish the processes needed to create and maintain the complex hydrologic attributes and resource values of the prechannelization ecosystem (Toth 1993).

Eutrophication of Lake Okeechobee and the WCAs

In response to increasing concern over water quality, the SFWMD (December 1981) formally embraced water quality as a goal of Lake Okeechobee management. The technical plan accompanying the statement of policy established phosphorus and nitrogen input goals to Lake Okeechobee in order to prevent eutrophication (Frederico 1981). The plan also set programs that involved (1) best management practices for the dairy farms north of the lake, (2) establishment of a holding area in the Holey Land for the discharge of water from the EAA, and (3) continuance of restrictions on backpumping into Lake Okeechobee and diversion to the WCAs until the Holey Land project was completed.

By 1985 concern by environmental groups and members of the press regarding the progress of the district's water quality program led to a state-initiated review process referred to as LOTAC I and II (Lake Okeechobee Technical Advisory Council/Committee). The Department of Environmental Regulation (DER) review and report (1988) confirmed that phosphorus concentrations in the lake had doubled over the past decade and that if this trend continued, the lake might become hypereutrophic. Backpumping from the EAA for water supply purposes had not been totally discontinued. DER, mirroring the district's technical analysis, recommended that phosphorus loading be reduced 40% by changing land use practices and diverting runoff. The report acknowledged but did not recommend action on the finding that diversions of water to the WCAs from the EAA were generating negative impacts in the WCAs, and potentially in the park. Since the Holey Land had been removed from consideration as an areawide storm water retention site by efforts of the Florida environmental groups, the report recommended that the use of flow ways (i.e., man-made marsh with nutrient uptake) be given further analysis.

In 1986, just as the DER report was being published, a 312-km^2 algae bloom appeared on Lake Okeechobee. The bloom created a political fire storm that precipitated legislative action (SWIM ACT of 1987) and created new WMD planning requirements; established a mandatory 40% phosphorus reduction target for Lake Okeechobee by 1992; and initiated another panel of independent technical experts (LOTAC II) to investigate the adverse effects of past diversions on indigenous wildlife and vegetation in the WCAs. The ensuing report (DER 1988) concluded, based on SFWMD research, that the diversion of nutrient-laden waters from the EAA had seriously damaged the biological integrity of the Everglades and, if left unchecked, could eventually damage the park. Native sawgrass and periphyton communities were being replaced by pollution-tolerant taxa. One bright spot from the final committee report (1990) was the conclusion that Lake Okeechobee was not in imminent biological danger.

The threat of eutrophication to the park and the Art Marshall Loxahatchee National Wildlife Refuge (WCA-1) contained in the LOTAC reports was more than the environmental community was willing to take. During the summer of 1988 rumblings of an impending lawsuit against the state or "Big Sugar," the major industry in the EAA, began to surface. Finally, on October 10, 1988, much to the district's surprise, U.S. Attorney Dexter Lehtinen (with the support of park superintendent Mike Finley) and Art Marshall NWR Superintendent Burkett Neely filed a lawsuit against the director of SFWMD, John Wodraska, and the secretary of DER, Dale Twachtmann. The principal charge was that the state had violated water quality standards and breached water quality agreements between the state and federal governments. The suit lasted more than 2 years and cost the state and federal governments tens of millions of taxpayer dollars. The agricultural community spent undocnted millions as well.

By the summer of 1991 a settlement was reached between the state and federal governments. The agreement put the state under a strict timetable for nutrient reduction and appeared to place considerable responsibility for the future of the Everglades water quality and quantity in the hands of a five-member Technical Oversight Committee (three to two federal over state representation) with ultimate jurisdiction resting with a federal court judge if the settlement was violated. The state legislature followed suit by passing the Everglades Protection Act (1991). The act required the district to apply to the DER for permits for the EAA pumps. It also required farmers to obtain permits from the district for drainage of their lands in the EAA. In response to these state and federal actions, the agricultural community filed a flurry of legal challenges to the state and federal settlement. Farmers claimed that the U.S. attorney was never authorized to file the suit. In addition, they contended that the settlement was negotiated privately and in violation of the Florida Administrative Procedures Act and the National Environmental Policy Act. In an effort to avoid years of protracted litigation the parties agreed to hire a mediator. On July 13, 1993, Secretary of Interior Bruce Babbitt and representatives of the district, state, and federal governments and agricultural interests announced agreement to a statement of principles outlining a framework for a mediated Everglades cleanup and restoration plan. Basically, the mediated plan called for 40,000 acres of stormwater treatment areas (STAs) to be created to filter agricultural runoff. That plan provided the basis for state legislation (The Everglades Forever Act of 1994) backed by Governor Chiles that ended the litigation between industry and the government and requires sugar and vegetable growers to pay between $222 million and $322 million over 20 years. The total price tag for the cleanup project is estimated to be $700 million.

However, the dispute over how much cleanup is "enough" and who gets to decide has not abated. Under the Everglades Forever Act the interim water quality goal for cleanup actions is 50 ppb phosphorus, with the state responsible for setting the number. Ironically, the same attorney (Dexter Lehtinen) who brought suit to have the state enforce water quality standards and begin the cleanup has recently brought suit on behalf of the Miccosukee tribe of Florida in conjunction with the Friends of the Everglades to halt operation of the Everglades Nutrient Removal (ENR) Project that was designed to test (a fundamentally adaptive strategy) the validity of the STA concept. The following quote from an editorial in the *Palm Beach Post* (June 16, 1994) captures the underlying conflict between the warring factions:

> there's this sentence in a Friends of the Everglades news release: "The ENR should not be allowed to operate until it is proven that no further harm to the flora and fauna will be caused. . . ." That's like the assertion farmers used to make—and some still make—that we do nothing until it is proven that they are polluting the Everglades. *Insist on enough "proof" and you will guarantee that nothing is done* [emphasis added].

Other Voices

As the 1980s began, Art Marshall carried the mantle of high priest for the environmental community in Florida. Following 10 years of incubation and coalition building, the Marshall plan was incorporated into public policy, and Governor Graham crafted his "Save Our Everglades" program from Marshall's clay and cast. Even the SFWMD director, Jack

Maloy, admitted that the Marshall concept of dechannelizing the Kissimmee was an "idea whose time had come" (Blake 1980). Marshall continued to build upon his vision that the "balkanized" system could be repaired.

Just before the "Repair the Everglades" Plan received the political acceptance that it deserved, Marshall went off on a new controversial tangent—the broken "rain machine." He seemed more comfortable when people were taking exception to his ideas. In 1982, after Florida's record drought, Marshall hypothesized that development and drainage over the past 80 years had disrupted the normal rain cycle. Gannon (1978) contended that significant orographic changes on the Florida peninsula scale were occurring that would lead to less rainfall in the long term. The rain-machine theory was roundly criticized at the time and has attracted little scientific attention since then.

Art Marshall died in 1984. No one of Marshall's stature and foresight has emerged to take his place. But a coalition of state and national conservation groups have picked up the torch and continue to champion and elaborate on many of his ideas. One of Marshall's greatest concerns was the lack of a systems perspective within the region and a lack of appreciation for the Kissimmee/Okeechobee/Everglades ecosystem. *Systems* was a key word for Marshall, as cited by Boyle and Mechum (1981):

> If you don't synthesize knowledge, scientific journals become spare-parts catalogues for machines that are never built. Until isolated and separated pieces of information are assimilated by the human mind, we will continue to rattle around aimlessly.
>
> We don't have any agencies in state government that consider whole ecosystems, and that's a major problem. If we don't change our way of thinking, we're going to see the collapse of our systems and the resources they've provided.

The spirit of Marshall's systems approach from a technical perspective endures in a consortium of scientists who work in the Everglades. In the midst of unparalleled interagency and intergovernmental warfare that has evoked comparisons with "WWIII," "Vietnam," and the "Roman Coliseum," an informal collegium of scientists and engineers has emerged and continues to persist. It grew out of an idea about writing a "white paper" on scientific issues in the Everglades. In 1987 several scientists from differing organizations (Steve Davis, Bill Robertson,

Lance Gunderson, and John Ogden) were concerned that research findings about the system were not being published and that no synthesis that could provide guidelines for restoration had surfaced or seemed under serious consideration by any of the major resource agencies in the region. The white paper idea turned into a symposium, and the symposium evolved into 2.5 years of workshops, simulation modeling, and publishable manuscripts involving some fifty technical professionals. The major conclusion of this informal collaborative effort is that enough is known about the Everglades ecosystem to rehabilitate it. Tinkering has not worked, nor have singular, quick-fix structural solutions, but composite policies will work. That is, integrated sets of structural and operational changes can be devised to satisfy restoration goals and provide alternative uses for water. Moreover, there must be more than one set of composite policies, so the region will not be dependent on one set of *unforgiving policies.*

Interpretive Analysis

In response to the management dilemmas faced in the Everglades during the 1980s, Walt Dineen, chief biologist for the SFWMD, lamented, "Our biggest problem is that we are faced with a diversity of interests, many in direct conflict with each other. We are constantly walking a tight rope" (Yates 1983). During the 1980s the "tightrope" snapped, sending the district and much of the region politically into free fall. The premise of the 1970s that Florida water institutions could be "perfected" and could solve complicated biophysical and political problems was dealt a punishing blow.

The district seems to have been a victim of what has been termed the "Uncle Jack" syndrome—trying to assimilate, gloss over, or patch up ever-widening water policy schisms in the region. The district's breadth of authority and the scope of responsibilities were not enough to overcome the fundamental lack of shared values among the powerful water interests in the Everglades. Attempts to "balance" competing objectives without forsaking long-standing sociopolitical moorings proved impossible. The SFWMD and DER attempts to redress the water quality problems caused by agriculture around Lake Okeechobee had to be frequently bolstered by additional political will in the form of specific directives from the governor, state legislature, and U.S. Department of Justice.

In a series of *New York Times* articles (Judd 1991) the SFWMD and its sister agencies (water management districts) were painted as fiefdoms for the power elite within their respective regions. These elites were accused of hiding real intentions under the guise of "regional flexibility to deal with localized problems." The *Times* was critical of the political elites for attempting to shake free of state oversight and emasculating selected state mandates that threatened their vested interests.

However, references in the articles to the "good ol' boy" network and "sacred cows" barely pierce the surface of more fundamental problems in the region. As Marshall often chided, Florida's institution for water management failed to live up to the expectation of being a systems-level manager. In fairness to the SFWMD, the institution, though remarkably innovative for its time, lacks the capacity to steer the entire region. The district had system responsibility but not the authority or the political consensus necessary for region-wide action. There is a fundamental mistrust among levels of government. The region's power elite was extremely polarized over the proper role of the agency on environmental issues (Penn and Schoen 1988). Furthermore, the district's resource management counterparts at the federal level (ENP, corps, and FWS) function under very different mandates and sources of authority. Traditionally, consistency and concurrence among and between these state and federal agencies have been the exception rather than the rule. Furthermore, the urban "rim communities," which could well hold some of the trump cards in the struggle over Everglades restoration, have been barely visible in the political debates of the past decade.

Within state government the SFWMD competencies became trapped between the avalanche of state environmental policy demands that were attempting to transform a federal flood control system and well-entrenched political and economic interests that had built their fortunes and political constituencies around an increasingly balkanized and degraded natural system.

A consortium of Everglades scientists and engineers has concluded (Holling et al. 1994) that the design of composite (i.e., holistic, system-wide) solutions is needed but that this will require *long-term commitments and collaboration* among rival factions, including many affected publics (e.g., Native Americans, sports fishermen, urban residents); the vested economic interests; and the leaders of the various water-related local, state, and federal jurisdictions. Few, if any, precedents exist for such an undertaking. As Martha Musgrove, editorial board member *Miami Herald,* lamented over the the prospects for real solution in the Everglades, "Alas, scientific consensus—when and if it can be achieved—

does not necessarily lead to political consensus or action. If the Everglades . . . are to be preserved, state and federal officials must reach accord—on a *remedy*, not on blame" (*Miami Herald*, January 4, 1991).

This period of water management has been filled with paradox. At times the water wars erupted into political chaos, but the state's Save Our Everglades program has somehow endured. The significant gap between public expectation/policy demands regarding environmental restoration and institutional performance remains. As Governor Chiles stated in December 1992 upon receiving presidential recognition for his administration's Save Our Everglades program, "We have not saved the Everglades. There is a lot more that needs saving. . . . [We must] step forward and continue down that path." Progress in rehabilitating the Kissimmee/Okeechobee/Everglades system has been slow but steady. There are real signs of hope and despair. Major initiatives for a significant reduction in phosphorus input into Lake Okeechobee and the Water Conservation Areas are under way. However, the fight to restore Everglades hydroperiod, a decidedly more intractable issue, must still be waged. Although FWS objections failed to block a more ambitious rain-driven operating strategy, adaptive strategies to improve water deliveries to the park require further attention in order to become operational. Despite attempts at almost every level to derail the project, efforts to restore the ecological integrity of the Kissimmee River continue to muster the necessary political and financial support. At a more fundamental level, although decision makers are bombarded with "facts" and "pseudo-facts" related to ecosystem restoration at the political level, the scientific community seems to be quietly moving toward a common understanding of the requirements for ecosystem restoration.

Strategy and Relevant Myths

Marshall was a superb strategist and myth maker. His ideas of repairing the Everglades inspired governors and other leaders within the state to think and do what many considered impossible. Since his departure the political struggle over the fate of the region's major ecosystem has intensified. Although small pieces of the puzzle are moving into place, an overall strategy or set of hypotheses capable of restoring the ecosystem and sustaining the other sectors of the regional water economy has yet to crystallize. Strategy appears to be in a state of flux and ripe for fundamental reframing. But the various factions seem unwilling to let go of the past to embrace a collective future. System-level options and innovations must be generated. More voices need to be heard (e.g., Native Americans) and issues still remain unresolved. Political solutions have failed as a catalyst for a new and broader dialogue among the interests in the region.

In summary, south Florida remains on the "edge of history"—a petri dish where the fundamental conflicts among population, development, and ecosystem services remain unreconciled (Ehrlich and Ehrlich 1990). The region is bereft both of long-range options and innovations and of viable myths upon which to build a sustainable future. Mobilizing the capacity within the region to address these system-level concerns is paramount. The question remains of who will put together the new understandings and new ideas that will attract sufficient political support to trigger action. The fundamental challenge of sustainability is yet to be solved. The attempt for certitude is illusory. People, not water resources, are both the problem and the solution. As Marshall foresaw decades ago, it is inconceivable that people think you can take a water-dominated ecosystem and turn it into a dry vacation wonderland and/or an agricultural bread basket and still maintain that system's integrity.

Barriers and Bridges

Recognition of the barriers and bridges that have arisen in the past is critical to meet the goal of renewing ecosystems. Ecosystem revitalization is a subset of a broader goal of sustainability of the coupled human and natural systems. A barrier is an impediment or obstacle to renewal or sustainability; a bridge is an enhancement of steps to the same goal. The bridges and barriers are intimately linked to and dependent upon the multiplicity of goals that are derived by humans and their assumptions and expectations in attempting to control or adapt to a turbulent environment. Bridges and barriers (table 3.2) appear to fall into several categories that relate to how: (1) humans view nature, (2) human institutions develop, and (3) institutions are coupled with ecosystems.

Views of Nature

Perhaps one of the most fundamental perspectives of nature is how humans view their role in the management of systems. There are at least two distinctly different views—one where humans are part of the natural system and one with humans removed and different from the system.

TABLE 3.2
Barriers and Bridges that Have Arisen in the Management of Water Resources in the Everglades

Argument	Barrier	Bridge
Matching of scales	Fixed boundaries in time and space	Boundaries appropriate to issue
Time scales of ecology and institutions	Institutions unable to adaptto ecological change	Ecological change incorporated into ecological changes
Space scales	No institution with systemwide perspective or authority	Informal collegia, interagency collaboration
Values	Conflicts at local scale	Hierarchy of values, multiple values coexisting at broad scale
Decision rules		
Constitution	Explicit organization, lack of interdependency	Hierarchy of rules; simple, robust rules within and between levels
Policy		
Implementation	Operational, lack of interaction	
Paradigms of stability	Equilibrium goal = control	Dynamic, evolutionary goal = resiliency
Paradigms of connectivity (institutions)	Single-purpose	Hierarchy + shadow domain, meshing organization
Paradigms of uncertainty	Problem too large,too complex; Internal myopia	Simplification External perception
Paradigms of learning (institutions)	No learning possible, single-loop type, feedback delay	Douuble-loop type, paradigms reviewed
Paradigms of change	None	Social science, four-stage, Holling four-box

The school of thought that views humans as part of the system and that led to the hands-off or laissez-faire management style had its genesis in the United States in the writings of George Perkins Marsh (1864). The hands-off approach fostered the concepts of natural control, whereby the biophysical system was left to self-organize generally after humans had altered some fundamental components. In Yellowstone

National Park, for example, complex predator/prey relationships were significantly altered by removing wolves and cougars, which resulted in unpredictable population changes in a variety of species. The first settlers and their actions in the Everglades were initially not linked by researchers to observed variation in the natural system.

The intensive management strategies developed from the "we are different" view have encouraged extensive control and exploitation of resources strictly for human consumption. This dichotomy between man and nature tended to separate thinking from doing to the exclusion of much error detection. Management strategies focused on the "realization of explicit intentions" (Mintzberg 1990) and ignored the unexpected (a barrier). This school of thought had its roots in the conservation movement that became embodied in public policy during Teddy Roosevelt's administration through men like Gifford Pinchot (U.S. Forest Service) and Frederick H. Newell (U.S. Geological Survey) (Hays 1959).

Resource matters were thought to be basically technical, so it was argued that foresters and engineers, not politicians, should make the key decisions. This style of management by a technocratic elite has dominated resource management during this century. Technological (a.k.a. "scientific") management is highly centralized. It assumes strict values, sets rigid goals and the means for achieving these goals, and is decidedly resistant to feedback. Technological management has glorified large public works, and it has nurtured excessive regulation and policy fragmentation that accelerated conflicts and thwarted learning among sectors of government and society. The large amount of capital (money and people) invested in these projects and programs, and the associated high cost of change have been a barrier to renewal. In the Everglades the view of nature that stresses control has led to an ever-increasing set of adverse ecological impacts and hence created a barrier to system renewal.

Botkin (1990) provides compelling arguments that neither approach (hands on or hands off) is working. Botkin (1990) points out that a useful bridge is that there are "both natural and unnatural changes and rates of change. . . . The key to a new, wise management is to accept change that is natural in kind and frequency, to pick out the melodies from the noise." Clark (1986) used the metaphor of a garden, to recognize the pervasive impact of humans in ecosystems and to redefine management in terms of the questions, "What kind of gardens do we want, and what kind can we get?" In essence, humans cannot be removed from the system, and the "hands-on" approach must be recast to incorporate alternative views of nature as resilient and evolving (Holling 1987). The implications of such alternative propositions are substantial and signal a profound shift in management approach to embrace uncertainty and manage adaptively.

Holling (1987) characterized three views of nature in terms of stability: equilibrium, dynamic, and evolutionary. The first cohesive plans for controlling water levels in the Everglades (Cut 'n Try and Turning Green Lines into Red) were based on an equilibrium-centered point of view of the system. That is, the goal of management was to achieve a lower equilibrium level for hydrologic regime by using extensive infrastructure (physical structures plus operational guidelines) to reach the desired regulation stages. The guaranteed minimum flows to Everglades National Park are another example of this view: Deliver a set amount of water that was determined from an assumption and model of equilibrium.

Recently, in attempting to recover and restore the viability of the system, dynamic paradigms were accepted as part of management. The development of a statistical formula (MacVicar 1987) that metered deliveries to Everglades Park as a function of upstream rainfall and allowed for variation in the regulation schedules of the Conservation Areas signaled a modification in assumptions to dynamic stability. Associated with this shift in paradigms is the change of management goals from one of control to one of resiliency; that is, water levels are allowed to fluctuate as they have for 5000 years, without constraints that have led to brittleness or loss of resiliency in the ecosystem. Accepting the view of dynamic and evolutionary systems is a bridge to achieving revitalization goals, because a more natural variation is critical to maintaining diversity and stability of these systems.

Understanding ecosystem dynamics is another bridge to the renewal of systems. The dynamics of many natural systems are captured in the four-phase heuristic of Holling (1986, 1992, and chapter 1 of this book). Each box represents a distinct phase or stage within which the system has a distinct function and structure. In the exploitation stage opportunists or pioneer species dominate in an environment of intense competition. As the system develops, conservation strategies appear as evidence in climax forests, where structure and biomass are greatest and where connections and organization (diversity) among units are tightest. The stages represented by exploitation and conservation boxes are alternative ways of stating the process of ecosystem maturation portrayed by Odum (1969). The phase of creative destruction occurs in the form of recurring physical disturbances, such as fires, hurricanes, and floods in the Everglades, and some of these effects are internally generated by the mature elements of the ecosystem. The renewal phase is the critical phase in reconfiguration, as organization in this stage will result either in a restart of the cycle, utilizing the existing components of the system, or in alternative configurations, using new components. The recycle through the system is consistent with a dynamic stability domain, whereas the exit and fundamental reconfiguration is characteristic of evolution or devolution into an alternative stability domain.

The successional dynamics of Everglades vegetation conform to this model. Fires, hurricanes, floods, and droughts are disturbances that occur frequently enough to cause cycles and periodicities in the vegetation structure to vary from about a decade (in the sawgrass marshes) to about a century (in the swamp forests) (Gunderson 1994). Water management and changing land uses have impacted these vegetation dynamics, as shown by the drowning of tree islands in WCA 2 (Worth 1987), conversion of sawgrass to cattails (Davis 1990, 1991), and losses of landscape heterogeneity (Davis et al. 1994). These all reflect changes into deteriorated, less resilient systems and hence are impediments to sustainability.

Institutions

In the history of management in the Everglades ecosystem, the number of institutions, either governmental or nongovernmental organizations, has increased dramatically. The institutional barriers and bridges are related to the actions within each institution and the transactions among institutions in response to demands from their respective task environments. For the purposes of this study, the actions of individual institutions are a function of a number of factors, including the designated authority or purpose of the institution, the rules of how decisions are made, the values implicitly and explicitly embodied by the institution, and capacity to innovate and learn.

Public decisions are made through a hierarchical rule system (Ciriacy-Wantrup 1976; Ostrom 1990). The broadest-scale rules are defined as a constitution, which defines how rules are made about collective choices. Mini-constitutions in Everglades water management include the enabling legislation for Everglades National Park (1934) and the Central and Southern Florida Project (1949). Policy embedded in organizations interprets the authority of an institution and mirrors formal or informal collective choices. Implementation requires a more refined set of rules to achieve policy goals. Operational rules are how policy is implemented and reflect the most defined (in space and time) set of rules. One example of these nested sets of rules involves water deliveries to Everglades Park. The authority for deliveries is vested in the corps and SFWMD by the U.S. Congress, which has twice (1970 and 1984) stated the policy on how water is delivered. Implementation was derived from these public laws, in the form of development of a statistical formula that predicts water flow as a function of rainfall and ambient water conditions (MacVicar 1987). Operational rules determine how the water is physically delivered (i.e., through which S-12 structure water is delivered).

The barriers that have developed in the rule-making system deal with interaction of rule sets within the hierarchy. Maintaining a strict dichotomy between policy formulation and implementation across institutions at different levels thwarts feedback and undermines learning. Within institutions the translation of action across layers (policy, which results in implementation, which leads to operations) can result in a gross distortion or perversions of policy intent. One example is the minimum delivery act (PL 91–181), which was intended to ensure a piece of the water pie to the park. (The park was guaranteed at least 415,800 ha m/yr.) The resulting operational procedure ended up delivering this amount on a regular monthly basis, without regard to the variation in rainfall and to undesirable impacts to the biota of the park.

Bridges appear as recognition of hierarchies, with simple, robust rules for decision making at each level and the appropriate nesting relationships. The rule-making system should anticipate failure and hence stress resiliency (a.k.a. mutually reinforcing processes that are robust to transformation) and requisite variety. The analogy of watch construction (Simon 1962) is a powerful example in which the system of rules is self-contained and resilient across hierarchical levels. That is, if the watchmaker is interrupted during construction, the work must be repeated not from the beginning, but from the last level completed. Hierarchies are an integral part of our system of governance, with states and regions nested within the larger federal system. Based on the Everglades experience, however, the current configuration of resource-related institutions and of asymmetrical power among them does little to cope with the increased complexity and interdependence of the social and natural systems of the region. This pan-institutional deficiency has contributed significantly to regional resource degradation and left unresolved questions of jurisdictional boundary setting (e.g., whether management strategies should be set on jurisdictional boundaries or on natural system boundaries). Without a robust form of combined state and federal governance organized at the regional level, top-down policy demands have created a labyrinth of jurisdictional issues and accelerated policy fragmentation and gridlock. Other pitfalls include the interpretation of decision rules to satisfy the self-serving interest of particular agencies or vested interests at the expense of resource protection and the concomitant loss of public interest/trust accountability.

One major impediment to the revitalization of the natural system has been the difficulty of institutions to learn (i.e., adapt to change) from experience and new information or knowledge. Certain structural problems (i.e., paucity, redundancy, and complexity) related to the experiences that institutions encounter have been recognized as significant barriers to the detection and correction of error (Levitt and March 1990). Learning takes at least two forms: single loop and double loop (Arygris and Schon 1978). Single-loop learning involves existing paradigms that are reinforced through new information and experience. Double-loop learning consists of existing models that are to some degree reframed, updated as new knowledge, understanding, and experience are gained. Double-loop learning occurs at each hierarchical level, but at different speeds. Constitutional changes of nation-states occur on the order of centuries, policy changes on the order of decades, and operational changes occur on the order of years.

In the Everglades experience, fundamental learning has not been a critical component of the institutional mode of activities. In one sense, much of the SFWMD is geared to producing information (Lynne and Burkhardt 1990), although this information is generally used in the single-loop sense to build on existing agency, not ecosystem-level competencies. The double-loop learning has been achieved through mechanisms like the AEA process (i.e., scientific consensus on Everglades Restoration goals) and provides a bridge for systems-level understandings (Environment Canada 1982).

Michael (1991a) discusses how the myths developed and maintained by institutions impede learning because of an unwillingness to embrace the uncertainties inherent in managing large, complex systems. The lack of double-loop learning or adaptive strategies (Holling 1978, Walters 1986) has been attributed to the denial or the fear of failure by policy makers or the technical staff. There may be certain societal taboos associated with the acknowledgment of uncertainties intrinsic to complex problems (Michael 1991b). Based on the Everglades experience, bureaucracies are simply not challenged to question their own purpose or actions, nor are they rewarded for doing so. They are therefore ill equipped for more than single-loop learning. (Are they doing things right?) The bridges (i.e., double-loop learning) arose during major crises and have originated outside or at the boundaries of public institutions. Examples are the Allison Rewatering Plan of 1939, the Governor's Assembly (1972), the Lake Okeechobee Technical Committees, and the current AEA group of scientists. The critical point is that the deep learning (double loop) necessary during periods of profound change rarely happen inside public institutions as they are currently configured, but must originate from their periphery. A corollary example of this would be in two of the large research units within existing institutions in the Everglades (SFWMD and ENP). Over time, the institutions shape the focus of research to one that is "mission oriented," rather than basic research that is geared toward the questioning of fundamental assumptions of prevailing management policies and actions.

New understandings about the Everglades evolve through double-loop processes. They tend to accumulate in networks over a period of years up to a decade, resulting in sharp and fundamental shifts in view that are transferred into institutional and operational rule sets or operating premises. Alternatively, they may be summarily dismissed as heresy or perceived as irrelevant. Rule sets and operating premises are repositories for new understanding that have been learned by institutions and individuals working within them. New knowledge or understandings that are ignored or rejected may continue to be retained by individuals, in the shadow network, and by libraries, or they may be lost completely to the system.

Just as decision rules are hierarchical, the set of values (e.g., operating premises) imputed by humans on the system can also be viewed as a hierarchy (Schumacher 1977). Value systems that attempt to determine worth based upon a uniform denomination can result in terribly difficult mismatch of choices. This is evident in the "No Easy Answers" era, when management posed water allocation choices between "alligators and people." The recognition of a hierarchical value system allows for multiple values to coexist at the same or different scales. At the broadest scale, society imbues value on the natural portions of the Everglades, by designating it as a national park and world heritage site. Therefore what happens to the birds in Everglades National Park has national and international significance. Other values, such as the perceived maintenance of economic viability within the region, have resulted in urban residents subsidizing agriculture for at least the past 45 years (Blake 1980; Diamond 1990).

As alluded to in earlier discussion, management systems and their rules are irrevocably tied to prevailing technologies. Institutions that exhibit slow rates of change tend to become trapped in technological competencies that persist but become increasingly dysfunctional over time (Arthur 1984). Left unattended, these "competency traps" (Levitt and March 1990) can challenge long-term viability and function as a formidable obstacle to ecosystem renewal. In the Everglades, the C&SF Project has reached its design life, yet fundamental redesigns to enhance the entire Kissimmee/Okeechobee/Everglades system are not under serious consideration. However, the process may soon begin. In 1992 Congress authorized the corps to review the adequacy of the C&SF project "to determine whether modifications to the existing project are advisable due to changed physical, biological, demographic or economic conditions."

Perhaps the most apparent institutional barrier to enhanced ecosystem viability has been in the interaction or connectivity among institutions. As the eras of water management arose, institutions were created generally with a limited set of objectives. The SFWMD, the local sponsor of the C&SF project, is responsible for operation of the system for water supply and flood protection in conjunction with the corps. Although the project serves multiple objectives, the resulting decision processes often become compartmentalized within one institution. Even though the SFWMD has very broad state-level authority, gaps in both scope and purpose are inevitable. For example, the SFWMD was sued by its federal counterparts for not maintaining state standards of water quality. Fragmented agency action could have been overcome or significantly reduced by mutually reinforcing processes (i.e., shared responsibility) among the institutions that govern various components of the hydrologic and ecological system.

Such an arrangement for the Everglades region was contemplated by a distinguished group of naturalists, biologists, and ecologists in 1970. George Cornwell, Art Marshall, Robert Downing, James Layne, and Charles Loveless advanced several radical ideas for their time. The study team recommended the establishment of an interagency coordinating committee to oversee decisions involving the entire natural resource base in the region. From their perspective, if it couldn't be accomplished by "gentlemen's agreement," then appropriate legislation should be passed. The committee was envisioned as a *formal* mechanism for interaction—both day-to-day exchange of information and coordination of long-term operations and plans (Cornwell et al. 1970). In addition, the team focusing on water quantity and quality issues recommended (1) a reevaluation of regulation schedules, (2) restoration and preservation of natural hydroperiods, (3) maintenance of integrity of ENP flora and fauna ensuring adequate supply of quality water, and (4) continuous monitoring of nutrients, toxics, and heavy metals throughout the ecosystem.

This obscure but prophetic work and its insights into the nature of the ecological and institutional problems of the region remain a beacon to those seeking ways to renew the Everglades. The Everglades Coalition requested and Congress authorized a comprehensive evaluation of the region's water system. However, no voices since Cornwell and colleagues have called for a new institutional configuration, a truly regional form of governance, that would put the power and authority of the management agencies on an equal footing and require consistency and concurrence among state and federal plans and actions.

Interinstitutional cooperation in the case study (or interorganizational relations) does occur on a routine basis and has been accomplished through coordination, formal agreements (memoranda of agreements), informal shadow networks (consortium of scientists), nongovernment organizations (Everglades Coalition, FOREVERGLADES), ad hoc groups (SAGE), and meshing organizations (Technical Oversight Committees). The type of relationship is a function of the number of rationales, including necessity, asymmetry, reciprocity, efficiency, stability, and legitimacy (Oliver 1990). Some of the less formal networks fill niches unoccupied by institutions or where gaps exist between "fragmented services and a more highly aggregated functional system" (Schon 1971). A similar function is described by the "epistemic communities" that fostered increased cooperation management innovation in the Mediterranean basin (Haas 1990). As stated previously, informal problem-centered networks appear to be a significant spawning ground for new ideas and understandings.

Coupling Institutions and Ecosystems

Coupling of ecosystems and human institutions requires two approaches. One is to match models or paradigms of change of ecosystems with those of institutions. The other is to match the spatiotemporal scales of system processes with appropriate institutional form and function.

We view humans and their development as part of the ecosystem. Some systems, such as those in Everglades National Park, minimize human development and seek to sustain natural systems. These natural systems are becoming more and more difficult to renew or sustain, because ecological processes do not conform to the artificial boundaries imposed by the human system. Agriculture is intermediate in the development gradient, relying upon management of natural components. Other sectors, such as the coastal ridge urban concentrations, are intensely developed and dominated by human systems. One barrier has arisen due to the way the system has been partitioned for these land uses.

The compartmentalization, both physically and conceptually, is due in part to a lack of systems perspective. If the system can be recoupled, alternative views (reconciling man and nature) must become a fundamental premise around which management is structured. This concept is not new to the region and dates from almost two decades to work by Marshall (1972) and Odum (1977). This fundamental assumption is part of the AEA work in southern Florida initiated by Davis, Holling, and Walters in 1988. The myopia generated by partitioning can only be corrected through these types of approaches. Although the SFWMD has jurisdiction over the entire hydrologic system, rarely have perspectives of ecological impacts spanning the entire system been incorporated into its decisions. And for understandable reasons. The SFWMD has no authority inside Everglades National Park, and in Water Conservation Area 1 policies and actions are increasingly dictated by the needs of the refuge and FWS policy. Therefore, even though the scope of the SFWMD efforts at times may eclipse other partners-in-management, its span of control does not. Again, there is a lack of mutually reinforcing decision processes at the regional level. Bridges have been constructed by the development of informal collegia of scientists and managers during the AEA work in the late 1980s. Signs of greater collaboration among scientists are emerging, as demonstrated by the restoration goals and composite policies to meet those goals agreed to during the series of AEA workshops (Walters and Gunderson 1994). More formal mechanisms for collaboration are beginning to take shape (i.e., TOC and SAGE).

Other barriers arise from the improper matching of spatial domains of problems and solutions. Recognition of the appropriate spatial domain and boundaries of problems has resulted in solutions that enhance system function. This is shown in the recent water quality lawsuit, where the public marshes of the water conservation areas were used to treat nutrient-laden water from agricultural fields, resulting in a dramatic shift in biota. The solution calls for treating the water within the area designated as the EAA, in order to prevent any further degradation of areas that are protected because of their unique natural resources.

To couple ecosystems and management institutions, it is important to match not only the spatial domain, but also the temporal domain. Part of this matching involves agreement between models of change over time. The four-phase heuristic developed for ecosystems has also been applied to human systems (Holling 1986). Elgin and Bushnell (1977) describe the same dynamics of social systems: discrete stages comprised of high growth (r phase), greatest efficiency (K phase), severe diseconomies and systems crisis (Ω phase). Both models recognize a transformation stage (α phase), whereby the system either reinitiates with similar components or evolves (or devolves) into a dramatically different configuration. Similar dynamics are reported by Tushman and Romanelli (1985).

The four-phase model can be used to help explain much of the dynamics of the coupled ecosystem of the Everglades during the past century (figure 3.5). The earliest era (Cut 'n Try) reflected development from box 1 through box 2, where increasing structure in the form of canals reflected an attempt to control the system. The crises (creative destruction) that precipitated movement into a reconfiguration occurred as the result of the hurricanes of the mid-1920s (resulting in federal involvement in the system and the Hoover Dike around Lake Okeechobee) and the extremely wet year of 1947 (brought in the corps and the present infrastructure). The system underwent a major reconfiguration as it entered the second management period (Green Lines into Red) and has cycled through this incarnation at least three times. The cycles reflect changes in the internal management or how water is partitioned among the user groups and how institutional roles have changed. The system developed during the 1950s and 1960s as the canals and levees were being built. The first cycle occurred when the droughts of the 1960s resulted in a guaranteed water delivery to Everglades Park. The drought of 1971 prompted another cycle, this time an institutional reconfiguration that resulted in the formation of the SFWMD and that was characterized by the era of "No Easy Answers." The latest era, "Save

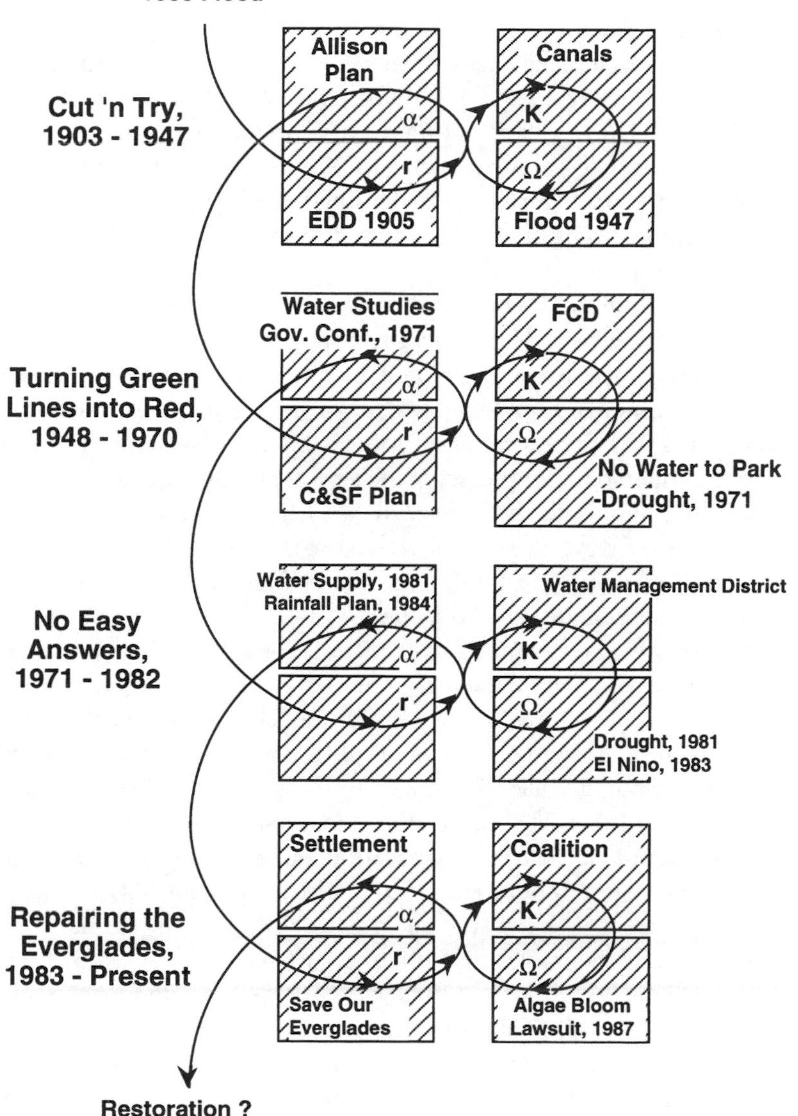

FIGURE 3.5

History of water management in the Everglades, as represented by the four-box model. The initial configuration is shown in the bottom set of boxes, the last major shift in the top set. Subsequent cycles through the boxes are listed sequentially, with corollary structures (box 2), crises event (box 3), and subsequent reconfiguration results (box 4).

our Everglades," reflects another cycle of the system and was brought about by crises in the early 1980s. The coincidence of high rainfall and fear of pollution in Lake Okeechobee resulted in a reconfiguration that included modified regulation schedules in the Water Conservation areas and the rainfall formula for making deliveries to the park. The last major cycle occurred with the settlement of the water quality lawsuit. It has triggered yet another major cycle focusing initially on water treatment areas, which would be used to minimize the impacts of water movement from agricultural to natural areas.

Adaptive strategies embrace these models of social and institutional change and have been adopted in attempting to meet restoration goals in two parts of the system. The rainfall delivery formula to Everglades Park is an example of passively adaptive management (Walters and Holling 1990). The restoration of the Kissimmee River and the Everglades Nutrient Removal project in the EAA are following adaptive techniques, in dealing with the uncertainties inherent in that complex project. Both may serve as preliminary tests that will ultimately support even "bolder strokes" in the future. The bridges summarized in table 3.2 are all facets of adaptive strategies.

Understanding the complex dynamics of ecological variation and human management is fundamental to the achievement of ecosystem restoration goals. The four-box or infinity model captures much of the dynamics of both ecologic and social-technological systems, as shown by work of Holling (1986), Tushman and Romanelli (1985), and Elgin and Bushnell (1977). The exploitation and conservation phases of these models have a rich history in application and are readily amenable to diagnosis and prescription. The creative destruction and system configuration phases have been studied in ecological systems, primarily in relation to large mesoscale disturbances, such as those caused by pests (Clark et al. 1979) and fires (Holling 1980). The least understood and probably the most difficult and critical portions to understand are the discontinuities and surprises resulting from the endogenous dynamics of human systems. Such concepts in social sciences have been studied, but rarely in the mainstream, because of the obsession with quantification and its presumed bearing on intellectual rigor (Mintzberg 1990).

The variables affecting both the ecological and institutional systems can be grouped in relation to the characteristic speeds or turnover times. Fast variables have time domains on the order of a year, intermediate-scale variables change over decades, and slow variables turn over on the order of centuries. In the ecological system, leaves and small fish are examples of fast variables, sawgrass plants and trees are examples of intermediate-speed variables, and landscape features (e.g., marl wetlands or tree island/wet prairie/sawgrass mosaic) are the slow variables (figure 3.2). The important or keystone variables in the institutional setting include fast variables, such as operational decisions and mass media input. Institutions appear to have characteristics of intermediate-speed variables. The mandated authority, personnel turnover time, and policy domains of such institutions as the corps, SFWMD, and NPS all change over decades. The spatial domains of these agencies cover large, regional areas correlated to the mesoscale domain of the ecological system. The slow variables that affect institutions are related to human myths, values, and mini-constitutions. The myths of nature, such as the equilibrium-centered stability regime, will spawn agencies with operating premises that emphasize a control over nature paradigm. Values also represent slow variables, with corollary spatial extent. The value of the Everglades National Park is international in scope, as reflected by its designation as a National Park, a World Heritage Site, and an International Biosphere Reserve. Indeed, some of the historic conflicts have been caused by a mismatch of values at different scales. One of these was the lack of Florida sponsorship of the minimum-delivery legislation in 1970. Other paradoxes reflect value conflicts at different scales. Although the international community views the Everglades as a unique physiognomic region, the local population seems less aware and has little experience-based knowledge of the system. Ninety percent of the regional residents have never visited the Everglades. Perhaps a more fundamental source of recurring conflict is the perpetuation of the myth that economic development and ecosystem preservation involve an either/or proposition. Recent work (Walters and Gunderson 1994; Holling et al. 1994) indicates movement away from win-or-lose competition within the region toward more win-and-win strategic combinations.

The past century has witnessed four management eras, each of which was precipitated by a crisis or surprise. These eras reflect evolutionary flips into radically new stability regimes, or de novo system configuration (represented by a new set of four boxes; figure 3.5). The exogenous surprises in the floods of 1903 and 1947 resulted in dramatically new sets of physical structures and operational procedures in the system. These structures and procedures were based upon myths that centered around a "quest for control"—that is, given enough money (which was available in 1947), the variation in water could be tamed. Since 1947 the physical appurtenances of the system have remained the same. The evolution of the system following the drought of 1971 reflected a change in institutional structure, whereby the South Florida Water Management District was given more systemwide responsibility. Although some adaptive strategies have been implemented and interaction among agencies has increased, the system appears perched on the cusp of a major reconfiguration of strategy because of ongoing changes that are systematically undermining the prevailing myth of "quest for control."

The myth of control has persevered through the past century and is characterized by a rational decision maker model that assumes that institutions can solve resource problems based on "objective knowledge" and the exploitation of technology in the name of progress. The successive eras of water management have been elaborations of this myth, with modifications of institutional and technological themes. The pursuit of control strategies has led to a tightly linked socioeconomic and natural system that is less resilient and more brittle. Successive attempts to reconfigure the system, with the control paradigm in place, will only reduce management options geometrically to the point where the principal options pursued will be dominated by cost-extreme measures using existing technologies.

The maturation of the system, within and between phases of state (as shown in figure 3.5), results in changes in problem definition and may lead to consequences that create or reinforce the next crises. Over the course of water management examined in this case study, the problem definition has changed from drainage to flood control to water supply to water quality to environmental protection to ecological restoration. This shift is due in part to "new" problems that are created from old solutions. The newer problems raise questions about the viability of large, irreversible projects, such as the C&SF project. The problem definitions shift solutions from one of controlling high-risk/low-frequency events to restoring broad-scale hydrologic and ecological functions under great uncertainty. The case study reveals that the presumptions about institutions and technologies, present but not apparent at each reconfiguration, set the stage for the next crises. Unrestrained growth, balkanization of the system, and faith in large, regional institutions for technological and scientific management are all myths that have created fundamental pathologies in the system. Recognition of failure is latent and only appears after the system has matured and the competency traps or pathologies become undeniable. Some deep pathologies, such as the fundamental incompatibility of the park and the

C&SF project, are ignored or denied, and remain unattended. Until these fundamental pathologies are dealt with, successive attempts at restructuring or reconception become self-limiting. Partial solutions to pathologies have short life spans. Successive iterations through the four phases tend to constrict and limit opportunities.

Two parallel sets of activities are viewed as tremendously robust bridges to ecosystem rejuvenation and renewal. The first is that adaptive management strategies are well in place, as illustrated by the rainfall delivery plan to Everglades Park and the Kissimmee River restoration project. These adaptive strategies are the only way of dealing with the uncertainties and inevitable surprises (both exogenous and endogenous). These strategies are designed to learn about the system from management experience and feedback. Recent syntheses (Holling et al. 1994) of existing strategies indicate that tinkering with the system (either physical structures or operational procedures) will not achieve restoration goals and that profound reform is needed. The second activity is the development of a shadow network, or informal collegia of scientists. The existence of a shadow domain at the periphery of institutions appears to be critical, not only in stabilizing revolutions such as in the box 4 reconfiguration phase, but also for advancing fundamental learning, strategy reframing, and innovation. The shadow domain in the Everglades has been present in the major transformations involving water management, that is, Allison's (1939) rewatering plan and Douglas's River of Grass (1947), which became the basis for the Central and Southern Florida Project. Art Marshall's vision that incorporated a system-wide perspective has guided the formation of the SFWMD and the movement to Save the Everglades. The recently formed informal consortium of scientists appears to have persisted in spite of bureaucratic and legal obstacles, but it still operates in an environment of rival, territorial institutions.

The new ecological system of the Everglades has been around for fifty centuries, while the epoch of intensive management by humans scarcely spans one century. Prior to this time, the dynamic, resilient wetland ecosystem was characterized as a broad, flat river of grass in harmony with the dramatic variation in mesoscale processes of water and fire. In the past 100 years the system has become home to millions of humans and boasts one of the world's largest public works projects. The accompanying management institutions appear to have become hopelessly gridlocked in a never-ending mode of decision paralysis and partial solutions. The adoption of adaptive policies and changing myths and values that embrace cooperation, renewal, and multiplicity of views are positive signs that renewal of the Everglades may already be under way.

Acknowledgments

This work was supported by the South Florida Water Management District and the Arthur R. Marshall, Jr., Chair in Ecological Sciences, University of Florida. The views expressed in this chapter are those of the authors and are not intended to represent those of the South Florida Water Management District. The authors would like to express their indebtedness and substantial reliance on the various historians of the south Florida region and the Kissimmee/Okeechobee/Everglades ecosystem. Without the diligent efforts of Dau (1934), DeGrove (1957), Dovell (1942, 1947), Johnson (1974), Carter (1974), and Blake (1980) this study would not have been possible.

4

The Chesapeake Bay and Its Watershed: A Model for Sustainable Ecosystem Management?

Robert Costanza and Jack Greer

The Chesapeake Bay is the largest estuary in North America and has been the subject of more scientific study and political wrangling than any other body of coastal water in the world. It has become clear that what happens in the bay is in large part a function of activities in the drainage basin, but the focus of most past studies has been narrow. We are only now beginning to develop a comprehensive picture of the bay and of its connections to a sprawling and often densely populated watershed—not only in ecological, but also in demographic, cultural, political, and economic terms.

In this chapter we first develop a historical and spatial perspective of human activities in the Chesapeake Bay watershed. Fundamental in gaining this perspective is conceptualizing the Chesapeake Bay watershed system as the combination of the drainage basin and the estuary itself. We have assembled and mapped past and present human activities in the Chesapeake watershed in order to gain this perspective.

The Chesapeake watershed comprises an ecological system whose beauty and productivity have led to high rates of human population growth and settlement. These high population growth rates have in turn, directly and indirectly, caused a troublesome infirmity, including declining fisheries, receding wetlands, vanishing seagrasses, and a devastated oyster industry. These trends have also led to a decline in the quality of human life. Traffic congestion, disappearing natural and agricultural areas, swelling landfills, and overtaxed water treatment facilities have all affected the tenor of life in the bay region. This chapter addresses the larger-scale terrestrial trends within the entire watershed that have produced such consequences, the barriers to effective environmental management capable of reversing these effects, and the bridges necessary to overcome those barriers.

In many ways efforts to manage the Chesapeake can be viewed as a "best-case scenario" for ecosystem management. Relatively early and widespread recognition of the bay's decline, combined with significant scientific analyses detailing both causes and solutions, helped lead to a broad consensus for remedial action. These aspects of the bay's management may serve as models for other areas, but one should also realize that significant barriers to implementation of the restoration strategy remain. These barriers include the absence of realistic, comprehensive, long-term goals for managing complex ecological and economic systems, the lack of systems for translating these goals into local incentives that can adequately change and guide human behavior, and the built-in but misplaced arrogance about our level of understanding the system.

Bridges that could help surmount such barriers include developing and applying (1) more realistic models of the ways ecosystems function, (2) more realistic acknowledgment of the severe limits of our knowledge, and (3) incentive systems that can translate this understanding (and especially the limits and uncertainty of this understanding) into local decision making.

A Summary of the Problem

Because of the special characteristics of the Chesapeake Bay and its watershed, it is, at one and the same time, extremely productive, unpredictable, resilient, sensitive to stress, and hard to understand and manage with traditional methods.

It is productive because it is a broad, shallow, large estuary where fresh water and nutrients running off a large watershed interact with seawater in very complex and unpredictable patterns. Its shallowness means nutrients are recycled easily and constantly, maintaining its high productivity. Its large watershed means that ample nutrients entered the system even before European settlers converted so much of the watershed to farms, towns, and cities. Because of its large and habitable watershed and its nutrient-rich waters, the bay from the beginning attracted many settlers, but, ironically, it also proved sensitive to the further addition of nutrients that settlement caused. In a further irony, because the bay has, for most of the last 300 years, remained a completely open-access resource, its tremendous productivity has been tragically overused and abused. Because of its complexity and unpredictability, the bay remains a relatively resilient ecosystem, and if we can devise ways to manage its use more intelligently and equitably, as well as a means of reducing the stresses we put on it, the estuary should recover and remain resilient and productive for a long time to come.

We can only do this by first understanding the history and current status of the bay and its watershed as an integrated system. This system involves interactions between various natural components between humans and the unique ecological life support mechanisms the bay provides. Clearly, we also need to develop a better understanding of human institutions—how they typically fail when dealing with resources like the Chesapeake and how we can fix them or perhaps replace them with more effective alternatives.

The Chesapeake Bay and Its Watershed

The early European settlers who arrived on the shores of the Chesapeake Bay came looking for a better life. Whether they came to Jamestown or Williamsburg in Virginia, or to St. Marys or Londontowne in Maryland, the Chesapeake Bay represented for them a new kind of promised land.

Settlers came to begin a new enterprise, and ultimately they founded a new nation. Even the Native Americans who greeted them had come from somewhere else, crossing a land bridge (where Alaska once touched Asia) many centuries before. But these Native Americans had preserved a lifestyle that had not taken a great toll on the natural environment or the countryside. They took oysters but did not destroy the oyster beds. They used logs for canoes but did not clear-cut the forests. The settlers who came from Europe, Africa, and Asia would have a much greater impact on the forests, streams, and rivers as they created what was to become a new country.

That new country often traced its beginnings to land grants from the kings of Europe, the undisputed powers of their day—land grants by the kings of England to such men as Lord Baltimore. Clearly, the country also had its origins in commerce, with the backing of businesses such as the Virginia Company (the equivalent of today's large international corporations, such as Exxon or Toyota). But the life the settlers made for themselves profited not so much from their European backing as from the riches of the land they had found.

Whereas explorers farther south often exploited gold from ancient South American cultures, the colonists of the Chesapeake Bay region found another kind of wealth: teeming schools of fish, hardwood forests, fertile soil, wild game. The riches of the Chesapeake were the riches of nature, and as the centuries passed, the bay region, with its large farms and plantations, became known as the "Land of Pleasant Living."

Of course living in this colonial milieu was not pleasant for all the people all the time. Indentured servants labored for years to gain their freedom. Slaves labored for a lifetime, and many were never free. War came to the Chesapeake region more than once, as Native Americans fought the settlers and slaves revolted against their masters. But the great wars were fought between the settlers and their European forebears and then finally among the settlers themselves.

As the nation suffered its pangs of growth and transition, the Chesapeake watershed repeatedly became a scene of bloodshed. The Revolutionary War's final campaign took place in the bay region, with the surrender of the British general Cornwallis at Yorktown on the shores of the York River. The War of 1812 saw the British return to the bay, with warships guarding its entrance and raiding parties making their way up bay rivers to attack cities—to burn Washington and Baltimore. A Baltimore lawyer by the name of Francis Scott Key watched the attack on his city and wrote a poem that eventually became "The Star-Spangled Banner."

The bloodiest battle of this nation's bloodiest war was fought in the Chesapeake watershed, along the edges of Antietam Creek. In the Battle of Antietam more than 24,000 Americans died, more than in any other battle, in any other war.

Through all this the Chesapeake Bay continued to provide. People took its productivity for granted, like the air or the rain, but changes were taking place in the Chesapeake, and the impacts of settlement began to be felt. The Chesapeake changed slowly at first, but it would never be the same.

Evolution of the Bay and Its Watershed

While humans were writing their own history of hardship and prosperity, war and peace, the Chesapeake Bay was evolving according to a history of its own. The bay's history is first and foremost geologic. Without the geologic and climatic rhythms of the earth—the shifting of the continents, the spread and retreat of glaciers—the bay would not have existed at all. As J. R. Schubel has pointed out, we should remember that the Chesapeake Bay we see today is only one of many that have come and gone over long geological epochs, growing and shrinking with the rise and fall of the sea (Schubel 1981, 1986).

The Chesapeake Bay we see today is a mosaic of estuaries that, taken together, comprise the largest single estuary in the United States. The bay is 193 miles long, is 3–25 miles wide, and with its tidal tributaries covers an area of 4400 mi^2, with some 8100 miles of shoreline (Tippie 1984). The drainage basins that feed it contain 64,000 mi^2 in six states, and three major cities lie on the banks of its tidal system (figure 4.1). Because of its size and location in temperate North America, the Chesapeake watershed represents an ecosystem of enormous historical, political, and economic significance.

This system formed when rising Atlantic waters drowned and eroded the mouths of the rivers that feed it. The Chesapeake Bay we see today began about 10,000 years ago when glaciers that had advanced as far south as present-day New York City finally began to recede.

For 90% of the recent geologic past, the sea has remained in a well-defined basin, with a sharp dropoff at the continental edge. But with the beginning of the Holocene Epoch, the glaciers melted and the sea rose. The overflowing sea covered the edge of the ancient oceanic basin, flooding the long, flat shoreline under a shallow sheet of water. That prehistoric shoreline, now covered with seawater, constitutes the continental shelf, an ecosystem whose physical and biological interconnections with estuaries like the Chesapeake Bay remain only partially understood.

The continental shelf in some ways represents an oceanic extension of the estuaries and coastal embayments of North America's east coast. Though covered by open ocean, the area is surprisingly shallow and flat. A vessel sailing from the Chesapeake Bay would find its depthfinder reading only about 40 ft or so—for 50 miles out to sea. At that point, the depthfinder would drop to more than 100 ft, but soon it would stop reading altogether because the vessel would have passed over the edge

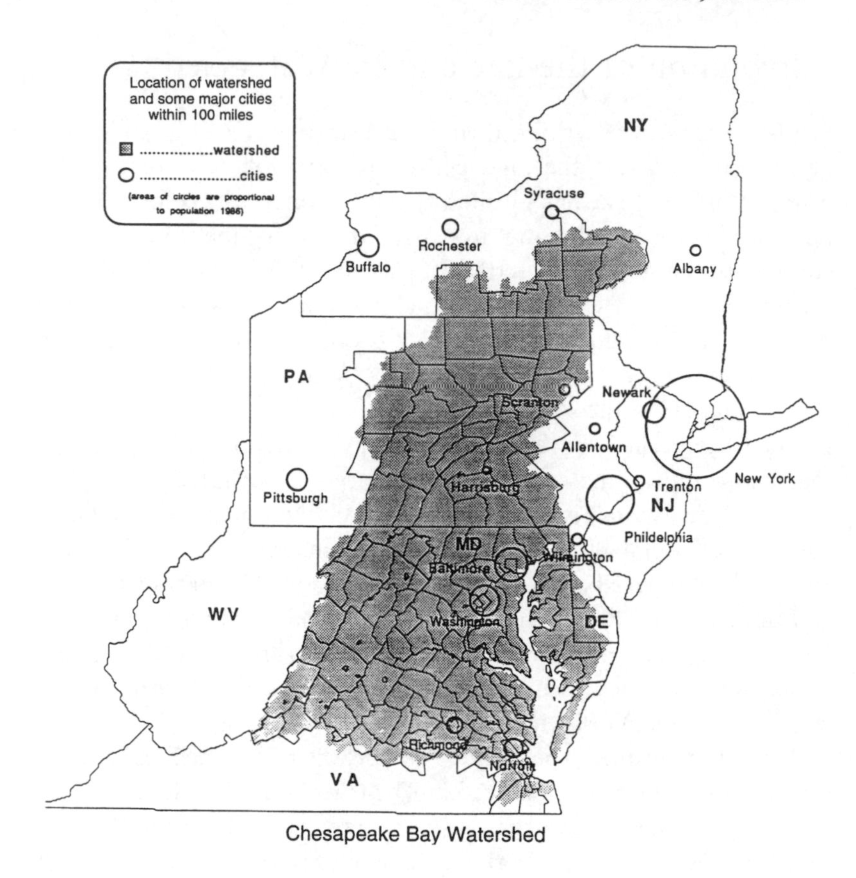

FIGURE 4.1
Map of the Chesapeake Bay Watershed.

of a great cliff, the perimeter of the oceanic basin. If the depthfinder's signal could reach that far, its numbers would rise rapidly to more than 1000 ft, then 2000, then 3000. About a 120 miles out of the bay, off the Virginia capes, the boat would be beyond the continental shelf, past the slope, and over the ancient sea bottom. The depth finder, if it could read that far, would say 9000 ft.

In the Chesapeake Bay the patterns are the same, even if the depths are much less. The Chesapeake, taken as a whole, is a *very* shallow body of water, with an average depth of less than 10 m. Many of its tidal flats may be only a few feet deep. Its deepest waters, 100 ft or more, follow an ancient channel carved by the river that once formed the liquid backbone of the bay, the Susquehanna. The Susquehanna, like the bay's other major tributaries—the Potomac, the Patuxent, the Rappahannock, the York, the James—still plays a significant role in the health and well-being of the estuary because of its capacity to deliver not only fresh water, but also nutrients, sediments, and toxic compounds to the bay.

The Susquehanna River, one of the longest in North America, begins in New York State and wends its way through Pennsylvania. Once it continued straight to the sea, emptying into the ocean basin at a point that is now well off the coast. Now the sea has come to meet it, rising and backfilling into Virginia past Norfolk, past Reedville, into Maryland past the Potomac, past the Patapsco. Tidal waters now reach as far north as Havre de Grace, where they flood a broad area only a foot or two deep called the Susquehanna Flats.

Though ocean waters have inundated the mouths of bay rivers for 10,000 years, the tributaries still exhibit very direct impacts on the estuary. Much of that impact is natural and desirable: the delivery of nutrients and the mixing of fresh river water with saline water from the sea. This natural mixing bowl has created a fertile feeding ground and nursery ground for fish and shellfish, such as the American oyster, *Crassostrea virginica,* a resource now facing catastrophic decline. It seems remarkable that a system that developed over the span of 10,000 years could change in a matter of a few centuries, perhaps even a few decades, as a consequence of the settlers that came to its shores. To understand the nature of this change, we must consider the different uses humans have made of the bay over the years and some of the problems that resulted from those uses.

Principal Uses—and Problems—of the Chesapeake Bay

As a Waterway

The settlers first used the bay to get off the ocean. According to historical writer Donald Shomette, the Spanish came even before the British, during the sixteenth century, and called the Chesapeake the Bahia Santa Maria. Regardless of nationality, the safe harbors of the Chesapeake clearly represented a boon to seafarers, particularly during an age when crossing the Atlantic in wooden ships was risky and, as Shomette and others have documented, too often disastrous.

Those who arrived safely could ply the relatively protected waters of the bay inland to the north and west, into the heavily wooded coastal plain. Even after founding settlements and towns, settlers continued to hold to the bay as their principal connection with the Old World, their supply line, and their chance for trade and commerce. The bay has continued as a major waterway to the present day, with ocean-going ships calling at Baltimore and Norfolk, which rank among the nation's busiest ports.

From the beginning, ships on the bay have probably caused some pollution: raw sewage, garbage dumped overboard, and river bottoms disturbed and sediment stirred up by anchors. But these impacts were quite small—until modern times. Garbage from an old square-rigger is one thing; flushing the oily bilges of a 500-foot freighter is something else. And while the colonial period may have seen a few watermen fishing or hunting for oysters and perhaps a few hearty souls rowing about for recreation, today the bay has become home to hundreds of thousands of recreational boats and thousands of watermen when the catch is good. And these modern boats rarely drift silently with the wind. Many of them have powerful engines and throw substantial wakes that beat against the shoreline and further erode the banks of rivers and creeks. These modern craft also use petroleum products—gasoline, diesel fuel, oil—that may find their way into the estuary, as do chemicals for cleaning and polishing, painting and varnishing, and antifouling agents to keep unwanted growth off of hulls. Because of their deleterious effect on marine life, some antifouling compounds, such as tributyl tin, have been banned from use in the bay, though Navy ships remain exempt.

As a Dumping Ground

All animals, whether humans or striped bass or blue crabs, continually cycle food, taking in nutrients and expelling wastes. When very few people lived on the shores of the bay, human sewage did not pose a significant problem to the health of the estuary. Even when large numbers of people crowded into cities such as Baltimore, the most serious problems related to sewage involved human health risks, primarily from bacteria transmitted through human feces, which can cause cholera and other diseases. Once modern sewage treatment began and modern medicine evolved, these waterborne diseases became less of a threat. Only in areas of the world where sewage treatment is not properly managed (as in underdeveloped countries) does human waste still pose a serious health problem.

But the Chesapeake Bay faces another problem from sewage, an environmental health problem: too many nutrients. Sewage is rich in nutrients, including phosphorus and nitrogen. These nutrients can overenrich the bay and make it too productive for its own good. Too many nutrients fuel the explosive growth of blue-green algae and other undesirable phytoplankton, which take away light from more desirable plants. When these algae finally die and drop to the bottom, they begin to decompose. As bacteria break down the fibers and minerals of the dead algae, they use up large amounts of oxygen, especially in deeper channels. This lack of oxygen makes it difficult for other animals that also need oxygen, such as oysters, fish, and crabs. Sometimes the waters have very little oxygen (hypoxic waters) or virtually no oxygen at all (anoxic waters), and fish and other sea life unable to escape often suffocate and die.

With expanding population growth, the Chesapeake has therefore suffered from receiving waters for human waste. It has also received the effluent from numerous factories and industries, and in some places its sediments bear significant traces of heavy metals, organic compounds, and other pollutants. The areas around Norfolk Harbor (the Elizabeth River) and Baltimore (the Patapsco River) have in particular been rated as "toxic hotspots." But according to most researchers, it is the overabundance of nutrients that has caused the most pervasive ecological decline in the Chesapeake.

As a Fishing Ground

Despite the other ways in which humans have used Chesapeake Bay, its reputation and its tradition center on its value as a fishing ground. In its heyday, the Chesapeake's shallow waters provided, acre for acre, more fish and shellfish than any other body of water in the world.

For many years, for example, Maryland's oyster harvest held steady at about 2 million bushels a year—nothing compared to the 15-million-bushel harvests of the nineteenth century, but enough to sustain a long-standing fishery and the bayside communities that depend on oysters to round out an annual cycle of working the water. Then, during the 1980s, oyster harvests in Maryland dropped to less than 2 million bushels, then less than 1 million, then less than half a million. Many feel that the bay's oyster industry has, for now at least, collapsed.

Other changes have also taken place in the Chesapeake's water trades. The blue crab has become king of the commercial fisheries, surpassing oysters as the bay's most lucrative harvest. Because of relatively low fuel prices during the 1980s and high demand for crabmeat, watermen hauling in blue crabs saw a 250% rise in profits during the 1980s. In many ways the blue crab, once an undesirable by-catch, has kept the bay's seafood harvesting industry alive.

Equally as dramatic, during the same time period as the blue crab's rise was the fall of the bay's striped bass fishery. Generally known as rockfish in the bay region, striped bass once traveled by the tractor-trailer load from ports like Rock Hall, headed for Lexington Market in Baltimore or Fulton Market in New York. Harvest figures for 1973 reached over 4 million pounds in Maryland alone. But a precipitous decline in striped bass stocks, coupled with an appreciation of the Chesapeake's importance as the Atlantic coast's major spawning ground, brought about stiff restrictions and then a total ban on fishing for (or even possession of) striped bass in Maryland in 1985.

A recent reopening of Maryland's fishery, based on promising surveys of juvenile stripers, has meant a short season for both commercial and recreational anglers, but it will be a long time, if ever, before the Chesapeake sees a return of the great harvests of 20 years ago. On the positive side, the strict management of the striped bass in the Chesapeake has, arguably, averted an irreversible decline of that species, and the careful allocation of stripers among recreational anglers, commercial netters, and charter boat operators may well serve as a model for other threatened fisheries.

The Bay and Its Watershed as a System

A Stressed Ecosystem

Because the Chesapeake Bay has been so productive, some have called it a food factory, or compared it to a powerful engine that runs on nutrients. But the bay is not a factory or an engine; it is an ecosystem. Instead of machinery, the bay is composed of living parts: animals, plants, and microorganisms that depend on each other. Take away or change some of these living parts, and the whole ecosystem feels the effects. It feels them in both direct and indirect ways, ways that are sometimes obvious and sometimes impossible to predict, that depend on the subtle interplay of the parts of the system at several spatial and temporal scales, and that determine whether the estuary is healthy and resilient or brittle to the breaking point.

This is not simply an abstract notion. Consider, for example, one of the bay's prized shellfish, the oyster. For years the reputation of the Chesapeake Bay oyster has spread across the country, and in the bay region oysters have meant money—some $20 million a year in Maryland alone during some years. As the oyster populations in the bay fell, largely because of disease and overfishing, many watermen and natural resource managers realized that this represented a significant loss of livelihood and a shrinking economic resource for the tidewater region. What many may not have realized, however, were the ecological effects of the oyster's demise.

Like the bay itself, the oyster bar is an ecosystem. The oyster is a gregarious animal; it prefers to grow in groups. Some scientists believe that young oyster larvae can actually detect certain chemical signals that draw them to other oysters (Weiner et al. 1993; Coon, pers. comm.). Because of this attraction, oyster larvae set and grow in clusters, ultimately forming large aggregations called oyster bars (or rocks or reefs).

In the Chesapeake today, after more than 100 years of harvesting by oyster tongs or dredges, most oyster bars lie low and scattered across the bottom of the bay. But during the early seventeenth century, when Captain John Smith first sailed into the Chesapeake, the oyster bars were said to reach from the bay bottom all the way to the water's surface. These bars actually formed oyster reefs, and as with coral in the tropics, these oyster reefs undoubtedly created rich habitats and functioned as underwater ecosystems.

Imagine for a moment oyster reefs stretching up both sides of the Chesapeake Bay, along the shallow margins. Fish and other marine animals would have gathered around the reefs to feed, and chances are that one could see these fish, because the water would be relatively clear. The water during those early years would be less murky than now for two reasons. First, because prior to European settlement and the introduction of intensive agriculture, the land surrounding the bay and its rivers was covered with forests, which protected the soil and prevented runoff. Second, the bay would be more transparent because the oysters themselves were actually cleaning the water.

Oysters are filter feeders. An oyster, when aggressively feeding, pumps about 50 gallons of water a day in order to filter out algae (also called phytoplankton), the tiny floating plants that serve as its primary food source. As oysters feed, they act like filters in a swimming pool, drawing out algae and clearing the water. Some researchers have estimated that the bay's once abundant oyster populations could pump through a volume of water equal to the entire Chesapeake Bay in less than a week (Newell 1988). Because oyster populations have dwindled to such low levels, it would now take a year or more for today's oysters to filter that same amount of water.

Disturb one part of the ecosystem, and the whole ecosystem changes. The bay now has too many phytoplankton. Not only does the water look murky, but this lack of clarity has meant a lack of light for rooted aquatic grasses. In many areas these grasses, which provide food for diving ducks and shelter for molting crabs, have died off as a result of too much algae, which not only cloud the water but also cover the submerged grasses with light-blocking slime. Overwhelmed, many grasses have disappeared, leaving large stretches of bay bottom bare.

In short, the bay has changed in a number of ways because of human beings.

First, we have harvested oysters and destroyed the oyster reefs. Most of the bay's oysters were taken during the end of the nineteenth century. During that time watermen hauled up in a single year what it would now take more than 10 years to harvest.

Second, we have increased the amount of algae in the water by adding more nutrients. These nutrients come from sewage treatment plants, from septic systems, from fertilized farm fields and residential lawns. These nutrients cause even larger blooms of algae, which die and decompose on the bay floor, a process that draws life-sustaining oxygen out of the water.

Third, we have increased the amount of sediments in the water. The bay naturally receives a heavy load of sediment, especially during heavy spring rains and storms, but human uses of the land have torn away the protective forests and left the soil to wash away. This increased runoff began many years ago when farmers increasingly cleared the land for agriculture. Soil erosion continues today, not only because of agriculture, but because of housing construction and other land development in the watershed. Every construction site has the potential to release tons of sediment into the bay and its tributaries.

Fourth, we have added new chemical compounds to the bay. Chemicals have become a part of our daily lives. Heavy metals like zinc and mercury from industrial uses, pesticides from farms and suburban lawns, cleaning solutions from households, and a host of petroleum products and other compounds all wash off the land or down storm drains and into the streams and rivers that feed the bay. We do not fully understand the effects of these chemicals. Scientists do know that, in significant doses, many of these compounds are toxic to fish and other marine organisms. What researchers are still trying to determine is the effect such toxic compounds have at very low levels in our waterways—levels so low that they may be difficult to measure.

Toxic compounds sometimes act together to create problems for bay organisms. For example, when rain becomes polluted with nitrous oxides (from automobiles) and sulfur dioxide (from coal-burning industries and electrical power plants), it becomes acidic. When this acidic rain falls on the bay, it does two things. First, it adds nutrients to the bay. Some researchers have postulated that as much as 25% or more of the nutrients entering the bay come from airborne sources. Second, acid rain causes elements like aluminum to leach out of the soil. If, for example, this leaching occurs in a tributary where fish are spawning, it can kill delicate larvae. Some evidence exists that this double threat from acid rain and aluminum has hurt the reproduction of striped bass.

Such problems have not gone unnoticed. The decline of striped bass, the Maryland State Fish and a popular sport and commercial fish up and down the Atlantic coast, has caused an uproar. But other effects of toxic compounds on the bay's plants and animals that may not be as visible to the public may go largely unnoticed. As one researcher has said, it may be that the bay has a giant "headache" caused by toxic compounds, and we just don't realize it.

There is one more thing to say about the Chesapeake Bay's ailing ecosystem. The bay serves as something of an indicator for the health of the entire region.

Since the bay lies at the base of an enormous watershed—some 64,000 mi^2 of mountains, foothills, and coastal plain—it gathers much of what is put on the land or poured into streams and tributaries throughout the area. Because it supports such a rich and productive ecosystem, the bay provides ample opportunity for us to witness changes and trends, such as significant decreases in the populations of animals and plants.

The history of the Chesapeake Bay tells a story about how the way we live affects the environment we live in. Clearly, the bay is not simply a pool of water where oysters and crabs grow. It is the most visible part of a vast network of plant and animal life. Human beings rely on this environment as much as any animal; we also bear a special responsibility, since our actions have an impact on the ecosystem greater than that of any other creature that walks the earth.

Population Growth and Change in the Bay Watershed

During the Revolutionary War, when George Washington traveled through Annapolis, about 500,000 people lived in Maryland. In the two centuries that have passed since, Maryland's population has grown almost ten times. Now 4.7 million people live in the state, and according to predictions by the Maryland Office of Planning, another 800,000 or more will settle in Maryland by the year 2020. Maryland's growth illustrates change throughout the Chesapeake watershed. Today more than 14 million people live in the watershed; that number is expected to top 17 million by the year 2020.

Population changes in the watershed are shown in figures 4.2. The number of dots is proportional to population in each county and distributed randomly within each county. Between 1940 and 1986 the population of the watershed increased 87%. About 20% of this increase was due to net migration into the watershed, the majority of it to the areas surrounding Baltimore and Washington in Maryland and Virginia. The population of the watershed grew at an average annual rate of 1.6% between 1952 and 1972, almost the same as the U.S. average of 1.5% for the same period. But the growth was concentrated in the Maryland and Virginia portions of the watershed, which averaged 2.6% growth, compared with the remainder of the watershed, which grew at only 0.4%.

The most striking changes were in three areas: Richmond, the Norfolk/Virginia Beach area, and the Baltimore/Washington corridor. Growth can, in part, be attributed to increases in industry related directly and indirectly to the expanding U.S. government, as well as the increasing fashionableness of the Chesapeake Bay as a recreation area. The latter forces are amplified by the high immigration rates to the area.

These areas also illustrate the "urban flight" and "suburban sprawl" phenomena that have at once undermined the more natural and rural atmosphere that many originally left the city for and removed businesses and middle- and upper-income residents that served as a revenue base for the cities. The resulting deterioration of services and infrastructure

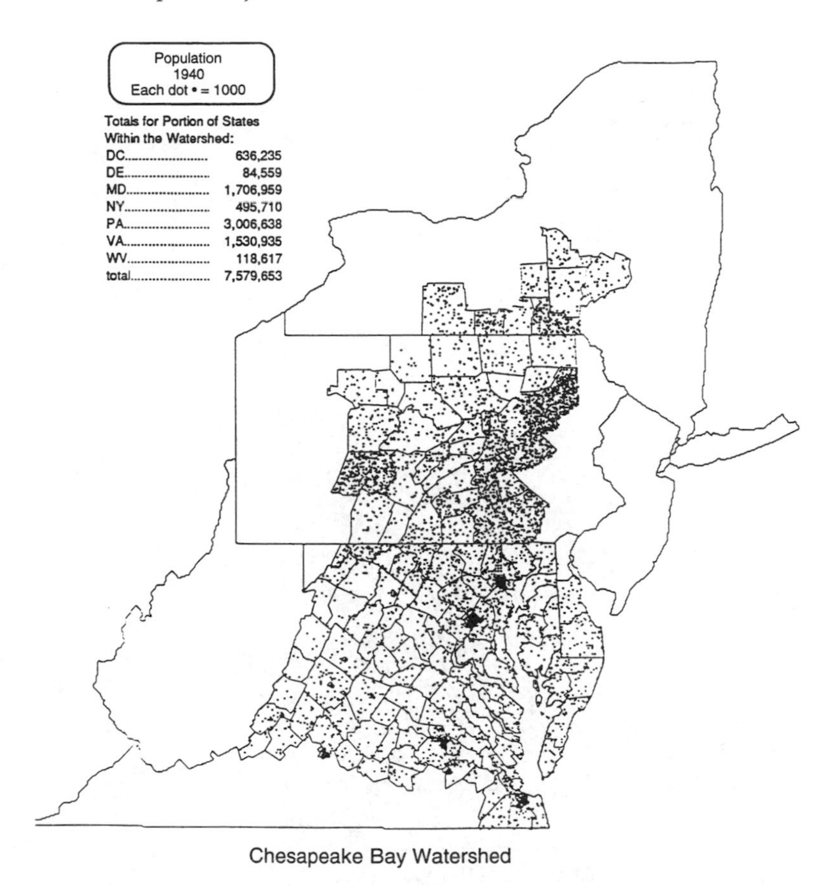

FIGURE 4.2a
Population in the Chesapeake Bay watershed. (a) 1940; (b) 1952; (c) 1972; (d) 1986.

worsens as people move farther and farther out. Increasing travel times required to reach work, worsening traffic conditions, and soaring property values are some of the resistive forces that quell the further spread of suburban development.

It appears that these forces may be approaching an equilibrium, at least for the time being. Between 1972 and 1986 the growth rate of the Maryland watershed population slowed to an annual rate of 0.9%, and in Virginia to 1.8%, and emigration rates from the cities have slowed. For example, Washington D.C.'s emigration rate decreased from an annual average of 1.3% in the 1960s to less than 0.8% in the early 1980s.

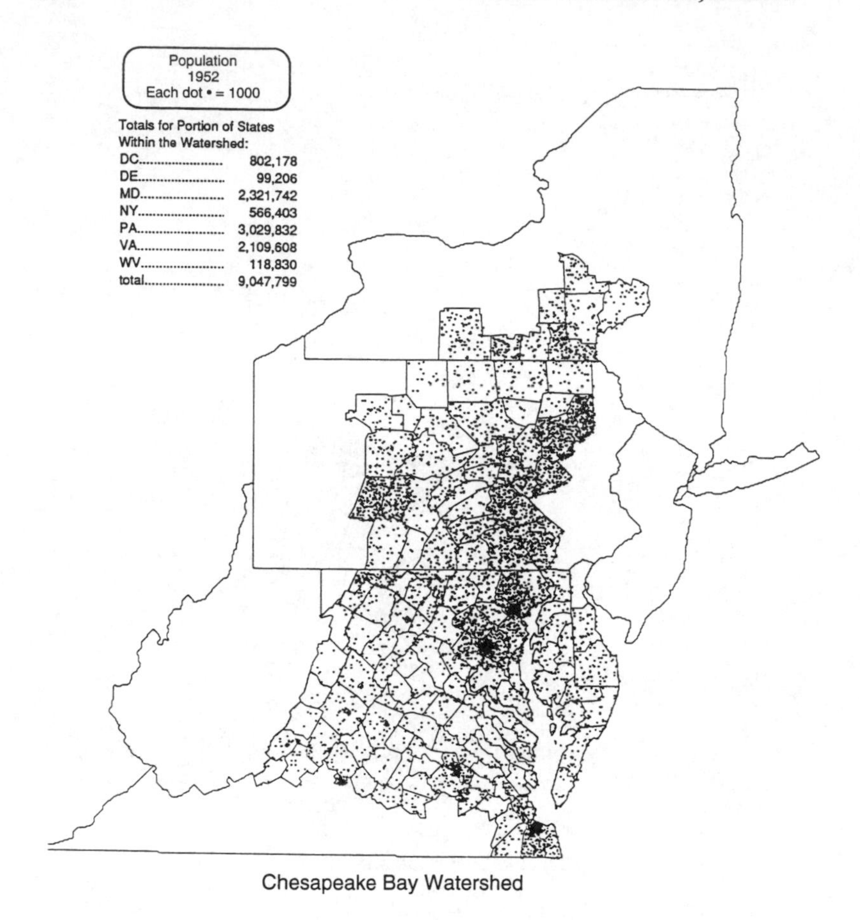

FIGURE 4.2b

In addition, the spatial extent of the sprawl appears currently to be limited to the counties immediately surrounding the cities in question (figure 4.2d). However, the information contained in these maps is only suggestive. Demographic trends can result from any number of factors. Variation in birth rates, cultural heritage, local versus long-distance moves, political climate, and zoning laws all muddy the waters. In addition, as sprawl and growth occur simultaneously, secondary economic centers inevitably spring up, initiating their own cycles.

As population has risen in the watershed, both land and water have felt the effects. Forest and farmlands have been lost to development, and, as described earlier, the Chesapeake Bay's underwater meadows of aquatic vegetation have died off, largely because of too many nutrients entering the estuary. As the underwater grasses have dwindled, widgeons and other diving ducks that grazed on the grasses have also gone. Rising population has also meant heavy harvesting of the Chesapeake's fisheries, which may already feel the strain of turbidity and disease.

Patterns of settlement and resettlement broadly affect land use in other ways as well, as shown by the percent change in land use maps in figures 4.3. For example, there has recently been a slight but noticeable increase in woodland in the New York portion of the watershed

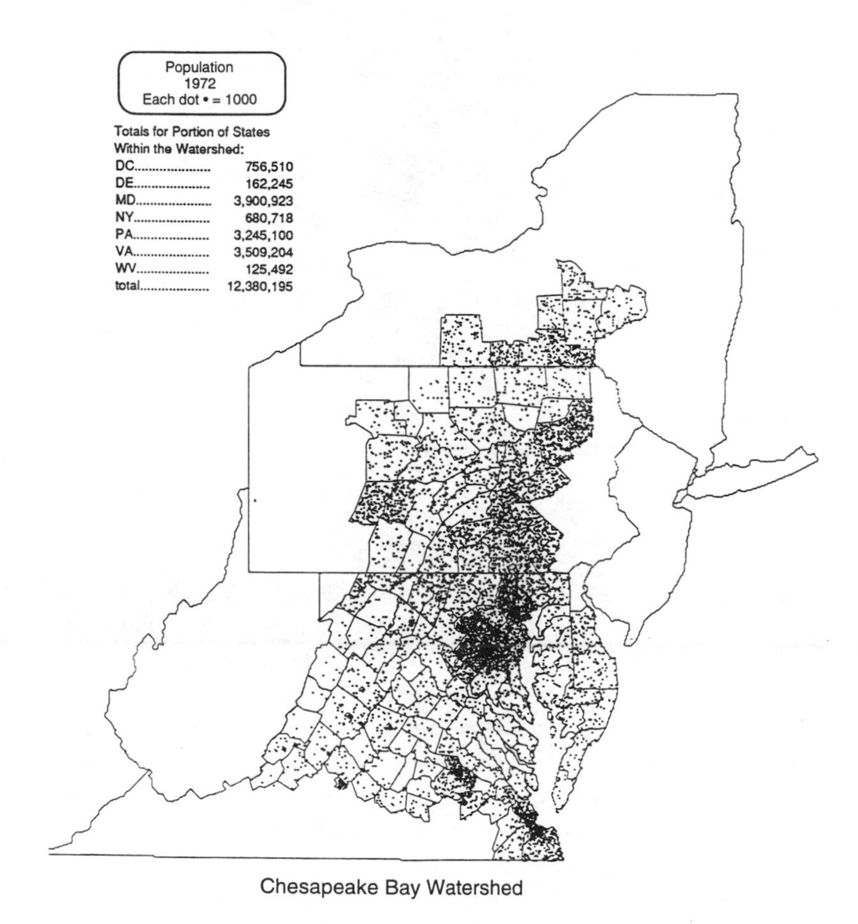

FIGURE 4.2c

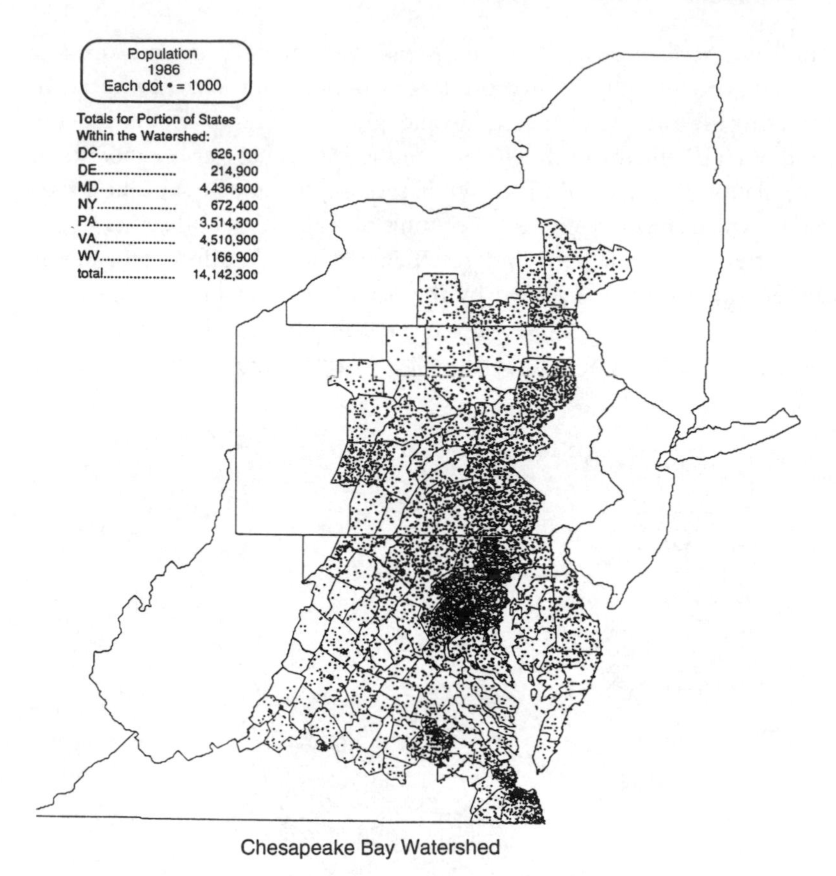

FIGURE 4.2d

(figure 4.3c). The fact that there was a concurrent loss of cropland and pasture (figures 4.3a and 4.3b) only indicates that there were net changes from fields to forests, but it does not explain why. In fact, the marked differences in changes in agricultural and forested acreage from one state to another strongly suggest that these are due to differences in state agricultural policies and tax laws. On the other hand, the pattern of urbanization is more straightforward in light of the population changes discussed earlier (figure 4.3d).

Increases in evidence of human activity accompany increases in population in both magnitude and distribution. Maps of manufacturers, energy consumption, housing units, water use, solid waste production, and air pollutants all closely resemble the maps of population. Changes in life-style have, however, caused accelerated increases in consumption and waste production. From 1952 to 1986 we have seen the following increases in the watershed: per capita energy consumption has gone from 567,000 btu/day to 744,000 btu/day; NO_x emissions by vehicles have gone from 1.0 lb/week per person to 1.7 lb/week per person; and

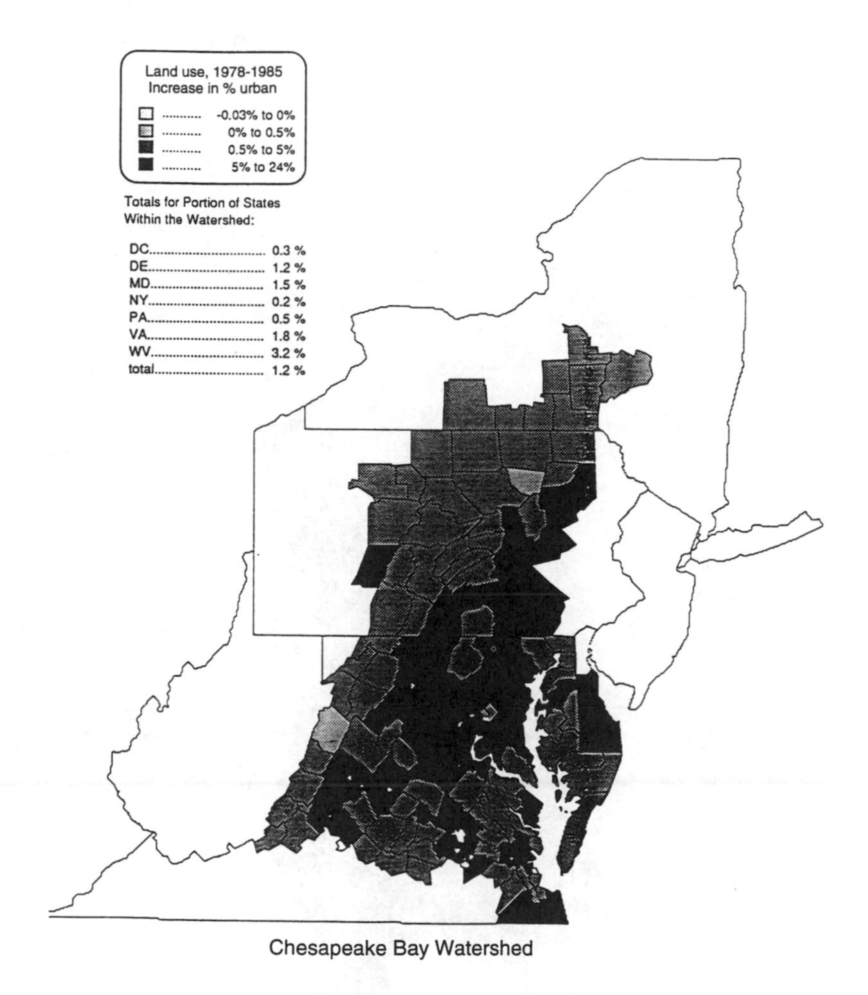

FIGURE 4.3a
Percent change in land use from 1978 to 1985 in the Chesapeake Bay watershed. (a) Urban; (b) cropland; (c) pasture; (d) woodland.

per capita solid waste production has risen from 2.2 lb/day to 3.7 lb/day. In contrast, public and industrial per capita water use has decreased in the same period for the watershed as a whole—from 334 gal/day to 278 gal/day. This is due in large part to the decline in total use in Pennsylvania during this period from 1.46 billion gal/day in 1952 to only 1.18 billion gal/day in 1986. In the Maryland and Virginia portions of the watershed, the per capita use rate held nearly constant at about 237 gal/day. This suggests that the former change may be due to changes

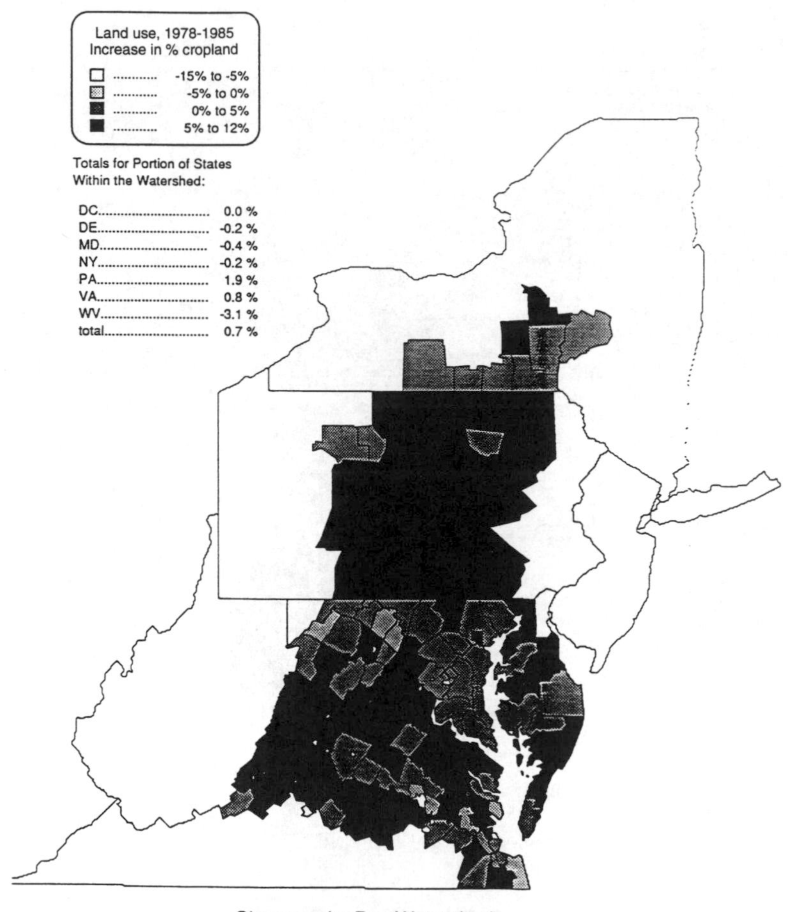

FIGURE 4.3b

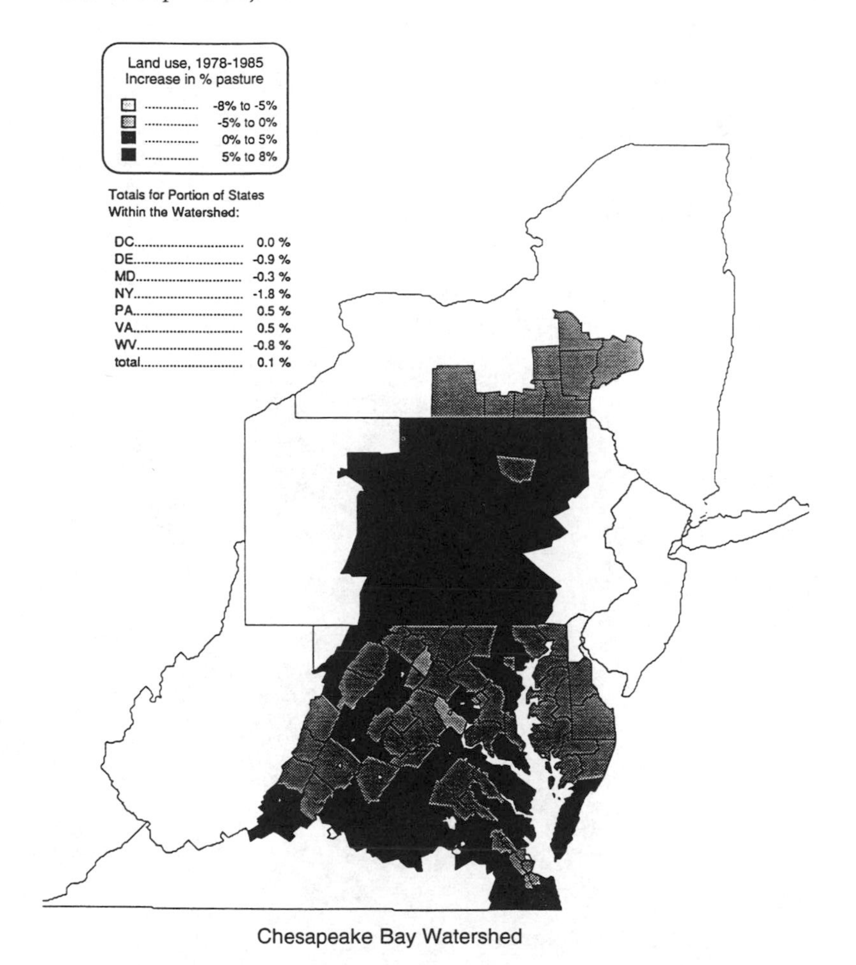

FIGURE 4.3C

in heavy industry in Pennsylvania, whereas the latter lack of change may be due to a relatively constant per capita public demand for water in a region that consists mostly of residency and light industry.

Agriculture

Changes associated with agriculture are tied to increases in population in their overall magnitude and to cultural and historical practices in their distribution and local magnitude. Heavily populated areas neces-

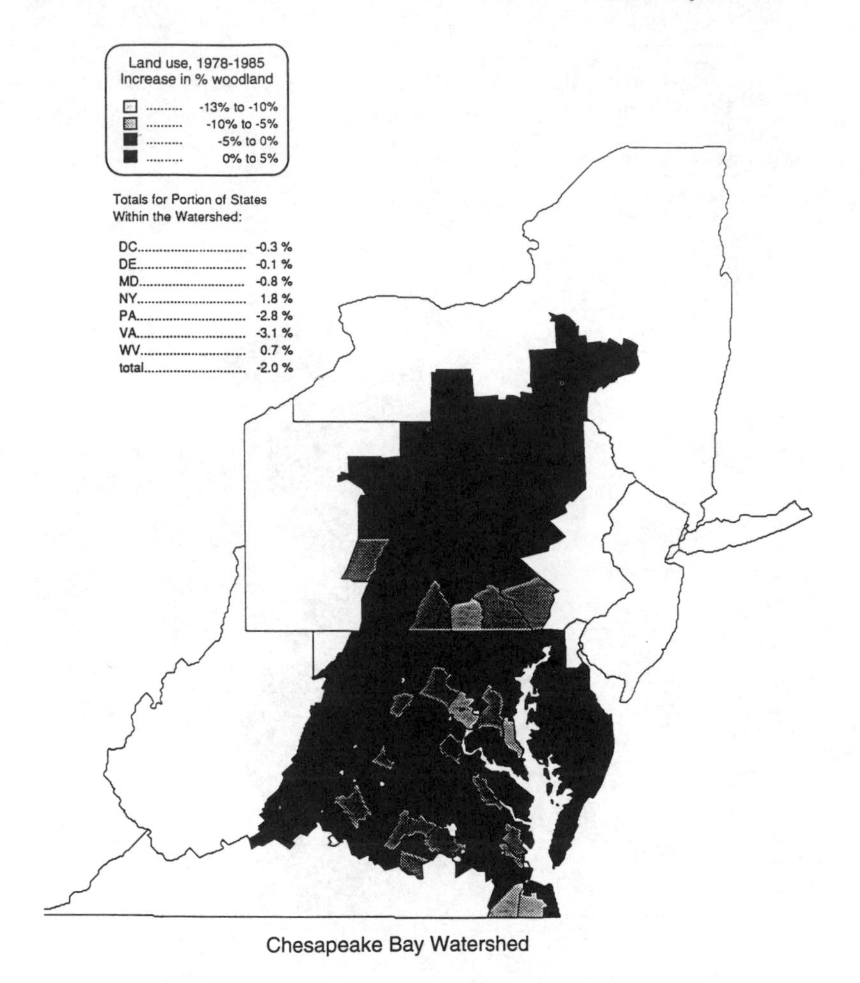

FIGURE 4.3d

sarily exclude agriculture. It is unfortunate that some of the best agricultural land in the country, which was the basis for the initial local growth, is rapidly being converted to residential developments, industrial parks, and shopping malls as rising property values make farming unprofitable. It is not clear, however, whether agriculture itself is more benign to an ecosystem such as the Chesapeake Bay than urban and suburban development. Both load the system with wastes and nutrients while consuming "natural" areas that could have absorbed some of that load.

Whereas total farm acreage in the watershed has decreased from 23.2 million acres in 1954 to 14.4 million acres in 1987, cropland has only decreased from 10.5 million acres to 9.0 million. Meanwhile, from 1954 to 1987 the average farm size increased from 126 acres to 190 acres. This means that larger percentages of farms are devoted to crops (64% versus 45%) and that less lies fallow, pastured, and wooded. Operations are larger and more intensive. Irrigation has increased from 40,000 acres to 180,000 acres, and fertilization rates have increased from about 210 lb per cultivated acre to 250 lb per cultivated acre. Figures 4.4 show that pesticide use, which was almost nonexistent in the early 1950s, rose to

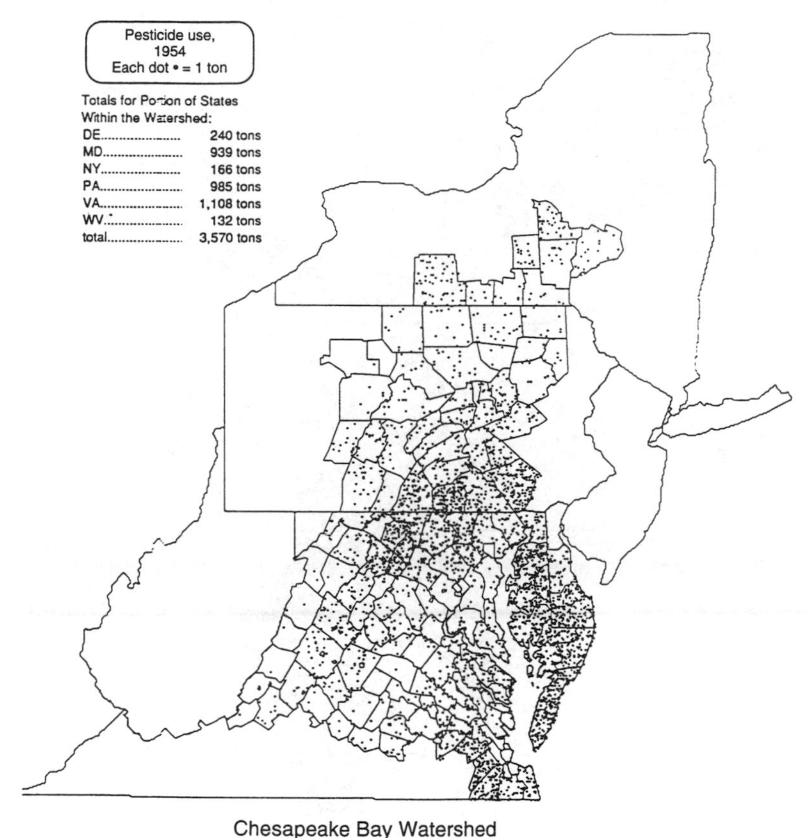

FIGURE 4.4a
Pesticide use in the Chesapeake Bay watershed. (a) 1954; (b) 1974; (c) 1987.

15,000 tons in 1974 and then decreased to 13,000 tons in 1986. Of course this trend also reflects the greater specificity of the pesticides used in 1986.

If we examine the application of fertilizer nitrogen (figures 4.5), we see a pattern that follows the general distribution of farmland but with an overall increase in the intensity of use. Although the phosphorus content of fertilizers used in the watershed remained relatively constant over the years at around 10%, the nitrogen content rose steadily from about 7% in the 1950s to 15% in the mid-1980s.

The nutrient loading rates for the mainstem Chesapeake Bay are 78,700 tons/year nitrogen and 3600 tons/year phosphorus (Boynton et

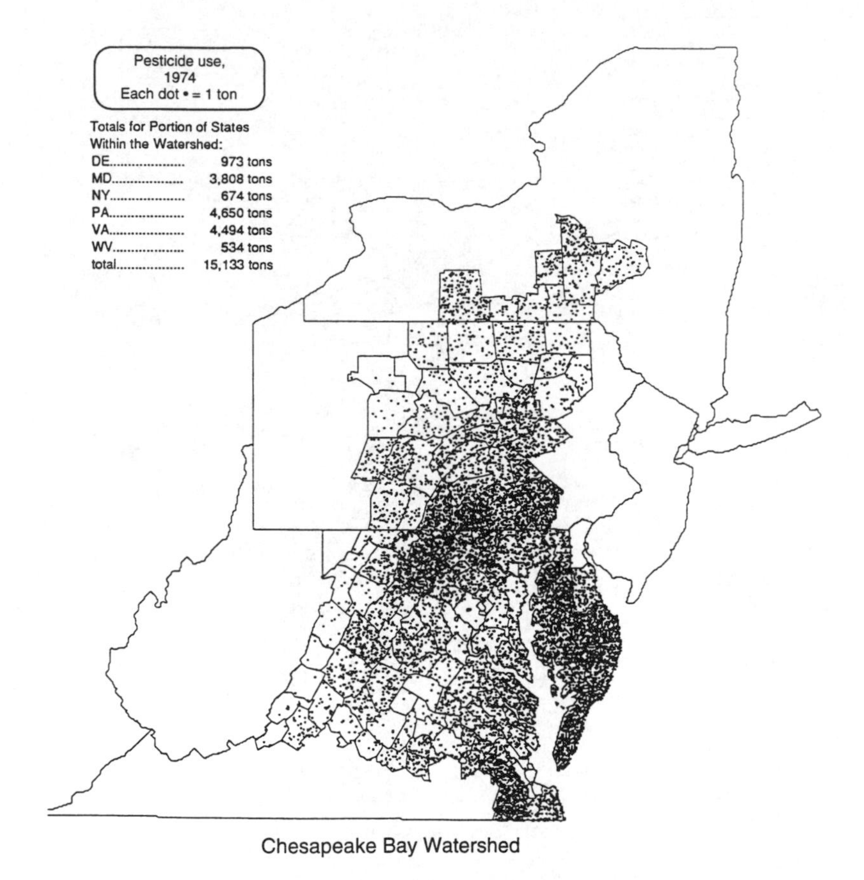

FIGURE 4.4b

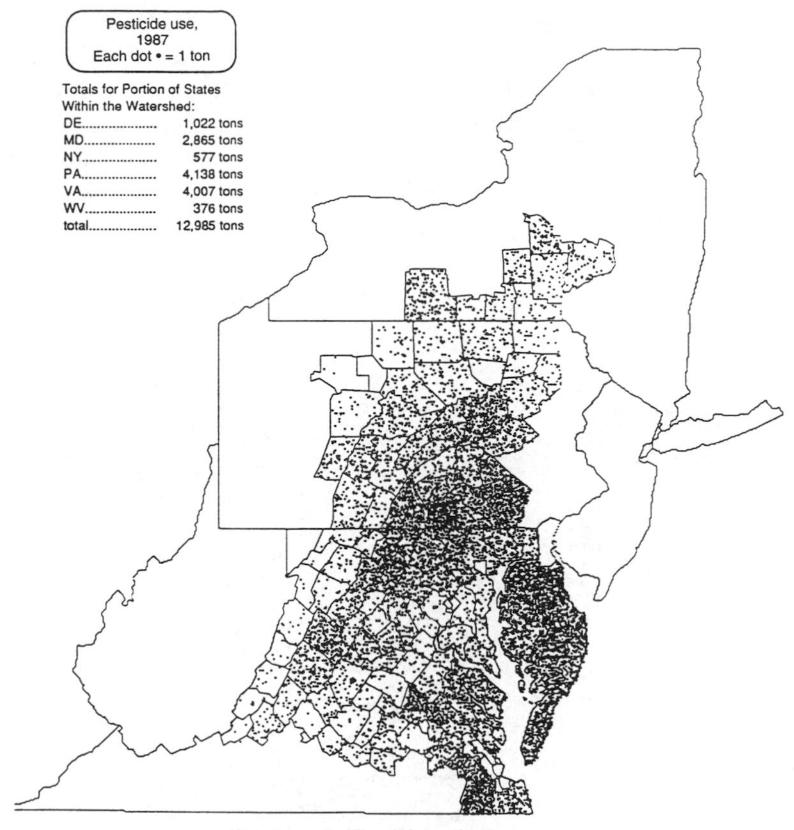

FIGURE 4.4c

al. 1994). If one assumes that the loading rate is proportional to the amount of nutrients produced and applied within the watershed, then this implies that phosphorus loading has increased around 4% to 10% and nitrogen loading has increased somewhere between 34% and 125% since 1954. This would not only be a great change in amount of nutrients but (and this may be even more crucial) a great change in the relative proportion of nutrients. The increased concentration of nutrients in specific localities, such as Lancaster County, means that this problem may be even more exaggerated in some parts of the system.

The Chesapeake Bay has undergone very rapid population growth, with the accompanying environmental impacts. We have mapped some of these changes as they are reflected in the characteristics of the bay's watershed. The impacts of these activities in the watershed on the bay itself are known to be large, but their specific interconnections are only now being investigated. The Chesapeake Bay has 200,000 people living in its drainage basin for every cubic kilometer of water in the bay. (The Baltic Sea has 4000 people/km^3, and the Mediterranean has 85 people/km^3, by way of comparison.) Even if all these people were minimizing

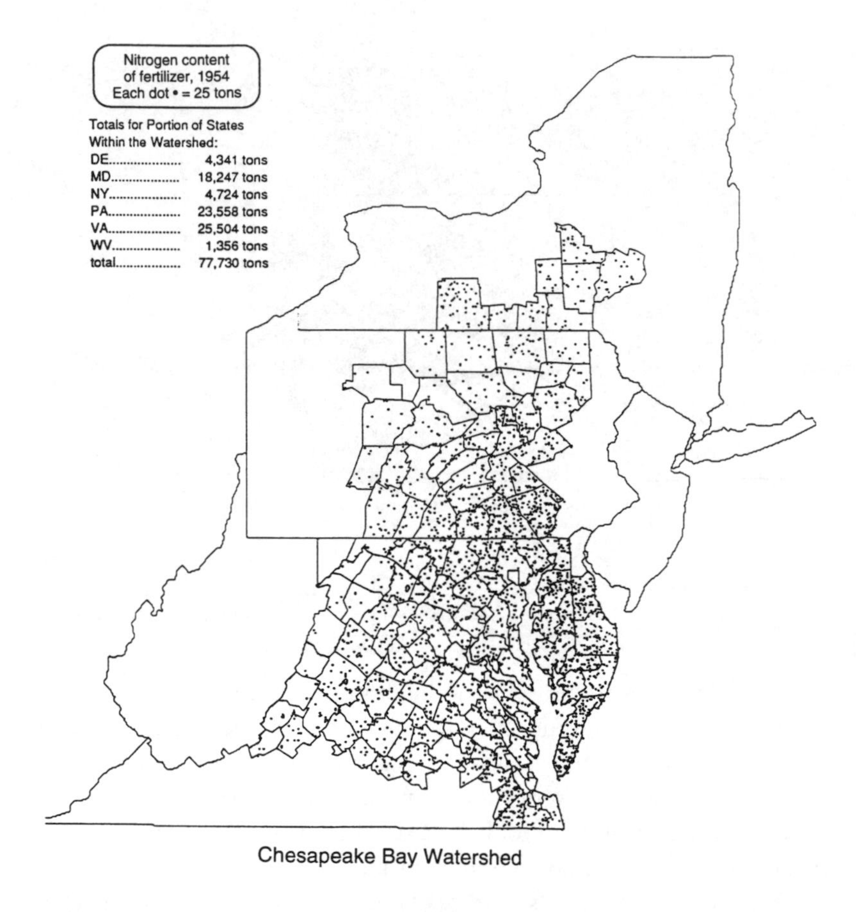

FIGURE 4.5a
Nitrogen content of fertilizer use in the Chesapeake Bay watershed. (a) 1954; (b) 1974; (c) 1987.

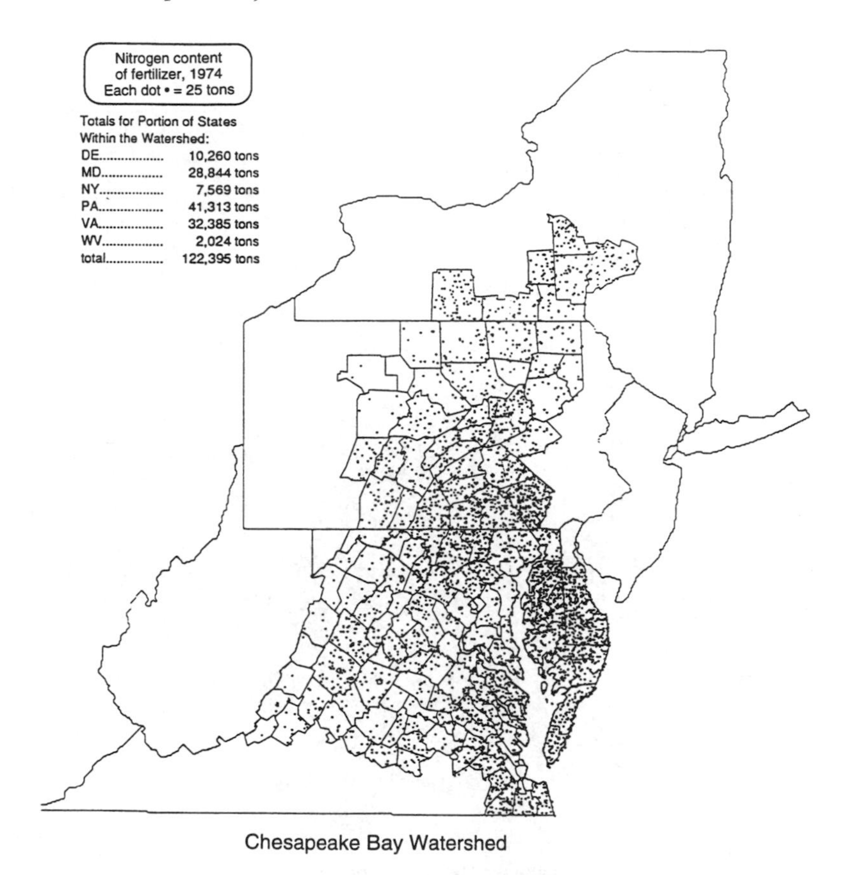

FIGURE 4.5b

their environmental impacts (which they are not), their sheer numbers are daunting to a system as sensitive as the Chesapeake. If these numbers continue to increase as they have in the past, the prospects for America's largest estuary seem bleak (table 4.1).

Evolution of Chesapeake Bay Management

Efforts to understand and manage the bay ecosystem in a comprehensive way began to gain momentum in the late 1960s, along with the general increase in environmental awareness in the United States. Several major studies of the bay have been undertaken, and the bay is

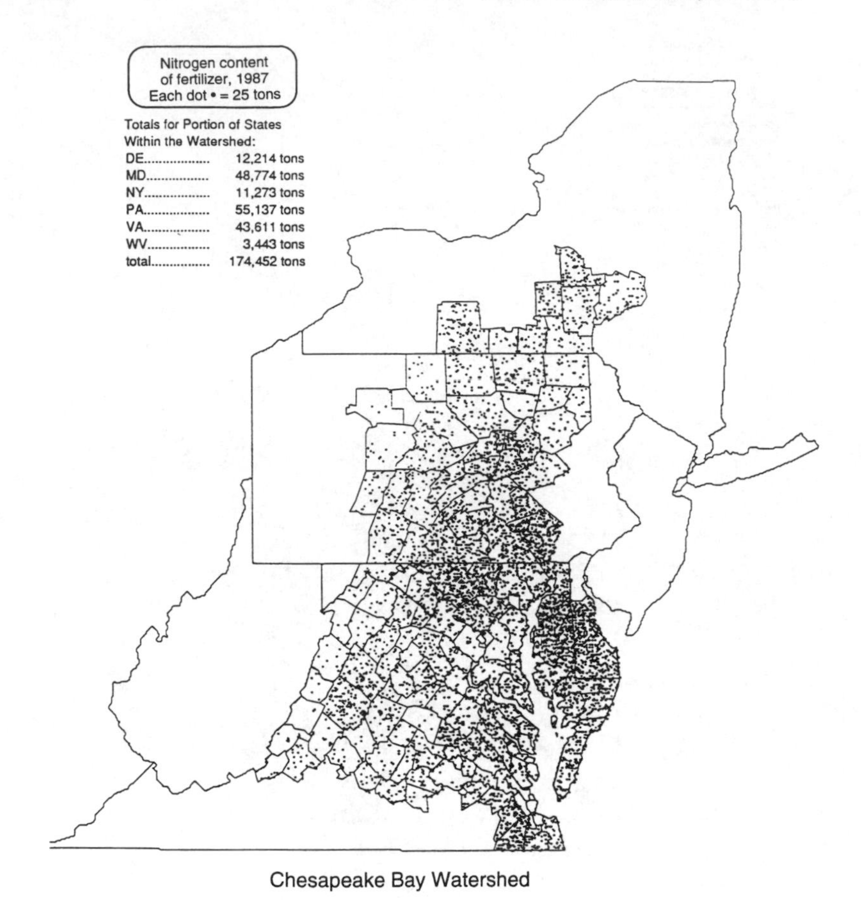

FIGURE 4.5c

arguably the most intensively studied single ecosystem in the world. It is also generally acknowledged that the Chesapeake Bay system is a "best-case scenario" of the integration of science, government, the media, and private environmental interest groups toward the goal of rational and effective ecosystem management. As such, it is worthwhile examining in some detail how this situation came to be, how well it has worked, whether it is an appropriate model for application to the management of other ecosystems, how it might continue to evolve and improve, and what major improvements still remain elusive.

Although, as we have already described, the bay has always been an ecosystem of concern for its inhabitants, a recent chronology of bay

TABLE 4.1
The Chesapeake Watershed's Population at a Glance

State or District	1940	1986	2020 (projected)
District of Columbia	636,235	626,100	626,100
Delaware	84,559	214,900	215,000
Maryland	1,706,959	4,436,800	5,496,600
New York	495,710	672,400	700,000
Pennsylvania	3,006,638	3,514,300	3,854,500
Virginia	1,530,935	4,510,900	6,229,800
West Virginia	118,617	166,900	166,900
Total watershed	7,579,653	14,142,300	17,288,900

Sources: Bureau of the Census, 1952, 1972, 1988, Maryland Office of Planning.
*Only those portions of each state within the watershed are included.

"management" should probably begin in the year 1965. In that year the U.S. Army Corps of Engineers conducted a major study of the bay, and two important pieces of federal legislation were passed: the Federal Water Pollution Control Act and the Federal Rivers and Harbors Act. By this time scientists and citizens alike had begun to notice significant declines in some key indicators. The water had become more turbid, algae blooms were occurring more frequently, and it was becoming harder to catch certain kinds of fish.

1965–1976: Era of Shared Experience and Raised Consciousness

Signs of decline are frequently ignored, but what separates the Chesapeake from other, more remote ecosystems is a broadly shared experience of the bay by a large cross section of the population. The bay was and is everyone's back yard. Government bureaucrats from Washington sail and fish on it side by side with scientists, watermen, local politicians, and everyone else. This common experience resulted in a shared commitment to protecting the ecosystem that is often missing from attempts to protect more isolated ecosystems that, however valuable, few people experience as part of their daily lives. In 1966 the Chesapeake Bay Foundation was established as a nonprofit conservation organization committed to help "Save the Bay" through public education and advocacy. "Save the Bay" bumper stickers began to appear everywhere.

Also significant was the role of the media in publicizing the bay's plight and subsequent efforts to "save" it. In 1976 William Warner published his Pulitzer Prize–winning book about crabbing on the bay, entitled *Beautiful Swimmers.* Tom Horton, then of the *Baltimore Sun,* served as a key environmental journalist in the effort to bring the bay's problems before the public eye (Horton 1989; Horton and Eichbaum 1991). His job was made easier by the fact that many in the region had observed very directly the problems he described in his reports. Another defining event occurred when Maryland State Senator Bernie Fowler waded into the waters of the Patuxent river near Broom's Island and discovered that he could no longer see his feet when chest high in water, as he could as a boy growing up in the area. Even compensating for the fact that his chest and toes were farther apart, the water quality in the Patuxent had obviously deteriorated significantly. This kind of shared experience got politicians off the dime and made the "save the bay" program much more than an academic exercise. The problems were, in any event, daunting, and included the sheer number of people living in the region. By 1970 there were 3.9 million people living in Maryland, and 12.4 million had made their home in the watershed.

1977–1983: Era of Intense Scientific Analysis with Political Backing

Once a consensus began to form that the bay was indeed in trouble, a phase of detailed applied scientific analysis was initiated. Unlike many purely scientific studies, this effort had political backing from the start and was aimed directly at providing answers to two fundamental questions: (1) what was responsible for the bay's decline and (2) what should be done about it? In 1975 Senator Mac Mathias toured the bay and requested that the EPA commence a study of the bay's problems. In 1977 a 6-year, $26-million EPA-funded study got under way. This study was unique both in its size and in its goal of supporting ecosystem management. In 1978 the Chesapeake Bay Legislative Advisory Commission (CBLAC)—an intergovernmental coordinating committee—was created to develop a method for managing the bay's resources. In 1980 the Chesapeake Bay Commission was created as an outgrowth of the CBLAC. The commission, composed of ten state legislators from Maryland and Virginia, was responsible for developing a cooperative arrangement between the two states to clean up the bay.

Meanwhile, population in the region continued to increase. By 1980 there were 4.2 million people living in Maryland, and 13.4 million in the watershed. In 1983 researchers completed the EPA-supported study, and government officials, scientists, and citizens began to consider how to implement their findings.

1983–Present: Era of Implementation and Monitoring

The next 9 years saw an era of implementation, combined with a unique effort to monitor and model the effects of implementation plans. In December of 1983, the first Chesapeake Bay Agreement was signed by Maryland, Virginia, Pennsylvania, Washington, D.C., and the EPA. The agreement, considered a compact among all the parties, committed them to establish (1) a Chesapeake Executive Council to oversee the implementation of coordinated plans; (2) an implementation committee, appointed and overseen by the Executive Council, to coordinate technical matters and evaluate specific management plans; and (3) a liaison office at the EPA's regional laboratory in Annapolis, Maryland.

The states acted quickly to begin the restoration process. Maryland, for example, in 1984 approved thirty-four legislative initiatives, focused into seven state programs known as the Maryland Chesapeake Bay Program. Among these, the Maryland Critical Area Law restricted residential and urban development in all areas within 1000 ft of tidal water. In 1985 the state of Maryland released its Restoration Plan, aimed at "restoring and maintaining the bay's ecological integrity, productivity, and beneficial uses and to protect public health." By 1987 a new Chesapeake Bay Agreement was put forward and signed by Maryland, Virginia, Pennsylvania, Washington, D.C., the EPA, and the Chesapeake Bay Commission. This agreement extended and expanded the 1983 compact and called for more precise goals to be reached by specific deadlines. (For example, a 40% reduction of nitrogen and phosphorus is to be achieved by the year 2000.) In 1988 the "Year 2020" panel released its visions for the future, detailing the environmental impacts of increased population growth within the watershed and making recommendations to avert further deterioration of the bay's water quality. By 1990 there were 4.7 million people living in Maryland, 14.5 million living in the watershed. In 1991 the Chesapeake Bay Growth and Preservation Act was presented before the Maryland legislature. This was the first bill to address statewide growth management and to focus on human population growth as the bay's number one problem. This bill did not pass, but in 1992 a much weaker growth management bill did pass the Maryland General Assembly.

It was not until the implementation era that the full extent of the bay's problems began to be realized, and in particular the necessity to treat the entire bay watershed as a system. Nutrient reductions required changes in activities in the watershed. Although point sources of nutrients, such as industries and sewage treatment plants, were relatively easy to control, it soon became evident that nonpoint sources, such as residences and agriculture, were responsible for a large part of the problem and were much more difficult to control. The growing population of the watershed itself came to be recognized as a primary cause of the bay's problems. The bay is now primed for new approaches to management that can go beyond the traditional command and control methods (which were relatively successful with point sources) to complete the implementation phase on nonpoint sources. This will be discussed in greater detail later.

Monitoring

The development of a comprehensive monitoring program was essential to moving the management of the Chesapeake forward. All estuaries are in constant flux, and the Chesapeake Bay, the nation's largest estuary, is no exception. With 150 rivers and streams feeding the bay and several thousand miles of shoreline, the bay system continually responds to nature's changing conditions: rains, tides, drought, shoreline erosion, storm runoff. The bay also responds to human activity—discharges of nutrients from waste treatment plants, fertilizer runoff from farmland, toxic chemicals in industrial outfalls, fish and shellfish harvesting. How do we distinguish between natural and human impacts on the bay? And why do we need to?

The Chesapeake's vital signs—its temperature, dissolved oxygen, pH, water clarity—are indicators of the bay's environmental health. These and other indicators, such as the presence (or absence) of underwater grasses or the numbers and diversity of fish, measured week by week, month by month, and year by year, give researchers and resource managers insight into the health of the bay by tracking changes in it.

Keeping track of these changes is the job of monitoring. To restore the health of the Chesapeake best, we need as complete an understanding of its ecology as we can get. Bay monitoring programs, funded by the state and federal government, help chart emerging trends, enabling us to see whether the ecosystem is responding to programs designed to reverse the decline of water quality or populations of fish or shellfish.

Government agencies in the bay region (e.g., the U.S. Environmental Protection Agency, the Maryland Department of the Environment, the Maryland Department of Natural Resources) are cooperating by regularly monitoring many different water quality factors, including toxic chemicals, metals, and nutrients. They are also monitoring the abundance of key fish and shellfish species, the presence of diseases in fish and shellfish, underwater grasses, and waterfowl. Especially important are samplings of dissolved oxygen and chlorophyll as indicators of whether overgrowths of algae are occurring. (Since virtually all plants, including algae, contain chlorophyll, if high levels of chlorophyll emerge, then scientists can assume that high numbers of algae are present.) Interpretation of these data can help scientists and resource managers determine whether management programs to restore living resources to the bay—such as underwater grasses, striped bass, oysters—are working or whether they need to be revised or new strategies must be attempted.

But the bay is enormously complex, and monitoring is too costly for government agencies to take all the samples necessary for a thorough understanding of the ecosystem. For this reason citizen monitoring programs in the bay are becoming increasingly helpful. Students and volunteers in citizen organizations have for several years been regularly monitoring streams and rivers throughout the watershed. With assistance from organizations like the Alliance for the Chesapeake Bay and Save Our Streams, volunteers learn how to take water quality measurements and how to record the data so that they will be reliable and useful for data managers.

Each week volunteers head for an assigned site along the shore of a stream or river, where they take samples and compile a number of measurements: pH (a measure of alkalinity or acidity), dissolved oxygen level, salinity, temperature, and turbidity (a measure of how clear the water is). These data are then entered into the Chesapeake Bay data base, along with data gathered by state and federal monitoring programs. Collected over a period of years, these measurements of water quality will enable resource managers to analyze and describe the bay's shifting ecosystem better. They will also help us to assess just how well our efforts to curb pollution are working.

Modeling the Bay and the Watershed

To begin to put the monitoring data together into an understanding of how the complex pieces of the watershed and the bay interact, we need to synthesize and integrate our knowledge in the form of models of various kinds. A model is simply an abstract representation of the system of interest that we can manipulate to aid our understanding. For example, architects build cardboard models of their projects to aid the design process. The mental picture we conjure up when someone mentions the Chesapeake Bay is also a model. More recently, hydrologists, ecologists, economists, and other scientists have begun to build mathematical models of various parts of the Chesapeake system that can be simulated on computers. These range from hydrodynamic models of the water flow dynamics in the bay itself (Dortch et al. 1988; HydroQual 1987, 1989) and other estuaries in the Chesapeake system (Hwang 1990) to models of water and nutrient runoff from the watershed (Summers 1990) and integrated models of the ecological and economic components of the system with spatial articulation (DeBellevue and Costanza 1991).

Since ecosystems are being threatened by a host of human activities, protecting and preserving them requires understanding the direct and indirect effects of human activities over long periods of time and over large areas. Computer simulations are now becoming important tools to investigate these interactions and in all areas of science. Without the sophisticated global atmospheric simulations now being done, our understanding of the potential impacts of increasing CO_2 concentrations in the atmosphere because of fossil fuel burning would be much more primitive. Computer simulations can now be used to understand not only human impacts on ecosystems, but also our economic dependence on natural ecosystem services and capital, and the interdependence between ecological and economic components of the system (see, for example, Braat and Steetskamp 1991; Costanza et al. 1990).

Several recent developments make such computer simulation modeling feasible, including the accessibility of extensive spatial and temporal data bases and advances in computer power and convenience. Computer simulation models are potentially one of our best tools to help understand the complex functions of integrated ecological economic systems like the Chesapeake watershed.

But even with the best conceivable modeling capabilities, we will always be confronted with large amounts of uncertainty about the response of the environment to human actions (Funtowicz and Ravetz 1991). Learning how to manage the environment effectively in the face of this uncertainty is critical (Costanza and Perrings 1990; Perrings 1991).

To use quantitative computer modeling effectively to understand and manage complex ecological economic systems like the Chesapeake watershed, we need an integrated, multiscale, transdisciplinary, and pluralistic approach. Moreover, we need one that also acknowledges the large remaining uncertainty inherent in modeling these systems and develops new ways to deal with this uncertainty effectively.

One important addition to our suite of models of the Chesapeake watershed is the "four-box" model of system development presented in this book's first chapter. This is discussed in more detail later.

Barriers and Bridges to Improved Management

Social Traps

A fundamental reason for system collapse and mismanagement is thought to be the inherent mismatch in many systems between the fundamental time and space scales of the ecological system, on the one hand, and the human institutions developed to manage it, on the other. When it comes to human institutions, one critical feature is the incentive structures that these institutions produce. These incentive structures often lead to behavior that is directly counter to the long-term health of the whole system, and often even to the stated goals of the institution itself. How does this happen and how can we fix it?

This process of short-run and local incentives getting out of sync with long-term and global goals has been well studied in the last decade under several rubrics (cf. Hardin 1968; Axelrod 1984), but the one we like best is John Platt's notion of "social traps" (Platt 1973; Cross and Guyer 1980; Teger 1980; Brockner and Rubin 1985; Costanza 1987). In all such cases the decision maker may be said to be "trapped" by the local conditions into making what turns out to be a bad decision viewed from a longer or wider perspective. We go through life making decisions about which path to take based largely on "road signs," the short-run, local reinforcements that we perceive most directly. These short-run reinforcements can include monetary incentives, social acceptance or admonishment, and physical pleasure or pain. In general, this strategy of following the road signs is quite effective, unless the road signs are inaccurate or misleading. In these cases we can be trapped into following a path that is ultimately detrimental because of our reliance on the road signs. For example, cigarette smoking has been a social trap because by following the short-run road signs of the pleasure and social status* associated with smoking, we embark on the road to an increased risk of earlier death from smoking-induced cancer. More important, once this road has been taken it is very difficult to change to another (as most people who have tried to quit smoking can attest).

The elimination of social traps requires intervention—the modification of the reinforcement system. Indeed, it can be argued that the proper role of a democratic government is to eliminate social traps (no more and no less) while maintaining as much individual freedom as possible. Cross and Guyer list four broad methods by which traps can be avoided or escaped from. These are education (about the long-term, distributed impacts); insurance; superordinate authority (i.e., legal systems, government, religion); and conversion of the trap to a trade-off (i.e., correcting the road signs).

Education can be used to warn people of long-term impacts that cannot be seen from the road. Examples are the warning labels now required on cigarette packages and the warnings of environmentalists about future hazardous waste problems. People can ignore warnings, however, particularly if the path seems otherwise enticing. For example, warning labels on cigarette packages have had little effect on the number of smokers.

The main problem with education as a general method of avoiding and escaping from traps is that it requires a significant time commitment on the part of individuals to learn the details of each situation. Our current society is so large and complex that we cannot expect even professionals, much less the general public, to know the details of all the traps. In addition, for education to be effective in avoiding traps involving many individuals, *all* the participants must be educated.

Governments can, of course, forbid or regulate certain actions that have been deemed socially inappropriate. The problem with this ap-

---

*This particular positive reinforcement has in the last few years begun to turn into a negative one. As smoking becomes less socially acceptable we should expect the number of new smokers to fall and many old smokers to escape the trap. But the process of escape is much more difficult than the process of avoidance.

proach is that it must be rigidly monitored and enforced, and the strong short-term incentive for individuals to try to ignore or avoid the regulations remains. A police force and legal system are very expensive to maintain, and increasing their chances of catching violators increases their costs exponentially (both the costs of maintaining a larger, better-equipped force and the cost of the loss of individual privacy and freedom).

Religion and social customs can be seen as much less expensive ways to avoid certain social traps. If a moral code of action and belief in an ultimate payment for transgressions can be deeply instilled in a person, the probability of that person's falling into the "sins" (traps) covered by the code will be greatly reduced, and with very little enforcement cost. On the other hand, the problems with religion and social customs as means of avoiding social traps are that the moral code must be relatively static to allow beliefs learned early in life to remain in force later and that it requires a relatively homogeneous community of like-minded individuals to be truly effective. This system works well in culturally homogeneous societies that are changing very slowly. In modern, heterogeneous, rapidly changing societies, religion and social customs cannot handle all the newly evolving situations, nor can they deal with the conflict between radically different cultures and belief systems.

Many trap theorists believe that the most effective method for avoiding and escaping from social traps is to turn the trap into a trade-off. This method does not run counter to our normal tendency to follow the road signs; it merely corrects the signs' inaccuracies by adding compensatory positive or negative reinforcements. A simple example illustrates how effective this method can be. Playing slot machines is a social trap because the long-term costs and benefits are inconsistent with the short-term costs and benefits. People play the machines because they expect a large short-term jackpot, whereas the machines are programmed to pay off, say, $0.80 on the dollar in the long term. People may "win" hundreds of dollars playing the slots in the short run, but if they play long enough they will certainly lose $0.20 for every dollar played. To change this trap to a trade-off, one could simply reprogram the machines so that every time a dollar was put in, $0.80 would come out. This way the short-term reinforcements ($0.80 on the dollar) are made consistent with the long-term reinforcements ($0.80 on the dollar), and only the dedicated aficionados of spinning wheels with fruit painted on them would continue to play.

The Chesapeake Bay managed to avoid many of the social traps that have afflicted many of the other ecosystems described in this book. It did this because it managed from the outset to achieve a broad dialogue between the interest groups involved and a broad consensus on the long-term goals of ecosystem restoration. This guaranteed that long-term, common goals were foremost, and short-term, narrow interest group goals did not drive the process. One way to achieve this result in other systems is to employ the methods of policy dialogue and consensus building described earlier in this chapter and in chapter 1. But even after this consensus is achieved, the system has trouble implementing its long-term goals unless these goals are effectively communicated to the narrow interest groups. Implementation of the bay management strategy is now stalled on this problem, especially for nonpoint sources. Later we describe some new suggestions for solving this problem by modifying the short-term incentive to make them more consistent with long-term goals.

Building Bridges with Incentives

In the context of the theory of social traps, the most effective way to make global and long-term goals consistent with local, private, short-term goals is somehow to change the local, private, short-term incentives to be consistent with the global and long-term goals (Costanza 1987). These incentives are any combination of the reinforcements that are important at the local level, including economic, social, and cultural incentives. The bridges are the social and economic instruments and institutions that we must design and build in order to bridge this gulf between the present and future, between the private and social, between the local and global. As we have seen in chapter 1, one of the most important inconsistencies is the underdevelopment of the "creative renewal" pathway that can lead to system collapse in the long run. To avoid this trap we need to reinforce creativity to the appropriate degree by creating the proper set of incentives.

An innovative instrument currently being researched to manage the environment under uncertainty is a *flexible environmental assurance bonding system* (Costanza and Perrings 1990). This variation of the deposit/refund system is designed to incorporate environmental criteria and uncertainty into the market and to induce positive environmental technological innovation. It works in this way: in addition to charging a firm directly for known environmental damages, an assurance bond equal to the current best estimate of the largest potential future environmental damages would be levied and kept in an interest-bearing escrow account for a predetermined length of time. In keeping with the precautionary principle, this system requires the commitment of resources now to offset the potentially catastrophic future effects of current activity. Portions of the bond (plus interest) would be returned *if and only if* the firm could demonstrate that the suspected worst-case damages had not occurred or would be less than originally assessed. If damages did occur, the portions of the bond would be used to rehabilitate or repair the environment, and possibly to compensate injured parties. By requiring the users of environmental resources to post a bond adequate to cover uncertain future environmental damages (with the possibility for refunds), the burden of proof (and the cost of the uncertainty) is shifted from the public to the resource user. At the same time, firms are not charged in any final way for uncertain future damages and can recover portions of their bond in proportion to how much better their performance is than the worst case.

Deposit/refund systems, in general, are not a new concept. They have been applied to consumer policy, conservation policy, environmental policy, and other efficiency objectives. They can be market generated or government initiated and are often performance based. For example, deposit/refund systems are currently used effectively to encourage the proper management of beverage containers and used lubricating oils (Bohm 1981).

Environmental assurance bonding would be similar to the producer-paid performance bonds often required for federal, state, or local government work. For example, the Miller Act (40 U.S.C. 270), a 1935 federal statute, requires contractors performing construction contracts for the federal government to secure performance bonds. Performance bonds provide a contractual guarantee that the principal (the entity doing work or providing service) will perform in a designated way. Often bonds are required for work done in the private sector as well.

Performance bonds are frequently posted in the form of corporate surety bonds that are licensed under various insurance laws and that, under their charter, have legal authority to act as financial guarantee for others. The unrecoverable cost of this service is usually 1% to 5% of the bond amount. However, under the Miller Act (FAR 28.203–1 and 28.203–2), any contract above a designated amount ($25,000 in the case of construction) can be backed by other types of securities, such as U.S. bonds or notes, in lieu of a bond guaranteed by a surety company. In this case the contractor provides a duly executed power of attorney and an agreement authorizing collection on the bond or notes if the contractor defaults on the contract (PRC Environmental Management 1986). If the contractor performs all the obligations specified in the contract, the securities are returned and the usual cost of the surety is avoided.

Environmental assurance bonds would work in a similar manner (by providing a contractual guarantee that the principal would perform in an environmentally benign manner) but would be levied for the current best estimate of the *largest* potential future environmental damages. In most cases, these bonds could be administered by the regulatory authority that currently manages the operation or procedure. For example, in the United States the Environmental Protection Agency could be the primary authority. In some cases or countries, however, it may be better to set up a completely independent agency to administer the bonds.

Protocol for worst-case analysis already exists within the EPA. In 1977 the U.S. Council on Environmental Quality required worst-case analysis for implementing NEPA (National Environmental Policy Act of 1969). This required the regulatory agency to consider the worst environmental consequences of an action when scientific uncertainty was involved (Fogleman 1987).

Strong economic incentives are provided by the bond to reduce pollution, to research the true costs of environmentally damaging activities, and to develop new innovative, cost-effective pollution control technologies. The bonding system is an extension of the "polluter pays principle" to the "polluter pays for uncertainty as well" or the "precautionary polluter pays principle" (4P) (Costanza and Cornwell 1992). It would allow a much more pro-active (rather than reactive) approach to environmental problems because the bond is paid up front, before the damage is done. It would tend to foster prevention rather than cleanup by unleashing the creative resources of firms on finding more environmentally benign technologies, since these technologies would also be economically attractive. Competition in the marketplace would lead to environmental improvement rather than to degradation. The bonding system would deal more appropriately with scientific uncertainty.

The 4P approach has several potential applications. We'll describe three very diverse relevant ones: growth management, toxic chemicals, and global warming. All of these are high-stakes, high-uncertainty problems for which effective management mechanisms do not currently exist.

The traditional approaches to growth management have centered on zoning and other forms of land use restrictions. While planning and zoning are better than totally uncontrolled growth, the approach leaves much to be desired, and one can certainly argue that they have not improved environmental conditions. The 4P approach suggests a flexible impact bond system in addition to regional planning. An initial impact bond would be posted by a developer that is large enough to cover the worst-case environmental and economic impacts of the proposed development. The developer would be refunded portions of the bond to the extent that the possible impacts did not occur. Innovative developers who could design projects with lower environmental impacts (e.g., by using porous surfaces or by not developing in ecologically sensitive areas) or lower economic impacts (e.g., by locating near mass transit or by helping to extend mass transit systems) would be directly rewarded by refunds of their impact bonds. Developers who defaulted on their bonds would do poorly in economic competition with their more innovative competitors, and their bonds could still be used to pay for the impacts they did cause. Impact fees have been tried before, but in most cases they have been flat, inflexible, one-time fees that offered no incentive to developers to produce anything but the standard fare. They also generally covered only a small fraction of the real impacts of the development. A flexible impact bonding system would solve these problems and help manage growth in a rational yet flexible way, without taking the right to develop away, but merely imposing the true costs of that growth on the parties that stand to gain from it, while providing strong economic incentives for them to reduce their impacts to a minimum.

Another particularly difficult environmental management problem is the control of toxic chemicals from both point sources (like factories) and from nonpoint sources (like agriculture and urban areas). Toxic chemicals can be damaging to ecosystems and human health in extremely low concentrations. Literally thousands are in common use, and there is enormous uncertainty about their cumulative and individual impacts. The standard approach is to develop lists of toxic chemicals and standards for their allowable concentrations in the environment. But because the list is so long and there is so much uncertainty about where to set the safe standards, as well as uncertainty about who is producing and releasing what quantities of which chemicals and how they interact once they come in contact in the environment, this approach has not been very effective.

The 4P approach suggests a flexible toxic chemical bonding system. Here the bond would be sized according to the best current estimates of the worst-case damages from the release of the chemicals. Refunds would be based on the extent to which each potential polluter was performing better than the worst case. This system would give polluters strong incentives to reduce their releases by recycling and more efficient use. Farmers could no longer afford to "overuse" agricultural chemicals just to be sure they were killing all pests. Industries could no longer afford to release new chemicals with poorly known impacts into the environment. Individual homeowners would pay a high price for using potentially dangerous chemicals on their lawns and would find cheaper alternatives, which, under the bonding system, would be more environmentally benign.

This system would be designed to compliment other regulatory schemes and would be self-policing, would be self-funding, and would provide strong economic incentives to correct environmental problems for which there are few good management alternatives.

Finally, the problem of global warming is probably our most severe current example of a high-stakes, high-uncertainty problem. A CO_2 tax has been proposed to deal with this problem using economic incentives, but current thinking on the tax does not incorporate uncertainty. The 4P approach suggests a CO_2 bond (with the size of the bond based on the *worst-case* estimates of the magnitude of future damages), rather than a tax (with the size of the tax based on much more uncertain estimates of *actual* future damages). In this way the efficiency advantages of economic incentives can be reaped without suffering the costs of unknown, future damages.

There are several other potential applications of the 4P system. Any situation with large true uncertainty is a likely candidate, and these situations abound in the modern world. To deal with these situations we have to broaden our understanding of science and change our approach to environmental management accordingly. We can only expect science to provide the envelope of our ignorance. Given that envelope, we must encourage policy makers to plan for the worst while providing incentives to produce the best.

Admittedly, the administrative details of how the 4P system would function need to be worked out, and these details would need to be different for the various applications of the system. These details are critical to the success of the proposal and need to be given careful consideration. There are several examples of well-intentioned performance bonding systems that have failed because the details were not given enough thought. For example, mine reclamation bonds have, in some cases, been set so low that it was cheaper for the mining companies to default than to reclaim the site. It is of prime importance that the bond be large enough to cover the worst-case damages so that malfunctions like this do not occur. We also do not wish to imply that the assurance bonding system is a general panacea that can completely replace our existing regulatory system. The bonding system is best thought of as a complement to management schemes of regulations and incentives that is better suited to handle those situations where uncertainty is high.

We think that working out which situations are most appropriate for the 4P system, and the administrative details of the system for particular applications, can be most effectively addressed via policy dialogues involving all the stakeholders affected by the system under consideration, including the environmental, business, regulatory, and scientific communities. Policy dialogue techniques involve the use of impartial third parties in structured meeting environments to bring together people with different views and find areas of consensus or agreement. Just such dialogues are currently in the planning stages, and we hope to use this process to develop detailed protocols and administrative procedures for pilot studies of the 4P system. Experimental work is also under way (Costanza and Cornwell 1991) to understand more fully the underlying incentive structures in the 4P system and to help in their effective design and use.

What can we learn from what happened in the Chesapeake? In many ways the Chesapeake was and is a "best-case scenario" for ecosystem management, but it still has a long way to go. What follows is one possible synthesis with an eye toward determining what may be extrapolated to ecosystem management problems in other areas.

1. A necessary first step for effective action is the *creation of a broad consensus* about both the essentials of the problem and common goals shared by all interest groups. In the Chesapeake this was relatively easy because a large percentage of the population had *direct experience* with the bay and could directly *perceive its decline.* There was also a broad preexisting consensus on the common goal of protecting the bay ecosystem and reversing this decline. In most ecosystem management cases,

these features are missing. There may be small groups of people who are directly affected by the decline of a particular ecosystem, but not a broad cross section, or the impacts may be so subtle and distributed that they are hard to perceive without some sophisticated tools. Achieving a consensus on a common goal across interest groups in ecosystem management is even more difficult, since in many cases interest groups are in direct conflict over management goals. Consider, for example, the spotted owl controversy in the Pacific Northwest. These general characteristics often lead to severe "social traps." It is therefore necessary in most cases to take special steps to achieve a broad consensus. One of the most effective tools in this regard is the Adaptive Environmental Management (AEM) techniques developed by Carl Walters and Buzz Holling (Walters 1986) or, using different terminology, the methods of Environmental Dispute Resolution developed by Gail Bingham and others (Bingham 1986). It is achieving this first step that often proves the biggest barrier, which, if not surmounted, can lead to the cycles of building and destroying bureaucratic structure described in chapter 1. The Chesapeake was able to avoid most of the lurking social traps by maintaining a relatively broad consensus and effective dialogue between interest groups.

2. A second step seems to be achieving broad consensus on the *details of the problem and the methods of solution.* In the Chesapeake this was effected by a large, relatively coordinated, EPA-funded study. In general, funding for detailed studies and the development of action plans is difficult to assemble. But if step 1 can be achieved, step 2 has a chance.
3. The third step of *implementation of the remedial action plans* follows directly from steps 1 and 2. A key here is holding the coalition together long enough to reach this stage and finding the resources to effect the plans. In the Chesapeake Bay (which we have already indicated is a "best-case" scenario) implementation is proceeding, but there are many rough spots. Now that the real extent of the problem and the magnitude of resources necessary to fix it are known, the coalition is showing strain. This may well be the point when things fall apart. To effect a stable solution, it is important that the short-term, local incentives that drive the system away from its long-term goals are

corrected. Pollution taxes, environmental assurance bonds, and other incentive-based instruments are effective ways to do this, as are continuing education and dialogue on the problems of the bay ecosystem and the goals of sustainable ecosystem management. If the Chesapeake can successfully implement this final stage, it will truly be a model of sustainable ecosystem management worthy of emulation.

Acknowledgments

We'd like to thank the thousands of scientists, environmental managers, politicians, and citizens who have worked to restore the Chesapeake Bay. We hope that the final steps can be taken to ensure a sustainable ecosystem. In particular, we'd like to thank W. Boynton, M. Kemp, J. Bartholomew, and J. Barnes for helpful comments on earlier drafts.

5

Deliberately Seeking Sustainability in the Columbia River Basin

Kai N. Lee

To succeed, sustainable development must originate in political choice and be carried into institutional transformation. This chapter explores the question of weaving sustainability into the institutional fabric of a large ecosystem, the Columbia River Basin in the Pacific Northwest region of the United States. Crises in that region's political economy in the 1970s have prompted an ambitious attempt to rebuild salmon populations in the midst of the largest hydropower system in the world. An institutional structure that realigns but does not supersede existing authorities is emerging, together with a shared perception of the possibilities and conflicts implicit in managing resources whose requirements are partly incompatible. The goal is an ecologically sustainable salmon population coexisting with an economically sustainable hydropower system. An optimist sees in the still incomplete story of the Columbia basin a social system searching for a path to that goal of dual sustainability; a pessimist sees resistance to the changes needed before sustainability can be realized.[1] This chapter takes the view of an analyst looking for larger lessons from a case pregnant with possibilities.

Rising in the Canadian Rocky Mountains and flowing 1200 miles through the Pacific Northwest, the Columbia is the fourth largest river in North America, draining an area that includes parts of seven U.S.

states and two Canadian provinces (figure 5.1). The river's average annual stream flow of 141 million acre-ft is more than ten times that of the Colorado (Kahrl 1978). The Columbia's high flows and extensive drainage have made it ideal for colonization, first, by fish and wildlife (Wilkinson and Conner 1983), as the glaciers retreated at the end of the last ice age and, much later, by dam-building humans.

Well into the nineteenth century the Columbia River Basin was a wilderness. Because it is a major spawning ground and nursery of the

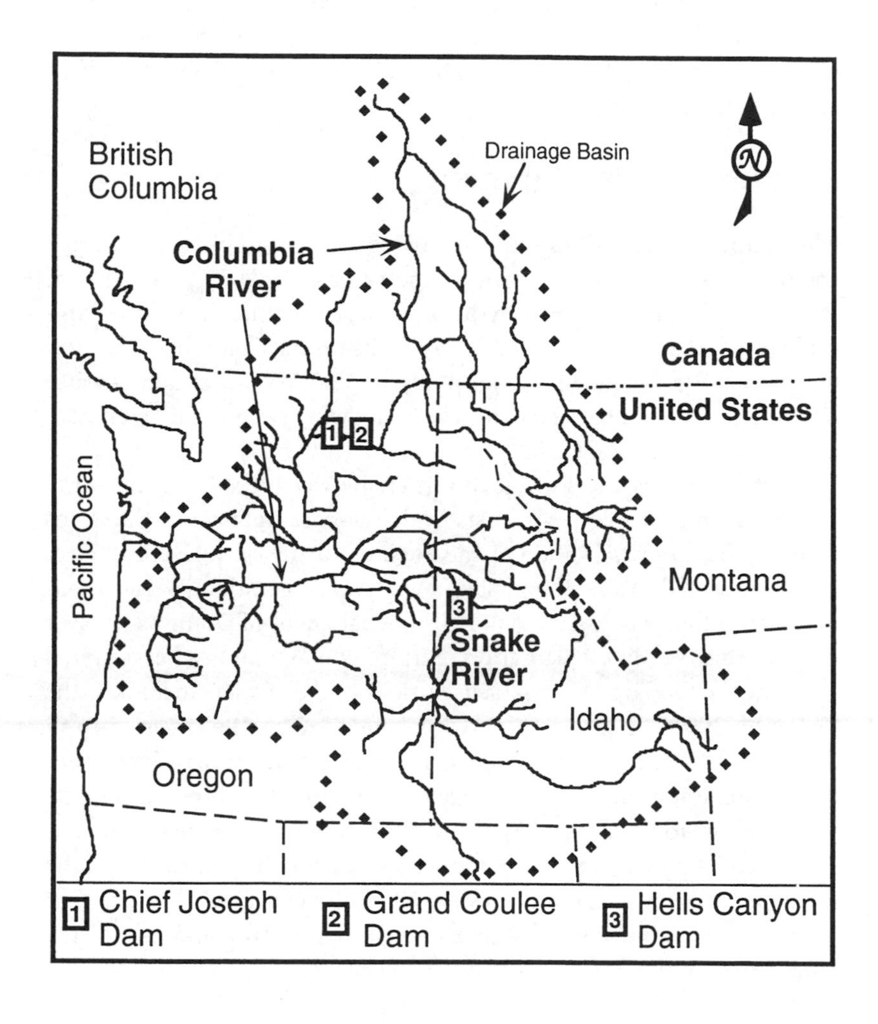

FIGURE 5.1
Drainage basin of the Columbia River (Lee 1993).

Pacific salmon (*Oncorhynchus* spp.), the Columbia's biological web reaches far into the North Pacific Ocean, where the fish mature for 2 to 4 years before returning to their native streams to reproduce. Before European settlement, this ecosystem supported a population of perhaps 50,000 Native Americans (Schalk 1986), whose world centered on the yearly migrations that brought 10–16 million salmon back to the river (Northwest Power Planning Council 1987). Harvested by spear, net, and boat, these fish provided both food and trade goods for the people of the river basin. The Native American tribes lived in a long-run ecological equilibrium, which fluctuated between bad times and good, but endured over many human generations. This original Columbia civilization lasted until about 1850.

Industrial Development

The second human civilization to invade the Columbia basin turned the river into a factory. The basin's nineteen major dams, together with more than five dozen smaller hydroprojects, constitute the world's largest hydroelectric power system. Today dams on the Columbia River and its tributaries generate on average about 12,000 MW from falling water (Northwest Power Planning Council 1991), which is more than enough power to run New York City.

Built largely by the U.S. government between 1930 and the early 1970s—a time of low labor costs and low-cost financing—the dams fostered the industrialization of the Pacific Northwest with cheap electricity marketed by the Bonneville Power Administration (BPA), a federal agency that is now part of the U.S. Department of Energy. The river basin has also become a plantation of more than 3 million acres, watered by some of the largest irrigation works on the planet, including the Columbia Basin Project anchored at Grand Coulee, the largest dam in the United States. Industrial and agricultural development have built the population centers of the Northwest: Portland and the Willamette Valley of Oregon, Boise and Spokane in the upper watershed, as well as Seattle and Puget Sound. BPA remains the economic keystone of the regional economy, and the agency's power sales contracts, together with the water rights that control where water flows on croplands, shape the landscapes of the Pacific Northwest about as decisively as does the weather.

The Native Americans who lived in wilderness have given way to a population of 9 million, more than 100 times the aboriginal level. That increase in population, by two orders of magnitude, reflects a fundamental change in the relationships between people and the environment. The domesticated river provides power and irrigation while also eliminating its once legendary floods; serving as an inland waterway navigable by tug and barge for 500 miles from Astoria, Oregon, near the river's mouth, to Lewiston in central Idaho; affording world-class windsurfing in the Columbia Gorge; and, last but not least, supporting sport and commercial harvest of salmon and other fish and wildlife. Development's dominant theme has been economically efficient management through engineered control.

The industrial Columbia is a multiple-purpose marvel, the product of a national government that saw its central role as the control of nature for economic ends. President Franklin Roosevelt called that role the New Deal, and it was carried forward by Senator Henry M. Jackson and other regional leaders for two generations, transforming the landscape and the people who lived on it. The Columbia became a river, as the historian Donald Worster (1985) put it, that died and was reborn as money,[2] its many functions ranked by their economic value: power first, then urban and industrial uses, agriculture, flood control, navigation, recreation, and finally fish and wildlife. The inferior position of fish and wildlife is evident in the decline of the annual fish runs of 10–16 million in the preindustrial era to 2.5 million by the late 1970s.

As these numbers imply, however, the Columbia has not died entirely. Thus there is hope that, just as the wilderness gave way to the power plant, so a new Columbia may rise, a river whose watchword will be *sustainable* multiple use.

Changing the Rules

Sustainable multiple use is not just a hope. Three intertwined crises have redrawn the rules by which humans attempt to govern the Columbia. Beginning in 1969, Indian tribes of the Pacific Northwest have reasserted their legal rights to harvest fish, under treaties concluded with the U.S. government in the mid-nineteenth century. Today more than $100 million is invested annually in fish and wildlife mitigation. The second change is a large shift in the price of electric energy—crystallized in a crisis over the development of nuclear power—that has raised the importance of husbanding the river's low-cost electricity by innovative and successful efforts to improve energy efficiency. Third, environmental awareness among the voters of the Pacific Northwest—a by-product of the rapid urbanization[3] brought by the harnessing of the Columbia's riches—has supported innovations in institutional relations unimaginable under the industrial order. The changes in fisheries, energy, and political consciousness came together in the Northwest Power Act of 1980, a statute—sponsored by Henry Jackson, a leader of the industrial river basin—that has spurred a search, still in progress, for a new, perhaps sustainable Columbia, a place that would be neither wilderness nor power plant, but an ecosystem requiring active management.

The wilderness the Native Americans knew is gone. Their world was an integral fabric whose natural time scale was the human generation. That cloth has been cut; management by preservation, permitting nature to set the terms on which its constituent species will equilibrate, is no longer possible. Some have questioned whether management by preservation is possible today, even in unpeopled parks and biological preserves (Chase 1987). Yet following the profit motive to its logical end point by increasing energy production as long as its revenues outbid the competing claims of irrigation and other uses is unacceptable. A sustainable Columbia River implies a culturally, economically, and ecologically viable relationship between people and the environment they inhabit. Sustainability would be likely to yield less than the maximum achievable short-run profit, and it would be likely to involve humans in the landscape more than is contemplated in the popular notion of pristine wilderness (Tietenberg 1992).

These statements of what sustainability is not do not specify what sustainability is, nor how to get there. That is why the Columbia basin is searching for a sustainable balance between the electricity that is its most economically important resource and the salmon that are the most emotionally compelling symbol of its natural integrity. The search involves both policy and politics, a combination called *social learning* in the following discussion, in which an idealistic approach to science is combined with a pragmatic approach to politics. It is, inevitably, an uneasy combination.

A Search for Sustainable Management

In 1980 the U.S. Congress passed the Northwest Power Act. This legislation recognized the three crises of fish, power, and environmentalism and created a public arena in which those questions would be worked on. The act was designed to solve a set of social problems by technological means. As demand for power grew during the 1970s, more power plants seemed necessary to utilities. They proposed federal legislation to enable them to finance more plants in 1977. However, citizen activists, whose voices were growing steadily in power and influence, argued that energy conservation could meet the demand for power at lower environmental and economic cost. The search for compromise took more than 2 years. Toward the end of that search, the Indian tribes and fishermen who had fought over the salmon made common cause, demanding that the damage to the Columbia's fish runs be repaired. Rather than choosing among these partially conflicting claims, Congress sought to accommodate them all. The result was a complex law, whose implementation has taken turns unanticipated by those who fashioned its compromises.

To the utilities the major challenge was to build new generating plants to augment the limited supply of hydropower. Pressed by the Bonneville Power Administration during the 1970s, utilities launched five nuclear power plants sponsored by the Washington Public Power Supply System (WPPSS), a public utility consortium based in Washington state but drawing upon the creditworthiness of more than 100 utilities throughout the Northwest. The Northwest Power Act was intended to buttress these arrangements, but it came too late. Plagued by cost overruns, high interest rates, and swiftly falling demand for power in the 1980s, WPPSS completed one plant, mothballed two, and canceled the other two.

Even as Congress hurried to preserve the low-cost power of the Northwest, the costs of new power plants began to come due,[4] and rates increased rapidly. The rise was all the more dramatic because of the low historical base from which it started. From 1979 to 1984 the Bonneville wholesale rate increased more than 700%, and the retail price of power followed, more than doubling on average. At the same time, high interest rates, together with a worldwide economic slowdown triggered by the oil crisis of 1979, depressed the Northwest economy, hitting its energy-intensive industries with gale force. In the rural hinterlands, layoffs and skyrocketing utility bills stirred rebellion.

By 1982, as the Northwest Power Act was in the early stages of implementation, the expected power shortage that had motivated its enactment had evaporated. Demand was far below expectations because the economy was in recession. With rates rising rapidly, conservation gained plausibility. Instead of a deficit, there was a surplus of power through the 1980s. Instead of a financing mechanism to build new power plants, the Northwest Power Act became the blueprint for a laboratory of energy and environmental conservation. By the end of the decade energy conservation had produced the equivalent of a small coal-fired power plant at roughly half the cost. More important, the success of conservation meant that official plans to meet growing electricity demand now took energy efficiency as the preferred alternative.

The claims of Indian tribes posed another threat to the region's industrial order. In 1855 treaties between the U.S. government and the Northwest's Native Americans created reservations within which the native peoples agreed to live while retaining rights to fish, hunt, and gather roots and plants over a territory well beyond the reservation boundaries "in common with" the settlers. That language would reverberate more than a century later. At the time it seemed not to concede much—it afforded the Indians their traditional livelihood, and there was plenty to share "in common."

Beginning in 1969, after the settlers and their descendants had transformed the landscape and obliterated many of the fish runs, the Northwest tribes filed, and won, lawsuits to claim their treaty rights. The immediate result was to reallocate shares of the salmon harvest, since "in common with" meant that Native Americans were entitled to harvest half the fish. Such a drastic and sudden curtailment of a fishing industry already in decline struck hard at commercial and sport fisheries that had ignored the Indians since the treaties were signed. After a decade of hard feelings, as the lawsuits made their way to the Supreme Court—where the treaty claims were affirmed—both non-Indian and tribal leaders realized that there was only one option all could abide: to rebuild the salmon populations so that there would once again be enough for all to take "in common" without battling one another for the right to kill off the stocks forever. Although this reality was articulated by Tim Wapato, the politically astute negotiator who led the Native Americans of the Columbia basin, no one knew how to rebuild the salmon runs; it was clearly going to be expensive, however.

The Planning Council

The government could legislate and tax, but it could not make kilowatts or fish. A law could not solve the problems of salmon or power, but it could arrange for their solution over time. The Northwest Power Act used a familiar strategy of governance, defining a new process, so that an array of choices could be made without further appeals to Congress or the courts. So far the strategy has been successful: After an initial flurry of litigation over the meaning of the act for power sales contracts and other legal matters, judicial activity has ceased, and despite concerns over endangered salmon, there is little prospect of new legislation.

The centerpiece of the new process is the Northwest Power Planning Council and the two plans it has promulgated—each several times—with wide public involvement. Chartered by the four Pacific Northwest states of Idaho, Montana, Oregon, and Washington, the council is composed of two members from each state, appointed by the governor under procedures established by state law. Under some circumstances the council has the unusual authority to restrict or redirect the actions of federal agencies. The council is in effect an interstate compact, a form of government organization that shares both state and federal authority.

Power Planning

The council's primary task is to formulate a plan to guide electric power development, including energy conservation. Three versions of the plan have been issued, the most recent in 1991. The plan's central premise is regional cost effectiveness, planning that minimizes costs across the Pacific Northwest's many utilities, a rule that the fragmented industry would not naturally follow. As noted earlier, the plans have induced investment in energy conservation, making this approach credible and feasible among the region's utilities, regulators, and energy consumers. About 300 MW of new demand has been met by existing supplies, because that quantity of existing demand has been eliminated by an investment of $600 million in energy-efficient technology; the cost ratio is about half of that required to build a coal-fired power plant in this size range.

The acceptance of the power plans has had a more subtle cultural influence, raising the legitimacy of a rational response to the uncertainties brought by large-scale change in the cost of energy. Rising costs affected different utilities in different ways, because power from the Columbia's dams was a large fraction of operating expenses in some cases but not in others. Some utilities were growing rapidly, and thus needed additional supplies of power, whereas others were not. Institutional fragmentation would have been the natural outcome of this situation, and all utility managements have moved toward greater vigilance of their own self-interest in dealings with BPA and other utilities. The presence of the council and its regional planning process has made regional information a public resource, however. Information in the form of demand forecasts and publicly debated plans to offer energy efficiency programs or to build generating resources has made it possible for utilities and citizens to estimate their own self-interest in ways that were not feasible before. In an industry in which the economic fate of nominally independent entities is actually coupled together by investments in transmission networks, dams, and other large facilities, making decisions in public and on a rational basis is helping to redefine the social function of utilities and the energy they provide.

The Columbia River Basin Fish and Wildlife Program

In an attempt to address the disruptive potential of Indian treaty rights litigation, Congress included in the Northwest Power Act directives meant to provide fish and wildlife "equitable treatment" in comparison to hydropower. In response, the Northwest Power Planning Council adopted an ambitious Columbia River Basin Fish and Wildlife Program in 1982, subsequently amended in 1984, 1987, and 1992, calling for a broad spectrum of mitigating activities.

Implementation of the council's program is funded from revenues that the Bonneville Power Administration charges its electric power ratepayers. The council has determined that losses of salmon and steelhead due to hydropower amount to between 5 and 11 million adult fish per year (Northwest Power Planning Council 1987). The result is an effort to rehabilitate fish and wildlife on an economic scale unheard of in natural resource management, with an economic cost of over $130 million per year. Since its reformulation in 1987 into a program that is explicitly "systemwide" in conception, the Columbia basin program has organized mitigation activities into the three principal points at which human activities intersect the salmon life cycle—at the point of harvest in the oceans and rivers, with hatcheries and habitat conservation at the time of birth, and during migration through the dams, when the fish are juveniles bound for the sea or are returning as adults.

Harvest of Pacific salmon is now being regulated by the states and tribes of the Pacific Northwest and by Canadian and U.S. governments, both to conserve and rebuild fish stocks and to assure fair apportionment of the catch. The regulations, determined annually, implement the terms of a treaty between the United States and Canada signed in 1985, as well as the treaties governing relations between Indian tribes and the U.S. government.

Enhanced production of fish is in progress, by artificial means, such as hatcheries, and via protection and improvement of natural spawning grounds. Several new hatcheries will be built in the basin to supplement the more than 100 artificial production facilities now operated by state and federal governments. The existing hatcheries are mostly in the lower river, below the traditional fishing grounds of the Columbia basin Indian tribes; the new facilities will be upstream, where traditional tribal harvest sites are located. The present intention is to use the hatcheries to raise fish only until they can survive in the wild. Juveniles would then be put into streams where they would imprint the smell of their adopted waters at the time of migration. This way the fish should return as adults to these streams, rather than to the hatchery. If enough adults do so, a natural spawning run will be reestablished, independent of the hatchery.

Natural spawning habitat is being improved primarily by reopening fish passages that were blocked by earlier human usage. For example, in the Wenatchee River of eastern Washington, Dryden and Tumwater Falls dams, originally built without fish ladders, have been equipped with passage so that migrating adults can now reach the habitat blocked by the dams.

The council has also identified 40,000 stream miles of "protected areas" where small hydroelectric projects should not be built. The council advises the state and federal agencies responsible for hydroproject licensing, particularly the Federal Energy Regulatory Commission, which is directed by statute to take the council's program "into account at each relevant stage of the decision-making processes to the fullest extent practicable" [U.S. Congress 1980, sec. 4(h)(11)(A)]. As a result, even though the council has no explicit regulatory authority over landowners, its legal influence over the federal commission effectively protects salmon habitat against hydropower development, a leading threat.

The most costly and controversial elements of the Columbia basin program are intended to enhance the upstream and downstream migration of anadromous fish in the main stem of the Columbia and its principal tributary, the Snake. In some years more than 80% of migrants from the upper Snake and Columbia are individually marked before being transported in barges to the river's estuary. In an effort to protect the fish that are not transported, Congress has also appropriated more than $30 million annually to install screens and carve bypass channels in dams to deflect young fish away from power turbines. Most ambitious of all, perhaps, the river's flow has been altered to benefit fish migration, at an annual cost of more than $40 million in lost power revenues. The revenues are diminished because water is released in the spring and summer, when it benefits fish, rather than being held back until autumn or winter, when higher prices can be obtained for the power.

The key measure, known as the water budget, re-creates the spring snow melt or freshet, providing a substantial volume of water to flush migrating juveniles to the sea. The water budget is a more generous compromise for the Columbia than for the Snake, because the upper Columbia discharges more water and has substantially more storage in its upstream dams. In practice, even the water dedicated to the water budget in the Snake River drainage has often been unavailable. In 1987 the council staff analyzed the relative abundance of salmon stocks in the Columbia basin and discovered five that were so depressed that their biological viability was clearly in jeopardy (Nehlsen, Williams, and Lichatowich 1991). Four of the five were in the upper Snake River drainage. By 1990 all five stocks were the subjects of petitions for listing as endangered species under the federal Endangered Species Act (Volkman 1992). In response, the council convened a year-long "salmon summit" that tried to negotiate a consensus approach to benefit stocks under extreme pressure. The result, adopted by the council in 1992 as an amendment to its fish and and wildlife program, forms the starting point of a recovery plan now being formulated by the National Marine Fisheries Service under the provisions of the Endangered Species Act.

Biological Uncertainty

The policy problems of the water budget illustrate the problem of searching for a sustainable way to manage the river. The biological benefits of the water budget are hard to see, in part because of its small size in comparison to natural fluctuations. Figure 5.2 shows a compilation of data from the Snake River on the relation between the volume of river flow and average travel time for migrating juvenile fish. This handful of measurements—the black dots in figure 5.2—constitutes the principal justification for losing $40 million per year in power revenues.

Each data point represents a measurement of how fast a typical juvenile salmon moves down the river, depending on how much water the river is carrying. Like riders on an escalator, the fish should go faster when the river flows faster—that is, when the flow level is higher. Therefore the data points should trend down to the right; the higher the flow, the shorter the travel time. The downward-pointing straight solid line

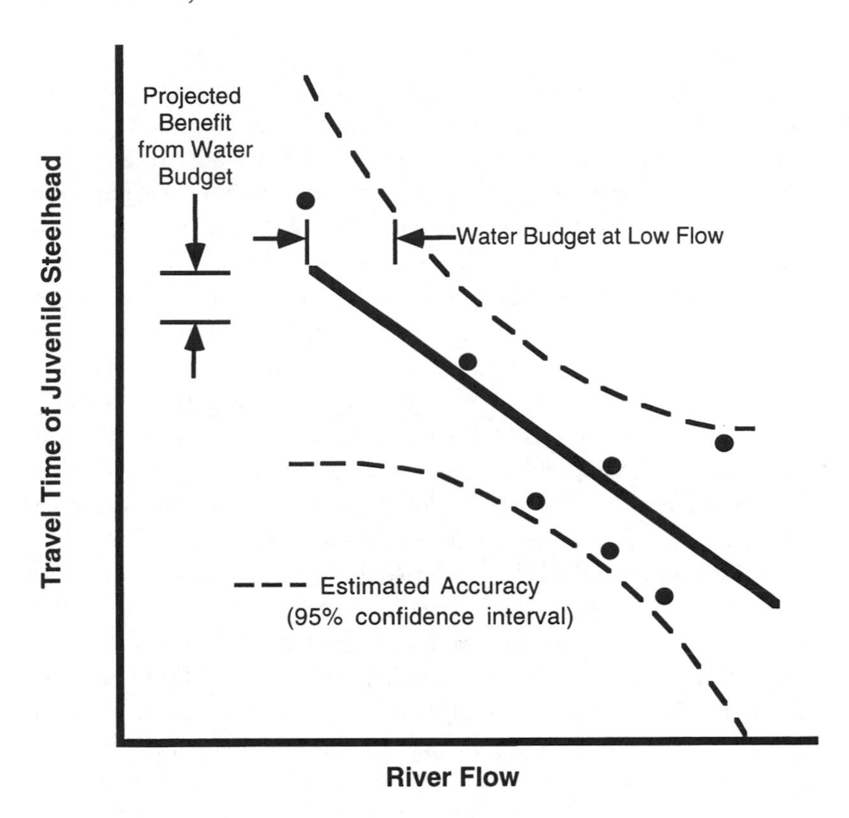

FIGURE 5.2
Effect of the Snake River water budget on juvenile salmon migration. Both vertical and horizontal scales are logarithmic. For consistency, the flow at Ice Harbor Dam is used as the standard measure of flow. The water budget in the Snake River (horizontal arrows) is superimposed on the measured relationship between river flow and migratory fish travel time. The projected reduction in travel time (vertical arrows) as a result of the water budget lies well within the measurement error (curved lines). Therefore the effectiveness of the water budget in speeding juvenile fish downstream is difficult to prove. (Northwest Power Planning Council, after Sims and Ossiander.)

represents the spring runoff in quantitative terms—the biological benefit, measured in reduced travel time, plotted against economic cost, measured in river flow. The water budget, shown as an increase in flow in the figure, should therefore produce a biological benefit.

The concept is straightforward, but there are complications. First, there are few measurements. Each point represents a sizable investment of research time and technical effort—marking, releasing, and recapturing thousands of migrating juvenile fish at dams hundreds of miles apart, to see how long it takes them to migrate downstream. At most a single data point is measured in a year, not only because of cost, but because different levels of river flow are needed to see how fish and flow relate to each other. Each year one sees no more than one Columbia—a wet one this year, perhaps, a drier one next, and so on.

Second, although the relationship shows the expected trend—the higher the flow, the lower the travel time—the observations do not lie along a single line; there are fluctuations and "noise" in the data. The relationship between travel time and river flow is affected by other factors, many of them unmeasured. For example, the condition of the fish when they start migrating can make a large difference, but trying to pinpoint the health and readiness of thousands of fish the size of a human finger is too expensive and time-consuming to be practical.

Third, the uncertainty is large relative to the size of the water budget. The effect of the water budget is largest when the flow is lowest, at the left side of the graph, because the effect of a fixed volume of water is largest when the underlying flow is lowest. Yet even under these conditions, the *change* brought about by the water budget lies well within the dotted lines, indicating the range of uncertainty in the available data.[5] The biological effectiveness of the water budget, even if it were fully implemented, would be difficult to observe, even in low-water years, when one would expect it to be most helpful.

For these reasons the idea of dramatically increasing the volume of flow of the Snake River during the salmon migration season was resisted fiercely by utilities and farmers whose irrigation waters would be drained at the start of the growing season. Instead, the new salmon strategy devised by the council experiments with an operating method called "drawdown": The reservoirs behind several dams are lowered, forcing the river to flow in a narrower channel; a smaller augmentation of the volume of flow is then able to produce a higher flow velocity. The drawdown method is now being studied to see how well it works.

Budgeting for Conservation and Fish

Oddly, it is easier to pay for sustainable management of the Columbia because of the failure of nuclear power in the Pacific Northwest. The wholesale cost of electric power soared more than 700% in the early

1980s (Bonneville Power Administration 1988), largely to pay for nuclear plants never completed. Thus the revenue stream is much larger than anticipated, and the percentage needed to pay for conservation and fisheries has been relatively small. The cost of the fish and wildlife program consumes about 1.5% of the Bonneville Power Administration's annual budget of roughly $3 billion.

The search for sustainability in the Columbia basin proceeds, accordingly, under conditions where budgetary limitations are only a secondary consideration. This condition is clearly unrepresentative of attempts to carry out sustainable development generally, so the Columbia River case should be regarded as no more than a proof of principle—a demonstration that the serious pursuit of sustainability can be launched.

A problem in the Columbia that is more representative is the large number of hands on the steering wheel. The Columbia River Basin Fish and Wildlife Program is implemented or significantly influenced by eleven state and federal agencies, thirteen Indian tribes, eight utilities that operate major hydroelectric projects in the Columbia drainage, and numerous organized interests ranging from agricultural groups anxious to protect water rights to flyfishers impatient for the return of wild fish stocks. If the river is to revive in any sustainable sense, it will have to be managed with a stability, durability, and awareness of biology rare in human affairs.

Mindful of the complex institutional repercussions of the changes it was making, the council adopted the concept of adaptive management in its fish and wildlife program in 1984, and expanded the idea into a process called *system* planning in 1987. System planning was intended to institute an experimental approach to implementation. The recasting of the search for sustainability brought about by the petitions under the Endangered Species Act has made clear the fragility of an experimental approach and the central role of political conflict in social learning.

Social Learning

The Columbia basin experience illuminates the *social learning* that is needed to search for sustainable development. Today humans do not know how to achieve an environmentally sustainable economy. If we are to learn how, we shall need two complementary sorts of education. First, we need to understand far better the relationship between humans and nature, that is, *adaptive management*—treating economic uses of nature as experiments, so that we may learn efficiently from experience.

Second, we need to grasp far more wisely the relationships among people. One name for such a learning process is politics; another is conflict. We need institutions that can sustain civilization now and in the future. Building them requires conflict, because the fundamental interests of industrial society are under challenge. But conflict must be limited because unbounded strife will destroy the material foundations of those interests, leaving all in poverty. Bounded conflict is politics.

This combination of adaptive management and political change is *social learning.* Social learning explores the human niche in the natural world as rapidly as knowledge can be gained, on terms that are governable, though not always orderly. It expands our awareness of effects across scales of space, time, and function. For example, we pump crude oil from deep within the earth and ship it across oceans; we burn in a minute gasoline that took millennia to form; with petroleum and its end products we foul water, soil, and air, overloading their biological capacity. Human action affects the natural world in ways we do not sense, expect, or control. Learning how to do all three lies at the center of a sustainable economy.

Adaptive Management

There are two critical elements in the transition to sustainability: biological uncertainty and institutional complexity. In seeking a path from the unsustainable vitality of industrialism to a sustainable order, learning from experience is the only practical approach. Without signposts the path to sustainability is easily lost. Consider some of the difficulties on the way (Hilborn 1987):

- *Data are sparse.* It is difficult to observe the state of the ecological system and the human economy interacting with it. Measurements of the natural world, such as the size of migrating populations, are inexact at best, and natural systems often yield only one data point per year (e.g., river flow).
- *Theory is limited.* Reliable observations are few, and theories of natural environments do not permit deductive logic to extrapolate very far from experience. Also, the perturbations caused by humans are frequently both large and unprecedented in natural history, so that it is unclear what theory is applicable.

- *Surprise becomes unexceptional.* With limited theory comes poor knowledge of the limitations of theory. Predictions are often wrong, expectations unfulfilled, and warnings hollow.

A general strategy has been devised to deal with natural resources under these conditions. The approach is called *adaptive management,* a term coined by C. S. Holling and co-workers at the International Institute for Applied Systems Analysis (Holling 1978; Walters 1986). Their work is built on a simple, elegant idea: If human understanding of nature is imperfect, then human interactions with nature should be experimental. That is, policies should be designed and implemented as experiments probing the behavior of the natural system. Experiments often surprise and scientists learn from surprises. So if resource management is thought of from the outset as an experiment, surprises are opportunities to learn rather than failures to predict. Adaptive management holds the hope that, by learning from experience, one can reach and maintain a managed equilibrium efficiently, with a resilience able to persevere in the face of surprise (Clark and Munn 1986).

Adaptive management originates in a comprehensive ecosystem perspective, in which the interactions among the components of the natural environment are highly structured, and the behavior of the system as a whole is consequently rich in surprise. Proceeding from a base of careful observations, experimental interventions into this interacting system provide insights into its dynamic character—insights, such as the long-standing belief that diversity reinforces stability, that are helpful even when they are not universally valid, useful even when one cannot rely implicitly on their quantitative implications. The adaptive perspective begins from a scientific viewpoint, and its continuation into the realm of action is informed more by the observational interest of a naturalist or astronomer than by the manipulative tendency of the engineer or entrepreneur.

Adaptive management is ecologically rooted in two more specific ways. First, the adaptive perspective is linked to biological time scales, because the effects of experimentation on a population often become visible only when measured over generations. For salmon this implies times of 5 years or more—a long interval in a governmental world, where senior policy officials serve terms shorter than the salmon life span. Second, the adaptive approach focuses on populations, not individuals. Failures are often fatal for individuals but rarely for populations.

There is, accordingly, a greater willingness to experiment when the unit of concern is the population.

Even if whole populations are being managed, however, the decisions are made by individuals. Put into governmental terms, a policy maker who regards each choice as an opportunity to succeed or fail may be reluctant to venture into the chancy—if realistic—terrain of adaptive management. Though the theory emphasizes the value of learning from failure, it requires individuals with a high tolerance for risk to carry it out. As in economics, where the theory emphasizes the benefits of competition, the risks facing individuals can be imposing.

Although virtually all policy designs take into account feedback from action, the idea of using a deliberately experimental design, paying attention to the choice of controls and the statistical power needed to test hypotheses, is one rarely articulated and usually honored in the breach. It is for this reason that the explicit adoption of an adaptive policy in the Columbia River Basin is noteworthy.

Negotiating Consensus

Adaptive management responds to biological uncertainty, but it is not clear how the adaptive approach can work in the presence of institutional complexity. That many interests have stakes in the transition to sustainability is hardly surprising, but finding and maintaining a balance among disparate and often noncomparable considerations, such as irrigation and tourism, is evidently a political task, one that may not be consistent with the rational pursuit of knowledge through adaptive management.

Because control over large ecosystems is fragmented, the search for a sustainable economy requires extensive social interaction: sharing analytical information, such as simulation models and data bases; identifying trade-offs and coalitions for joint action; and learning from surprising outcomes. These interactions are ways to negotiate shared agendas that individual organizations cannot achieve by themselves.

The central role of negotiation emerges from the surprising blend of technocracy and consensus building that has gained visibility and favor among natural resource managers over the past decade (Amy 1987). In cases previously characterized by lengthy litigation and embittered conflict, informal negotiations have produced plans of action acceptable to traditional adversaries: tribes and state governments, environmentalists and developers, and resource managers and harvesters. Although wary of advocacy in the guise of science, the parties have found it possible to use technical analyses and have invented measures to assure the political and scientific credibility of analysts and their findings. The negotiated agreements have included joint oversight mechanisms, because unforeseen circumstances are to be expected during implementation. As a social process, the negotiations have sought to achieve and maintain the measure of consensus necessary for experiential learning to occur. Thus consensus-building negotiations have created the open political environment that is necessary for adaptive management.[6]

Consensus building is central to sustainable development as well, because the natural systems being managed cross the spatial and functional boundaries of existing human institutions. Without a comprehensive perspective, the fragmentation of jurisdictions promotes abuse of the environment, because individual institutions seek to achieve purposes that often turn out to be incompatible with sustainable use of the whole; this is the tragedy of the commons (Hardin 1968). Yet implementation requires a decentralized, fragmented perspective, because decisions are carried out by parties whose responsibilities are narrow compared to the breadth of the analytical tools used by planners (Baskerville 1988).

The complexity of both human and natural systems is high enough to outrun anyone's ability to command from a central vantage. Building consensus by negotiation can link central perception to decentralized action. Consensus may also improve long-range plans to rehabilitate ecosystems.

Remedial actions require consensus when they encounter problems of economic sustainability. Damages from past actions are a sunk cost: The value of the resource has been taken by the exploiter and is no longer available to pay for remediation. The damaged ecosystem also contains hidden opportunities, since the ability of natural systems to recuperate is often uncertain. In that circumstance strict cost–benefit estimates are likely to undervalue the worth of rehabilitating the ecosystem, especially if it is difficult or impossible to fund rehabilitation from the profits of exploitation (Regier and Baskerville 1986). When these conditions occur, a negotiated consensus reflecting a mandate for rehabilitation is needed to justify expenditures. Moreover, past damages may have altered the political environment by driving out a group of resource users (such as the Native Americans of the Columbia basin). In that event, rebuilding a sustainable suite of uses may require that points of view silenced by earlier misuse be actively sought out. Consensus building has been strategically important in the Columbia River Basin, where a central agent finances decentralized actions, no one of which meets a narrow cost/benefit test, even though their cumulative impact may be economically sound.

A consensus that fosters learning both facilitates and benefits from an open political setting. By lowering the barriers to participation and, in effect, organizing their own political environment, planners can negotiate and sustain a pluralist, competitive political setting in which disparate considerations can continue to be weighed as learning goes forward.

Planning and Politics

This consensus-building approach can be seen in the work of the Northwest Power Planning Council (Evans and Hemmingway 1984). Because important matters are at stake in development projects, a wide spectrum of interests is motivated to participate in the planning. Barriers to participation should be low at the outset and can be kept low by the planners. Established relationships are usually weak at the beginning of a development project, and because there is often substantial uncertainty about how the links among different interests will be changed, it does not require much previous experience to become an effective player. Where external support is important to the implementation of the plan, however, planning must turn to the outside world.

Backed by a legal mandate to keep the public informed and involved, the council lowered barriers to participation and judged its success by its credibility with the public. The council's first chairman, Dan Evans, a popular and well-known figure who had served three terms as governor of Washington state, led the way with an open political style. Evans made a special effort to approach the Native American tribes, whose legal battles on fishing rights began to be fought while he was governor. More generally, the council approached organizational and opinion leaders both in and out of government and consciously developed a constituency for implementation of its energy plan and fish and wildlife program. Support for the council came almost entirely from organized groups because of the complexity of the council's plans. Despite the wide popularity of efforts to protect and enhance the fish runs, the work of the council has not been visible to the public at large. Instead, the council cultivated a reputation for well-informed, evenhanded judgment among organized interests. This low visibility was a liability when the Endangered Species Act petitions were filed and critics of the council's approach sought drastic change (Blumm and Simrin 1991). Yet the strength of the institutional network fostered by the Northwest Power Act became apparent during the salmon summit negotiations: The broad scope of the conflict over how to respond to the decline of the salmon runs required adjustments from a wide spectrum of powerful economic interests. Although the council did not develop a consensus to which all agreed, it did formulate a salmon strategy that no one disagreed with so strongly as to seek to upset it; moreover, the strategy has the backing of the four Northwest state governors, a considerable degree of support given the potential for economic disruption.

This kind of planning constitutes an institutional style: gathering information from sources throughout the basin and subjecting the data to public review as a prelude to a public process of priority setting. Building an institutional structure for sustainable development in the Columbia River basin has been based on several conditions:

- Commitments in law, reinforced by political support, to preserve and enhance environmentally valued resources.
- Explicit recognition of ecological, economic, and social uncertainties.
- Acceptance of conflict as an indispensable element of social learning.
- A commitment to act on the basis of knowledge.
- Adequate funding.
- An institutional process open to experiential learning, including conflict.
- A systems orientation.

Changing Myths

The emergence of social learning as an approach to the Columbia can be understood as a change in governing myths, a shift in the way people imagine their place in the natural landscape they inhabit. Table 5.1 rearranges the chronicle to bring out this change in myths—a change only partially perceived by the actors themselves.

From the construction of the first dam on the main stem of the Columbia in 1930 until the enactment of the Northwest Power Act in 1980, the Columbia Basin was an industrializing economy, in which natural resources were treated as economic assets. The transition be-

TABLE 5.1
Elements of the Shift from an Industrial Order to a Search for Sustainability in the Columbia River Basin

Element	Industrial (1930–90)	Searching (Since 1980)
Myths	Economic return is social objective	Sustainability is social objective
	Control of nature through engineering and mastery of fluctuations	Adaptation to nature through management and learning; acknowledgment of uncertainty and natural time scales
	Energy of primary economic importance	Low-cost energy, obtained through energy conservation, economically important Environmental quality, including healthy salmon populations, socially important and worth a significant (but not unlimited) measure of economic sacrifice
	Electric power producers and irrigators dominant	Fish and wildlife interests play a legitimate role in constraining exploitative use
	Management should be centralized	Decisions should be negotiated in open processes
	Politics should be subordinated to rational, technocratic management	Political conflict is inherent in renegotiation of social priorities. Political conflict is crucial for the recognition of errors. Negotiated consensus is nonetheless necessary if scientific study is to be possible over times of biological significance.
Key institutions	Bonneville Power Administration, operators of major dams	Northwest Power Planning as a forum for negotiation among existing institutions and interests
	Power sales contracts Water rights	Native American treaty rights Endangered Species Act
Key events	Building of dams and irrigation system	Indian treaty rights litigation
	Commercial salmon harvest	Nuclear crisis
Key people	Franklin Roosevelt, Henry Jackson	Tim Wapato, Dan Evans

yond industrialism toward a search for sustainability is marked symbolically here by the passage of the Northwest Power Act. Yet, like all social transitions, the process has been punctuated by conflict and crisis, and its outlines are still taking shape. Indeed, if an ecologically and economically sustainable order is achieved, it will not be visible for some time, simply because sustainability implies stability in long-term averages, which cannot be established rapidly by definition.

Social learning entails wrenching changes in beliefs. These are not obviously matters of ideology in the usual sense, for acknowledgment of uncertainty and natural fluctuation is neither liberal nor conservative. Rather, the beliefs that have yielded to changing circumstances have to do with whether humans can control nature, or whether they can prosper by making intelligent use of limitations on control. The idea is implicitly economic and cognitive: The cost of control may be unaffordably high; human efforts should seek a satisfactory level of influence upon the forces of nature and human activities should avoid irreversible damage to natural systems.[7]

Human institutions are the channels of change. Organizational routines and ways of allocating the attention of decision makers shape their perception of and response to surprising changes. To say that institutions are important is not to say that institutional processes produce orderly change, however. Because social learning is a conflict-ridden process, choices are characteristically made in an anarchic fashion, in which institutionalized patterns of behavior generate outcomes that can seem irrational to those participating in the processes (Kingdon 1984). Certainly the utility leaders who sought to preserve nuclear energy as a viable means of supporting the growth of an industrial society did not imagine that their labors would lay the economic foundations for an ambitious attempt to rebuild salmon. Yet the much higher electricity rates that followed the nuclear crisis made salmon rehabilitation affordable.

For the same reasons, the events and people whose actions would cast long shadows are difficult to identify in real time. Thus the philosopher Hegel spoke of the owl of Minerva: History's shape is discernible only at dusk, when the battle is over. This does *not* mean that one must be resigned to an existential fatalism, however. Because institutions and the beliefs they embody are influential, it is paradoxically more important to assure that the basis of those beliefs is well founded. Only then can actions taken in the press of incompletely controlled change be guided by reliable knowledge. That is why adaptive management, the testing of experience against the organized skepticism of science, is an essential element of social learning. The political challenge is to support adaptive management through the turbulence of change. Dan Evans's most significant contribution as founding chairman of the Northwest Power Planning Council may therefore be his dictum that, in a situation of high uncertainty, "the best politics is no politics." Similarly, Tim Wapato's quiet investment in biological training for the Native American field staff who began to take their places alongside the non-Indian resource managers responsible for the salmon may turn out to be his most durable legacy.

Three Caveats

The challenge of a sustainable Columbia River is no different from the challenge of sustainable development generally. Can humans endure on this planet? Nobody knows. It is clear that continuing the exponential increase in resource use of the past 150 years will have serious effects on the global climate. Yet stalling the rush toward the inhospitable greenhouse raises profound questions of economic justice and strains the ability of the international system to maintain order. Although it is necessary to proceed adaptively in the search for sustainability, learning from experience may be neither sufficient nor feasible in the transition ahead. Three caveats are in order.

1. There is the problem of conceptual tractability. The natural systems to be managed sustainably are inherently complex, and their complexity exceeds both traditional human comprehension and the institutions that have managed portions of them in the past. Complexity is a barrier to sustainable development.
2. Moral viability is another problem. Perhaps only rich, stable nations can afford sustainability and make the transition with some semblance of political consent. The concept of sustainability comes with no guarantee that it is attainable or that, if feasible by some quantitative measure, it will be politically and morally palatable.
3. Social learning emphasizes the interest of the population, not that of the individual. Belief systems that value individuals may find little comfort when learning comes at a high cost in suf-

> fering. The long delays inherent in many aspects of global sustainability also limit the utility of adaptive methods, since the signals of success or failure may come back too slowly to inform action. These are the same conditions one encounters in social welfare policy: slow or incomprehensible feedback, combined with urgent, undeniable individual needs. Thus far, social welfare programs have been more anodyne than cure.

The experience of the Columbia River Basin points to two unorthodox paths for study and reflection. First, look to the industrial economies for examples of sustainability. Sustainable development may be like the demographic transition: Nations rich and stable enough to be able to experiment with different modes of living may discover the viable alternatives.

Second, the strategic importance of uncertainty on the path to sustainability must be considered. The adaptive approach offers a conceptually sound way to deal with uncertainties in the natural system and with the complexities of institutional structure. Thinking in terms of whole systems while acting through fragments ("think globally, act locally") requires an explicit organizational and political strategy. Social learning is such a strategy.

Taking sustainability seriously is a question of governance. There are promising leads, but time is short and resources are dwindling. Learning what does not work is a cost of finding ways that will. Minimizing that cost preserves humanity's already limited ability to pursue sustainability with justice and mercy.

Acknowledgments

I am grateful to C. S. Holling for guidance and inspiration.

NOTES

1. Lee (1993) provides more extensive discussion of social learning. Detailed discussion of the Columbia Basin experience in energy conservation and salmon rehabilitation, with full references to the literature, is in Lee (1991a, 1991b); see also Northwest Power Planning Council (1987, 1991).

2. Worster was commenting on the Colorado River.

3. On the importance of rising economic welfare in the formation of

environmental consciousness, see Hays and Hays (1987), who describe environmentalism as part of the "history of consumption."

4. In traditional utility practice, the cost of a new power plant is not counted as part of the "rate base," or invested capital, until it goes into service. Ratepayers are consequently shielded from the cost impact of utility managers' investment decisions until power is delivered; by that time it is generally too late to affect the project. In the Northwest the costs of the first three WPPSS nuclear plants were financed by Bonneville, but the cost impacts were delayed by the financing arrangements until 1979. Since then, even though only one of the plants is in service, the costs of all three are being recovered through Bonneville's rates.

5. The dotted lines in the figure reflect a 95% confidence level; that is, there is a 95% probability that the true trend line will fall within the dotted lines. A lower confidence level would allow the dotted lines to fall closer together. The fact that the observed data are already close to the dotted lines, however, indicates that a more relaxed confidence level would not change the conclusion stated in the text. No statistical manipulation will make the noise in the data go away; only a longer series of observations can do that.

6. Negotiated consensus may not be a necessary precondition for adaptive management. The simultaneous emergence of negotiated settlements to natural resource disputes and of adaptive management appears to be a historical accident, though it is clear that both draw upon ideas "in the air," including a bias for consensus as a management style and a commitment to use science in decision making despite conflict.

7. For a general formulation of the procedural rationality reflected in this approach, see Simon (1983).

6

Barriers and Bridges to the Restoration of the Great Lakes Basin Ecosystem

George R. Francis and Henry A. Regier

The Great Lakes are in a class by themselves. Their size and the fact that they are the world's largest set of freshwater resources assure this. The 766,000-km^2 drainage area of the Lakes, which serve as a huge headwater region for the St. Lawrence River system, is home for some 38 million people. It is also the location of the urban and industrial heartland of North America. This heartland sprawls over the southern portion of the basin while, economically, the northern portion serves as a vast resource-based "hinterlands" for the heartland and beyond (table 6.1). The governing framework for the basin is set by the two constitutional federalisms, which through historical compromises some two centuries ago, extend to the middle of four of the lakes and their connecting channels (rivers). This regional ecosystem is equivalent in its geographic scale, human population, economic base, and institutional complexities to a medium-sized industrialized nation-state.

The Great Lakes have inspired many superlatives from those who have written about their natural features and aesthetic vistas (Ela and King 1977), the history of resource use (Waters 1987), and the sagas of commercial shipping over the years (Stephans 1930; Willoughby 1961; Havighurst 1975). Unfortunately, the urban/industrial heartland and some of the resource extraction industries in the basin have taken their

TABLE 6.1
Some Geographic Data on the Great Lakes Basin Ecosystem

Dimensions	
Land drainage area	521,830 km^2
Water surface	244,000 km^2
Shoreline length	17,017 km
Water volume	~23,000 km^3
Population	
About 38 million (some estimates of 40 million)	
Settlements	
With population of more than 100,000	31
With population of more than 1 million	6
Largest cities: Chicago and Toronto	
Land Use in Basin	
Forests	49%
Agriculture	39%
Residential	7%
Other	5%
Total water usage, 1990*	
Withdrawals and instream uses	3.7 trillion liters/day

*From Great Lakes Commission, Regional Water Use Data Report, 1990. Rest of the data are from Krushelnicki and Botts (1987). Economic data are compiled by state/provincial jurisdictions and cannot easily be disaggregated and recompiled for some regional economy within basin boundaries.

toll upon the Lakes. Superlatives have also been incorporated into expressions of environmental or ecological concerns about the conditions of the Lakes' aquatic ecosystems (Burns 1985; Ashworth 1986; Weller 1990).

The response to these problems has led to extensive and growing binational cooperation aimed at ecosystem redevelopment or renewal. Early initiatives began more than a century ago. The most concerted and sustained efforts have unfolded over the past three decades, beginning in the early 1960s with a small binational network of limnologists and fishery biologists sharing their interests in the biophysical phenomena of the Great Lakes. This was soon followed by growing public outrage over the degradation of the waters and by consultations among officials in the two federal governments about what might be done. Now binational cooperative endeavors involve all levels of government, coalitions of citizen and interest groups, a much expanded set of research networks, and multiple centers for initiatives to promote remedial measures to reverse degradation and for preventive measures to maintain improved conditions.

Along the way, the challenges from the Great Lakes have led to a growing sense of a shared major bioregion, to a depth of understanding of some biophysical phenomena of aquatic ecosystems that is acknowledged worldwide (Ragotzkie 1990), and to the emergence of ecosystemic thinking that casts new light upon how humans must begin to adapt to and understand the ecosystems of which they are a part.

This story is sketched out later. The "tri-partite model" that runs through all the case study accounts is certainly discernible, although somewhat diffused by the scale of human efforts vis-à-vis the scale of the entire Great Lakes Basin Ecosystem, and long response times inherent in some of its processes. Barriers are more obvious than bridges to the restoration efforts. The similarities with other case study examples arise mainly from systemic features in the institutions for governance, which lead to many of the same kinds of ecological problems everywhere they operate.

Structures and Processes

Issues of scale are pervasive for any account of the structures and processes in the natural and human systems in the Great Lakes Basin Ecosystem, and in descriptions of the ecological problems that have been identified as the ones to address.

The Natural System

Abiotic Components

The present-day morphology of the Great Lakes Basin is the result of glacial processes, most recently, those from the Wisconsinan period. Beginning some 75,000 years before present (yBP), what is now the Great Lakes Basin became buried under up to 3 km of ice. During the late Wisconsinan period, from 23 kyBP to 10 kyBP, the basin was covered with six major lobes of ice. Beginning about 15 kyBP, the glaciers began to recede, with retreats and periodic readvances over a period of some 6000 years. The ice had retreated entirely from the basin by about 8000 yBP (Hough 1962; Denton and Hughes 1981; Karrow and Calkin 1985; Fulton et al. 1986).

As the glaciers retreated, the precursors to the present Lakes were formed with separate drainages south into the Mississippi drainage basin and southeast toward the Atlantic. Isostatic rebound of land from the retreating glaciers results in a gradual north/south tilting of land, currently estimated to be in the order of 30–50 cm/century upward along the northeast shore of Lake Superior. Drainage, only to the Atlantic, occurred northeastward. The basin's morphology/bathymetry and drainage patterns are thought to have taken their present form some 3000–2000 yBP. The changes yet to come, in part through climate warming, are speculative, and the duration of what may well turn out to have been another interglacial period is unknown.

The underlying geology of the basin is composed of the ancient pre-Cambrian Shield formation of metamorphosed rock, which has been exposed by glacial abrasion around the northern portion of the basin; fossil-rich limestone and dolomite formations dating back to the Silurian period of the Paleozoic (ca. 425–400 million yBP) that are exposed along an arch that surfaces as the Niagara Escarpment and Manitoulin Island in Ontario, and again in northern Michigan and the Door Peninsula of Wisconsin; and glacial till deposited by the glaciers throughout the rest of the basin. The till deposits in turn have been molded by glacial meltwaters into moraines, drumlins, eskers, outwash sands, and kettle lakes. Soils on top of the pre-Cambrian shield are generally very shallow and acidic in wetland sites. Glacial till is a varying mix of silt, sand, clay, and cobble.

Figure 6.1 is a map of the present-day Great Lakes Basin showing the Lakes, their connecting channels (actually large, short reservoirs), and their outlet to the St. Lawrence River. Figure 6.2 is a cross-sectional profile of the Great Lakes system that notes the comparative elevations and depths of the Lakes.

The Great Lakes fluctuate over different time periods. These fluctuations may in turn be superimposed on larger and longer-term fluctuations over periods of three or four centuries. From geological and archeological evidence, the "high" water levels of the mid-1980s may in fact coincide with the long-term average level for the past 1500 years (Larsen 1987; NRC 1989). Seasonal variations also occur, and storm-driven seiches can cause variations in the surface levels within a matter of hours. A 4–5-m difference between water levels at the northeast and southwest end of Lake Erie occurred during one such event on December 2, 1985 (IJC/Project Management Team 1989). These fluctuations, combined with storm events, drive the dynamics of coastal wetlands.

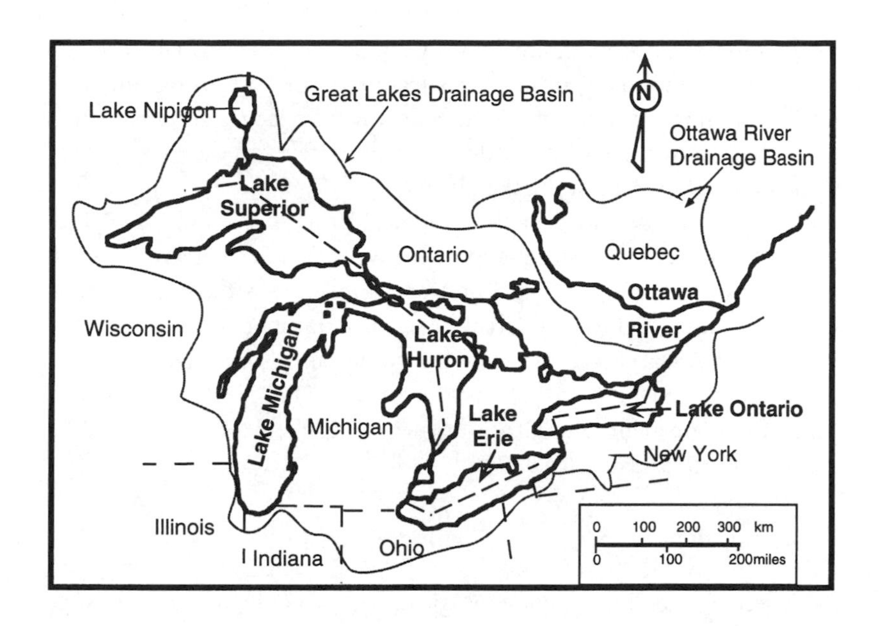

FIGURE 6.1
The Great Lakes/St. Lawrence Basin.

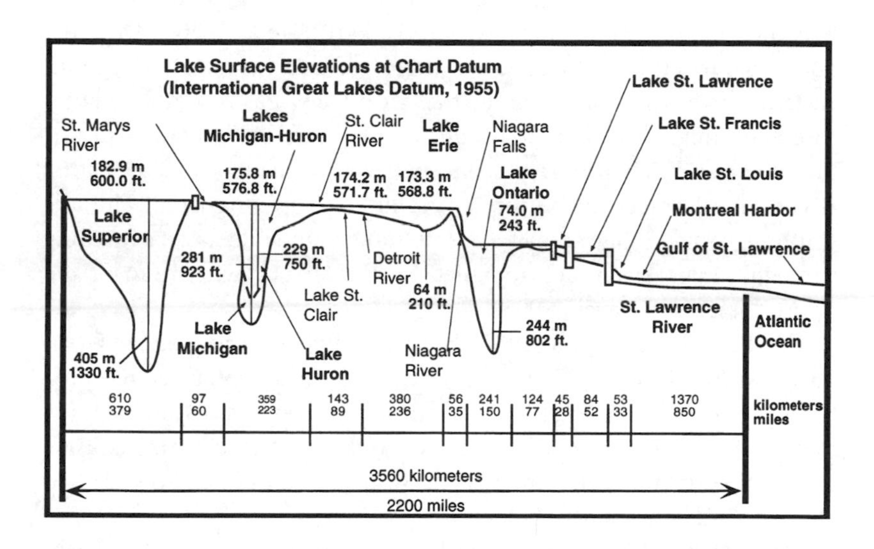

FIGURE 6.2
Schematic profile of the Great Lakes/St. Lawrence River.

There appears to be some uncertainty about the extent to which high-water levels affect the geomorphological processes of erosion and sedimentation in the lower Lakes which have sand beaches and clay bluffs along their shores. They have little observable effects on the rocky shorelines of Lake Superior and Lake Huron.

The bathymetry and basic limnological processes—such as surface currents, areas of up-wellings and down-wellings, thermal stratification during summer, and ice formation and movements during winter—have been described for the Great Lakes in a number of studies. Compilations of research are available for some lakes—for example, Lake Erie (Rathke and Edwards 1985; Boyce et al. 1988; NOAA 1989). The general climate regime is also known (Phillips and McCulloch 1972; Eichenlaub 1979; Saulesleya 1986).

Biotic Components

Evidence from geological, palynological, and archeological sources suggests that the immediate postglacial vegetation and associated other biota in the Great Lakes Basin were similar to the present tundra/taiga now further to the north. Deciduous forests and tall prairie grasslands gradually replaced the conifer forests in the southern part of the basin between 12,000 years before present (12 kyBP) and 8 kyBP. Delcourt and Delcourt (1987a,b) traced the history of forest change throughout eastern North America, including the Great Lakes Basin, and noted that forest taxa migrated differentially due to their different tolerance thresholds, life history characteristics, and patterns of barriers and corridors. A continuing migration of forest taxa in response to current climatic warming could, in one scenario, result in the temperate forest extending throughout the basin, accompanied by the virtual extirpation of the boreal forest (Hengeveld 1991).

Krushelnicki and Botts (1987) mapped thirty ecozones for the Great Lakes Basin. However, there is no one classification that is accepted throughout the basin for planning purposes. Agencies within each jurisdiction use their own land classifications, all of which vary considerably in scale and detail and in the relative importance accorded to climate, landforms, and vegetation.

The vertebrate fauna and vascular plants of the Great Lakes Basin have been extensively studied, albeit almost entirely with reference to their distributions within individual jurisdictions (i.e., United States or Canada, individual states and provinces, and other administrative subdivisions, such as counties). A few studies have adopted the basin as the geographic region of interest—for example, Hubbs and Lagler (1949) for fish, Burt (1972) for mammals, and Guire and Voss (1963), Marquis and Voss (1981), and Keddy and Sharp (1989) for Great Lakes endemic and disjunct plants. A joint project of the Nature Conservancy (U.S.) and the Nature Conservancy of Canada will, when completed, provide for a Great Lakes regional perspective to locate key biodiversity sites and potential sites for biomonitoring and to assess the conservation status of natural communities and species (Crispin 1991).

The fish of the Great Lakes have been studied for the past century, with attention given almost entirely to species of sport or commercial interest and to the "forage fish" that support them. Studies include compilations of commercial catch statistics since the 1860s (Baldwin et al. 1979); reviews of the changing composition of fish communities, mainly as a result of human activities (Wells and NcLain 1973; Berst and Spangler 1973; Christie 1973); and major syntheses of knowledge on the community ecology and management of economically important fish stocks (Loftus and Regier 1972; Colby 1977; GLFC 1980).

Socioeconomic Environment

Pre-European Period

There are four generally recognized periods of prehistorical inhabitation of the basin: the Paleo-Indian period, 11–7 kyBP; the Archaic period 7–3 kyBP; the Initial Woodland period, 3–1 kyBP; and the Late or Terminal Woodland period, from 1000 yBP until the period of European contact in the early seventeenth century (Wright 1972). The transition from one stage to the next is marked first by the appearance of pottery; then by some evidence of small-scale agriculture, distinct burial mounds, and particular ceramic designs; and finally by evidence of increasing reliance upon agriculture (by native groups in the southern part of the basin) and the appearance of villages with surrounding palisades (Brose 1976). The extensive use of Great Lakes wetland resources by native peoples has been documented by Raphael (1987).

Tanner (1987) provides a comprehensive and thorough review of the native peoples in the Great Lakes Basin from about 1600 to the mid-nineteenth century. She notes that the main factors influencing their history during this period were "territorial conquest by intertribal warfare, refugee movements, epidemics [of European origin], English–

French trade rivalry and wars, international treaties, Indian–white hostilities, intertribal alliances, encroachment by settlers, Indian Treaties ceding land to state and national governments, and imperfectly administered arrangements for reservation, removal, and land allotments" (1987). Some twenty-two different tribes belonging to the three main linguistic groups (Algonquian, Iroquoian, and Siouan) occupied the basin, and their total population was estimated to be about 60,000 in 1768. Many of the tribes maintained home villages from which families traveled for lengthy periods to seasonal hunting and fishing camps. Limited agriculture was practiced in the southern part of the basin by Iroquoian tribes who grew corn, tobacco and vegetables.

The remains of villages and hunting and fishing camps have been identified throughout the Basin and in many cases are protected as archeological sites. Native peoples themselves have been assimilated extensively into the dominant European culture, especially in the southern portion of the basin. Some remain as residents on the Indian Reserves established in the nineteenth century. In recent years, there has been a marked revival of interest among native people in their cultural and spiritual values, and this has been accompanied by political mobilization to challenge long-standing treaties or to exercise their right to resources under them. These rights are beginning to be realized under co-management arrangements, such as the Great Lakes Indian Fish and Wildlife Commission and the Chippewa/Ottawa Treaty Fishery Management Authority, both of which, in 1989, became signatories to the Strategic Plan for the Management of Great Lakes Fisheries.

European Period

European explorers first entered the Great Lakes Basin early in the seventeenth century. The rest of the century can be characterized as a period of further French exploration, development of the fur trade, and intermittent warfare among the Iroquois and other tribes and with the French. The French continued to exercise control over settlement and trade in the Great Lakes Basin until 1763, when it was ceded to the British under the Treaty of Paris, which concluded the Seven Years' War between France and Britain. Following the Declaration of Independence leading to the creation of the United States in 1776, another Treaty (the Definitive Treaty of Paris) in 1783 defined the boundary between Britain and the United States. It ran through the Great Lakes and remains to this day as the boundary between Canada and the United States.

The latter two decades of the eighteenth century were also ones of vigorous military campaigns against native tribes, partly as surrogates for continuing British and American rivalry, followed by rapid European settlement in the southern Great Lakes Basin. The Congressional Northwest Ordinance of 1787 promoted this settlement which led to the creation of the Midwest states by the early nineteenth century. In response to this, the British, particularly after the 1794 Jay Treaty which provided for the final removal of British militia from the U.S. portion of the basin to what is now Canada, promoted rapid settlement of upper Canada in what is now Ontario.

European settlement continued rapidly throughout the nineteenth century, leading progressively to extensive forest clearing for agriculture settlement throughout the southern and western portion of the basin, the development of resource-extractive industries (lumber, fish, and minerals), and the establishment of urban centers based on heavy industry, manufacturing, and trade. The banning of vessels of war in the Great Lakes under the Rush-Bagot Agreement of 1817 allowed water transportation throughout the Lakes to play a major role in servicing these developments and in the linking of trade to the Atlantic coast through the Erie Canal and the Rideau Canal in particular. Road and rail ties also developed rapidly.

New technologies such as the automobile (ca. 1900) and hydroelectric power generation at Niagara Falls (ca. 1890s) hastened the urbanization and industrialization processes. These developments continued throughout the twentieth century. They were paced by economically significant events such as the Depression of the 1930s, World War II defense production, the postwar economic boom from the mid-1940s to the mid-1960s, and the more recent economic restructuring of the Great Lakes Basin economy in both countries. Over the past 20 years there has been a relative decline in heavy industry and manufacturing, along with out-migration and an increase in service industries, including the "high-tech" knowledge-based economy of the present time. The restructuring occurred earlier in the U.S. basin economy, but it is now well under way in Canada as well.

At the present time, the Great Lakes Basin has a highly diversified and extensively integrated economy. The integration is evident in the transportation and communication networks, in electrical power sharing, in the continental "Auto-Pact" agreement of 1965, and in the elaborate structures of intercorporate ownerships and controls across all sectors of the economy. The 1989 "Free Trade" agreement between

Canada and the United States only served to acknowledge this development in the case of the Great Lakes Basin. The Council of Great Lakes Governors established a Canadian Trade Liaison Office in Toronto in 1990.

Socially, in terms of language and culture, the peoples in both countries in the basin are distinguished more by their similarities than by their differences, and the continuing immigration into the basin of people from many parts of the world is leading to similar multicultural urban societies throughout the Basin. A Great Lakes megalopolis (Leman and Leman 1970) is still forming around the lower Great Lakes and upper St. Lawrence Valley.

The Management Institutions

Formal Framework for Governance

"Governance," the exercise of authority and control through institutional structures and decision-making processes, is enormously complex in the Great Lakes Basin ecosystem. Governance entails not just the jurisdictions and their administrative agencies, but also the private sector, and various other nongovernmental organizations and local groups. Different configurations of agencies and other organizations are associated with particular ecological and environmental problems and with different resource uses.

The formal framework for governance over the Great Lakes is provided by the two constitutional federalisms of the United States and Canada, a framework composed of two federal governments, ten state and provincial governments, and many hundreds of municipal-level entities. Over the years, governing institutions have been created at hierarchical levels below the state or province but above that of cities, counties, and other municipalities. As a result, most citizens of the basin live under four levels of government. The number of governmental bodies whose responsibilities affect the Lakes is overwhelmingly large. For example, tallies have indicated that some 132 governmental bodies had (at one time) responsibilities concerning water quality for Ohio's portion of Lake Erie (Kent State University 1975), more than 650 governmental bodies have responsibilities for Great Lakes shorelines (Bulkley and Mathews 1973), and more than fifty agencies in both countries had responsibilities for fish and the fishing industry (Francis et al. 1979).

Binational Agreements

Within this formal framework for governance, a number of binational agreements have been reached over the years (table 6.2). They range from formal treaties and conventions to less formal "good-faith" agreements, have varying administrative arrangements for implementation in the cooperating jurisdictions, and were responses to some mix of perceived "crises," problems, or opportunities in the years immediately preceding their negotiation. Over half the agreements now in place have been reached within the past decade, reflecting the growing awareness of a shared bioregion and realization that cooperative arrangements are necessary to deal effectively with the problems that have been identified.

TABLE 6.2
Binational Agreements Concerning the Great Lakes

Boundary Waters Treaty, 1909
International Lake Superior Board of Control, 1914
International St. Lawrence River Board of Control, 1953
International Air Quality Advisory Board, 1966
Great Lakes Water Quality Agreement, 1972, 1978, 1987
International Great Lakes Levels Advisory Board, 1979
Air Quality Agreement, 1991

The Migratory Birds Treaty, 1916
The North American Waterfowl Management Plan, 1986
Eastern Habitat Joint Venture, 1989

The Niagara Treaty, 1950
International Niagara Board of Control, 1953

Convention on Great Lakes Fisheries, 1955
Joint Strategic Plan for the Management of Great Lakes Fisheries, 1981

St. Lawrence Seaway, 1959

Great Lakes Charter, 1985
Great Lakes Regional Water Use Data Base, 1988

The Great Lakes Toxic Substance Control Agreement, 1986
Great Lakes Protection Fund, 1988

Declaration of Intent (for the Niagara River and Lake Ontario), 1987
Lake Ontario Toxics Management Plan, 1989

Maritime Agreement for the Great Lakes–St. Lawrence System, 1991
[Declaration of Indiana]

The Boundary Waters Treaty of 1909 provides a framework for the resolution of disputes over water resources shared by both countries across the entire continent. It was stimulated by several disputes that had arisen by the turn of the century. Several were in the Great Lakes, notably issues of navigation and electrical power generation arising in the Saint Mary's River in the 1890s, the possible effects on navigation of the Chicago Drainage Canal, which first opened in 1900 to flush pollution in Lake Michigan at Chicago to the Mississippi drainage, and concerns about the impacts on Niagara Falls of new hydroelectric power generation using the Niagara River at the falls (Spencer et al. 1981).

The treaty established a six-person International Joint Commission (IJC) with powers to adjudicate on matters concerning water resource development schemes in one country that affected water levels or flows in the other country and to serve as a commission of enquiry for matters referred to it by the two federal governments; arbitration powers provided for in the treaty have never been used. The IJC has had a modest, although important, role in the Great Lakes almost since it was established in 1912, and especially throughout the past 25 years or so, as will be noted later. It remains by far the most studied of the binational arrangements pertaining to the Great Lakes (Dworsky et al. 1974; Spencer et al. 1981).

A comprehensive binational fisheries treaty was signed in 1908 in response to concerns about overfishing in the Great Lakes, but it failed to receive ratification in the United States. Almost half a century later the Great Lakes Fishery Commission (GLFC) was established under a convention signed in 1956. Concern about overfishing was joined with a perceived need to eradicate exotic sea lampreys and strengthen cooperation to foster the recovery of preferred fish stocks in the lakes.

Problems and Crisis, Triggering Events, and Responses

The Situation in Brief

Interrelated issues of physical restructuring, environmental degradation, and resource use in the Great Lakes have been a concern for at least a century, episodically until the 1950s and continuously ever since. Triggering events vary as did the responses. These are described later with reference to pollution, especially the phenomena of eutrophication and toxic contamination; fisheries, especially the efforts to control sea lampreys and restore the salmonid stocks; and lake level fluctuations and diversions. These are certainly not the only matters of concern about the Lakes, but they are the ones for which the greatest amounts of binational cooperative effort have been mobilized.

Responses in each case were formulated in terms of the proximate causes of the perceived problems from within the legislative mandate and particular expertise of designated lead agencies in each main jurisdiction. Initially, some of these agencies themselves had to be strengthened, and institution building went along with the development of programs to respond to the perceived problems. Although the binational coordinative arrangements of advisory bodies, task forces, or working groups allowed for more exploratory excursions into issues, follow-up was constrained to whatever the implementing agencies were willing and able to do. As the limitations of these structurally narrowed approaches to problem solving became apparent, critics pointed to the need for systems perspectives (usually talked about as an "ecosystem approach" in Great Lakes circles) to be taken with appropriate interjurisdictional and interagency coordination.

One result is that, over the past decade or so, ecosystem management has been much discussed but little practiced, especially when it became apparent that management of human uses of ecosystems must take precedence over further manipulations of nature. Although ecosystem language has been written into statements of policy and program intent, the institutional arrangements through which management programs are "delivered" for the Great Lakes are ill prepared for systemswide or ecosystemic strategies that would focus primarily on needed changes in human uses of ecosystems. Some reasons will be noted later. As time goes on, and the discrepancies between statements of intent and results become more widely apparent, there is a growing public perception of institutional dysfunction. This results in impatience, skepticism, and cynicism. Perhaps, however, this could be a force for transformation.

Pollution: Typhoid, Eutrophication, and Toxic Contamination

Three References and the First Result

Pollution became a continuing concern in the latter part of the nineteenth century as outbreaks of typhoid fever and gastrointestinal infections occurred in the cities and towns around the Great Lakes. One of the first tasks assigned to the International Joint Commission in 1912 was to investigate water pollution in the Great Lakes and elsewhere in water shared by the two countries. The commission had bacteriological surveys conducted to reveal the extent of pollution from untreated sewage, runoff, and industrial discharges. In 1918 the IJC reported severe near-shore problems, recommended that treatment works be constructed, and called for the commission to be given powers to oversee the work and to regulate pollution control in boundary waters. None of the recommendations were accepted. Instead extension of sewage outfalls further into the lakes and chlorination of drinking water were seen as less costly alternatives to sewage treatment (NRC/RSC 1985). Other initiatives had also been taken from about 1910 on to create public health services for communities around the Lakes (Denis 1916). These represent the beginnings of preventive measures for pollution associated with human pathogens.

In 1946 a similar reference was given to the IJC to review the extent of pollution in the "connecting channels" (i.e., St. Mary's River, Lake St. Clair, the Detroit River, and in 1948, the Niagara River) to determine if pollution was causing injury to health or property on either side of these rivers (i.e., the international boundary). In 1951 the IJC reported that from the Lake Huron to Lake Erie section there was serious pollution affecting health (e.g., 45,000 cases of dysentery in Detroit in 1926) and property (the water was unsuitable for industrial purposes). It proposed "Objectives for Boundary Waters Quality"; called for waste treatment measures for industry, for municipal sewage (primary and secondary treatment), and for combined sewer and stormwater overflows; and recommended that an international board be established to monitor progress and make further recommendations. Similar recommendations were made for the Niagara River (Dworsky 1988). The advisory boards were established. Although it is not clear what these boards were able to do, during the 1950s the jurisdictions began to develop better, though still only partial, pollution control and municipal waste treatment systems.

In 1964, in response to the continuing degradation of Lake Erie and Lake Ontario, the IJC was given another reference to investigate conditions in these lakes. Through technical advisory groups, the commission marshalled evidence on the extent of eutrophication caused by phosphorus loadings which came from various sources. In a 1970 report the commission called for a phosphate control program, new water quality objectives, and a new international board to coordinate programs and monitor their results. Given the growing public concerns over the environment generally, and the media coverage of the "death of Lake Erie," negotiations began between the two countries at about the same time as steps were being taken to establish the US Environmental Protection Agency in 1970 and the federal Department of Environment in Canada in 1971.

The negotiations resulted in the 1972 Great Lakes Water Quality Agreement (GLWQA). It contained provisions to establish common water quality objectives and a commitment to strive to meet these objectives using similar standards for pollution control, to monitor the lakes to assess results, to undertake studies of the water quality in the upper lakes and the extent of pollution from land use activities, and to establish a regional office (in Windsor) to help coordinate all this. The agreement was widely hailed as a major result, coming as it did 60 years after the first studies were made of Great Lakes pollution and a century or so after the first outbreaks of water-borne diseases from polluted water in the Great Lakes were reported. The jurisdictions had, however, an institutional basis from which they could now begin to address pollution issues on this larger scale.

Building on the First Result

As called for in the 1972 agreement, a review was made of the GLWQA (in complete bureaucratic secrecy) after the first 5 years. In 1978 the agreement was revised and extended with the following main provisions:

- A statement of purpose, "to restore and maintain the chemical, physical, and biological integrity of the waters of the Great Lakes Basin Ecosystem" (this statement was apparently adapted from US PL92–500, Sec. 101[a]) (Regier and France 1990).
- A commitment to a policy of prohibiting the discharge of toxic substances in toxic amounts and to the virtual elimination of any or all persistent toxic substances by adopting a zero-discharge philosophy (also from PL92–500).
- A definition of the Great Lakes Basin Ecosystem as the "interacting component of air, land, and water and living organisms, including humans, within the drainage basin of the St. Lawrence" upstream from the international boundary.

- Revisions to a set of water quality objectives, along with a list of some 350 "hazardous polluting substances" to be eliminated from the Lakes.
- A commitment to have municipal and industrial pollution control programs in place no later than 1982 and 1983, respectively.
- A reduction of total loadings of phosphorus entering Lake Erie and Lake Ontario to be agreed upon within 18 months (by about mid-1980).
- A decision that the Water Quality Board (composed of representatives from environmental regulatory agencies) would be the primary advisor to the commission (vis-à-vis the Science Advisory Board).

In 1987 an amending protocol to the 1978 GLWQA was signed. The processes for negotiation of this were slightly more open, mainly by including representatives from Great Lakes United (GLU), a binational, nongovernmental coalition, on the negotiating teams of both countries. This followed a set of public hearings that GLU held around the Lakes in 1986 (GLU 1987). The additions to the agreement were mainly in the form of specifications for

- Lake ecosystem objectives and ecosystem health indicators for Lake Superior, and a commitment to prepare comparable objectives and indicators for the other Lakes.
- Remedial action plans for degraded near-shore "areas of concern" and for the development of lakewide management plans.
- The development of watershed management plans to reduce nonpoint source pollution [9 years after the report from the Pollution from Land Use Activities Reference Group (IJC 1978) authorized by the 1972 GLWQA].
- The handling of contaminated sediments.
- Research, surveillance, monitoring, and control activities to resolve problems of atmospheric deposition of toxic chemicals.
- Identification of sources of contaminated groundwater.
- Biophysical, ecotoxicological, and mass balance studies of contaminants and their effects.

The current agreement is an impressive document, with fifteen articles and seventeen technical annexes (IJC 1988). It is meant to guide the most intensive binational efforts to deal with the degradation of the Great Lakes.

Eutrophication: Targetting Phosphorus

By the time of the 1972 GLWQA there was a strong consensus among limnologists that biologically active forms of phosphorus stimulate algal growth; this consensus was based on Swiss research that suggested this as early as 1905, with subsequent experimental and comparative tests in various countries over the decades. Phosphate enrichment leads to conditions associated with eutrophication, such as increased turbidity, aesthetic nuisance, clogged water intakes, taste and odor problems in drinking water, and oxygen depletion of bottom waters. In the Great Lakes, these conditions showed up most strongly in the waters of shallow Lake Erie and in the inshore waters of deep Lake Ontario. Both lakes were heavily loaded. Historical simulations and the use of phosphorus budget information suggested that there had been a big increase in land runoff into Lake Erie from 1850 to 1880 because of land-clearing and agricultural practices, a considerable increase in human waste loads into both Lake Erie and Lake Ontario from 1870 to 1970 as the population grew, and that household detergent was a major source of phosphorus entering these lakes from 1946 to 1970 (Aubert 1984).

One consequence of phosphorus discharges into Lake Erie was the deoxygenation of bottom waters, especially in the central basin, during the summer months. The phenomenon has been noted episodically first in 1929, again in 1943, in the early 1950s, in the 1960s and early 1970s, and on a more regular basis since the mid-1970s. The critical depletion rate of 3 mg/liter/month was being recorded by about 1961, and this rate was apparently maintained for a number of years later. Rosa and Burns (1987) documented the increase in oxygen depletion rates in the central basin from 1929 to 1980 and the substantial increase in the size of the anoxic area from the 1930s to the 1970s. These changes were associated with biotic changes, notably the reduction or collapse of fish stocks requiring colder bottom waters in the summer and the disappearance of burrowing mayflies (*Hexagenia*). There is circumstantial evidence that DDT and other persistent pesticides played a role in the demise of the burrowing mayfly.

Priority under the GLWQA, especially during the first 5 years, was given to reducing phosphorus entering the Great Lakes. The intent was to reduce nuisance levels of algae throughout the lakes and to restore year-round oxygen to the bottom waters of Lake Erie. A five-point phosphorus management strategy was subsequently developed, starting first with the elimination of point sources, as follows:

- Place a legislated limit on the phosphorus content of detergents to 0.5% by weight while substitutes are found by the soap and detergent industry.
- Construct or upgrade municipal sewage treatment plants so that those discharging 1 mgd or more effluents directly into the lakes would meet a phosphorus effluent standard of 1 mg/liter.
- Regulate phosphorus from industrial discharges "to the maximum practicable extent."
- Negotiate total loading reductions for phosphorus entering Lake Erie and Lake Ontario.
- Introduce strategies to reduce nonpoint sources of phosphorus entering the lakes from agricultural and other land use practices.

In 1973 the Canadian federal government legislated a limit of 2.2% by weight phosphorus content for detergents. Over the next 15 years (1973–1986) all eight Great Lakes states individually legislated the 0.5% limit. Canada has so far refused to adjust to the lower limit on the grounds that it is meeting its load reduction commitments under existing arrangements.

Municipal sewage treatment plants discharging directly to the Lakes had been brought up to standard by the mid-1980s, and most have been conforming to the 1-mg/liter standard of phosphorus in effluent discharges. An earlier intent to tighten up this standard to 0.5 mg/liter for plants discharging into Lake Erie and Lake Ontario seems to have been abandoned. Information on discharges from industrial sources is incomplete or not comparable, in part because it is included in general reports of industrial "compliance" with standards and regulations, and the meaning of *compliance* varies from one jurisdiction to another.

Loading reductions for phosphorus entering Lake Erie and Lake Ontario were negotiated by 1983 (3 years later than expected). It is reported that regulatory control of point sources will achieve the Lake Ontario target. Programs to encourage no-till or low-till agriculture are under way in various parts of the basin; their effects on the reduction of phosphorus runoff are still to be assessed.

Although portions of the central basin of Lake Erie still become anoxic in late summer, the oxygen depletion rate is reported to be much lower than that recorded in the 1960s and 1970s. Nuisance blooms of algae are now much less frequently encountered. There is some debate about the factors most responsible for the reduced eutrophication. Besides reductions in biologically active phosphorus entering the water due to the regulatory controls, the constant stocking of salmonids in the Great Lakes may be generating food web effects that result in higher populations of zooplankton to graze the phytoplankton. In addition, the recent accidental introduction of zebra mussels into Lake St. Clair in 1986 and their subsequent rapid spread in large numbers throughout Lake Erie (and beyond) have resulted in a considerable additional filtering of phytoplankton, to the point where the mussels have had a noticeable cleansing effect in waters at various local sites (Hebert et al. 1991). With the possibility that phosphorus may no longer be a major limiting factor to biological productivity in the Lakes, there is some interest in examining the role of nitrogen compounds (IJC/WQB 1989).

The long-standing commitment is that the biological productivity of the Great Lakes must be reduced to at least some mesotrophic (if not oligotrophic) ideal and that this conforms with proposed ecosystem objectives to be set under the GLWQA, the fish community objectives to be set under the Strategic Great Lakes Fishery Management Plan, and the commitment to restore and maintain ecosystem integrity. These are major policy decisions that could very well find political and public endorsement, but this has not been sought for the whole system. Public attention, however, is now devoted mostly to issues of toxic contaminants.

Toxic Contaminants and the Human Health Connection

A combination of environmental awareness with the development of laboratory analytical capabilities led to the increasing detection of toxic contaminants in the Great Lakes Basin ecosystem, starting in the early 1960s. The earliest studies documented the presence of pesticide residues in herring gulls and fish (Delfino 1979). PCBs were also soon discovered, and the extent of contamination in the bottom sediments of the Lakes by PCBs and heavy metals was first mapped by PLUARG

(IJC 1978) from data collected in the late 1960s. Mercury levels in fish led to the closure of family-owned commercial fisheries on Lake St. Clair in 1970.

The 1978 GLWQA contained an initial list of 271 "hazardous polluting substances" and 106 "potential hazardous polluting substances" with the intent of taking the necessary measures "to minimize or eliminate the risk of release" of these substances into the lakes. The lists were also included in the 1987 protocol, but the parties to the agreement have so far ignored the requirement to keep the lists up-to-date. The IJC regional office kept a compilation of all toxic contaminants reported from the water, the sediment, the fish, and the fish-eating birds, and in 1985, the IJC's Water Quality Board reported that this list exceeded 1000 (IJC/WQB 1985). This triggered a critical review of the evidence, and it was subsequently declared that only 362 contaminants had been reliably reported.

A number of these substances bioaccumulate and biomagnify in their passage through food webs. Evidence for this has been accumulating from the more degraded areas of the Great Lakes for the past 25 years at least, and it has been recently extensively summarized by Colborn and others (1990) and Environment Canada (1991). The bioaccumulation and magnification of PCBs from water to the eggs of bald eagles, for example, is from 15 to 25 million times (Environment Canada 1991). Impacts on biota include congenital birth defects in birds and turtles, fish with internal and external tumors, and reproductive failures and behavioral changes in colonies of fish-eating birds (Kubiak et al. 1989; Gilbertson et al. 1991).

The human health implications of this, although subsumed under routine toxicological analyses for individual contaminants, were not acknowledged for a number of years. However, evidence obtained from longitudinal epidemiological studies of children born to mothers who consumed fish from Lake Michigan suggested that the children had adverse behavioral changes resulting from interference with brain development during the fetal stage (IJC/SAB 1991). The most recent health concerns have come from evidence that exposure to low concentrations of contaminants does not strongly affect the health of adults of various species of biota, but it does noticeably affect the embryonic development and survival rates of their young. This intergenerational effect of exposures has raised considerable concern among members of the public in the Great Lakes Basin, especially given some recent documentation on the extent of their overall exposures to low concentrations of contaminants from multiple sources, especially foods.

Generally, there have been two kinds of responses to the quandaries posed by these findings and conclusions. Nongovernmental organizations and citizen groups have begun to mobilize to have the zero-discharge provision of the 1978 GLWQA implemented. Some have taken initiatives to think through the technical and legal strategies for achieving this (National Wildlife Federation and the Canadian Institute for Environmental Law and Policy 1991). Scientists call for more research to specify sources, pathways, distribution, fates, and the mass balances of these substances in the environment, including epidemiological studies of the more exposed groups in society, such as native people and other consumers of Great Lakes fish.

Regulatory agencies, after a period of several years of inconclusive attempts to set priorities for a short list of substances from among the 362 acknowledged to be in the ecosystem, have selected eleven critical pollutants for special "primary track attention" (IJC/WQB 1985). Selection was based on the fact that they are known to be present in the Great Lakes; are highly toxic and persistent; can bioaccumulate to levels that threaten human health and biota in the aquatic system; have had abatement or other corrective actions taken but are still present in unacceptable levels; represent a variety of sources, pathways, and uses; and may be members of larger chemical families, so that actions taken for one could be expected concurrently to control or apply to other substances with similar properties.

Measures taken for regulation vary somewhat among jurisdictions. In the United States, measures include setting state standards and placing particular requirements in each NPDES permit issued by the EPA. In Canada the main provisions come under the self-monitoring and load reduction requirements that will be required under the Ontario Municipal and Industrial Strategies for Abatement (MISA) program for eight industrial sectors, backstopped to some extent by regulations under the 1988 Canadian Environmental Protection Act and the Fisheries Act.

There is no comprehensive toxic substances management strategy for the Great Lakes Basin Ecosystem, although governments have been urged to develop one (IJC/WQB 1989; IJC 1990). Both countries announced new pollution prevention programs in 1991, but details for their coordinated application to the Great Lakes are still to come. Toxic discharge reduction plans for the Niagara River and Lake Ontario have been drawn up, but their implementation relies entirely on existing jurisdictional programs. In 1991 the federal state/provincial governments announced a "Binational Program to Restore and Protect the

Lake Superior Basin," responding to earlier IJC recommendations that this lake be a demonstration area for zero discharge and to a growing Lake Superior Alliance of nongovernmental organizations in support of this goal.

Since the 1972 GLWQA, various near-shore areas—such as harbors, river mouths, and the connecting channels—were flagged as problem areas not meeting agreed-upon water quality objectives. By 1985 the same set of areas was showing up time and again, indicating that routine regulatory measures were not working. Much more comprehensive remedial action plans (RAPs) were then (1985) mandated for these "areas of concern," which numbered forty-three in 1992; each requires multistakeholder involvement in planning and in plan implementation. Originally, it was expected these remedial plans could be completed within 2 years. However, the planning process has become more participatory and drawn out because of the need for stakeholders to develop some consensus on the preferred future for an area (or to decide "how clean is clean"), identify causes of the degradation and options for remedial measures, hold public consultations, and mobilize the political will to have agreed-upon plans implemented by different agencies responsible for different components of the plan. By early 1992 only five RAPs had been completed to the satisfaction of an IJC review process (IJC/WQB 1991). Implementation of these plans will be the real challenge.

There is no clear understanding of the overall current situation with toxic contaminants in the Great Lakes, although the trends for a few that were banned or heavily restricted in their use (such as DDT, dieldrin, PCBs, and mirex) are down from peak levels recorded during the 1970s. The GLWQA called for regular monitoring of the Great Lakes to assess progress and trends and to detect new problems. This provision was the most innovative one in the whole agreement. By the mid-1970s a Great Lakes International Surveillance Program (GLISP) was developed and debated. It was subsequently revised in 1986. It called for common baseline monitoring for all lakes each year and periodic intensive surveys for each lake every 3–5 years. Arguments over what GLISP should monitor have gone on to the point that, 20 years after the original agreement, there is still no consensus on the common elements for a basinwide GLISP. The main problems include meshing the need for data to enforce regulatory actions with data needed to understand ecosystemic processes, and the extent to which monitoring should be keyed to the forty or so substances for which scientifically supportable specific water quality objectives have been set, vis-à-vis monitoring for those that regulatory agencies decide are important because of jurisdictional requirements, even if objectives for them have not been negotiated for the entire Great Lakes.

Monitoring is thus being done by each agency for its own purpose. In 1988 Canada reported 28 federal and 34 provincial monitoring activities for the Great Lakes. The United States, with nine jurisdictions, no doubt has more. It is highly doubtful that these could all somehow add up to a coherent understanding of the Great Lakes Basin Ecosystem. Three of the proposed intensive surveys for individual lakes have been undertaken, but the results in each case were not published until at least 6 years later, when they would be more of archival interest (Rathke and Edwards 1985; Stevens 1988).

In 1989 the IJC/WQB reported that the programs called for in GLISP "are not being sufficiently conducted to permit the required assessment of the state of the lakes." The IJC has been complaining for at least the last 6 years about the lack of data that would allow an assessment of progress under the agreement, and the commission commented upon many quality problems with data it did receive.

The condition of biota in some of the most degraded areas, together with the human health implications of intergenerational effects, put conventional strategies of the regulatory agencies and their scientific research priorities into question. *Zero discharge* symbolizes prevention, not remediation long after the damage is done. At least one group of nongovernmental organizations in Canada has negotiated with industries a statement on toxics, including "sunset" policies for select substances (New Directions 1991).

Sea Lampreys and Salmonids

Fish Species and Associations

Fish species in the salmonid family thrive in rigorous habitats of aquatic ecosystems in geographic regions that have been scoured by continental glaciers in recent geological time. They prefer cold, clear, highly oxygenated waters and key the reproductive part of their life cycles to locales that are annually scoured by strong currents. At lower latitudes in their range salmonids thrive at higher altitudes in mountains and greater depths in lakes, as well as in streams fed continuously by cold spring water. At higher latitudes, where summer is short and surface waters do not become warm, salmonids also thrive in clear shallow lakes and slow rivers.

In such cold, oligotrophic ecosystems the salmonids tend to form a complex of species, including a complex of races within some species. The taxonomic complexity in larger lakes or river systems may approach that of the cichlid swarms in African rift lakes or the sculpin swarm of Baikal, noting that species segregation within the set of salmonid taxa is not as complete as that with the cichlids and sculpins of the old rift lakes.

Two centuries ago, each of the Great Lakes contained an interactive set of salmonid taxa, where the salmonids include charrs and other species of the Salmoninae subfamily and the ciscoes, whitefishes, and other species of the Coregoninae subfamily. Lake Superior may have contained as many as forty recognizably different salmonid taxa, most of which are conventionally lumped under the designation of lake trout. Lake Michigan may have had a comparable number, with a large assortment of coregonine taxa. The community of salmonid species dominated the aquatic association of the cold, oligotrophic parts of the Great Lakes. Biodiversity was high.

Salmonids have features that make them highly desirable to humans. Their flesh is preferred in part because of its firmness, which results from the vigorous life-style of these taxa, and because it is not bony. Externally, these fish have a simple, streamlined beauty. They are notoriously easy to catch by a variety of means, which makes them a favorite of fishers of all sorts.

Cumulative Effects of Cultural Stresses and Corrective Measures

The various attributes of salmonids gave them few preadaptations for coping with technologically advanced humans that invaded the Great Lakes Basin in large numbers two centuries ago. These human invaders interacted with the aquatic habits of salmonids in a number of ways, here sketched generically:

- Harvesting of biotic resources, with a preference for large salmonid taxa.
- Physical restructuring or removal of habitat features, especially in areas of rapids in spawning streams, where mill dams were built or on spawning shoals where rocks were removed.

- Loading of unwanted or sacrificed materials into streams so as to muddy, foul, and poison waters.
- Introduction or invasion of exotic species.

In each of the Great Lakes two centuries of industrial and commercial progress led to the extirpation of the great majority of the salmonid taxa that once thrived there. The extirpation of most of the vanished stocks was likely due to a combination of causes. The proximate causes seem often to have been unsustainable fishing practices and predation by the invading sea lamprey, though other causes likely contributed by making at least some stocks more vulnerable to fishing and lamprey predation.

The salmonid taxa and their habitats reached a low point in each of the lakes in the early 1960s. Fishermen had largely cast aside more conservationist practices of earlier decades, since they had little hope that stocks would recover if they were to practice better husbandry. Loadings of the lakes with various harmful substances reached a peak. Sea lamprey had invaded all parts of the lakes and had found numerous tributary streams in each that met their spawning requirements. Small exotic fish species, rainbow smelt and alewife, had been introduced into the lakes and were thriving in habitats once dominated by salmonid species. These small pelagics appeared to create a complex self-reinforcing process that exacerbated the eutrophication of the lakes through their trophic requirements and preyed upon young salmonids, thus suppressing them further.

Some fishery managers sensed that the lakes had become transformed under the assortment of human abuse into a kind of system that was more typical of shallow fertile seas of the continental shelf than of deep, oligotrophic lakes. They experimented with the introduction of several large marine salmonids that thrived in such marine systems, eventually suppressing the small exotic pelagics. These large exotic salmonids were supported by technically sophisticated fish hatchery programs that required measures (not fully successful) to prevent the introduction of diseases normal to these exotic species. Near the ends of the trophic network, because of biomagnification, these salmonids became quite contaminated by hazardous chlorinated hydrocarbons. These fish were generally reserved for the sport fishery, and anglers were advised not to eat their catch of large salmonids.

Meanwhile reduction of loadings into the lakes of various enriching and fouling materials undercut the trophic process that fed the small exotic pelagics. The pathological positive feedback loop (if that is what it was) was eventually ended. Fisherman concurred with tight restrictions on catches of salmonid species in order to permit recovery. Sea lampreys were controlled partially with a variety of methods related to their spawning and nursery habitats in streams. Various techniques have been used, including electric weirs and barrier dams, to block adults from reaching spawning streams, and releasing chemical lampricides into known spawning streams to kill larval lampreys. Some of the native salmonid taxa that had survived in the more remote, less accessible parts of these lakes began to reappear in some numbers. Attempts to reintroduce close relatives of extirpated stocks have also occasionally been successful.

What might be termed a delicate balance now exists between salmonids and sea lampreys. The sea lamprey's requirements for spawning habitat resemble those of the charrs—lake trout and speckled trout. This is also the case with habitats of juvenile and mature fish. Attempts to protect and restore charr habitats thus will likely also work to improve sea lamprey habitat.

The primary method heretofore for controlling sea lamprey has been to use lampricides that have some measure of selectivity for sea lamprey but that cannot be used in a fully selective way. Further, cost-effective control methods are not available for those larval sea lamprey populations in large rivers and along lake shores. There has long been speculation that large lake trout (e.g., greater than 4 kg, say) would not usually suffer mortality because of sea lamprey attacks. It is possible that a natural control measure would develop spontaneously for sea lamprey, though little evidence of this has been discovered.

From Fishery Management to Ecosystem Husbandry

There have long been two contending traditions related to fisheries: a pro-growth and often politically dominant faction that presupposed the inexhaustibility of the resource and a pro-sustainability faction that emphasized the risks of overexploitation. (There has been a similar split with respect to the assimilative capacity of the environment for pollutants and contaminants.) The 1954 binational Convention on Great Lakes Fisheries was in support of the pro-sustainability concerns; it made no explicit reference to environment or habitat issues and was long interpreted as relevant largely to direct exploitation and measures to control or mitigate overexploitation.

But fisheries researchers had long realized, at least tacitly, that all forms of human activities on land and in the water of the basin accumulated in the rivers and lakes, especially in the ecological association of fish species. This is reflected, for example, in the Canada Fisheries Act, which is the strongest piece of legislation in Canada with respect to an ecosystem perspective on pollution. Thus fisheries ecologists, together with naturalist ecologists, have been leaders in an ecosystem approach to the integrity of the whole basin (Regier 1992).

In 1980 a Strategic Great Lakes Fishery Management Plan (SGLFMP, pronounced *siggle-fump*) was endorsed by senior officials of the fisheries agencies of the eight Great Lakes states and the province of Ontario, the two federal agencies, and the Great Lakes Fishery Commission. Individual leaders in this initiative had also been active in similarly innovative work under the IJC auspices (e.g., the Pollution from Land Use Activities Reference). Thus there was an informal attempt to develop congruent interjurisdictional political commitments for environment and fisheries interests.

In 1992, again under the auspices of GLFC, a "Strategic Vision" with respect to fisheries was produced. It demonstrated further progress toward a commitment to ecosystem integrity as apparent in the Great Lakes Water Quality Agreement of 1978 as amended in 1987. The vision statement had three main features:

- Healthy Great Lakes ecosystems.
- Integrated management of sea lamprey.
- Institutional/stakeholder partnerships.

The commitment to healthy ecosystems emphasizes that reliance is to be on naturally reproducing stocks in self-regulating communities of indigenous species to produce sustainable benefits to society.

With respect to sea lamprey management, the vision statement does not repeat an earlier internal commitment within GLFC to the gradual phasing out of toxic chemical lampricides, but it does commit to a reduction in lampricide use and to further development of alternative, more socially acceptable control methods, consistent with a 1982 decision. (One of these methods, a seemingly innocuous 1-m-high dam inaccessible to lamprey during their spawning runs upstream, has already led to deaths of two canoeists and thus seemingly poses a greater risk to humans than the reasonably selective lampricide.)

The institutional/stakeholder part of the vision statement may signal a continuing interest on the part of federal fisheries agencies and private interests to devolve more responsibility to state and provincial agencies for finding ways to manage interjurisdictional fisheries problems. A gradual devolutionary process has been under way, and GLFC has been tacitly facilitating it in providing organizational help on issues that go beyond GLFC's formal mandate. Something similar has also occurred within the IJC family of transboundary institutions. The GLFC vision statement supports further collaboration between fisheries and environmental interests toward a goal of ecosystem health.

Of course each fishery agency has within it a sufficient number of bureaucratically constrained professionals to ensure that a broadening vision will only be addressed slowly and gradually. Many bureaucratic professionals take it as natural that the next year's programs should not deviate from those of the current year. A real challenge, as in any reform, is to find humane ways of enlisting such constrained people into the bigger challenge.

Lake Level Fluctuations and Diversions

Existing Controls and Diversions

During the latter part of the nineteenth century and the first half of the twentieth century, engineering works were undertaken to enhance the control over water level fluctuations or to increase diversions of water into or out of the Great Lakes (e.g., Yee et al. 1990). These include

- An upgrade of the old Illinois and Michigan Canal by the construction of the Chicago Sanitary and Ship Canal in 1900 (which allowed pollution to be diverted away from Lake Michigan into the Illinois River system), with the volume of the diversion limited (in 1980) to 3200 cfs (90 cms).
- The New York State Barge Canal, which, since 1918, conveys 1070 cfs (30 cms) of water from Lake Erie to various tributaries for Lake Ontario.
- Control over outflows of Lake Superior by construction of regulatory works at the head of the St. Mary's River by 1921

and a subsequent agreement to maintain the levels of Lake Superior to within a range of 1.5 ft (0.45 m).

- Construction of the Long Lac and Ogoki River diversions from 1939 to 1943 to divert 5600 cfs (159 cms) of flow from the Hudson Bay drainage into Lake Superior to generate hydropower near Terrace Bay, Ontario.
- Upgrading of the Welland Canal from 1929 to 1932 to provide navigation and power facilities resulting in an increase of 9400 cfs (270 cms) flow from Lake Erie to Lake Ontario.
- Construction of navigation and hydropower facilities at the international rapids section of the St. Lawrence River, which since 1958 (and as part of the St. Lawrence Seaway) has maintained water levels for Lake Ontario within an agreed range of 4 ft (1.2 m).

The Multiple Studies of Water Level Fluctuations

In 1964 the IJC received a reference to study factors that affect the fluctuations of water levels in the Great Lakes and to determine if actions to bring about a more beneficial range of levels were practicable. The reference came at a time of record low water levels. A number of possible regulation plans were examined, ranging from doing nothing to options for engineering works to achieve complete control over the levels and flows of the Great Lakes. By 1973 record high water levels were being recorded in all the lakes except Lake Superior, and control measures were modified for both Lake Superior and Lake Ontario to reduce damages to shore works and buildings in low areas of the coastal zones in the four lower lakes.

After 10 years of technical studies and twenty-two public hearings, the commission (IJC 1976) noted that the natural regulation of levels and flows in the Great Lakes was "very efficient" and that measures to achieve complete control would be too expensive compared to the benefits. It further noted that issues of the regulation of levels and flows must be based on the concept of basinwide benefits, not localized ones, and that an improved hydrologic network was needed to assist with lake level forecasting and to optimize operational procedures at existing control points: "Finally, the Great Lakes are viewed by the Commission as a massive ecological and economic centerpiece for the whole mid-continent, affecting its industry, its agriculture, its total lifestyle and character, and must be seen therefore as a whole, as a bi-national treasure and as a bi-national responsibility" (IJC 1976).

In 1977 the IJC was asked to study whether limited regulation of water levels in Lake Erie "would be in the public interest in both countries." Some possible regulatory works at the head of the Niagara River were examined. The study board convened by the commission to do the studies concluded that economic losses to commercial navigation, recreational boating, and hydropower generation would far outweigh any benefits. The main problem was the potential overloading of Lake Ontario and the St. Lawrence River, which would then need remedial channel enlargements (IJC 1981). The study board further recommended that the study be ended, that a public information program be developed, and that governments encourage coastal zone management to reduce flood and erosion damage.

The IJC was also asked in 1977 to examine the potential for dealing with extremes in lake levels by changing existing diversion flow rates and existing projected consumptive uses of water in the Great Lakes Basin. The commission (IJC 1985) subsequently recommended that governments not consider further manipulation of diversions as a way to manage lake levels, but that they instead take steps to ensure that better coastal zone management practices be followed to reduce flood and erosion damage. With regard to consumptive uses, the "most likely projection" for the year 2000 would be approximately from 5700 to 8400 cfs (161–238 cms). The commission noted the need for a better data base and regular monitoring of existing diversions and consumptive uses, and urged governments to undertake a joint review of their policies with a view to reducing consumptive water use. It also added that the "two Governments would be well advised at this stage to engage in broad but systematic discussion of their use of Great Lakes water before they are faced with any sense of crises, actual or imminent, and before any relationships deteriorate or become jeopardized" (IJC 1985).

In 1985 lake levels were high again, and following extensive shoreline property damages from winter storms, the IJC was given a reference in 1986 "to examine and report upon methods of alleviating the adverse consequences of fluctuating water levels in the Great Lakes–St. Lawrence River Basin." An elaborate plan of study was developed by November 1987 to address the perceived need to take a comprehensive systems perspective on key issues, including principles and strategies that should be agreed upon, governance, and public involvement. A continuing management process was deemed necessary in order to reconcile diverse interest groups and devise various combinations of on-site solutions at different locations around the Lakes. The organizational structure for the study consisted of a project steering committee with advisory groups, a management team, and five "functional study groups," each with subgroups.

In a phase I report (IJC/The Project Management Team 1989) the team noted that the issues to be addressed "include [recognition] that the essence of the nature–human complex is inescapably systemic; that an ecological dynamism deserves priority consideration before taking any action on water level fluctuations; that misperceptions and misunderstandings of the water fluctuations phenomenon and of our ability to affect it abound; and that the extant bi-lateral and hierarchical governance poses impediments to concerted and coherent collaboration."

The report posited that phase II of the study would address four collective objectives: "a set of binational principles as guides for decision-making; an overall strategy and general plan of action; improvements in governance; (and) refinement in understanding of critical aspects of the system."

The directives given by IJC for phase II were, however, somewhat different. They called for establishing a set of guiding principles, completion of a set of short-term technical studies, and evaluation of a range of management measures on a variety of type-specific sites throughout the basin. The technical studies are to include the feasibility of partial-to-total structural controls options, further review of the existing regulatory control plans at Lake Superior and Lake Ontario, and study of the impacts of fluctuating water levels on shoreline features and erosional processes. A new and different administrative structure was created for phase II. It included a study board; five working committees, each with a number of task groups; and a citizens advisory committee with members on each of the working committees. The final report to IJC was subsequently endorsed by the commission (IJC 1993).

Diversion Threats and Responses

There have been at least ten proposals over the past 30 years or so for substantial diversions of water out of or through the Great Lakes as the perceived solution to water shortages elsewhere on the North American continent. Most were made in the 1960s, when engineering megaprojects were more in vogue.

The most persistent proposal is the GRAND (Great Recycling and

Northern Development) Canal project. It would build a dam across James Bay, which would then be flushed out from a number of rivers flowing into it to become a freshwater reservoir. The water in this new reservoir would be pumped south across the Great Lakes/Hudson Bay divide through a series of pumping stations and reservoirs along the Harricanaw River. Once at the divide the water would flow south through the upper Ottawa River to be diverted again through the Mattawa and French River systems into Georgian Bay for storage and eventual redistribution through the United States. This project has been promoted intermittently since 1959. It has received some political endorsement (Bourassa 1985) but little in the way of feasibility studies.

The most recent studies in the early 1980s suggested that water from the Great Lakes could be diverted from Lake Superior into the Missouri River system for eventual transfer further west to replenish the depleted Ogallala and related aquifers in the High Plains of the central United States. Bulkley and others (1984) reviewed these suggestions and concluded that they were not feasible on economic grounds.

In 1985 the governors and premiers of the Great Lakes states and provinces signed the Great Lakes Charter, in part because of concerns raised by the GRAND Canal and Ogallala schemes. The charter endorsed a set of principles, including one to protect the "integrity of the natural resources and ecosystem of the Great Lakes Basin," pledged cooperation among jurisdictions to protect the water resources of the Great Lakes, agreed to the establishment of a common data base for the management of water use, and committed the signatories to a prior notice and consultation process for any new or increased diversion or consumptive uses of Great Lakes water exceeding 5 million gal/day (19 million liters/day). The data base is being maintained by the (U.S.) Great Lakes Commission in Ann Arbor. An interjurisdictional Water Resources Management Committee oversees the terms of the charter (WRMC 1990). Most jurisdictions have also passed legislation intended to prevent or control diversions out of the basin.

Control or Adapt?

The main issue underlying the constant studying of water levels for the past 25 years or so is a dispute between agencies and interest groups. One group seeks structural engineering works to secure near complete control over water flows; these interests until recently may not have realized that control of the levels of the lakes as well as the flows of the rivers is simply impossible, given the annual variations in precipitation over the basin. Others urge proper coastal zone planning and management as the most cost-effective means of reducing the exposure of people and property to the effects of fluctuations. The dispute also pits one level of government against others, since local and state/provincial governments have the main responsibility for coastal zone management, while the federal governments have jurisdiction over flow control structures.

Politically, the issue rises and falls with the water levels. The lengthy, repetitive studies of various aspects of lake level fluctuations and possible means of control serve to "ride out" the extremes until the lakes readjust to levels perceived to be more normal. The studies over the past decade have on several occasions yielded recommendations that a much more ecosystemic and longer-term approach be taken to dealing with this issue, especially given the prospects of climatic change. But it has yet to happen.

The prospect of climate change adds some uncertainty (Cohen 1986; Croley 1990; Hartmann 1990). Current scenarios suggest that water levels in the Great Lakes will fall by 0.5–1.0 m or more on average over the next three decades or so and that the volume of water flowing through the St. Lawrence River could drop by up to 20% (Hengeveld 1991). If so, the low water levels of the mid-1960s could become the most frequent and "normal" levels for some unknown future period.

Emergence of an Ecosystem Perspective for the Great Lakes

University-Initiated Networks

Over the past 20 years, four series of meetings have been convened under the innocuous title "Canada–United States Inter-University Seminars" for the Great Lakes; the fourth is under way (1991–1993). They have been organized jointly by academics in both countries to bring together about fifty participants from universities, government agencies, and nongovernmental organizations from around the basin to discuss emerging Great Lakes issues and their implications for governance (Caldwell 1988; Francis 1990).

At the conclusion of the second series of meetings in 1977, the group recommended that governments give a Reference to the IJC (under the terms of the Boundary Waters Treaty) asking it to create a board on

Great Lakes Rehabilitation and Restoration to examine a range of issues relating to a rehabilitation and restoration goal systematically (versus the nearly exclusive preoccupation at the time with reactive limitations of particular problems such as phosphorus, a few chemical contaminants, sea lampreys, and salmonids). The proposed board was envisioned at the time to be similar in its scope to the board established during 1972–1978 to assess water quality problems associated with land use runoff, one of the most successful references carried out by the IJC in terms of the range of its work, its findings, and its consultation processes (IJC 1978).

Participants in these seminar discussions then took the proposal to the GLFC, asking this commission to endorse the idea and offer to work closely with the IJC on it, and to the IJC's Research Advisory Board (RAB). The GLFC instructed its board of technical experts to investigate the feasibility of ecosystem rehabilitation for the Great Lakes. This in turn led to a series of studies organized around stress-response analyses of ecosystem change (Francis et al. 1979; Harris et al. 1982; Francis et al. 1985). This general approach has subsequently been used by the remedial action plans (RAPs) being prepared for the designated areas of concern around the Lakes.

The IJC's Research Advisory Board prepared a report on taking an "ecosystem approach" to Great Lakes matters and tabled it with the IJC in 1978. The notion of a Great Lakes Basin Ecosystem also appeared in the 1978 GLWQA. Subsequently, this general approach has been used to provide comments, critiques, and suggestions on agreement activities from both the Science Advisory Board (the RAB under its current title) and commission staff. Another follow-up from the seminar series was the organization of an IJC Workshop on Anticipatory Planning (IJC/SAB 1979) and continued promotion of the idea of establishing some kind of Great Lakes Ecosystem study board (e.g., Dworsky 1986; Allee and Dworsky 1990).

Ecosystemic Concepts in the Great Lakes Water Quality Agreement

The incorporation of the notion of ecosystem integrity into the statement of purpose for the 1978 GLWQA seemed to be mainly for symbolic or rhetorical purposes. It was not until a decade later, in 1988, that the IJC and GLFC convened a workshop to examine the concept from a number of perspectives (Edwards and Regier 1990). Generally, the scientific concept refers to ecosystems that are fully functioning in some vigorous, self-organizing way combined with human uses of them that are compatible with this healthy functioning. There is an element of political choice and preference with regard to particular ecosystem features people may wish to retain. The IJC/SAB in 1989 recommended that, "The people in the basin clarify the desirable attributes of ecosystems that would characterize integrity, including such factors as sufficient protected nature, carefully husbanded protected waters and lands, and beautifully maintained urban areas."

This incorporation of explicit values into a political notion of ecosystem integrity and the fact that the values often differ from those implicit in certain notions of "management" pose a challenge to the ready acceptance of ecosystem concepts in some circles and/or contexts (Regier 1990; 1992; Regier and Bronson 1992). To the extent the values are based on ecocentric views of the proper relationships of people with the ecosystems in which they reside and upon which they are totally dependent, a major cultural shift is entailed —that is, a shift from viewing the environment in a political, people-oriented context to seeing politics in an ecosystem context (Vallentyne and Beeton 1988). This in turn can generate a paradigmatic clash with the anthropocentric, utilitarian, and "resourcist" beliefs and values underlying the organizational structures and ethos of conventional management agencies. As fundamental contradictions, they become formidable barriers should this underlying difference become recognized.

In contrast to "ecosystem integrity," a less inclusive concept of "ecosystem health" appears currently to be more specifiable. The latter was exemplified by an interpretation given for Lake Superior in the 1987 amending protocol: Ecosystem health indicators for Lake Superior are stable, self-reproducing stocks of lake trout, free from contaminants that adversely affect the trout themselves or the quality of their harvested products; an average production of more than 0.38 kg of trout per hectare per year; and population levels of an amphipod crustacean *Pontoporeia* defined as a range for the number of individuals expected from standardized samples taken above and below a depth of 100 m. It was subsequently suggested (IJC/SAB 1989, 1991) that this interpretation be extended to all oligotrophic lakes or portions thereof, that similar objectives be expressed for mesotrophic lakes in terms of walleyes and mayflies (*Hexagenia*), and that healthy populations of bald eagles in numbers that flourished prior to the impacts of pesticides and other toxics be a general, overall indicator of well-being (Bets et al. 1990; IJC/SAB 1991).

Governments have not acted on the suggestions to extend the use of indicators of ecosystem health to the other lakes. Instead, under the 1987 protocol, governments (the parties to the agreement) took responsibility for setting the lake ecosystem and health objectives. As of mid-1992, none have gone forward, although a draft set for Lake Ontario has languished "in house" for several years (EOWG 1992). Some prime monitoring sites for ecosystem health in coastal wetlands have been identified by the nature conservancies of both countries (Crispin 1991). A framework for developing indicators of ecosystem health has been prepared by the Council of Great Lakes Research Managers (IJC/CGLRM 1991).

Various initiatives have come forward from nongovernmental groups to promote or elaborate upon the notion of an ecosystem approach. They include a major conference (Christie et al. 1986), the articulation of an ecosystem charter (RAAS 1989), and a revised statement on the ecosystem approach based on current scientific perspectives on ecosystem dynamics (Allen et al. 1991). An initiative is also under way to develop a state-of-the-environment reporting format for the Great Lakes (IJC/Task Force 1991).

Emergence of Organized, Binational, Nongovernmental Expertise

"Great Lakes Tomorrow" was created in the late 1970s as a spin-off from the Lake Michigan Federation to organize extension courses on Great Lakes issues in cooperation with colleges and universities around the lower Great Lakes and to participate in public hearings associated with these issues. Several successful courses were run, and some new entitlements to involvement by citizen groups in governmental decision processes pertaining to the Great Lakes were achieved before the organization folded in the late 1980s.

In 1983 a center for the Great Lakes was established in Chicago; it opened a Toronto office in 1985. The center undertakes policy studies on matters of interest to governors/premiers and business leaders. It has developed more recently into a facilitator for bringing stakeholders together to address issues or launch new initiatives, such as the successful Great Lakes Protection Fund. It currently has a Great Lakes Legacy project directed toward mobilizing support for the conservation of biodiversity and scenic areas around the Lakes, viewed as a kind of complement to the remedial action plans for the degraded "areas of concern." This initiative, in turn, follows up earlier work to review natural heritage conservation from the perspective of the entire Great Lakes (Smith 1987; Weller 1989; Crispin 1991). The center publishes a periodic newsletter, *Great Lakes Reporter,* that discusses issues and reports on activities in the basin.

In 1986 Great Lakes United (GLU) was formed as a loose coalition of citizen advocacy groups in both countries. About one-third of its 200 or so member groups are Canadian. Most of GLU's attention is directed toward issues of implementing the GLWQA (e.g., GLU 1991). It reports to the public through a newsletter, *The Great Lakes United.*

In response to the high water levels and winter storm damages during 1985–1986, riparian property owners founded the International Great Lakes Coalition to lobby for further regulation of lake levels. Membership reached a peak of about 20,000 but appears to have declined as lake levels returned to "normal" (or slightly below). Key members of the coalition are involved with the current round of studies under IJC's lake levels reference.

In 1991 two industrial-sector organizations were formed, the Council of Great Lakes Industries (CGLI 1991), and an association of nineteen maritime organizations committed to a new maritime agreement for the Great Lakes/St. Lawrence system (GLC 1991). The industries council, whose members have been subjected to environmental regulations for 20 years under the GLWQA, has come together at a time when nongovernmental groups are rallying around the zero discharge provisions in the agreement. The maritime coalition has emerged at a time when various interests are calling for more effective controls over spills into the Great Lakes and the enforcement of controls to prevent the introduction of exotic species from the ballast tanks of ocean-going ships.

The emergence of these "horizontal" networks of groups who involve themselves in Great Lakes policy and management issues (and in informal contact with academically based networks) can provide a counterbalance to the "vertical" division of labor among the established institutions of government, the corporate sector, and the science establishment. By bringing a diverse array of perspectives, kinds of knowledge, and experience to bear on particular issues, these networks could help foster a more holistic thinking about the context within which the issues must be understood and resolved. Whether this will serve as some kind of prerequisite for a transformation of institutional arrangements and capabilities remains to be seen.

Evolution of Ecosystem Management: An Assessment

The Tripartite Model

The tripartite model (Holling 1986) of the interactions among institutions, nature, and society—with the ultimate "pathology" evolving in the form of institutional rigidity, ecological brittleness, and societal dependencies and vulnerabilities—can be applied as an interpretive scheme for the Great Lakes, at least to a degree.

Management agencies did seek to control undesirable ecosystem changes by targeting particular features, such as phosphorus, individual (or individual "families" of) toxic substances, and sea lampreys, for reduction or removal from the ecosystem. The continual restocking of desirable species of top trophic salmonids has been carried out for decades with questions about the "carrying capacity" of the lakes' aquatic ecosystems deemed to be largely academic until about 5 years ago. For controversies over lake levels the strategy was different. Rather than agree to expensive engineering works to control fluctuations to the fine-tuned levels demanded by the riparian interests, governments commissioned the IJC to organize increasingly convoluted, drawn-out studies (almost continuously over the last 25 years). These serve to mollify some of these interests, with respite every time lake levels return to "normal."

Socioeconomic dependencies did build up as a result of these management responses, but they were somewhat diffused. Pollution control measures were to restore the "beneficial uses" of water, long interpreted publicly as making the Lakes "fishable, drinkable, and swimmable" again.

A much stronger dependency built up around sea lamprey control and salmonid restocking programs, as became evident with the growth of the sport fishing industry. Direct expenditures by anglers for Great Lakes sports fishing were estimated to be about $2 billion (U.S.) in 1985, resulting in a total economic impact within the Great Lakes region of some $2.3 to $4.3 billion, depending upon a range of assumptions (Talhelm 1988b). The direct expenditures by anglers were up from a little more than $1 billion estimated in 1980 (Talhelm 1988a). The economic impacts were shared among a diverse set of enterprises, many of them in "hinterland" locations where the business was eagerly sought.

In the case of lake levels, the struggle was to shift the dependencies from federal agencies to state/provincial and local governments, or onto the riparians themselves (i.e., to substitute proper hazard lands zoning and development controls for costly engineering works subsidized by the senior governments).

The profiles of the "subsequent crises" resulting from institutional rigidity, socioeconomic dependencies, and ecosystemic changes are only beginning to emerge, and in somewhat uncertain and ambiguous ways. The reduction of eutrophication opens up possibilities for human choice in ecosystem features that are preferred. The local area remedial action plans are demonstrating the processes whereby collective choice could be made in these kinds of decisions. Thus there is an emerging opportunity, but so far there have been inadequate attempts to seize it.

The toxic contaminants issue presents a crisis for a long-established strategy for dealing with pollution (i.e., by legislated standards for "end-of-the-pipe" effluents, applied at the sources of individual discharges). The impacts of biomagnified accumulations in biota demonstrate that water quality objectives relate to the wrong subsystem of the ecosystem. For many contaminants that have low solubility in water but that are readily transported when adhering to particles or contained in organisms, the fact that they can be detected in the water at all means there are far too many of them in the ecosystem. Generally, nothing of a corrective nature can be done about most of what is already in the environment. This established strategy must be complemented with one that puts high priority on preventive measures that place the burden of proof on the dischargers and moves toward "zero discharge" by sanctioning major changes in products and processes of production, and by gradually phasing out particular substances.

For the fisheries, the prospects of a discontinuation of sea lamprey control as both unworkable and too costly may appear severe in terms of some immediate rebound in population levels of lampreys and the likely political response among anglers and their outfitters. A user-pays policy of privatization for sea lamprey control paid for by anglers could be one solution, if costs were the only concern. Yet if lampreys did behave as a "pest" species, they would eventually be brought under some kind of natural control, as discussed, for example, by Christie (1991).

The possibility of exploring sea lampreys as a resource either for animal or human consumption (the recipes are known from the Baltic) merits consideration. It would require a considerable reevaluation of the thinking that has been institutionalized for more than 35 years. However, there may not be sufficient lamprey biomass for this to be economic. Lampreys do not school in open waters but do form spawning runs in more than 200 streams, with few of the runs being large.

Embedded Cycles and the Great Lakes Experience

Ecosystemic Cycles

The model of a nested set of embedded four-step (box) cycles [i.e., exploitation, conservation, creative destruction, and mobilization, and then the same process repeated (Holling 1986)] is intuitively attractive but difficult to apply unambiguously. Many of these cycles operate on scales far removed from the human scales of detection or experience. As a result, they would, in effect, be taken as near permanent background conditions at the large-scale and slow-moving end of the spectrum, or as "background noise" if detected at all at the small-scale and fast-moving end of the spectrum.

This concept could be applied at different scales. For example, at a geological scale, the Great Lakes could be seen as evolving toward some kind of conservation position at an unknown point in the current interglacial period. European settlement (ecological invasion) of the basin resulted in a major perturbation in this cycle, inflicting some forms of creative destruction, remobilization, and exploitation, especially of the terrestrial components of the basin ecosystem.

The four-step cycle could also be used to depict a general understanding of the kinds of changes that have been occurring in fish communities in the Great Lakes and the management responses to them. The pre-European "harmonic" fish communities were drastically disrupted by European settlement and have been replaced with "managed" fish communities. Three possible management goals—strive to restore former fish communities, rehabilitate the fisheries with comparable fish communities, and create artificial put-and-take fisheries—have each been pursued to a degree.

The time lags associated with problem detection and institutional response have been quite impressive in the Great Lakes. No doubt this reflects the additional complexity of geographic scale, the necessity for dealing with multiple jurisdictions, and the time required to mobilize binational commitments. Nevertheless, the examples of eutrophication, toxic contaminants, and sea lampreys suggest lags in the range of from 15 to 30 years between identifying a problem and developing a binational commitment to act on it (table 6.3). It can then take another one to two decades for results to appear, including some not anticipated (surprises).

The geological and human historical phenomena exhibit evolutionary change rather than cycles in the strict sense. Cyclic phenomena, such as population fluctuations, lake level variations, or patch dynamics of

TABLE 6.3

Time Course of Problem Detection and Institutional Response for Issues of Eutrophication, Toxic Contaminants, and Sea Lampreys in the Great Lakes

Eutrophication and phosphorus reduction	
*1929	Anoxia noted in the bottom waters of the Central Basin of Lake Erie
*1955	Anoxia increased and occurred quite regularly during the summer, at least locally in the Central Basin
*1961	Anoxia reached critical oxygen depletion rates
*1972	GLWQA
*1985	Phosphorus reduction strategy fully in place; improvements noted
*	Oxygen restored year round to bottom waters of Lake Erie
Toxic contaminants	
*1962	Herring gulls in Lake Michigan had high levels of DDT/DDE and experienced poor reproductive success
*1971	First fish consumption advisory issued upon discovery of PCBs in fish
*1972	Herring gulls started to be used as a spatial and temporal indicator of organochlorine contamination in the Great Lakes Basin
*1978	Niagara River–Love Canal emergency; 238 households evacuated
*1978	Revised GLWQA
*1990	United States and Canada agreed to develop a pollution prevention strategy for the Great Lakes
*1991	First major compilation by government of information on toxic chemicals in the Lakes and their associated effects, which also acknowledges human health issues (Environment Canada et al., 1991).
Sea lampreys	
*1890s	Sea lampreys adapted to Lake Ontario
*1923	First entered Lake Erie via Welland Canal
*1950s	Found throughout the upper Great Lakes
*1956	GLFC formed; sea lamprey eradication a major goal
*1958	Introduction of chemical lampricide to spawning streams
*1975	Lamprey abundance reported reduced by 90%
*1982	Formal acceptance of policy for integrated pest management for lamprey control
*1980s	Lampreys found spawning in St. Mary's River—a "connecting channel" (Regier and Goodier, 1992)
*1990	Larval lampreys found in Lake Superior shallows near St. Mary's River

terrestrial ecosystems, are embedded in the large-scale and longer-term evolutionary change.

A complementary interpretation can be sketched using the concepts of self-integration of living systems described by Bertalanffy. As Davidson (1983) noted:

> Bertalanffy's model of hierarchical order [has] four related concepts: As life ascends the ladder of complexity, there is a *progressive integration,* in which the parts become more dependent on the whole, and *progressive differentiation,* in which the parts become more specialized. In consequence, the organism exhibits a wider repertoire of behaviour. But this is paid for by *progressive mechanization,* which is the limiting of the parts to a single function, and *progressive centralization,* in which there are emerging leading parts (like the brain) that dominate the behaviour of the system.

Bertalanffy perceived that these processes occurred within the self-organization of biological (i.e., organismal), sociological, and ecological living systems. Because every living system is unique within its specific historical and environmental context, such an apparently simple set of universalistic statements cannot explain everything. Rapport and Regier (1992) applied this concept of a self-organizing ecosystem to depict the changes that occurred in the Great Lakes; formerly "healthy" natural/cultural ecosystems experienced cumulative disintegrative effects from the stresses of an industrial society, followed by partial system reintegration resulting from human corrective actions.

Organizational Cycles

Coordinated management actions taken under binational agreements could be interpreted as strategies directed toward achieving some desired set of ecosystem attributes along a trajectory between exploitation and conservation of the four-step aquatic ecosystem cycle. The strategies require several decades of persistent efforts, with results that are either uncertain, ambiguous, or yet to be achieved. Presumably, they also engender a number of shorter and smaller epicycles within ecosystems.

Organizational analogies to the ecosystemic cycles can be described (Holling, chapter 1). Leifer (1989) posed a similar model for organizational transformation (or "death" if it misses the transformation process at the right time). If an organization faced with a "point of singularity," "bifurcation point," or trigger event can quickly reframe issues and create some vision for itself, it may then be able to conduct some successful experimentation under new conditions, which then allows it to regroup and achieve some new stability and equilibrium with a capacity to adapt to further change. If it responds to some trigger event by denial, it may well be on the path of organizational decline and eventual "death" through abolishment or bankruptcy.

It is tempting (but with no sense of malice) to locate the lead agencies responsible for both the water quality and fishery management programs somewhere at or near bifurcation points with an inertia that threaten them with decline. The way forward, if Leifer's model is to be helpful, is to engage in a major reframing or revisioning exercise to redefine goals, objectives, problems, and opportunities, and then to start up various experiments to test how best to proceed. The urgent need is for social learning processes that would have to be accompanied by changed reward structures within the affected organization.

At the institutional level, there have been impressively long lead times between some first expressions of need for new institutions and their eventual creation. Calls for some kind of fisheries agreement for the Great Lakes date back to 1875. Twenty-seven commissions and conferences were convened to study the matter over an 80-year period prior to the establishment of the GLFC. A first call for the St. Lawrence Seaway came from an 1894 waterways conference, and 65 years later a binational agreement was reached to construct it. In contrast, the Boundary Waters Treaty came about in a record time of about 15 years (i.e., between the first call for some such agreement at an 1894 irrigation conference and the signing of the treaty in 1909).

Although potentially discouraging, it may be helpful to note these time spans associated with institutional changes. Various suggestions for even minor reforms to make the IJC more effective in its Great Lakes responsibilities have been successfully ignored or dismissed by governments and/or the commission for the past 20 years. It has been a scant 15 years since the call for an "ecosystem approach" to Great Lakes matters began.

It becomes highly speculative to consider what it would take to make the entire institutional and organizational overlay for the Great Lakes Basin Ecosystem functionally compatible with the dynamics of the ecosystem. The continual appearance and accumulation of "environmental problems" can be seen as symptoms of systemic dysfunctionalism in the relationships among society and ecosystem. The adaptive changes required can be specified only in general terms, such as the restoration and maintenance of ecosystem integrity and an ecologically sustainable society.

No conventional political ideology within the Canadian and American parts of the Great Lakes Basin now has an ethical and conceptual base that relates clearly to an emerging vision of "ecosystem integrity" and sustainability. Exaggeration of the role of purely selfish behavior, as in the theory of the "invisible hand" with respect to neoclassical free-market processes, predisposes a culture to behave disintegratively within itself, toward other cultures and toward nonhuman nature. As an allocative institution the free market may be a lesser evil than the "invisible mind" of a hierarchic centralized command bureaucracy, but the latter is not the only alternative.

Some opinion leaders forecast an integration in the 1990s of a number of North American reforms of the 1960s that have gone separate ways (i.e., reforms related to gender, race, the poor, war, and the environment). In each of these reforms, ethical principles—of what is inherently proper and improper with respect to the rights of individuals, peoples, or species—played important roles. Each involved attempts toward redemocratization rather than imposition of a new monistic ideology, and each of these reforms had to contend with the tyrannies of the "invisible hand" and of the "invisible mind." Both the free market and the centralized bureaucracy are recognized as important and valuable institutions, but it is crucial that their roles be accountable to the ethical principles of the emerging vision of ecosystem integrity and sustainability.

The integration of 1960s reforms in North America may create a basis for "green politics," as happened in Europe. A transnational version of green politics may now be emerging in the Great Lakes bioregion.

Myths, Barriers, and Bridges

The processes for communication and consultation opened up by the evolution of the horizontal networks and elaborate participatory processes also bring forth incompatible "myths," defined as partial captures of some reality that motivate people to do things. The relative ease with which these incompatibilities might be resolved will depend on how deeply rooted the myths are in the fundamental beliefs of their holders. Policy change struggles might reveal this (Sabatier 1988; Heintz and Jenkins-Smith 1988). Throgmorton (1991) outlined the challenge and pitfalls arising from the need to reconcile three primary audiences in policy disputes—scientists, politicians, and the lay public, with each viewed as its own "interpretive community" having little communicative effectiveness in dealing with the other two.

An "epistemic community" (Haas 1990) may only be comprised of individuals from the scientific community, whose members are otherwise affiliated with diverse organizations. Although this might enhance a lobbying effort, it does not resolve inherent problems associated with relations to the other communities. Previews of these kinds of difficulties over Great Lakes issues have been evident at the last two biennial meetings of the IJC to review problems and progress under the GLWQA. Public groups were encouraged to attend, and they subsequently actively challenged views of governments and scientists alike. Differences of perspective among interpretive communities (as well as among local geographic communities) are also endemic to the debates over lake level fluctuations and what should be done about them.

Myths

Myths that have come into play in debates over Great Lakes issues include the following, in no implied order of priority. They juxtapose the old with the new.

- Knowledge claims: nineteenth-century scientific method, applied strictly within existing disciplines, versus additional experiential knowledge based in part on intuition and the adoption of transdisciplinary perspectives and methods.
- Ontological status of "ecosystem": a metaphor or an interpretive concept that has no external reality of its own versus an entity that does have an existence (and history) of its own.
- Science and public policy: unilateral determination of policy by science versus science as subordinate to policy (as mandated science, *sensu* Salter 1988).
- Environmental problem solving: the view that the human ingenuity to invent will solve problems via "tech- fix" with appropriate market incentives versus the idea that solutions also require major change in beliefs, life-styles, and institutions.
- Social-choice mechanisms: the view that these, especially the market, can incorporate ecological knowledge and generate appropriate management strategies versus the idea that none of the known social-choice mechanisms can lead to "ecological rationality" (Dryzek 1987).
- Need for economic development: the view that this is necessary for societies to survive versus the notion of sustainable, steady-state societies (in terms of quantity) where improved quality is possible and is essential for the survival of humans and other living beings.

Barriers

The Great Lakes Basin Ecosystem has an overlay of human-imposed barriers of many kinds: international, jurisdictional, administrative, and proprietal. They in turn are overlain by organizations with other boundaries, such as those that fragment knowledge into professions and disciplines, each with special claims and privileges. Altogether, these act as constraints upon understanding ecosystem dynamics and impede the development of the intelligence systems needed for effective adaptive management at different levels of scale, including, when necessary, the entire Great Lakes Basin Ecosystem.

There is also a fundamental contradiction, becoming increasingly apparent, between the basic paradigm for understanding ecosystem dynamics and the paradigm that forms the foundation for the major institutions and "management" organizations. The contradiction centers on different worldviews about humans vis-à-vis ecosystems (i.e., between the view of humans as members of, and utterly dependent upon, complex, interdependent ecosystems related over a wide range of temporal and spatial scales and the prevailing anthropocentric view of human dominance over, and exemption from, many of the natural processes, combined with a utilitarian belief that only those components of ecosystems that are of use to humans matter and should be managed accordingly). The prevailing view is one of "instrumental rationality" that reduces wholes into parts to be analyzed or managed independently. It underlies the traditional natural and (some) social sciences, and the design of management organizations.

"Instrumental rationality" is expressed in organizations structured as hierarchical bureaucracies. Each such organization has core strengths in the form of facilities and technologies they control, along with the skills, beliefs, values, and commitments brought by their members to give the organization both a purpose and an identity. Examples abound in the institutions of government, the corporate world, research institutes, academia, and research funding agencies. The strong points in these kinds of organizations have inherent limitations and can become their major weaknesses at times of rapid or fundamental change.

The narrow range of professional expertise that dominates individual organizations (e.g., engineers and lawyers in environmental regulatory agencies, or biologists in fish management agencies) can become a major source of rigidity (as noted in the tripartite model mentioned earlier), effectively preventing organizational learning needed at times of change. Such organizations have a strong interest in making certain that new situations are not perceived or defined in ways that reduce the importance of their established expertise. Yet at times of change, when the old remedies are no longer working, the capacity to "reframe" problems and solutions is crucial. We suggest that this now confronts both the environmental regulatory and fish management agencies in the Great Lakes.

Evidence of this rigidity associated with a narrow range of expertise in Great Lakes organizations is anecdotal, but from our own participant-observer experience, it is both real and widespread. There are recurring struggles for control over management agendas involving engineers and biologists, and more recently, public health specialists. They are evident in debates about the ecosystem approach (unified information and management systems versus networking and adaptive management) or on indices of environmental quality that would reorient programs (health of biota vis-à-vis water quality objectives alone). Research agendas have been focused by governmental funding agencies largely on biophysical questions, and increasingly, ones that would support regulatory agencies and/or government policy initiatives.

"Safe" social science (e.g., to document the economic benefits of clean water or sport fisheries) is acceptable. Research that could raise contentious issues about policies, institutional arrangements, information and communications systems, and decision processes—all of which are crucial to the success of commitments expressed through binational agreements—is deemed not to be scientific, and hence beyond the terms of reference of governmental funding for Great Lakes work. A more positive interpretation of this phenomenon is that safe science on issues somewhat remote from the real difficulties provides a level playing field to serve mediational efforts among jurisdictions or agencies who otherwise might not be communicating at all.

The strong, nearly exclusive emphasis on the Great Lakes as hydrological systems (and the understandable fascination of scientists with all that goes on in them) has produced a wealth of particular information and data. This is evident in the conferences and publications associated with the International Association for Great Lakes Research (IAGLR) since the association was formed in 1966 and in occasional reviews of the thrust of ongoing research by the Council of Great Lakes Research Managers. At the same time, this huge volume of material remains virtually intractable to retroactive interdisciplinary syntheses, even if the inherent problems of interdisciplinarity could be resolved and funding for such work could be obtained. Yet the understanding of ecosystem dynamics implies syntheses.

The mandated science in particular has provided documentation of the symptoms of ecosystem degradation and some proximate causes, as these are manifested in the water or biota. This helps meet the needs of regulatory and management agencies (as these are driven by the legal system), but it does not get at issues of remediation or the prevention of such degradation arising from economic activities elsewhere in the basin or beyond. To the extent that science remains for the most part analyses of water or laboratory experiments that address "safe" topics, it can easily appear to the lay public that science has become a substitute for decision making. Major cuts in science budgets over the past decade and the decline in the "critical mass" of expertise in some well-established regional laboratories have brought little public protest in the Great Lakes region.

The Great Lakes organizations are driven by their policy and legal mandates, not by intergovernmental agreements. The agreements are not accountability-forcing documents, agency reporting arrangements are not keyed to them, and nobody is responsible or accountable for achieving ecosystem integrity or ecosystem health for the Great Lakes. Although it would be possible to develop from the GLWQA (or perhaps to include some amalgam of the SGLFMP and other main binational agreements) an evaluative framework based on declared goals, objectives, elements of strategy, and some schedules for activities that could then be used to evaluate the collective effectiveness of actions by a whole range of organizations, this has never been done.

The IJC, which has an assessment function under the GLWQA, complains about the lack of information on program results and on the fate of their recommendations to governments. Formal reporting arrangements have evolved into an elaborate system of published statements, recommendations, and rejoinders between IJC (and its advisory boards) and lead agencies in different jurisdictions (e.g., Canada-Ontario 1990). This could go on indefinitely. Some means must be found to introduce performance accountability at two levels, one (as now) at the level of individual organizations for the achievement of their own objectives and one at the systems level in terms of how the organization contributed to the achievement of ecosystem integrity, with the expectation that discrepancies between the two levels could lead to reforms within organizations rather than to the abandonment of integrity.

These institutional problems and inertias have generated another irritating response. Increasingly, spokespersons for the established bureaucracies are declaring that they cannot cope with Great Lakes problems because the public is ignorant. Hence those from the bureaucracies must devote time and resources to public education. Besides many reports, there are now many posters, brochures, fact sheets, and videos available to anyone who knows what to ask for, from which office. This is innocuous for the most part, but some of the information also deliberately shifts the responsibility on to the public and away from government and the corporate or other sources of problems. Fish consumption advisories (issued by all main jurisdictions, but with different definitions of safety) give consumption guidelines for fish likely to be contaminated. Although these are helpful, they put the onus on fisherpeople for problems created by others.

The major disjunction between systems operating at all levels—professional, institutional, and operational—is between the basin economy and the problems that show up in the aquatic ecosystem. Great Lakes research and management focus entirely on the latter. This gives a rich potential for "surprises," since so many of the industrial and other activities that can inadvertently generate surprise are not routinely considered to be within the purview of Great Lakes issues. This is true even for major users of the lakes. Only in recent years, for example, has attention been given to the issue of spills from commercial shipping (IJC/SAB 1979; IJC/SAB 1991) and to the role of ocean ships in bringing exotic species into the Lakes (IJC and GLFC 1990).

The nuclear industry is another example. Although past reports (IJC/WQB 1975) have noted the locations of nuclear installations in the basin, this served as background information for reports on the low levels of radionuclides in the Lakes, much of it residual from aboveground nuclear weapons tests in the 1950s. Most recently (IJC/SAB 1991), the issue of "siting nuclear reactors on drinking water reservoirs" was raised; this refers to the thirty-seven nuclear reactors in the basin, most of which draw their cooling water directly from the Great Lakes, and to the fact the some 20 million people rely directly upon the lakes as a source of domestic water supply. Note is also made of the unresolved long-term disposal of spent fuels. A nongovernmental group has drawn attention to nuclear facilities with a record of design or equipment failures, the legacy of a now defunct nuclear fuel reprocessing site, and large volumes of uranium mine tailings exposed to wind and water erosion not far from the shore of one of the lakes (Nuclear Awareness Project 1990).

Bridges

The human-imposed boundaries noted earlier are loosening or dissolving somewhat. Maps showing the entire Great Lakes Basin with the same amount of detail are superseding those that in the past detailed only half of the area, with only a sketch (or blank space) to depict whatever lay beyond the international border. Maps for daily weather forecasting in the media are increasingly being extended beyond U.S. national boundaries. Actors in Great Lakes issues, which some 25 years ago were largely confined to federal agencies and a network of natural scientists, now involve state/provincial governments, the private sector, and a growing number of citizen groups in both countries, with extensive network links for communication and cooperation among them. Isolated disciplinarians remain even more isolated, but only if they so choose. Transdisciplinary perspectives are now sought, with the recognition that they also must include some intuition and a commitment (at least implicitly) to normative goals. Awareness of the common interest in the well-being of this large, shared bioregion is growing extensively.

A long-discussed need to adopt "anticipate-and -prevent" strategies along with the "react-and -control" ones is gaining acceptance, in part through the publicity given to issues of sustainability. This is evident by the many community initiatives in recycling, energy conservation, wetlands conservation, stream and habitat restoration that, although necessarily local, can add up. The "zero discharge" goal, dismissed so derisively for a number of years, may be on the verge of being accepted by those responsible for achieving it.

There are many signs that the era of heavy industrial and manufacturing enterprises, long associated with the economic core of the urban and industrial heartland of North America around the lower Great Lakes for the past 50–75 years, may be coming to some kind of close or transformation. Ontario, through a mixture of economic recession, free trade and corporate restructuring, is currently being deindustrialized, much as were the rust belt Great Lakes states in the late 1970s and early 1980s. The promised new world of high-tech, knowledge-based services is developing, and there are many reports of technological innovations within the remaining industries. This does little for many of those victimized by the massive job losses in long-established industries. It may do more to relieve pollution stresses, especially those associated with the heavy industries. Initiatives are being taken to analyze the Great

Lakes Basin economy as a distinct, highly interdependent regional economy, within not only North America but also the world (Federal Reserve Bank of Chicago and Great Lakes Commission 1991).

Evolution of Adaptive Management: The Set of Postulates Applied to the Great Lakes

Our broad judgments about the postulates of bridges and barriers as they relate to the Great Lakes experience follow.

The chronology of resource management has been a history of crises and unitary responses, generating polarization and conflict. Generally, the tripartite model fits the experiences as noted. The time lags are such that the processes have not played themselves out fully, given the scale of human interventions in comparison with the scale of the Great Lakes Basin ecosystem. Institutional structures, as noted, inhibit easy reframing of problems and require organizational accountability for achieving the goals of the binational agreements.

A foundation of understanding begins to emerge following periods of crisis and polarization. The foundation of understanding began to emerge in the mid-1970s. This was in anticipation not necessarily of crisis, but of an underachievement or likely ineffectiveness of what was being done vis-à-vis what was perceived to be necessary in adopting an ecosystem approach to Great Lakes issues. This may have been reinforced by the emergence of social reform thrusts on several fronts from about 1968 on.

The nested sets of the four-step cycle concept of ecosystem dynamics is of heuristic value but requires considerable subjective judgment to apply. The underlying processes of change are more evolutionary than cyclical.

The main crises—arising from issues of toxic contaminants, links to human health, and the ignoring of zero discharge commitments undertaken in 1978—are only just now being acknowledged. The crisis in lamprey control is recognized, but with a determination to find even more effective eradication techniques. The new generation of experts and policy advisers is only now emerging, but its members still appear to be selected to maintain organizational continuity. The force for change is coming more from organized nongovernmental groups, some of whom are taking leadership initiatives to demonstrate alternatives to established governmental policies.

In conjunction with an emerging foundation of understanding, an informal collaboration developed and deepened among scientists in various institutions. To a degree this is true, but most of the scientific community remains discipline, technique, or budget bound, and through mandated science many are closely associated with (and defensive of) the policies that are beginning to fail. The networks of collaboration draw upon people from an increasingly varied background, with a result that scientists, especially those associated with government and industry, are likely to be viewed as institutionalized parts of the problems rather than as the source of solutions. The self-isolation of most academic scientists from anything that appears political contributes to this situation.

Underlying this, however, are different scientific mindsets based on different ontological and epistemological assumptions:

- Causally predetermined reality, closed system, old science.
- Chance-driven reality, stochastic system, current science.
- Emergent evolutionary reality, open system, new science.

Scientific and technical professionals commonly balance their work or iterate among these mindsets, but with no explicit strategy or procedure for doing so. In the Great Lakes region, collaboration among scientists is only just beginning to contemplate the implications of new science for a historically contextual and integrative understanding.

An informal consortium of concerned citizens, scientists, and business leaders was established. This situation may well be emerging, but the form it will take and the common-ground issues to be identified are not clear. A shared frustration with government seems to be the one thing that unites diverse groups. The new (September 1991) Council of Great Lakes Industries may take a lead with some of the citizen and scientist networks already in place. The best examples of consortia kinds of arrangements are to be found among some of the stakeholder groups associated with remedial action plans for "areas of concern" at the local level.

RAPs necessarily include government as major stakeholders. The RAP process, in effect, involves a redemocratization, consistent with a growing recognition that more than shallow reform is involved.

Catalysts triggered political action that resulted in system reconfiguration and renewed commitment to broader-scale ecosystem management. Not yet, but some ideas have been put forth about how this might, or should, occur—for example, evolving a learning society (Milbrath 1988), striving for "unified management" (Allee and Dworsky 1990), maintaining flexibility and adaptiveness to system-generated change, restructuring and transformation as discerned from a nonequilibrium systems dynamics perspective (Slocombe 1990), waiting for a major catastrophe to catalyze reform (Caldwell 1991), or promoting a general redemocratization.

The preceding events resulted in a new institution characterized by systemwide change, adaptive in nature and adequately funded. Not yet. It would be interesting to know if threshold conditions are being approached. This kind of institutional change, first dimly perceived in the late 1960s, will obviously not come about with the relative speed of say, the Boundary Waters Treaty (i.e., 15 years after the need for it was first noted). Presumably, however, it will occur more quickly than the Great Lakes Fisheries Convention, 80 years after the need for such an agreement was first noted.

Acknowledgments

The Donner Canadian Foundation provided financial support for this work.

7

The Baltic: The Sea of Surprises

Bengt-Owe Jansson and Harald Velner

What is the Baltic Sea? An optimist's picture would include a glittering sea framed by a blue sky with schools of white sails tacking back and forth between rocky green islands. Pessimists would describe it as one of the most polluted seas in the world, filled with turbid water, green algae, and toxic fish. In fact, the Baltic is a large, diverse, semienclosed sea covering an area one-thousandth of the world's total ocean surface. In 1985 it yielded 1 million tons of seafood, or 1% of the total world catch. The catchment area is four times as large as the sea; it houses 10% of Europe's population and accounts for nearly 15% of worldwide industrial production (Kaasik 1989). All these views are facets of the Baltic Sea, a sea that has provided many surprises in a long history.

The history of Baltic Europe, defined as the sea and its catchment area, is one of change characterized by waves of migration and invasion, with the centers of commerce moving from place to place through time (Zaleski and Wojewódka 1972). The area has always been politically unstable; the recent dramatic breakup and reassembly resulting from *glasnost* resembles a pattern similar to the period prior to World War II. The political importance of the Baltic Sea during the cold war period was based on its geographical structure as a gate to the open ocean rather than on its natural resources. So not only has the natural system in this low-salinity area been highly dependent on the connections to the ocean, but the political system has been as well. What crucial historical events have created the present Baltic Sea? Are there strong couplings between the surprises in the natural system and developments in the socioeconomic/political system? Have human events triggered the natural events or vice versa? Are there examples of man learning from history?

The present story represents an attempt by one systems ecologist and one technical scientist/international administrator (the latter, until recently, living on the east side of the Iron Curtain) to examine the interplay between the systems of nature and the systems of man. We recount recent history, but in a time of dramatic change. It is not yet possible to penetrate fully the development in the countries of the East European countries prior to the liberation. Events previously unknown to the West, and often embarrassing, are uncovered nearly every day. Our description of these events is therefore a result of mainly looking through two pairs of eyes only—a Swedish pair and an Estonian one. It is hoped that as more reliable data become available, our analysis will stimulate broader, more extensive, and detailed descriptions of these extremely interesting and dynamic events.

Postulates

Our natural world consists of both changes and stability. Cycles, like the sequence of days, each characterized by rhythmic changes of light, or the succession of years, each incorporating seasonal changes, we call "warm and sunny" or "cold and dark," thus turning change into stability. Pulsing is characteristic of many ecosystems, established by basic processes of production and respiration or consumption in natural systems. Holling (1986) has systematized the world of pulses into four phases: *exploitation* (a buildup using released resources), *conservation* (increasing complexity and size), *creative destruction* (e.g., fires, storms, pests), resulting in release of basic material, used in *reorganization,* which prepares for the next round of pulses. This cycle holds for natural systems, such as the Baltic Sea, but is it true also for the human system? Humanity's enormous potential for thriving in a world so diversified in time, space, and structure is the result of a constructive mind and access to huge energy sources. The failures to shape a sustainable society probably can be traced to some fundamental gaps in the mind's potential network of associations: an apparent inability to learn, except the hard way. Three signs of this imperfection stand out when analyzing environmental crises.

One concerns humanity's seeming inability to understand, absorb, and adjust to the pulses of the natural system. Society acts as if the present conditions are eternal, with no built-in mechanism for rapid change ("release"). Large fishing fleets with high maintenance costs are built when the fish come in, only to be immobilized and depreciated when the schools are gone or overexploited. Humans appear to have little sense of time and space and therefore seem unable to adjust to or manage fluctuating resources.

The second sign concerns the inertia that people display while absorbing and reacting to a new understanding of the natural system. Changes in the natural system are usually characterized as "natural" and humanity's crucial role is therefore downplayed or denied.

The third sign is our inability to grasp the systems network and realize the fundamental feedback loops, which then appear as "surprises." In an attempt to check overproduction in Swedish agriculture, the government remunerated farmers for fallowing their fields, releasing nutrients that overfertilized the inland waters and the Baltic Sea, a process they were busy trying to check by another set of laws.

These imperfections are also related to myths of the environmental crises—concepts that intuitively and uncritically have been generally adopted and used as starting points for management but later proved disastrous because the concept was wrong. Are there many myths in the Baltic story?

If this is at least a part of the explanation to the man/nature conflict, how does it fit with the Holling four-phase infinity cycle? To test this a set of postulates can be formulated. Their presentation will allow a running comparison with the following description of the evolution of the man/nature complex in the Baltic area. The six postulates are

- The chronology of resource management has been a history of crises and unitary responses, generating polarization and conflict.
- A foundation of understanding only begins to emerge following periods of crisis and polarization.
- In conjunction with an emerging foundation of understanding, an informal collaboration developed among scientists in various institutions.
- An informal consortium of concerned citizens, scientists, and business leaders was established.

- Catalysts triggered political action that resulted in system reconfiguration and renewed commitment to broader-scale ecosystem management.
- The preceding events resulted in a new institution characterized by systemwide change, adaptive in nature and adequately funded.

During the reviewed sequence of history there are situations, people, and decisions that prove to be either bridges for adaptation to sustainable development or barriers that delay this evolution. Before we identify these components or attempt to invalidate the postulates, we describe the coupled systems of man and nature in the Baltic.

The Baltic Europe System

The Natural System

The marine component of the Baltic Sea constitutes one of the largest low-salinity, brackish-water areas in the world, nearly twice the size of the American Great Lakes (Voipio 1981; table 7.1). Shallow and narrow entrances located at Kattegat and Skagerrack limit exchange with the North Sea (figure 7.1), resulting in a residence time of the total water mass on the order of 25 years. Freshwater inflow from large rivers in the north, combined with the heavier saline water from the North Sea, create a halocline, a sharp density gradient, at about 80 m depth between the upper, slowly outflowing, fresher surface water and the heavier, saltier inflowing deep water. Negligible tides contribute to the stability

TABLE 7.1
Some Oceanographic Data of the Baltic Sea

Statistic	Measure
Area	415,000 km^2
Length (N–S)	1300 km
Width (W–E)	1200 km
Average depth	60 m
Maximum depth	459 m
Sill depth	17 m
Volume	21,700 km^2
Residence time of water	25 years
Total drainage area	1,641,650 km^2

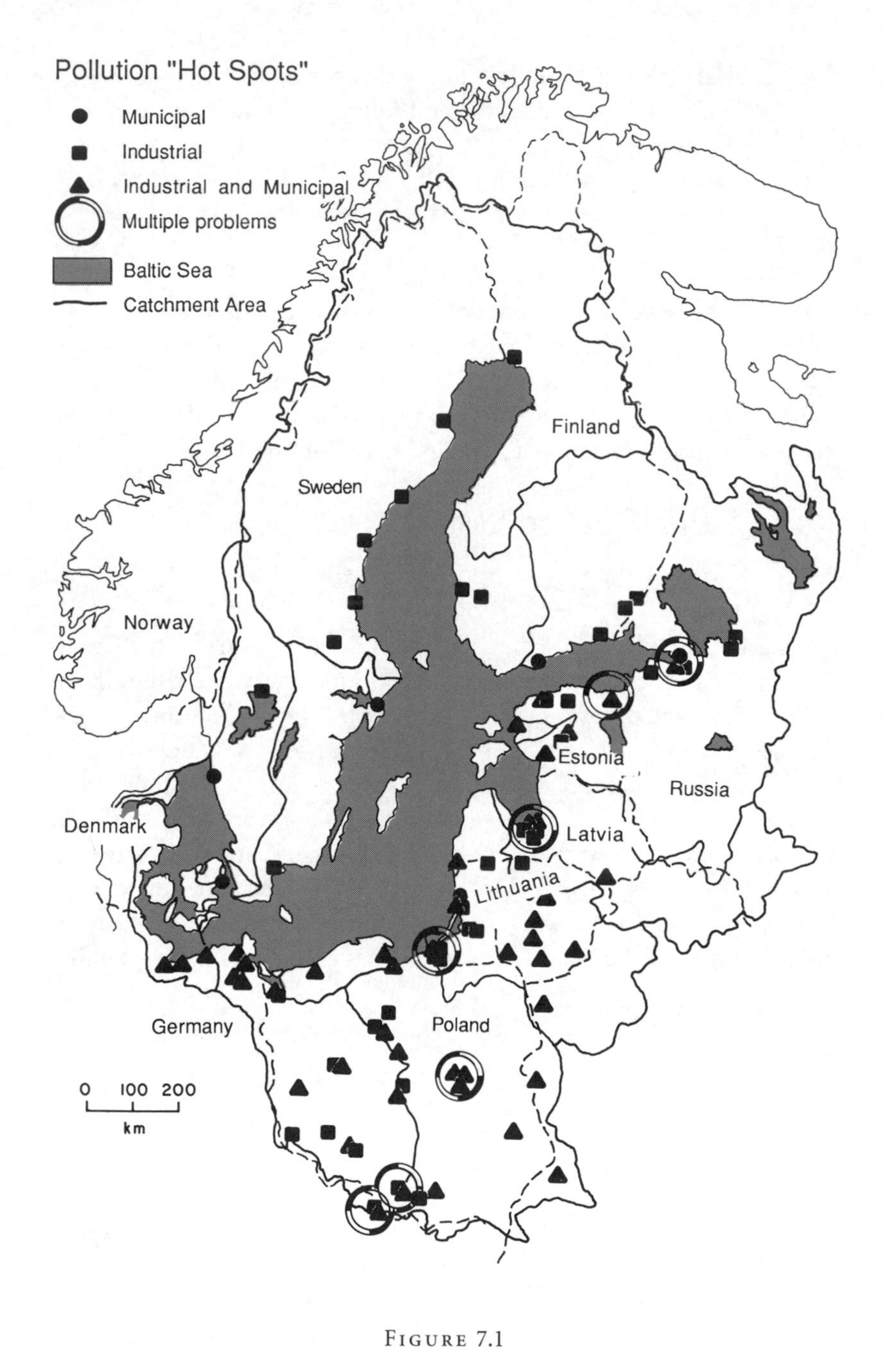

FIGURE 7.1
The Baltic Europe: drainage area, nations, basins, and pollution hot spots (based on HELCOM Task Force, pers. comm.).

of the stratification. Oxygenation of the bottom water takes place through the inflow of North Sea water; the surface water is oxygenated mainly through the mixing work of wind. The small but steady inflow through the straits is insufficient for maintaining either salinity or oxygen in the deep water. Large pulses of water, generated by the interplay of high- and low-pressure systems over the Atlantic, are crucial for the Baltic. The large-scale hydrodynamics are stochastic in time and size. The volume of saltwater intrusion is on the order of hundreds of cubic kilometers or less, and the effect of a medium-sized inflow through the sound is only felt in the middle of the Baltic after 8–9 months. So the system reacts like an internal combustion engine to spark, the difference being that the sparks come intermittently and the initiated processes of the engine thus decline to various degrees before the next ignition.

The biological system consists of a low number of freshwater and marine species (Jansson 1980). Freshwater plants, such as reed belts, pond weeds, and millfoils, dominate the shallow, muddy areas. The surface zone of rocky shores in the extensive archipelagoes (figure 7.2) are colonized by the freshwater alga *Cladophora,* whereas deeper zones are dominated by familiar North Atlantic algal taxa, such as the brown, marine bladder-wrack *Fucus vesiculosus* and red algae. The dominant macro-invertebrate in the Baltic is the blue mussel (*Mytilus edulis*), which forms extensive mats on deeper hard bottoms, where light is insufficient for seaweed.

The Baltic Sea could be described as a massive trap for organic matter. The soft sediment bottoms form the spatially dominant subsystem with a poor macrofauna comprised of the Baltic mussel (*Macoma balthica*), amphipods (*Monoporeia affinis* and *Pontoporeia femorata*), and the bristle worm (*Harmothoe sarsi*). The organic particles make the sulfur cycle important. A rich meio- and microfauna breaks down the organic material. The end product of the sulfur bacteria is the poisonous hydrogen sulfide, which, depending on the frequency of the North Sea inflows, sometimes covers more than one-third of the total bottom area.

The primary production of the microscopic, pelagic algae is similar in magnitude to the ocean (Elmgren 1984) and forms the basis of energy flow in the ecosystem. A heavy pulse of diatoms in the spring, nearly half of which sink to the bottom, provides nearly half the annual food supply for the soft bottom community. The weather during spring is crucial to the development of food for fish, because storms decrease the primary production of the algae. Following the peak in summer pro-

FIGURE 7.2
The rocky archipelagoes dominate most of the Swedish and Finnish coastlines and are a typical Baltic feature. The enlarged hard-bottom areas are covered by seaweeds, absorbing the nutrients coming from land, serving as spawning and feeding areas for fish and recreation areas for citizens. This view is from Östergötland Archipelago (O. Norman, Sv. Aero-Bilder, Stockholm).

duction, when the water is devoid of nitrogen, a bloom of the bluegreen, nitrogen-fixing algae *Nodularia spumigena* appears. These blooms sometimes cover the whole surface of the Baltic and draw large amounts of nitrogen from the atmosphere into the sea. In this manner the system compensates for the low limiting levels of nitrogen that exist during most of the year. The winter levels of nitrogen and phosphorus in the open waters have more than doubled since the 1960s (Nehring et al. 1990) and are approaching the eutrophic regime of the Vollenweider scheme (figure 7.3f).

The Baltic fish population consists of a mixture of freshwater and marine species; pike and perch can be caught in the same gillnet as flounder and cod. Cod, herring, sprat, and salmon are the most important commercial species. Cod and salmon have been overfished to the point where the open sea is currently closed part of the year. In

1985 the total fish catch was in the order of 1 million tons, but it has dropped rapidly (figure 7.3g). Fish stocks have decreased 30% since 1970 (Thurow 1989). The decline in the herring population, an effect of overfishing, has switched to an increase lately, but the declining cod population has a more complex background. Cod eggs develop while floating in saline water (salinity 14 0/00) (Nissling and Westin 1991) and under good oxygen conditions. These conditions normally occur in the deeper troughs of the Baltic: the Gotland Deep, the Gdansk Deep, and the Bornholm Deep (figure 7.3). Because of bottom water stagnation and the eutrophication of the Baltic, sufficient oxygen levels at the proper depths have only occurred intermittently during the past 15 years at the Bornholm Deep. A few successful single-year classes, such as occurred in 1985, will comprise the total population, much like the intermittent pulses of the refreshing North Sea water maintain the salinity of the Baltic.

The Baltic functions like a large funnel, receiving the runoff of water and particles from the surrounding landmass and the atmosphere. Some of this is filtered off and recycled in the rim of coastal vegetation. Some of the material is incorporated into the living world of the open waters for periods of weeks to months before settling onto the deeper bottoms. At the bottom, some material is consumed by living organisms, and when eaten by fish, is returned to the upper layers for a new cycle. Some material is broken down by the activity of organisms, including bacteria, and as inorganic, low-molecular-weight products returned to the faster flow of the biogeochemical cycles. Most of the material is transferred out of rapid circulation in the system and deposited as geological material. Thus the turnover time of sediments, even the top layers, is very long, occurring over decades rather than years. The absence of tides and the resulting strong stratification make the cold Baltic an effective nutrient trap, much like a continuously stocked but seldom cleaned refrigerator.

The Human System

The Socioeconomic Environment: A History of Battles for Space, Resources, and Governance

The lengthy birth of the Baltic Sea began when it was more or less cut off from the ocean after the last glaciation. Around 8000 B.C. the Baltic Ice Lake spilled into the ocean, but it took a long time before the Baltic

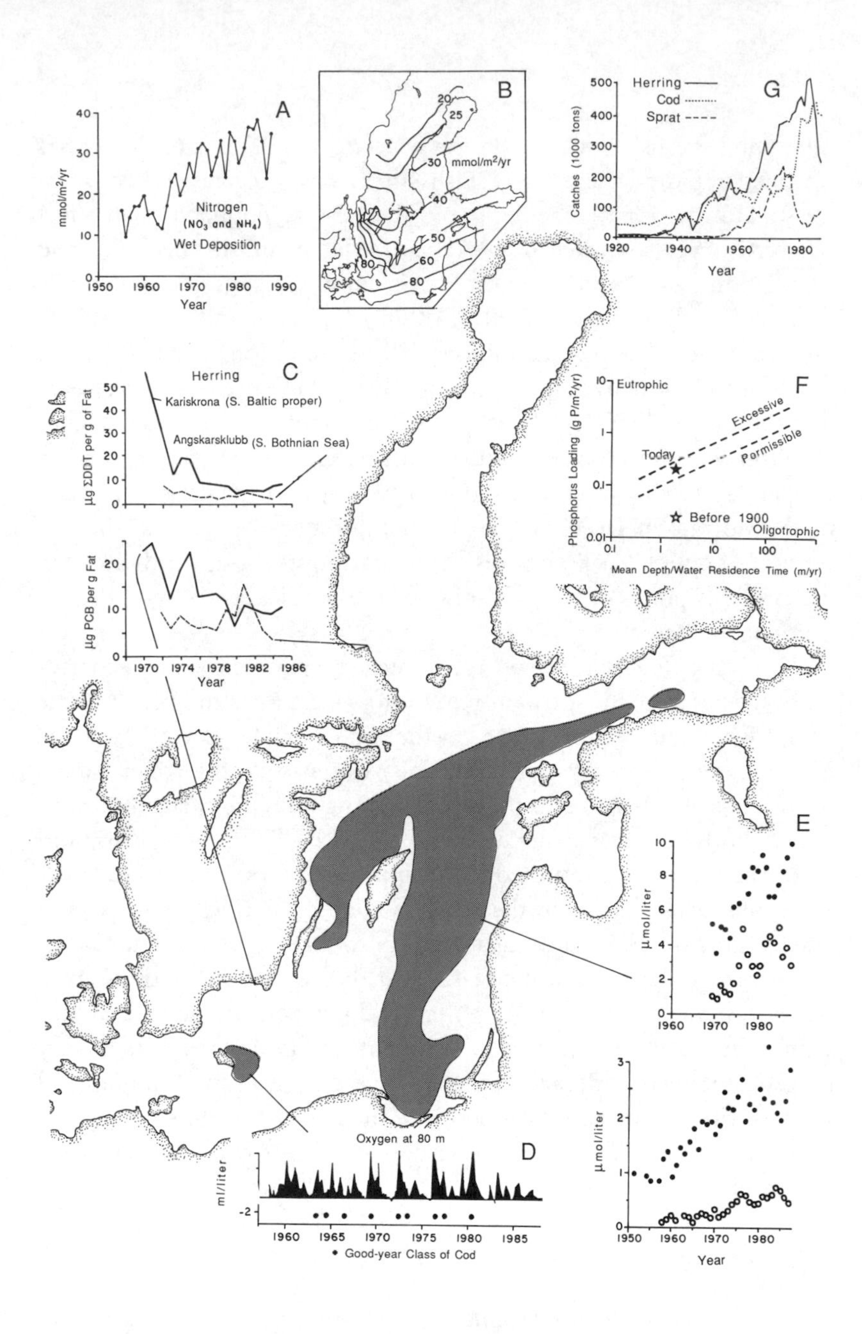

FIGURE 7.3

The changing Baltic Sea. The surface salinity isoclines are in parts per thousand (o/oo). The shaded bottom area corresponds to bottoms frequently with oxygen concentrations less than 2 ml/l. (A) Wet deposition field of inorganic nitrogen compounds (NO_3, NH_4) in mmol m^{-2} yr^{-1} (Stockholm Environmental Institute 1990). (B) Wet deposition of atmospheric nitrogen (NO_3, NH_3) in central Sweden in mmol turned into the largest brackish-water body of the world. As the increasing availability of natural resources increased because of technical development, the social conditions for the people living on its shores improved.

During the ensuing centuries waves of people pulsed through the Baltic Europe system, competing for space, resources, and governance. At the beginning of the Christian Era, the southern shores of the Baltic were occupied by German tribes, who were later displaced by Slavs. In the eastern Baltic region, Finns, Estonians, and people of Fenno-Ugric origin settled in the north, and various Indo-European tribes settled in the south. The Baltic area has a long history of commercial contacts and trade outside the region. The sea was and is an important means of communication linking peoples and civilizations. During the ninth to the eleventh centuries, the Vikings controlled the Baltic Sea, playing a constructive role in the formation of modern northern Europe, founding Denmark and Sweden in the west, and Novgorod in the east. Poland's independent existence originated in the tenth century, as a union of a number of Slavic tribes in the south.

Following the Viking Era the economy of the Baltic region was controlled by the Hanseatic League, a group of North European, mainly German, port cities that had a monopoly on the east/west trade. Herring, grain, fur, and wood were among the commodities shipped to the west. The Hanseatic League cooperated with the Teutone Order of German Knights, a powerful state along the southeastern Baltic coast, and allied with the Danes, who controlled the Straits (the Belts and the Sound). Denmark's collection of tolls after the fifteenth century from ships passing through the straits was a constant source of conflict.

By the early thirteenth century, Estonians and other Baltic tribes (except the Lithuanians) had been converted to Christianity by force and reduced to a state of serfdom (Christiansen 1980). The colonization of the eastern Baltic brought the Finns, Estonians, and Latvians into the

---

m^{-2} yr^{-1} (Stockholm Environmental Institute 1990). (C) Body burden of DDT and PCB in Baltic herring (Bernes 1988). (D) Oxygen content in the Bornholm Deep and successful spawning of cod (from Bernes 1989). [E] Trends of phosphorus (P) and nitrogen (N) concentrations (mol/l) in the surface (open circles) and 100-m depth water (filled circles) of the Gotland Sea (Elmgren 1989). (F) A Vollenweider diagram showing the trophic status in relation to phosphorus input for past and present Baltic (Larsson et al. 1985). (G) Fish catches (1000 tons) in the Baltic during 1900s (Bernes 1988).

cultural sphere of the West. The line along the Narva River and Lake Peipsi (Peipus) became the cultural dividing line between the Roman Catholic west and the Russian Byzantine east. The Lithuanians remained independent until the end of the 1400s, when they were converted to Christianity and brought into the Western sphere by Poland. The Polish/Lithuanian Union's victory over the Teutonic Order of Knights in 1410 at Tannenberg made it the largest and most populated state in the Baltic region.

After the decline of the Hanseatic League between the fifteenth and sixteenth centuries, Baltic trade fell into the hands of the Dutch merchant marine (Lisk 1967). Sweden acquired sovereignty over the Baltic until defeated by Russia in the Great Northern War in 1721. However, Baltic trade continued to be dominated by the Dutch and the English. With the onset of the industrial revolution in the mid-1900s, the Baltic region began to decline as a center of European commerce and Russia became the strongest power in the Baltic region.

In the Nordic countries, at the end of the eighteenth century, human populations increased rapidly because of a decrease in infant mortality. The higher populations increased the exploitation of natural resources and led to large-scale migrations, including the transatlantic emigration during the latter half of the 1800s (Odén 1980).

Following World Wars I and II, the political map changed radically. After World War I, Finland, Estonia, Latvia, Lithuania, and Poland became independent states, raising the number of littoral states to eight. During this period the states of the region developed highly specialized industrial activities. This period lasted only 20 years, until the Molotov/Ribbentrop treaty in 1939 and World War II. Following World War II, Estonia, Latvia, Lithuania, East Prussia, and Finnish Carelia were incorporated into the Soviet Union (U.S.S.R.) and Germany was divided. The U.S.S.R. became the strongest Baltic state and repeatedly tried to make the Baltic a *mare clausum,* while the NATO countries, equally persistent, argued for the Baltic as a *mare liberum.* During this period the former Baltic republics were colonized by Russians. In 1939, for example, 25,000 Russians lived in Estonia, whereas now more than 600,000, or 40%, of the population are Russian. This was a period of rapid industrialization and a transformation in the economic infrastructure. Three states—U.S.S.R., Poland, and the German Democratic Republic (GDR)—formed the Warsaw Treaty Organization, where a centrally planned system of managing their own national economics had been adopted. The Federal Republic of Germany (FRG) and Denmark belonged to NATO, whereas Finland and Sweden maintained a neutral foreign policy.

The postwar industrialization increased the load of pollution, prompting a series of international agreements regarding the Baltic. In 1972 delegates to a United Nations conference decided to strengthen environmental activity in the Baltic Sea area and find a political means for international collaboration. In 1973 FRG recognized GDR as a state, opening the door to multilateral cooperation among all seven littoral states. Two conventions, one on fishing and conservation in the Baltic Sea and the Belts (1973) and one on the protection of the marine environment of the Baltic Sea area (1974), were signed. These treaties led to a rapid evolution of supportive national legislation in each of the seven Baltic littoral and to cooperation among the states in many relevant areas. In 1975 the final act of the Conference on Security and Cooperation in Europe was signed by thirty-five states in Helsinki, Finland. The participating states, including United States and Canada, declared that "problems relating to the protection and improvement of the environment will be solved on both a bilateral and multilateral, including regional and subregional, basis, making full use of existing patterns and forms of cooperation."

Much has happened in the Baltic region since the 1970s. M. Gorbachev's glasnost and perestroika policies of the early 1990s served as a "switch" for the flow of material and information, and offered possibilities to change the Eastern European political and socioeconomic situation radically. The Berlin Wall was destroyed and the two Germanies were reunited. Estonia, Latvia, and Lithuania declared independence. The Warsaw Treaty Organization was disbanded. The process of democratization in Russia is now under way and Russia has adapted and become a member of the European Community. The Baltic Sea area is again totally open for communication and joint activities; a new period of exploitation is beginning.

Resource Use Patterns in the Baltic Catchment Area

The exploitation of the Baltic has gone through many cycles. The first natural resources to be exploited were iron, copper, timber, tar, hides, fur, hemp, flax, and fish. One consequence of this was the decimation of populations of larger fur animals. Shipping, navigation, and capital provided the prerequisites for overconsumption (Odén 1980) and led to the first overuse of natural resources during the fifteenth and sixteenth centuries (Odén 1980). Larger environmental effects have occurred since the early 1900s, and are related to discharges from various industries, such as sawmills, pulp mills, and chemical and manufacturing plants. Urbanization paralleled this process and increased rapidly after the innovation of the sewage system, based on the myth that the self-cleaning capacity of water is infinite (Lundgren 1974).

The nine littoral states—Denmark, Estonia, Finland, Germany, Latvia, Lithuania, Poland, Russia, and Sweden (figure 7.1)—were inhabited by 290 million people in 1989 (table 7.2). Only a small part of Russia (St. Petersburg, Kaliningrad) and Germany (Schleswig-Holstein, Mecklenburg) drain into the Baltic Sea. Approximately 80 million people live in the drainage area, with population densities ranging from 15–20 per km² (in 1989).

Crops and grass areas dominate the drainage areas in the southern Baltic; forests dominate the boreal coniferous biome in the northern areas. Pulp and paper production, chemical, and metal-working industries dominate the industrial sectors, but the importance of the Baltic as a natural resource lies mainly in fishing, navigation, and tourism.

Fishing is, and has always been, one of the most profitable trades for the coastal populations. Total annual catches have increased from 50,000 tons at the beginning of the century to almost 1 million tons in 1985 (figure 7.3g). Sprat, herring, and cod dominate the harvest, whereas the highly rated salmon and eel are strongly overfished. Be-

TABLE 7.2
Demographic Data for the Baltic Littoral States (1990)

Territory	**Area** (10^3 km²)	**Catchment Area** (10^3 km²)	**Population** (10^6)	Population in Catchment Area (10^6)	**Coastline** (km)
Denmark	43	31	5.1	4.4	7500*
Estonia	45	45	1.6	1.6	3780*
Finland	338	304	5.1	5.0	19000*
Germany	357	24	77.3	2.4	615
Latvia	65	64	2.65	2.6	~500
Lithuania	65	65	3.7	3.7	99
Poland	312	312	38.0	38.0	524
Russia	1708	326	~150.0	9.0	~500
Sweden	450	470	8.3	~ 8.5	2500
Baltic total	3383		291.9		
World	14,8800		5162		

*Includes archipelagos.

tween 1975 and 1985 the total biomass of fish declined by almost 30% (Thurow 1989), and the decline continues in some stocks because of overfishing and pollution. In ancient times fluctuations of fish stocks were common, resulting in changes in large-scale trading patterns. One of the reasons for the demise of the Hanseatic League was a decline in herring stock because of overfishing (Boczek 1989). During the twentieth century, however, other factors have contributed to the fluctuations. The salmon population is dependent on extensive stocking with artificially reared smolts, since most of the original spawning sites in the larger rivers have been affected by the construction of hydroelectric plants. Eutrophication in the Baltic has favored pelagic fish, such as herring and sprat (Schulz 1970). As previously mentioned, oxygen levels at a depth of 80–120 m are often too low for the successful spawning of cod. This is due to the stagnant conditions of the bottom water and the increase in oxygen-consuming material resulting from fertilization. The fish stocks in the open sea are regarded as a common resource for exploitation.

Navigation and the freedom of commerce have been the subject of legal discussions for centuries. Following World War I, the 1919 Treaty of Versailles stipulated complete freedom of access to the Baltic Sea for ships of all nations. This open sea policy has been stressed frequently up to the present. The Baltic was recognized as a special area by the International Maritime Organization (IMO) 73/78, which means that special protection measures by ships should be taken during navigation and in harbors. For the marine environment the biggest danger is the bulk transportation of oil and chemicals by tankers. During the last 15 years more than ten big tanker accidents have occurred, the biggest one in 1981, when 16,000 tons of crude oil from the *Globe Assimi* polluted the marine environment, including the beaches near Klaipeda, Lithuania. The number of merchant ferries and sailing boats traveling from one Baltic harbor to another is growing each year. The role of the navy, however, is rapidly decreasing and will probably be small in the future, because of changing strategic weapons.

Tourism as a resource is based on the diverse and beautiful coastal area of the Baltic (figure 7.2). Sandy beaches in the southern and southeastern areas; rocky archipelagoes in the north; the beautiful, historical towns of Lübeck, Visby, Tallinn, and Riga are some of the attractions. More than ten shipping companies and various air and railway companies transport tourists and businessmen from state to state and from town to town around the Baltic. Recently, the number of tourists from the United States, Japan, and other countries outside Europe who have visited the Baltic littoral countries has exceeded 10–15 million per year. The midsummer and Christmas trips to Lapland (the midnight sun!) are among the most popular excursions.

The archipelagoes have always been loved by sailors. Now the new wildlife coastal areas (some of them former U.S.S.R. military areas), including biosphere reserves, protected areas, and wetlands, can be used by specialized tourist groups. The curative mud medical centers in Kemeri (Latvia), Kuressaare (island of Saarema, Estonia), and Haapsalu (Estonia) have been well known in Eastern Europe from the beginning of the nineteenth century. The basis for tourism is a well-protected and managed environment.

Management Institutions

The management institutions for the Baltic span hierarchies ranging from international agreements to local entities. The international arrangements are governed by the Convention on the Protection of the Marine Environment of the Baltic Sea Area, signed in Helsinki, Finland in 1974 by seven of the littoral states (Denmark, Finland, GDR, FRG, Poland, Sweden, and the U.S.S.R.). The convention came in force in 1980 and the executive agency—the Baltic Marine Environment Protection Commission (HELCOM)—opened in Helsinki. The aim of the Convention is to ensure that all parties take appropriate measures to prevent and reduce pollution and to protect and enhance the marine environment of the Baltic Sea. The convention was the first international agreement to prevent or limit both waterborne and airborne pollution from the land or ships. In 1992–1993 a new convention text, covering the entire catchment area of the Baltic, was signed by Denmark, Estonia, Finland, FRG, Lithuania, Latvia, Poland, Russia, Sweden, and the European Economic Community (EEC). According to the Ronneby Ministerial Conference (1990 in Sweden), the Joint Comprehensive Environmental Action Program (JCP) should be carried out in collaboration with four international banks: the World Bank, the European Bank for Reconstruction and Development, the European Investment Bank, and the Nordic Investment Bank. The program, approved in 1992 by the Baltic Sea Environmental Declaration, aims to facilitate restoration of the ecosystem of the Baltic Sea via concerted efforts and activities within the whole catchment area. This is the first environmental managing program to be supported by international banks.

Most of the Baltic Sea states are members of different global and/or European conventions connected to environmental management in the Baltic region (IMO 1973 and 1978, London and Bern 1979, Geneva 1979, Basel 1989.) and are concerned with marine and airborne pollution, and conservation of European wildlife and natural habitats. The bilateral cooperation between different Baltic states has been very important. The Joint Soviet/Finnish Working Group on Protection of the Gulf of Finland, started in 1968, provides a good background for further cooperation between the Baltic states (Helsinki Convention 1974). The cooperation between Sweden and Finland with regard to the Gulf of Bothnia also has a long history and may help in the management of the pulp-paper and chemical industries for the benefit of the environment.

A number of nongovernmental organizations (NGOs) have been formed to address issues of the marine environment in the Baltic region. The Baltic Marine Biologists (BMB) was founded in Rostock, GDR, in 1968; the Conference of Baltic Oceanographers (CBO) held its first meeting in Helsinki in 1957; and Greenpeace International was established in Lewer, UK, in 1971. The oldest NGO is the Conference of Baltic Hydrologists (the "father" of CBO), founded in 1925, although its work was interrupted by World War II and its last conference took place in Berlin in 1938. The New Hanseatic League (head office in Lübeck, Germany) and the Association of the Baltic Sea Towns (Gdansk, Poland 1990) perform important tasks to protect the environment from pollution by cities. The NGOs perform many roles: doing scientific research, preparing background documents for environmental management purposes, and acting as pressure groups that push governmental authorities to implement promptly the recommendations and decisions adopted by the intergovernmental organizations.

In 1973 the Baltic Fishing Convention (the "Gdansk Convention") was signed by six Baltic states. Another group—including FRG (1977), EEC (1984), Estonia, Latvia, Lithuania, and Russia (1991 and 1992)—later joined the convention. The major aims of the convention are to prepare recommendations regarding total allowable catches and to coordinate relevant scientific research. In 1978 the Baltic was divided into fishery (or equivalent) economic zones, subject to intergovernmental decisions. The International Baltic Sea Fishery Commission in Warsaw (within the International Council for Exploration of the Seas [ICES]) was created to establish the appropriate level of fishing effort for the various fish stocks and to serve as secretariat of the convention. In the last several years the commission has often exceeded the catches recommended by ICES leading to overfishing of some stocks. Because of short-term economic pressure, the optimal catch levels have not been achieved.

The activities of environmental management in the Baltic Sea region are connected with other major regional sea projects, such as Chesapeake Bay, the Mediterranean Sea, the North Sea in close cooperation with the United Nations Environmental Program (UNEP), Joint Oceanographic Commission (JOC), World Meteorological Organization (WMO), and several others, which are also observer organizations for HELCOM.

Environmental issues in the Nordic states and Germany are handled by the ministries of the environment under different subdivisions. The management of fisheries, navigation, and protection of the marine environment belongs to other authorities.

The political situation in the Baltic area has changed drastically during the last few years. The Soviet Union collapsed and three republics—Estonia, Latvia, and Lithuania—regained their independence. Russia is part of the Baltic community represented by St. Petersburg and Kaliningrad (former Königsberg) districts. The two former German states, FRG and GDR, united in 1989. The environmental management policy in these countries therefore has two versions, one old and one new. Eastern Germany is managed by the Federal Ministry of the Environment Protection of Germany according to its rules and responsibilities. Poland and the three Baltic States have had to rebuild their Environmental Management systems, taking into account the international legislative acts concerning the Baltic Sea region under HELCOM, EEC, and others. In the Russian Federation all responsibility is given to the new Ministry of Environmental Protection, which has to manage through the local district departments of St. Petersburg and Kaliningrad.

Estonia is one example of the eastern Baltic states. The environmental authorities include the Ministry of the Environment and the district environmental departments, each of which has different functions and responsibilities. The districts are relatively independent units of local governments. They are responsible for issuing licenses and permits for the use of natural resources, permissible amounts of pollutants, and control over polluters. The Forest Department and the Fisheries Department are responsible for their respective problems. A very important issue for Estonia is the management of biosphere and nature reserves and wetlands. Matsalu Nature Reserve is close to the internationally known West-Estonian Inlands Biosphere Reserve. The goals of Estonia's environmental policy are to provide the conditions for stable development and to safeguard the state's environmental and natural resources. The development of environmental legislation, quality standards, policies and strategy of management of natural resources, environmental protection activities, and efficient economic measures are priority objectives. The framework for environmental legislation is provided by the "Law on the Protection of Nature in Estonia" (1990). A number of specific acts are being developed. Some of these acts are based on legislation of the Republic of Estonia that was in force before 1940; another part was approved during the Soviet occupation (1940–1991) and is based on corresponding U.S.S.R. regulations and acts. These acts and regulations correspond to those used in the regional conventions (HELCOM and others) and in the European communities.

Environmental management in the Baltic Basin cannot be successful without international cooperation between governments and nongovernmental organizations. In the next section we examine how these organizations have responded to surprises in the natural system.

Surprises and Responses

Events in Nature

The Beating of the North Sea Pulse

Heavy storms and large supplies of fish are examples of surprises that can be traced back to the twelfth century (Voigt 1983). A scientific registration of the dynamics of the Baltic Sea began toward the end of the nineteenth century. Since 1900 the Fishery Board of Sweden has maintained a continuous registration of the main hydrographic variables, constituting the longest existing record of its kind in the world; it is available in the archives of ICES. The more dramatic events, in terms of both seriousness and scale, appear to occur when an industrialized society is emerging and growing with increasing speed around the Baltic Sea.

Through the 1950s the problems of the Baltic were restricted to the fisheries and to the usual difficulties of utilizing a natural resource of fluctuating and poorly understood quantity. In Sweden concern for the water resources first focused on the freshwater system following the detection of extensive pollution. A strong movement for cleaning up the lakes and rivers, and a corresponding need for strengthening education and research within limnology, swept through the country. The campaign was enthusiastically and skillfully led by Professor Wilhelm Rodhe, a long-time president of the International Society for Limnology.

Pollution concerns were extended into the coastal areas of the Baltic as the restricted bays and inner archipelagoes, such as the Stockholm Archipelago. This was the scene when Dr. Stig Fonselius from the Swedish Fishery Board of Sweden published his "Hydrography of the Baltic Deep Basins III" in 1969. Fonselius (1969) showed that stressed conditions of the Baltic Sea were not restricted to the coastal areas but were typical for the stratified parts of the open Baltic Sea! His analyses of the long hydrographic series in the Baltic proper indicated at least eight stagnation periods between 1957 and 1969, when hydrogen sulfide was formed in the bottom water. The oxygen content in the deep basins below the halocline had decreased from 1902 to 1967 and was near zero. The longer time series of measurements showed some thirty stagnation periods, characterized by low oxygen concentrations, hydrogen sulfide formation, and bacteria as the only life forms. Almost one-third of the total Baltic Sea bottom had been affected. The main reason was traced to the intermittent inflows of North Sea water, interfoliated with the stagnation periods. Dickson (1973), through data simulations, showed the flows to be a function of the distribution of high- and low-pressure systems over the Atlantic. The Baltic ecosystem is thus driven by a pulse that is as difficult to predict as next month's weather!

The pulses of North Sea water influence the rest of the system. Phosphorus, previously bound in insoluble form in the oxygenated upper layers of the sediment, is released during the oxygen-free periods to the bottom water and stimulates plant growth in the upper water layers. The nitrogen compounds are transferred by bacteria to nitrogen gas at low oxygen levels and escape to the atmosphere. The inflowing water from the North Sea contains, depending upon the season, animals and plants that develop in the now oxygenated bottoms in the Baltic. A classic example is the big inflow of 1923, which brought high saline water to Kattegat, inducing spawning of the usually sparse haddock. The larvae were transported to the southern Baltic and gave rise to a ten-fold increase in haddock catches during 1925–1926 before the remaining fish returned to the North Sea to spawn. Haddock are not normally found in the Baltic because the low salinity does not allow for successful spawning. The effects of this surprise-induced condition are further strengthened by the increase of organic matter in the form of dead, microscopic plants and animals falling like rain from the dense clouds of plankton in the surface layers, which have been stimulated by the increasing inflows of nitrogen and phosphorus from the atmosphere and land.

The Mercury Problem

Parallel to the debate on "dead bottoms," as they were called by the newspapers (although, in fact, they are crowded with millions and millions of sulfur bacteria), mercury emerged as another crisis. The problem started in the catchment area, where Borg in 1958 warned about high concentrations of mercury resulting from its extensive use in agriculture as a fungicide. Subsequent work (Borg et al. 1966) linked concentrations to periods of spring and autumn sowing (another pulsing!). Fish and other organisms were found to contain high amounts of methyl-mercury, and Johnels and colleagues (1967) first showed increased levels of mercury in fish from the Baltic—pike from the Stockholm Archipelago. An extensive mapping of the mercury content in different organisms in the Baltic ecosystem, especially fish and birds, was undertaken (Ackefors 1971). Pike, which was used as a test organism, had a concentration of mercury 3000 times higher than that of the water. Fish-eating birds, such as osprey showed even higher values, and the mercury content in hair of humans who consumed a lot of fish was in some cases fifty times higher than normal.

The main source of the mercury increase in the Baltic system was the paper and pulp industry, where phenyl mercury was used to prevent slime formation and as a fungicide to improve storage conditions. Other related industries (chlorine alkali) were also responsible for the increase in mercury. But mercury was put into circulation through other activities in society: electrical installations, dental hygiene, diffuse atmospheric fallout, and garbage dumping.

By the early 1970s mercury levels were down to the values of the turn of the century as a result of bans on its use. However, other dangerous substances were on the environmental scene—the chlorinated pesticides.

The Chlorinated Pesticides

In 1966 the Swedish chemist Sören Jensen, when analyzing pesticide residues, was able to identify some previously unknown substances as polychlorinated biphenyls (PCB) occurring with DDT and its metabolites (Jensen 1966). Samples were taken of different organisms in the marine food web (blue mussels; herring; plaice; cod; salmon; gray, common, and ringed seals; guillemot; white-tailed eagles; and heron) along the Swedish Baltic and west coast. Jensen and colleagues (1969) found a roughly a tenfold content of DDTs and PCB in the Baltic organisms compared to corresponding species on the west coast. The top consumer of fish, the white-tailed eagle, showed an increase of as much as 100 times.

Clear signs of these effects became apparent. The population of white-tailed eagle declined, especially along the Swedish Baltic coast (figure 7.4). Out of ten pairs only two showed a successful hatching. Further analyses showed that DDT caused thinning of the eggshells (Odsjö and Edelstam 1972), which was one reason the hatching of the birds failed. In fact, the species was close to extinction. Other fish-eating birds with very high levels of DDT and PCB did not show a decrease in numbers. Levels of these compounds in seals were ten times as high as in populations in the Netherlands, Great Britain, and Canada. Apparently, the toxins were transferred to the pups via the mother's milk and perhaps via the blood.

The Oil Spills

During the 1960s another threat darkened the skies of the blue, glittering Baltic Sea, oil spills from the increasing fleet of freighters that maintained the flow of energy and material between Baltic Europe and the world market. In addition to ten big accidents, many small spills were registered (Simonov 1990). The first of the few spills investigated occurred when the the Finnish tanker *Palva,* carrying Russian crude oil, ran aground in the southern part of the Archipelago Sea in May of 1969. The 150 tons of oil soon covered an area of 200 kmFD (Heino 1972). In October 1970, 1000 tons of medium crude oil spilled in the southern part of the Stockholm Archipelago and wiped out nearly the total fauna of the small bay where some 400 tons of the oil had accumulated (Notini 1978). In both of these cases, the effects were moderate. In the *Palva* case, no effects were seen a year later (Pelkonen and Tulkki 1972), and the heavily affected small bay south of Stockholm recovered after 4 years (Notini 1978). The number of spills increased, however, in Swedish waters to some 200 accidents per year (Landner and Hagström 1975) and inspired the scientific community to begin experimental studies. Lindén (1974) showed that low concentrations of dispersants decreased hatching of eggs and deformed larvae of herring. Sophisticated studies of the effects on the ecosystem were made by Ganning and Billing

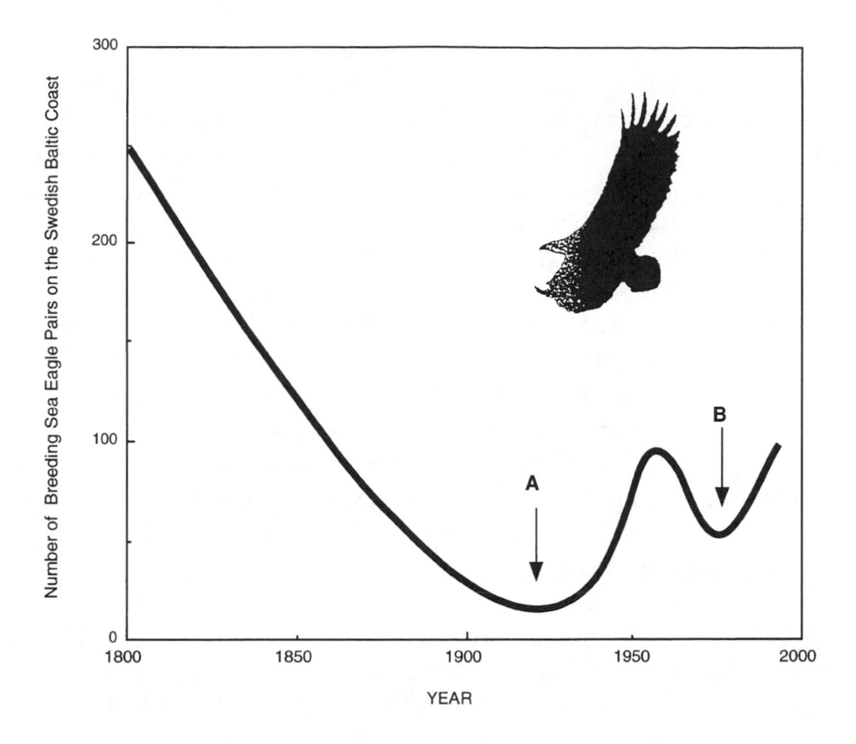

FIGURE 7.4
Fluctuation of the population of the white-tailed sea-eagle (*Haliaeetus albicilla* L.) on the Swedish Baltic coast. The number of breeding pairs declined rapidly during the 1800s due to hunting, as the sea-eagle was regarded as a strong competitor of the huntable wildlife fauna. After protection in 1924 (arrow A) the population increased only to be "taken by surprise" in the middle 1950s, when the effects of accumulated DDT caused a rapid decline. The Swedish Society for Conservation of Nature (SNF) started in 1971 (arrow B) the campaign "Project Sea-Eagle" to save the population—for example, by winter feeding with toxic-free meat (slaughter, spill, etc.). DDT was banned in 1975 and the population is now steadily increasing. This is an excellent example of how humans are able to change an undesirable development in a large and dynamic system when citizens and governments work in concert. (Data from Dr. Bjorn Helander, Swedish Society for Nature Conservation, Stockholm.)

(1974), who measured the community metabolism of the bladder wrack (*Fucus vesiculosus*) and showed that the net primary production decreased at oil concentrations found in sheltered areas and that the effects of the detergents used to disperse the oil were far more toxic than the oil itself.

A real test case occurred in October 1977, when the Russian tanker

Tsesis went aground on the rocks close to the Askö Laboratory, the marine field station of Stockholm University. Scientists from Askö, in collaboration with the Swedish Water and Air Pollution Research Institute and the American Spilled Oil Research Team of NOAA (Lindén et al. 1979), were able to follow the impact of 1100 tons of spilled oil. Since most water fowl had already migrated south, the minor amount of oil that was spilled was not treated with detergents, which kept the effects small. The pelagic community recovered within a month; the seaweed and their fauna recovered after little more than a year. The soft bottoms, however, which had received a fair amount of oil through sedimentation, were clearly affected 6 years later, with clams instead of crustaceans dominating one population (Hansson 1984). The understanding from these studies and other surveys improved the emergency response system and lessened the threat posed by oil spills.

Eutrophication Again!

In 1980 the Laholm Bay in Kattegat, on the Swedish side of the Baltic (the transition area between the Baltic and the North Sea), experienced an extensive kill of the bottom fauna. The foul-smelling sediments were full of dead mussels, clams, worms, and other bottom animals as a result of low oxygen levels and the formation of hydrogen sulfide. Hundreds of tons of clams washed ashore, and no fish were to be found on the usually very productive bottom areas. The ensuing investigation confirmed the first hypothesis that the cause of the event was excessive agricultural fertilization from the land surrounding the bay. The stratification of the bay, with the outgoing nutrient-rich water from the Baltic on top and an in-going colder and saltier water mass at the bottom, usually gave lower oxygen values at the end of the summer, because of the increased decomposition in the heated water. Unexpectedly, vigorous blooms of tiny algae, dinoflagellates, lit up the dark autumn evenings with their sea fires, and this extra organic material was sufficient to initiate a period of anoxia.

The Seals Are Becoming Extinct!

A British expert on seals, O. Hook, visited the northern Baltic in 1962 and 1963 and was struck by the absence of gray seals on the ice except for the carcasses of dead animals shot and recovered by Finnish fishermen on a few large ice floes between Gotland and Estonia. The seals compete with the fishermen for cod, salmon, herring, and eel. Hence large numbers of both gray and ringed seals have been killed over the years. They do eat a lot of fish. The total seal population during the 1920s was estimated at 200,000 animals, and it consumed as much herring as the fishermen do today (Jansson and Zuchetto 1978; Folke and Kautsky 1989)!

The suspicion arose that increased amounts of mercury, DDT, and PCB found in seals may have contributed to the drastic population decrease (Hook and Johnels 1972). Subsequent research confirmed this. In Bothnian Bay only 27% of the ringed seal females were pregnant, whereas rates of 80% to 90% are typical for areas with low pollution levels (Helle et al. 1976). Four months before the normal pupping seasons half of the nonpregnant females showed higher levels of both DDT and PCB, along with uterine scarring, indicating resorption or abortion of embryos. The incidence of lesions, uterine stenosis, and occlusions increased in gray and ringed seals (Bergman and Olsson 1985). Experimental studies on other mammals (Lundberg and Kihlström 1973; Kihlstrom et al. 1992) pointed to PCB as the responsible factor. During the 1980s other morphological effects from PCB poisoning were detected: deformation of flaps and jaws, loosening of teeth, and bleeding of the soft tissue in the mouth, indicating disturbances in the endocrine system. The ban of DDT and PCB has had the intended results; the seal population is now slowly recovering—too fast according to the fishermen and too slow according to the scientists.

Ringed seal populations have not fared as well. The population in the heavily polluted Riga Bay (10,000 animals in the late 1950s) is now down to a few hundred (Kuresoo and Leibak 1992).

The Petroleum Hydrocarbons: Traffic Leaves Its Fingerprints

Most of the previous oil pollution research has been concentrated on the more conspicuous events—accidental oil spills or intentional blowing of tanks to get rid of dirty ballast water. During the mid-1980s the focus was on the diffuse spill from urban areas containing combustion products from cars, industries, house heating, and runoff substances from ferry and leisure boats (Broman et al. 1987). Broman and Ganning (1985) used the water-filtering mussels and clams as indicators of this flow of substances and demonstrated a striking gradient from the inner part of the Stockholm Archipelago with mean levels of up to six times those typical of more open waters. A special case is that of the daily ferry traffic between Sweden and Finland through the Archipelago Sea, which has devastating effects on the shores and the bottoms because of erosion of the bottoms and algal belts, resuspension of sediments, and spills from the boats (Rönnberg 1981).

Heavy Metals: A Heavy Problem?

Although heavy metals had been recognized as pollutants for many years, they became an important issue in the 1980s as studies revealed their presence in many top consumers. For example, herring from the Baltic had about twice the cadmium concentrations as those in the Kattegat, whereas concentrations of copper, lead, and zinc were similar between the two sites (Odsjö and Olsson 1987). The reason for this is still unknown, but the focus on the discharges from Swedish industrial plants uncovered significant reactions in the animal communities outside the plants: greatly impoverished fauna, skeletal deformities, and disturbances in the physiology of the bottom fish (Bengtsson et al. 1985). Studies of the metal content in migrating shore birds (such as sandpipers) pointed to cadmium as a possible future threat for aquatic birds and migrating birds as a vehicle for transporting hazardous substances from sea to land and from industrialized areas to remote territories (Blomqvist et al. 1987).

The well-developed pulp mill industry in the Gulf of Bothnia became a target for multidisciplinary studies on the effects of the effluents on the natural system (Södergren et al. 1988). Bleached mill kraft effluents caused severe disturbances in the physiology of fish and to critical processes such as reproductive ability and immune defense (Larsson et al. 1988). The affected area outside the mill turned out to be far larger than previously thought. The important brown alga *Fucus vesiculosus* was shown to be very sensitive to pulp mill effluents (Lindwall and Alm 1983; Lehtinen et al. 1988).

War Pollution

In addition to the surprising and far-reaching degradation of the environment around former U.S.S.R. camps in the previously occupied countries, where even the groundwater is seriously affected by the leakage and discharge of oil and gasoline, fishermen often report that bombs and containers holding chemicals such as mustard gas are dreaded catches in their trawls. In fact, according to Directive 28 of the Potsdam

Conference, about 12,000 tons of toxic chemical warfare agents—including 50% mustard gas, 20% arsenic-containing substances, and 20% irritants—have been dumped northeast of the island of Bornholm and southwest off the Latvian coastal town of Liepaja (Anon. 1993a). Unsealing may last from 10 to 400 years from time of dumping. Mustard gas and "arrinie oil" will remain as jellylike leaking substances in the sediment, whereas insoluble substances, such as adamsite and chloracetophenon, will remain in the sediment.

Beware of Bathing!

Between the two world wars the southern coasts of the Baltic provided summer residences for the northern European continent as well as Swedes and Finns. Many families owned summer houses along the Estonian and Latvian coasts. Around the beaches in Poland and Balticum (Estonia, Latvia, and Lithuania) large continental hotels still bear witness to the Happy Thirties and the health care culture. Why shouldn't this natural resource flourish again now that the borders are open? One has a strange feeling when strolling along the miles-wide, dazzling sandy beaches at Sopot in Poland, Riga and Jurmala in Latvia, and Pärnu in Estonia, which are washed by an inviting clear blue water, and reading the signs warning against swimming, even against sitting in the sand, because of the risk of infection and skin diseases. Improved wastewater treatment is badly needed! Recently, some progress has been made, for example, in Riga and Pärnu, which now have functioning sewage treatment plants.

Nuclear Power Plants and Radioactivity

The more or less proven fact of the unsafe state of the large number of nuclear reactors in the former U.S.S.R. was exemplified by the Chernobyl catastrophe, which affected both the Baltic Sea drainage area and its ecosystem. Monitoring ^{131}I levels in the brown seaweed *Fucus* spp. from several coastal stations in cold temperate waters of the Northern Hemisphere 18 days after the accident yielded the highest value—31.7 Bq/g dr wt—in *Fucus* from the Askö Laboratory area (Druehl et al. 1988). The increasingly free release of information regarding both the past and present conditions of the plants indicates the radiation problem, including the safety of the reactors and the storage of nuclear waste, as one of the biggest environmental problems in Europe.

Killer Algae and Seal Deaths

Along the Swedish west coast, including the Kattegat, since the 1970s, blooms of dinoflagellates and mass invasions of jellyfish (both nuisances to fishermen and tourists) have increased. In 1988 a new type of bloom appeared, dominated by the microscopic alga *Chrysochromulina polylepis.* It moved like a hurricane, coming from the German Bight, sweeping along Denmark, Sweden, and northward along the Norwegian coast, wiping out red algae, starfish, mussels, clams, and fish on the bottoms down to a depth of 10 m (Rosenberg et al. 1988). The mussel cultures in the fjords were struck heavily in the first stage, before a warning system incorporating surveys from air, sea, and land made a transport of the mussel rafts to the open sea possible. Catastrophe meetings with ministers, director generals, scientists, and federal representatives followed the event and guided the field work of sampling and quantifying the bloom. The field laboratories used as centers for the operation were crowded with media people, coming from every direction with every type of transportation and effectively blocking telecommunications between the scientists. The bloom declined in 10 days, leaving behind barren sea bottoms, dead mussels, angry fishermen, and exhausted scientists.

At the same time the seals began to die along the Swedish west coast and in the southern part of the Baltic. A similar fish kill, but of much smaller dimensions, occurred in the Finnish part of the Archipelago Sea in 1990 (Lindholm and Virtanen 1992).

Events in the natural system thus washed over the Baltic society during the 1960s, 1970s, and 1980s like waves of dirty water, making people gasp for breath, ask for explanations, and shout for immediate action from people other than themselves. Who were those "other people" and how did they react?

Responses in Society: Barriers or Bridges?

Scientists as Alarm Clocks and Bees in the Environmental Bonnet

Our analysis of the very confused and erratic reactions of society to natural events is simplified to exemplify our postulates. Reading a newspaper during the 1970s was like diving into a jungle of headlines announcing the findings of new and stronger toxics in the environment, new candidates for the title of worst threat to the Baltic, interviews with scientists in white coats or diving suits, reports of new organizations for analyzing the pollution threat, and repeated accusations from citizen organizations of governmental passivity. Society did not respond to events in the natural system in a simple way. Society was itself responsible for these surprises in blind beliefs in an infinite and stable environment that could absorb and neutralize any interference. It responded also to its own reactions in debates on the realism of the predictions. We can here give only a few examples of this process, and mainly from our own countries.

The "Dead Bottoms" Stir Up

Fonselius's report in 1969 on large, anoxic, bottom areas in the Baltic had an immediate effect on scientific society. The first one to react constructively was Professor Svante Odén, the "father of acid rain." He was trained as a soil scientist at the School of Agriculture, Uppsala, to integrate natural events and material processes in a dynamic system like the soil. He used his inventive and original mind to adopt to the larger-scale challenge. I (B.O.J.) vividly recall how, listening to the radio on Christmas Day in 1967 at our Askö Laboratory field station, I jumped at hearing that Svante Odén, together with the director general of the Board of Technical Development, had received a grant to prepare a research program to explain the Baltic environmental conditions. I phoned him immediately, saying, "We have worked in the Baltic for 10 years and want to participate!" His response was, "I know; you are already in."

Odén organized a handful of scientists from Swedish universities and a colleague from the Institute of Marine Research, Helsinki. The ensuing discussions resulted in a research program to combine process studies and mass balance computations with synoptic mapping of chemical and biological components. The main aim was to find out if the present condition of the Baltic was a natural trend or the result of human activities. The program was met with heavy criticism, some saying that the goal was too vague and the project an overly ambitious and general program for marine research. One scientist from a governmental board wrote to the Natural Science Research Council (NFR), which had allocated the money, complaining that the project was too large, since it covered most disciplines except the fisheries. The responsible persons were mostly engineers with no previous background in Baltic research, no contact had been made with the governmental laboratories, and the character of the project was too applied for the Natural Science Research Council. Odén's action and his public arguments—including his undiversified statements concerning the acidification of the lakes in western Sweden caused by airborne, industrial pollution—had seriously harmed the reputation of the council. The project belonged to the Environmental Protection Board, to be carried out in collaboration with concerned offices and institutions, said the governmental scientist.

The Odén group fought for independence to shape the project and carry it through with money from research-financing organizations. The controversy resulted in what somebody called the "night of the director generals." Representatives of the government, together with the heads of the Environmental Protection Board, the Meteorological and Hydrological Institute, the Technical Board of Development, the Geological Survey, and the Fishery Board, decided that the problem was an affair for governmental institutions. In late December 1968 the government commissioned the Swedish Environmental Protection Board (SEPA) to review and coordinate the research and investigation activities in the Baltic in collaboration with concerned institutions. In view of the of the size of the project in terms of time, space, financing, and importance to society, this decision was historic.

The Swedish Environmental Protection Board tried to take a firm hand from the beginning, and we scientists balked. We felt the program was originally ours, now presented with a few, but unacceptable, changes. The discussions with the SEPA were performed in the traditional, formalistic way with polite tours and obscure comments, although we all knew we had the same negative opinion. I (B.O.J.) could not stand the dissimulation, spoke out, and caused considerable embarrassment, which polarized future negotiations. We were asked to present proposals for respective suggested research tasks and did so. When the irritations continued, the Askö Laboratory returned its grant.

In 1970 an American political journalist saw the Baltic's pollution problems as an interesting playground for scientific and political interaction, interviewed the participants of the debate, and published an article in the international journal *New Scientist* titled "The Lesson of the Askö Lab" (Jameson 1971). The article described how the laboratory had withdrawn from the collaboration and wanted to work on its own, following the theories of the American ecologist Howard T. Odum of the University of Florida, with whom the director of the laboratory and his scientist wife were going to spend a year. Even though the schism had been resolved and the universities were working with the governmental institutes on a draft program, the article strongly upset the Swedish Environmental Protection Board (SEPA), which demanded a published comment. This came (Jansson 1972), and the Baltic study continued, coordinated by the Environmental Protection Board, which sought a balance between basic research and monitoring studies.

The SEPA program was successively enlarged, with the Fishery Board responsible for monitoring offshore hydrography and bottom fauna, the SMHI following the water transport and the ice conditions, and the universities studying the coastal biological processes. The offshore program was based on the "Baltic Year," CBO's international field experiment in 1969–1970. The program developed into the marine part of a nationwide Program of Environmental Control (PMK), which so far has reported on the changing situation in the Baltic Sea and delivered invaluable data for the basic ecological studies at the universities. International evaluations have given this program a reputation for quality and adequacy.

Rebuild the Sewage Plants!

The role of sewage plants in the cleanup of the Baltic had been closely followed for a number of years in the Himmerfjärden project. The bay of Himmerfjärden receives treated wastewater from the most developed sewage plant in Sweden, receiving wastewater from the Stockholm region at a total of some 250,000 person equivalents. The plant made the Stockholm water cleaner (you can now fish for salmon in the middle of the town and not only catch some, but even eat them) and made Himmerfjärden more eutrophic. The reaction of the bay was followed by the Askö Laboratory from 1975, supported not only by the SEPA and the sewage plant but also by a citizen's group of people living around the bay and very concerned about its future. This long time period has resulted in several important studies regarding the basic response of marine systems to nutrient enrichment (Larsson and Hagström 1982; Rudestam et al. 1992). Project Pelag, at the University of Helsinki, studied the dynamics of the pelagic system to changes in the nutrient levels in coupled field measurements and experiments (Anon. 1990a). A considerable amount of data on the Baltic ecosystem has been collected, but we are still not sure of the limits of its resilience. At this writing, an international review board has just evaluated the marine part of the Swedish Environmental Monitoring Program.

The Laholm Alarm

The Laholm Bay in the Kattegat turned out to be the alarm clock of the Swedish west coast. A lively debate was surfacing in the mass media, in the scientific community, and in the managing institutions. This debate forced the SEPA to initiate a new project, "Eutrophication in the Marine Environment," and to allocate money for case studies on the Swedish west coast and in the Baltic (Rosenberg et al. 1990). The alarm rang several times during the 1980s, each time waking the authorities into new action (figure 7.5). The minister of energy and environment, Birgitta Dahl, took great personal interest in the problem, preparing the coastal counties for a tough period of measures to check the nutrient flows from land to sea. Action groups north, south, and west were established and given the task of identifying the main sources and coordinating a regional monitoring of the nutrient flows. The SEPA promptly devised an action plan (Swedish Action Program Against Marine Pollution) that reported on the actual conditions in the Swedish marine waters and that contained concrete measures for substantially cutting down the levels of phosphorus, nitrogen, toxic substances, and heavy metals (Anon. 1987b). Reduction of nitrogen in coastal sewage plants, rules for the storage and use of manure, regulation of effluents from pulp mills, and so on were proposed, and the costs of each measure were estimated. This was followed by a second plan, "Marine Pollution '90" (Anon. 1990b), an updated version of the first plan, including the decisions taken by parliament in 1987. The discharge area of the Laholm Bay, with its intensive agricultural activities, was turned into a test area for working out agricultural practices and ecotechnological methods to decrease the nutrient flows to the sea.

Algal Blooms and Seal Deaths Build Marine Centers

By the end of the 1980s, Swedish marine science had spent three decades struggling for adequate resources to match the issues confronting the scientific community. The glorious days of Swedish oceanography during the 1930s and 1940s with O. Pettersson, V. W. Ekman, and the *Albatross* expedition could reasonably be construed as a swan song. Perpetual proposals for increased maintenance, new positions, and large vessels had produced nothing but new committees. Pressed between an agitated public and a deteriorating environment, the reports of the successive committees became increasingly concrete, and in 1988 the gov-

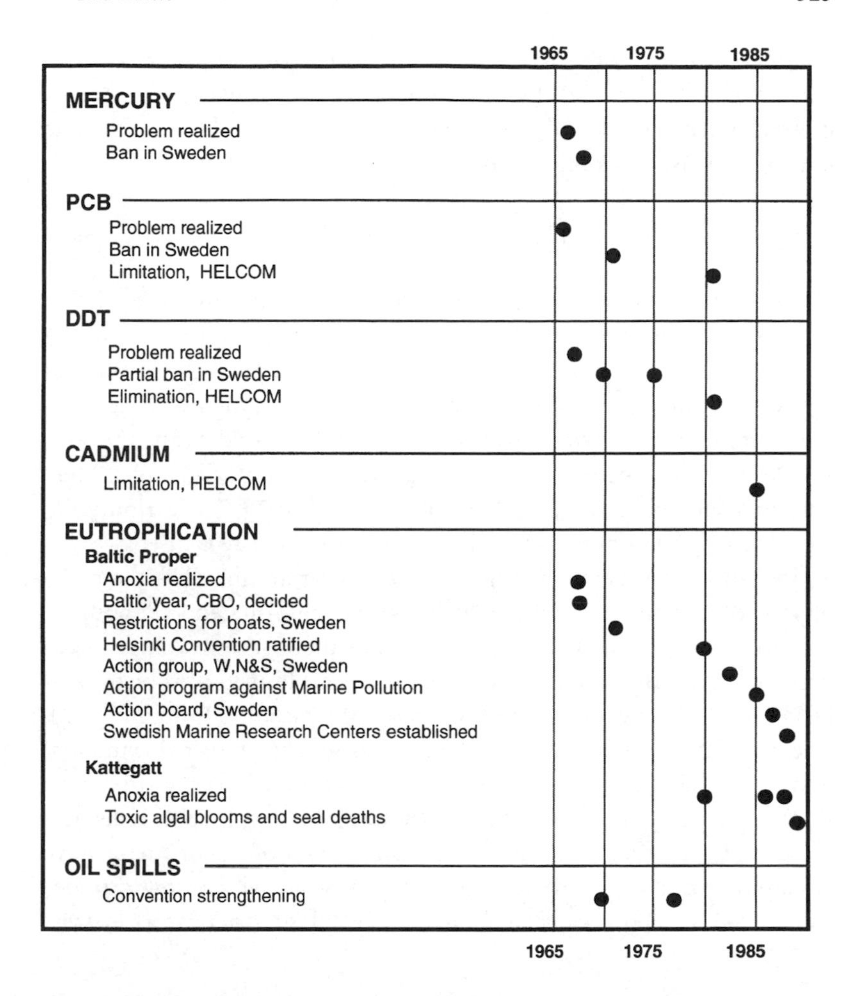

FIGURE 7.5
Some of the main "surprises" and their approximate location in time. Note the increasing activity after 1980 within the eutrophication sector both in the Baltic, reflecting the last stagnation period in 1976–1993, and in the Kattegatt.

ernment decided to strengthen marine activities at three of its universities: Umeạ, Stockholm, and Gothenburgh. A draft of the design for this new program had just been completed (Jansson 1988) when the algae started to bloom and the seals to die on the Swedish west coast. Following the public cry for action, the program was adopted by the House of Commons, and the Umeạ, Stockholm, and Gothenburgh Marine Research Centers were born. The national centers are responsible for interdisciplinary research and higher education, long-term control over marine waters, and the dissemination of information to governmental organizations and the public on environmental matters. The marine expansion, so much longed for, came at last!

Our Known Toxics: The Tip of an Iceberg

Since the first alarms regarding high levels of toxic substances in the Baltic's biota, poisoning of the environment has been hot news in the media and of great public concern. As scientists from governmental agencies became involved in the mapping activities, the alarming findings were met with swift action in Giftnämnden, Folkhälsan, and Livsmedelsverket in the form of bans and health recommendations. Mercury was banned in 1968, DDT in 1975, PCB in 1976, TBT (tributyltin, an extremely poisonous antifouling substance) in 1990. Intensive scientific work identified one toxic substance after another, but the complexity of certain substances, such as the chlorinated compounds, proved enormous (Wulff et al. 1993). As a result of administrative measures, the levels of mercury in the Baltic are no longer a problem, and DDT and PCB have been at low levels since the early 1980s (figure 7.3c.). The Swedish SEPA has a long-standing committee for environmental toxins and has successively created project areas for crucial problems: PETOS, environment and cellulose (Södergren et al. 1988), seals and toxics (Olsson et al. 1992). The decreasing levels of major toxins have apparently resulted in the recovery of gray seals, which now number some 3000, or about twice as many as existed 10 years ago (Almkvist 1982).

Reactions in the East to the U.S.S.R.

After World War II, concern for the environment was first made public in the Soviet Union in the early 1960s. In 1962–1965 an all–Soviet Union project on water supply, water management, and water protection, including the Baltic Sea Subarea, was finalized. The Environmental Management Program, including a forecast of the water quality of the coastal sea and inland waters up to 1980, was carried out. The implementation of this project in the 1980s indicated that the seriousness of the environmental situation had been underestimated and that the recommended protection measures had been only partly realized (30% to 40%).

The project of the Baltic Subarea was actively discussed by scientists, ecologists, engineers, chemists, and members of the Nature Protection Society. For the first time the public realized the dangerous trends in the environment and the strong need to push officials to implement the plans. The Nature Conservation Society was founded in 1966, in Estonia, the first in Soviet Union. By the end of the 1960s, scientists actively cooperated across borders as the CBO, BMB, and bilateral working groups began working and signaled a thaw in the political landscape. The working group on the protection of the Gulf of Finland (comprising Estonian and Russian scientists from St. Petersburg, together with Finnish and, later, Swedish scientists) exerted the pressure. The scientific cooperation was formalized on an official level by the Soviet, Finnish, and Swedish governments. As a practical outcome of this cooperation, the sewage project of St. Petersburg and Tallinn was initiated. The implementation of the project began, but progress has been very slow. Today, after more than 20 years, the sewage treatment plants in both cities are still not completed.

Ecology Marries Politics in Estonia

Estonia was the first state in the Baltic area in which, in the early 1960s, Nature Protection Societies were formed. The societies were "underground" centers acting for the cultural and, later, political independence of Estonia. Lahemaa, the famous national park in northern Estonia, was founded to protect the coastal nature and wetland area. The annual summer meetings of the Nature Protection Society have been important steps on the way to independence.

The real catastrophic nature of the environmental pollution was clearly presented at the end of the 1970s in Poland by the Ecological Club movement. The scientists involved in this movement advised and forced officials to save the environment. It is important to remember that the Ecological Club movement cooperated with *Solidarnost* and actively participated in the political fight for democracy.

At the end of the 1980s, an "ecological war" was waged by the Estonian environmentalists and intellectuals against phosphorite mining in the north of Estonia. In the beginning the ecological movement increased political activity, and with the help of the Estonian public it became successful. The mining project prepared by the Moscow ministry was stopped. It is important to know that the potential phosphorite mining area in Estonia is one of the biggest in northern Europe. Phosphorite deposits cover an area of more than 1000 km^{2,} and the reserves are estimated to contain about 750 million tons of P_2O_5. The ore's content of P_2O_5 is 15% to 20%, but it also contains relatively large amounts of harmful components, especially heavy metals. Thus the Estonians succeeded in protecting nature, and the phosphorite deposit was saved for future exploitation using advanced technology.

The "phosphorite war" in 1987 signaled the starting point for a "Singing Revolution" in Estonia, and later in both Latvia and Lithuania. The People's Front in August 1991 succeeded in regaining Estonia's independence, which had been lost to the Soviet Army in June 1940. From the spring of 1988 the people and the renewed local administration acted hand in hand.

The Estonian "Singing Revolution" in the spring of 1988 (figure 7.6) marked a starting point for continued changes in the Soviet Union as support for Gorbachev's *Perestroika* and *Glasnost* continued. These changes meant more openness in environmental statistics and management, and rejection of the old philosophy that "Soviet is best." The crash of the Soviet Union and its socialistic single party and planned economy was a logical endpoint.

The definitive independence of the Baltic and other Eastern states will be reached only after the final departure of the former Soviet Army. Unfortunately, an ecological disaster in the military's former station areas has been uncovered. Soil and groundwater close to the military airfields in many cases are heavily polluted by oil and gasoline. How nature can be saved and who should pay for the restoration of the damaged ecological systems are difficult questions. Russia and other East European states are moving toward close cooperation with the European Community and the Nordic countries in environmental protection activities. This means that environmental criteria and standards may be normalized and that EC standards should be applied. For the Baltic states cooperation with western and eastern neighbors will be increasingly important. For example, in the Gulf of Finland, where Estonia and Finland are located, more than 75% of the total load of pollution is discharged from St. Petersburg by the river Neva.

So far Russia has used the harbors of Finland and the Baltic republics, but as it does not want to be "under pressure," it has decided to build a new international harbor in the mouth of the Luga River, near Viborg in the eastern Gulf of Finland. Does this represent a new barrier or a new bridge for Baltic Europe?

Figure 7.6

One strong nationalistic feature of Estonia, Latvia, and Lithuania is the special fondness for singing. During the early days of *perestroika,* with the escalating tension between the stationed Soviet troups and the native citizens, people demonstrated for their rights by singing in huge choirs for hours. An example of this "singing revolution" in 1989–1991 is this picture from Tallinn (Jaan Röömus, Tallinn, Estonia).

The effective realization of environmentally oriented joint projects requires a good understanding between the parties and cooperation on both intergovernmental and nongovernmental levels. It is important therefore that the reborn Baltic states—Estonia, Latvia, and Lithuania—will be member states in different conventions. The membership fees should not be barriers for the Baltic states' participation in common projects.

The Emergence of an Ecosystem Approach

From Pieces to Puzzle

It is probably fair to say that the picture of the Baltic Sea as a total ecosystem—the coupling of nutrient dynamics to the production of seaweed and plankton communities, which in turn fed the deeper, soft-bottom communities—was first painted at the Askö Laboratory. A fairly detailed report of the development of that picture would therefore give good examples for our discussion on barriers and bridges in the Baltic Basin. [A broader outline of the evolution of ecology as a scientific discipline in Sweden is given by Söderqvist (1986). It has a considerable bearing on the theme of this book.] The director of that laboratory (B.O.J.) provides a historical description that cannot be claimed to be totally objective. Founded as a station for field education and field research, the first years of activity at the Askö Laboratory were devoted to studies of single species and their ecology. The first step toward an ecosystem approach was taken by Ann-Mari Jansson, through her study of the food chains in the *Cladophora* belt, the typical green algal belt that lies just below the water surface, inhabiting a rich microfauna, which turned out to be an important nursery for crustaceans (Jansson 1966). In a proposal from 1964 to the Natural Science Research Council (NFR), the goal of the common research of the laboratory was formulated as the study of the metabolism of energy and matter of the Baltic Sea. The *Cladophora* study substantiated the abstract talk about energy by measuring the energy content in various parts of the food web with a bomb calorimeter. The ecosystem approach was later manifested in the studies of the total metabolism of a rock pool by Ganning and Wulff (1969), using the methods of H. T. Odum and Hoskin (1958). The proposal to the Natural Science Research Council in 1968 asked for continuous support to analyze the Baltic ecosystem, mainly the seaweed and soft-bottom communities.

The next proposal to NFR, in 1969, noticeably changed the previous research policy of the Askö Laboratory: "the separate research problems begin to overlap. The time is ripe for larger team works, with biological production in the Baltic as a unifying theme.... the research at the laboratory has converted from its first zoological phase to a general biological one." The proposal put the focus on the energy flow between the algal belts, the soft bottoms, and the pelagic communities, stressing the importance of the transport of organic matter through sedimentation. At the same time we applied to the recently established Swedish Environmental Protection Board for support of field investigations in a pollution gradient from the small town of Trosa to our field station in the outer archipelago. The study had been going on for several years on the initiative of a couple of environmentally concerned young students, first without financial support, later also involving the botanist Inger Wallentinus, focusing on the distribution of algal species.

Light from an American Spark

The research report for 1969 was a broad documentation of the activities at the Askö Laboratory, where now also a geological group under Dr. Rolf Hallberg had started experimental work on the flux of phosphorus, nitrogen, and sulfur between sediment and bottom water (Hallberg 1973). The publication list also included an article for the NFR Yearbook on the Baltic coastal waters as an ecosystem. This was actually the result of a process, triggered by the great American ecologist Howard T. Odum. I (B.O.J.) can even date the moment of the transformation. Sitting in an airplane, reading Olson and Burgess's book *Marine Pollution and the Marine Environment,* I was struck by Odum's chapter on the Texas coastal systems, both by the overall view he presented and by his energy circuit language. Here was what I had been searching for when trying to summarize the different studies in our coastal area!

In 1970 Howard Odum visited our laboratory thanks to a grant from the Swedish Board of Marine Technology, whose chairman, Jan Zeilon, had met the Odum brothers in the United States and, impressed, had asked if we were interested in bringing H. T. Odum to Sweden. He arrived, accompanied by an "interpreter," the biological director of the U.S. National Science Foundation, Harve Carlson, who translated not only the American dialect into our school English, but also Odum's condensed ecological discussion into our level of understanding. Of course, Odum was particularly interested in Ganning-Wulff's community metabolism studies, and the following year Wulff was invited to spend the summer with him, learning the energy circuit language and how to use analog computers. When Odum returned to Sweden in 1971, invited to the Nobel Symposium on the Chemistry of the Ocean, we again visited at the Symposium of the Baltic Oceanographers in Helsinki, where Odum gave a talk on marine systems and, following the traditions of the these symposia, participated in the evening parties on the research vessels, where he had animated discussions with East German, West German, and Finnish scientists. Symbolically, it was a meeting between two worlds: one claiming the sea was governed by physics and chemistry, with some, although economically important, elements of biology, and the other seeing the biosphere as flows of energy and matter operated by living systems marvelously described in *Environment, Power and Society,* which had just appeared (Odum 1971).

Dynamics and Energy Flow in the Baltic Ecosystem

At the beginning of this year the meteorological conditions were favorable for a bigger inflow of oxygenated, salty North Sea water after the previous stagnation in the Baltic. To prepare for this event the SEPA announced a 1-year grant for following the supposed "changing hydrographic situation of the Baltic Sea." The main actors of the previous discussions between the Odén group and the SEPA were summoned, including, among others, the chairman of the Biological Division of the NFR, professor Erik Dahl, and representatives of the SMHI. The Askö Laboratory had at that time, as stated earlier, resigned from its previous SEPA grant, but I (B.O.J.) was asked to participate by the NFR. After discussions back and forth, Dahl suggested that the task of designing a research program be given to the Askö Laboratory, because of its broad overview of the Baltic environment. A draft of ecosystem structure was discussed within a small ad hoc group with representatives of the main governmental agencies. Acknowledged by the SEPA, a couple of young, promising graduate students were contracted and a dynamic 1970s had begun.

Parallel to this a proposal entitled "Energy Flow in Baltic Ecosystems" was sent to a nongovernmental research foundation, but it received no funding. Within the NFR, however, the Ecological Committee, probably inspired by the IBP programs, was looking for larger projects to launch, and we were asked to formulate a proposal to compete with other themes. We were lucky with our "Dynamics and Energy

Flow in Baltic Ecosystems," which now started to grow. It was to be followed by another large-scale project, "The Coniferous Forest," and for the financing of both projects the committee's dynamic secretary Bengt Lundholm began intensive lobbying, including editing a small book on the projects to be presented to each member of the Swedish parliament. This proved to be effective and the scientific 1970s was dominated by the two projects, to the pleasure of the participants and the dismay of many representatives of other fields of ecology, who strongly felt that too much money was spent on systems ecology. Financed by money from the Swedish NFR and the American NSF, I (B.O.J.) and my wife AnnMari spent a year with Howard Odum in Florida learning system ecology modeling and simulation on analog computers. As is obvious from the preceding description of the tours with Odén and the SEPA, this was at the time of the article in *New Scientist* and during the first year of the large-scale project Dynamics and Energy Flow in Baltic Ecosystems. To motivate the Baltic Project scientifically for an international scientific audience I was urged to write a fuller description of it: "The Ecosystem Approach to the Baltic Problem" (Jansson 1972), sometimes called "The Blue Book." The description and analysis were strongly influenced by Odum's systems ecological thinking. The booklet was received with interest not only in the scientific community in and outside the Baltic area but also in administrative circles, as judged by the fact that its formulations often found their way into official documents.

International Collaboration

An interdisciplinary project had been launched in 1971 at the Institut fur Meereskunde, Kiel, Germany (Hempel 1975) as one of many by the Sonderforschungsbereich. Its aim was to understand the interaction between the seabed and the water column, including an ecosystem project for the Kiel Bight. Close collaboration was established with the Askö project. The strong and still existing ties between Kiel and Askö have been important for the development of Baltic science, thanks to Professor Hempel's experience and high standing in international marine science administration.

Even the international organizations were active early. "Recalling that the pollution of the Baltic in recent years calls for coordinated action by the surrounding nations," ICES and the Scientific Committee on Oceanic Research (SCOR) established a working group in the fall of

1971, under the coordination of Professor Bert Bolin, with one representative from the involved institutions: International Institute of Meteorology, Stockholm; Institute of Marine Research, Helsinki; Department of the Baltic Sea, Tallinn; Institute of Oceanography, Warnemunde, GDR; Fishery Board of Sweden, Gothenburgh; Institut fur Meereskunde, Kiel; and Askö Laboratory, Stockholm. The report from the group dealt exclusively with problems of the Baltic as a whole and strongly emphasized international collaboration. The research program presented in 1973 contained both a description of the total Baltic ecosystem, with submodels for the main types of the coastal zone, and specific research aims, such as the study of excessive eutrophication, toxic substances in the food chain, oil spills, and dumped solid wastes. It culminated in a specific research plan, including large- and small-scale water exchange, continuous stations in the open sea, biological monitoring and experiments, and determination of toxic substances in the food web in different areas (Anon. 1974). The year 1975 was set up as the Baltic Year, when all the parts of the total program would be run simultaneously. Thus the ecosystem concept was established also in the international organizations in the early 1970s. The Swedish Environmental Protection Board initiated the coordination of methods for determining nutrients, toxic substances, and oil with a large number of scientists from all the Baltic countries in 1974, which was to be followed by many similar efforts in the collaborative work on the Baltic.

Evolution of Ecosystem Management

Systems Approach to Management Problems

Pollution in the Baltic has reached the point that damage may become irreversible. The reasons for this environmental degradation are both natural and manmade. The large catchment area; the semienclosed cold water body, heavily stratified without tide; and a water residence time of several decades make the Baltic a very sensitive sea. The intermittent pulses of oxygenated North Sea water seem to come in long intervals, leaving periods of stagnation and low oxygen levels in between. Man's urban areas, industries, and agriculture have discharged pollutants directly through rivers, which act as large-scale collectors of wastewater from various sources in the different drainage basins and unload them in wetlands, coastal lagoons, bays, and gulfs. A large number of cities have dumped their untreated or insufficiently treated wastewater directly into the rivers and coastal areas. Runoff from agricultural areas, including areas with intensive livestock husbandry, is a major contributor to the high nutrient load on the Baltic. Many pollutants are transported from the industrial areas of Europe to the Baltic Sea by the atmosphere. Accumulation of toxic substances in fish, birds, and seals is probably the most conspicuous sign of the closed nature of the Baltic ecosystem. Since the 1900s the loads of phosphorus have increased eightfold and nitrogen has risen fourfold (Larsson et al. 1985). The *Fucus* algal belt has been displaced by 3–4 m in areas far from point sources of pollution (Kautsky et al. 1986). The concentrations of chlorinated hydrocarbons have decreased considerably (Olsson and Reutergård 1986), but new toxic substances are continuously emerging (Anon. 1990b). The need for multinational ecosystem management is overwhelming.

However, we have come a long way toward an ecosystem perspective. For example, addressing the pollution of the coast outside urban centers in Sweden was a continuation of the cleaning process of lakes started during the 1950s. Individual scientists have produced strong trigger effects in addressing the large-scale pollution of the Baltic. As mentioned earlier, Stig Fonselius, "Mr. Baltic," started the era of chasing the cause of the "dead bottoms" in the Baltic in the mid-1960s, and Svante Odén gave the problem the proper space and time scale. The ecosystem approach was born at the Askö Laboratory with the description of the Baltic Sea Project (Jansson 1972), thanks to the group of young scientists working within the same geographical area, which made an interdisciplinary approach possible. The trigger was provided by Howard Odum of the University of Florida, whose energy flow theories became the backbone of the Baltic Project. The simultaneous monitoring project for the coastal area, commissioned by the SEPA, and the necessity of pooling available resources transferred the ecosystem concept to the monitoring program. Although originally started without the aim of solving pollution problems, the basic research project, "Dynamics and Energy Flow in the Baltic Ecosystem," sponsored by the Natural Science Research Council, became the reference base for more patchy and restricted investigations elsewhere, as it identified and secured the main ties among the various subsystems. The offshore measurements were carried out as before, by agencies in the different countries, in Sweden enlarged by support from the SEPA.

While the concept of the Baltic Sea as a single, huge ecosystem rapidly gained acceptance, the technical side of the pollution problem still continued along traditional trends. The Swedish government had been the first in the Baltic area to understand the necessity of decreasing the pollution entering the sea and to have the economic ability to do so. The program for building biological sewage purification plants, which started in the 1960s in Swedish fresh waters, was extended to the Baltic coast, incorporating techniques for precipitating phosphorus. Finland soon followed, but the loads of pollution from the southern and eastern coasts of the Baltic continued to increase. A large part of the pollution, especially nitrogen, emanates from sources outside the Baltic region and therefore calls for long-range efforts to reduce transboundary loads.

Banning Solves the Problem?

When the steering committee of the Helsinki Convention, HELCOM, started to work in 1980, it mainly followed the recommendations for action by the contracting parties, based on the information of the state of the sea and its pollution loads. The preparatory work was done by special working groups in which the best scientists and experts from the littoral states, including NGOs, prepared assessments of the Baltic marine environment (1980, 1986, 1990) and guided the Monitoring Program of the marine environment. Over the decades many dozens of recommendations for decreasing different sources of pollutants have been prepared. Very hazardous substances, such as DDT, PCB, and PCT, were totally prohibited. Attempts were made to minimize the introduction of such noxious substances as mercury, lead, oil, and radioactive material. Dumping and pollution from exploration and exploitation of the resources of the seabed were also restricted. Unfortunately, there was little success in stopping marine eutrophication processes. The contracting parties approved the maximum permissible load by municipal sewage water discharges: (BOD [biological oxygen demand] *15 mg/liter, total phosphorus* 1.5 mg/liter) but could not yet decide on nitrogen limits (15, 12, or 8 mg/liter).

The decisions of the commission are reached unanimously with preliminary approval by one of the permanent advisory bodies: scientific, technological, maritime, and combatting (spillage) committees. Unfortunately, the commission has had no power to enforce its decisions upon its member states; such enforcement was the responsibility of the individual countries. In most of the member nations the decisions of the commission were implemented by national Baltic Sea Committees established for this purpose. The main activities of the commission,

TABLE 7.3
Land Uses in the countries of the Baltic Sea Drainage Area*

	Land Use					
Country	Urban Area	Forests	Arable Land†	Water Bodies (Lake Surface)	Marshes, Swamps, Wetlands	Other
Denmark	15	12	60	13		
Estonia		39	30	6.2	20	4.8
Finland	2	51	7	10	27	3
Germany	3.9	15.2	72.3	3.5		5.1
Latvia	6.2	43.2	39.8	1.5	4.9	4.4
Lithuania	2.5	27.9	49.2	1.5	6.4	12.5
Poland	6.2	28.4	60.1	2.6	17	8.1
Russia	2	55	12	17	13	1
Sweden	1.1	48.8	6.2	8.1	12	23.8

*Values are percentage of land use types within a country.
†Includes grasslands and greenfields

including its scientific and technical results, are published in the *Baltic Sea Environment Proceedings,* now in its fortieth edition.

Achieving the goals would be impossible without close cooperation between the commission and other international organizations and such observers as UNEP, WMO, IMO, and the Paris and Oslo commissions. The most important and supportive groups have been the CBO (Charlottenlund, Denmark), BMB (Lysekil, Sweden), and ICES (Copenhagen, Denmark). The guidelines for a Baltic monitoring program and the periodic assessments have been prepared in close collaboration. Most important contributions have been made by experts preparing the first (Anon. 1987a) and second (Anon. 1993b) Baltic Sea Pollution Load Compilation. All this material could give advice on how to manage ecosystems and how to determine the priorities for the Baltic Sea area (tables 7.3–7.6).

Science Versus Management

Except for some groups working with the effects of increased nutrient levels in the coastal area, such as the outstanding Waern group (Waern 1973) and the long-term monitoring of the Finnish IBP–PM program (Luther et al. 1975), Baltic marine science in the 1960s had little involvement in pollution problems. Most researchers were not interested

Table 7.4a

Division of the Baltic Sea Drainage Area Among the Baltic States and the Baltic Subregions (km^2)

Country	Bothnian Bay	Bothnian Sea	Archipelago Sea	Gulf of Finland	Gulf of Riga	Baltic Proper	Western Baltic	The Sound	Kattegat	Total
Finland	146,000	39,000	9,000	—	—	—	—	—	—	301,300
Russia	—	—	—	276,100	23,700	15,000	—	—	—	314,800
Estonia	—	—	—	26,400	17,600	1,100	—	—	—	45,100
Latvia	—	—	—	3,500	48,500	12,600	—	—	—	64,600
Lithuania	—	—	—	—	16,500	48,800	—	—	—	65m300
Poland	—	—	—	—	—	311,900	—	—	—	311,900
Germany	—	—	—	—	—	12,600	10,950	—	—	23,550
Denmark	—	—	—	—	—	1,200	12,400	1,700	15,800	31,100
Sweden	131,000*	180,100	—	—	—	84,900	—	2,600*	71,600*	470,200*
Total	277,000	219,100	9,000	413,300	106,300	488,100	23,350	4,300*	87,400*	1,641,650*

*Including Norwegian drainage area.

Table 7.4b
Division of the Baltic Sea Drainage Area Between the Noncontracting Parties

State	Subregion	Discharge Through	River Basin	Drainage Area (km^2)
Belarus (BY)	Gulf of Riga	Latvia	Daugava	25,800
Belarus	Baltic Proper	Lithuania, Russia	Nemunis, Pregel	46,900
Ukraine (UA)	Baltic Proper	Poland, Lithuania	Vistula, Neumunas	11,000
Czechoslovakia (CS)	Baltic Proper	Poland	Vistula, Oder	8,500
Norway (NO)	Kattegat	Sweden	n.i.	n.i.
Norway	Bothnian Bay	Sweden, Finland	n.i.	n.i.

TABLE 7.5
Annual Pollution Load Data for the Baltic Sea (1987)

	Pollutant		
Territory	**Organic** (10^3 t BOD_7)	**Total Nitrogen** (10^3 t)	**Total Phosphorus** (10^3 t)
Denmark	160	69	10
Estonia	49*	10*	0.6*
Finland	250	70	4
Germany	36	20	2.4
Latvia	46*	7*	0.9
Lithuania	43*	8*	0.9*
Poland	393	110	19
Russia	176*	33*	5
Sweden	362	129	7
(Byelorussia)	(6)	(0.5)	(0.1)

*Data from 1991 HELCOM Task Force Joint Comprehensive Program.

TABLE 7.6
Annual Pollution Load Data for the Baltic Sea by Source (1990)

	Pollutant		
Sources	**BOD_7** (10^3 tons)	**Total Phosphorus** (10^3 tons)	**Total Nitrogen** (10^3 tons)
Rivers	1010	30	336
Urban areas	207	11	81
Industries	155	1.8	12
Total	1372	42	429

in applied questions, and BMB debated for several years if they should address such topics in their symposia. The main focus was on basic problems in brackish water ecology, although the ecosystem concept was received with interest, and the suggestion from the Askö Laboratory at the symposium in Stockholm 1971 of a number of working groups for standardizing methods for quantitative sampling as a preparation for the "future work on a common model of the Baltic" was enthusiastically accepted. The effects of the stagnant periods were recognized as an ecosystem problem; therefore, understanding the system is the key to its solution. Few, if any, car owners would leave their broken car to a person who does not understand how an engine functions. The BMB scientists felt that HELCOM at the beginning appointed advisers who were not always familiar with how an ecosystem works, and even less with what should be done to a disturbed one. This changed the attitude toward the more applied Baltic problems, and BMB offered to advise HELCOM.

Eutrophication: Good or Bad?

Besides the evolution of ecosystem management in the Baltic Sea, the scientific community had its own internal battles. One was fought over the reason for the "dead bottoms." Some people from the fisheries denied the role of eutrophication, saying that the cause is the natural stratification of the Baltic and the intermittent flushing of North Sea water. The Baltic fisheries have benefitted from the increasing nutrient loads; in their view the only problem with the Baltic is the increase in toxic substances (Otterlind 1986). This was strongly opposed by scientists from the universities, who showed that the water exchange by itself could not explain the lack of oxygen. Increased oxygen consumption in the seabed was important, and only eutrophication could explain this (Shaffer 1979; Stigebrandt 1983). On the other hand, data showed that in the oxygenated surface layer, the biomass of the macroscopic bottom fauna had increased four times since the 1920s (Cederwall and Elmgren 1980).

At the end of the 1970s, when the authorities thought the eutrophication of the Baltic was under control, since the reduction of phosphorus at coastal sewage plants had resulted in improvements in the receiving areas, the debate started again, triggered by the decrease of oxygen and accompanying fishkills in the Laholm Bight in the Kattegat. The public became very concerned, which was made worse by the many catastrophe reports in the news media. The scientists on the west coast feared that their coast was dying, and their Baltic colleagues countered by saying that oxygen depletion had existed in the Baltic for decades and over much larger areas, a position supported in the 1981 report from HELCOM (Melvasaalo et al. 1981). As a result of a SEPA-initiated investigation, a special program area was formulated in 1980, "Eutrophication of the Marine Environment," and projects both at the Swedish west coast and in the Baltic were formulated for a period of 6 years, started after international reviews and delivering valuable data for the understanding of marine systems dynamics (for summaries see Rosenberg et al. 1990). And just as in the early 1970s, they were attacked by the same fishery biologist and with the same arguments: Oxygen depletion has always taken place during periods of low water exchange; eutrophication is good for the fisheries; and the loss of secondary production in the deeper bottoms is compensated for by an increase in the shallow areas. The biologist continued to argue that the scientists' alarming reports should be seen as lobbying for increasing financial support and that the real problems were the toxic substances. But in the international arena, the UNESCO/IOC expert group on marine pollution judged the eutrophication to be the largest threat, as did ICES for Kattegat and Skagerrack.

These projects had a strong ecosystem basis. An ICES/CBO/BMB project, "PLEX," initiated by Dr. Bernt-Ingemar Dybern, of the Fishery Board of Sweden and the driving force of the BMB, addressed the important patchiness problem (the tendency of organisms to appear in patches or clumps rather than to be evenly distributed). The patchiness was striking when studying the remote sensing images of the bluegreen algal blooms in the Baltic proper. Research vessels from all Baltic countries simultaneously studied the plankton distribution within a fixed area in the middle of the Baltic (Dybern and Hansen 1989). Initiated by the Marine Committee of the SEPA, two large-scale projects are coupling the pathways and decay of a few characteristic toxic substances with the major energy and material flows—one studies the Baltic (Wulff et al. 1990), the other the Kattegat/Skagerrack (Rosenberg et al. 1990). In spite of the early findings of toxics in Baltic biota, few studies of the cycling in the system of hazardous substances have been made, although the ecosystem studies have offered excellent opportunities for sampling the crucial flows. Within the international WOCE-program a study of the hydrology and meteorology of the Baltic Basin, BALTEX, has been launched (Raschke 1992).

All these projects together clearly show how scientific interest in the Baltic initially focused on its hydrography and flora and fauna, then successively widened its scope via production studies and ecosystem analyses to large-scale, international projects. The need for further integration of studies in the catchment area (land use/water quality) and in the marine environment has also been stressed by the COPERNICUS project within the European Rector's Conference (CRE), where it was argued that the universities should increase and intensify education in the basic, multidisciplinary areas, such as systems ecology, ecotoxicology, ecotechnology, and ecological economics (Jansson 1990). The important link between ecology and economy was demonstrated for the Baltic system in the 1970s (Jansson 1976) and will certainly prove to be a potent weapon in the redeveloping the Baltic Sea.

P or N?

For their fixation of solar energy (i.e., their growth), plants need carbon, which they have in excess in the water, phosphorus (P), and nitrogen (N). Light is of course limited during winter, when P and N are increasing because of the decomposition of organic matter. Phosphorus is usually the limiting substance in freshwater systems, which has led to its reduction as the standard method used in sewage plants. In the early debates of the stagnant bottoms in the Baltic, the classical, limnological phosphorus trap model dominated the discussions: When they are oxygenized, sediments bind and store phosphorus, as it is made insoluble by iron and manganese; when they are deoxygenized, phosphorus is released to the bottom water. Waern and Pekkari (1973) were outstanding in showing the intricate shift from phosphorus to nitrogen limitation in the Stockholm Archipelago when going away from shore, and Sen Gupta (1972) was virtually alone in his early arguments for nitrogen as the major limiting nutrient. Accumulating results from field measurements and enrichment experiments now point to nitrogen as the usually limiting nutrient in the Baltic proper and to phosphorus as the limiting factor in the Bothnian Bay. This agrees with the common and well-founded rule of thumb that phosphorus is limiting in fresh water and that nitrogen is limiting in the sea. The importance of using the nitrogen-to-phosphorus ratio was beautifully demonstrated by Niemi (1979) for the typical bluegreen algal blooms in the Baltic proper.

Triggered by the Laholm incident, however, the debate started all over again. A lone but strong and frequent voice representing the county administration pointed to the difference between the cycles of the two nutrients. Phosphorus stays in the sea, switching between the sediments and the water phase, depending upon the oxygen concentration of the water. At low oxygen concentrations bacteria reduce nitrogen to nitrogen gas, which rises into the atmosphere. The amount of nitrogen, therefore, is automatically regulated, and we need only bother about phosphorus. The marine scientific society countered, saying that experiments with plankton showed that stimulation occurred with nitrogen enrichment, not with phosphorus enrichment. Even the full-scale experiment at Himmerfjärden Bay, where the loads of total phosphorus and phosphate were allowed to increase five- and ten-fold, respectively, from January through October, did not result in an observable increase of the primary production or increase of phytoplankton biomass (Elmgren and Larsson 1987). Besides, even if the nitrogen levels were eventually regulated by the amount of phosphorus, the time lag between the stimulation of algal growth by the nutrient load and the final transfer of nitrate to nitrogen gas coincided with accumulating, decaying, oxygen-consuming algae, a nuisance to both man and marine organisms.

Cut the Loads by 50%!

Discussions within the scientific community were sometimes carried on in the mass media. Given the close connection to environmental practices, the media often confused the management organizations. The scientist became a person who, seemingly, could never give a straight answer. In some cases the alternative solutions were obscured by economic consequences, such as the far more expensive reduction of nitrogen compared to phosphorus or the body burden of toxic substances affecting the sale of fish.

Decisions had to be made, however, and during the 1990s activity at the international level began at last to increase. HELCOM was busy preparing an analysis of the Second Pollution Load Compilation adopted in 1993 (Anon. 1993b; tables 7.4–7.6). Following the late, but positive, decision to also include the drainage area, the analysis is based on demographic data both on the new constellation of countries (table 7.2) and on land use patterns in the drainage area (table 7.3). It also includes information long sought after on the anthropogenic impact in the southeastern Baltic Area countries (figure 7.3). For example, the analysis shows a very unsatisfactory environmental situation in the St. Petersburg and Kaliningrad areas.

The ministers responsible for environmental protection in the seven Baltic littoral states met in February 1988 in Helsinki under the auspices of HELCOM. Expressing concern for the present and future state of the Baltic marine environment, they formally promised to reduce the charges of phosphorus, nitrogen, heavy metals, and toxic and persistent organic substances by about 50% by 1995. The commission was asked to take the necessary control actions as matters of priority in its future work.

Input Management Instead of Output Management

During the recent decades, "end-of-pipe pollution control" approaches have helped to reduce the direct discharge of some pollutants from point sources, especially when the "Best Available Technology" (BAT) has been used. Unfortunately, too often the problems resulting from the pollutants were not solved, since the materials were merely transferred from one medium to another. Often the point source problems were simply transformed into non–point source problems. A preventive approach, based upon the elimination of the problems at their sources, is far more satisfactory. All phases of the life cycle of the pollutants must be addressed, with the objective being to prevent or at least minimize short- or long-term risks to both humans and the environment (UNEP 1989). The Cleaner Production approach is an ongoing process geared toward a minimal waste and risk status. Governmental and private corporations are challenged to address all aspects of the life cycle of their products, including the selection and production of raw materials, the product design phase, the production and assemblage of the final products, and the management of all used products at the end of their life. The emission control technology is the last as well as the most important step in the process. An example is the handling of municipal sewage systems. The Cleaner Production approach also should be used in attacking the diffuse sources of pollution in the agriculture and forestry sectors.

In the south and southeastern Baltic area, the most important issue will be to decide whether a polluting industry stands any chance of surviving future competition. When an industry has the potential to survive, the first step should be a transition to a cleaner technology, fulfilling the EC standards of pollution load. In the eastern Baltic countries, because of a lack of equipment, many purification plants have been under construction for years (e.g., Tallinn, Riga, and St. Petersburg since the beginning of the 1970s). These projects should be finalized as soon as possible. Unfortunately, the eastern Baltic states have very little local funding or foreign exchange to invest. They need soft loans from abroad for joint ventures to produce new equipment. Considering pollution on a regional scale, investment funds are far more effective where the pollution loads are largely uncontrolled. As a result, levels of the investment should be highest in the south and southeastern parts of the Baltic area, where more than 70% of the total pollution load is located.

The "Greens" Cry: "Back to the '50s!"

In many informal meetings, the "Greens" expressed a desire to restore the Baltic Sea to the conditions of 1950, which they considered to be a satisfactory state. A rough estimate of the total cost of restoration is $600 billion (US), comprising $350 billion for the reconstruction of industry, agriculture, and so on, and $250 billion for fulfilling the requirements of the "end-of-pipe" control system. If this program were to be realized during the next 50 years, the annual investment would be $12 billion (US). The biggest investment, 25% to 30% of the total, preferably would be used in agriculture and to remove nutrients from the sewage systems. Such large investments probably could only be realized by the FRG for the former GDR. The FRG hopes to achieve this in 15–20 years (Wulff et al. 1990). For the whole Baltic region such an investment seems unrealistic. The best we can hope for is a return to the marine system of the 1970s.

Promises for the Future

The need to establish a realistic and specific plan for the restoration of the Baltic Sea was expressed by the prime ministers at the Baltic Sea Conference in Ronneby, Sweden, in September 1990, where the Baltic Sea Declaration was adopted. The declaration calls for a concrete, action-oriented Joint Comprehensive Program (JCP) to restore the Baltic Sea to a state of sound ecological balance. To implement the JCP, an ad hoc high-level task force was established within HELCOM. The task force consists of representatives of the contracting parties to the Helsinki Convention of 1992 (adopted by the countries in April 1992). The JCP was reviewed by the Helsinki commission in 1994. International financing institutions such as the European Bank for Reconstruction and Development, the European Investment Bank, and the World Bank are hiring consultants to gather information for prioritizing areas that require special attention. The JCP was prepared on the basis of the national plans of the countries concerned, material provided by HELCOM, and the results of pilot feasibility studies by the consultants nominated by the banks as executing agencies. This was done in a very short time and adopted by the diplomatic Conference on the Protection of the Marine Environment of the Baltic Sea Area in April 1992 in Helsinki, Finland.

In a 1992 declaration the high representatives of the governments participating in the diplomatic conference approved the strategic approach and principles reflected in the JCP. As a background to this program, the use of precautionary principle, the best environmental practice, and the best available technology when designing measures to eliminate pollution and improve environmental management were adopted. The international financial institutions have promised to support the implementation of JCP and to mobilize national and international financial resources.

Investment priorities are determined primarily by the character and magnitude of their impact on public health and ecology, by economic feasibility, and by their effect on the Baltic Sea. The most cost-effective measures are those actions focused on "hot spots" (areas of acute environmental concern) where pollution has the greatest impact on the sea and its catchment area. The program will focus on 132 hot spots. The task force has identified ninety-eight actions appropriate to key hot spots, out of which forty-seven have been identified as priority hot spots (figure 7.1). The cost of these actions is estimated to be $6.5 billion (ECU). The price of implementing the entire long-term program is estimated to be at least $18 billion (ECU) over a 20-year period. The program will be implemented in two phases: the first phase (1993–1997) is estimated to cost $5 billion, the second (1998–2012) is estimated at an additional $13 billion.

Nearly two-thirds of the hot spots are located in Eastern countries, with a total population of about 60 million. A highly problematic macroeconomic situation makes ordinary loans and commercial financing difficult. The polluter pays principle cannot always be applied. In the northeast area of the Baltic the Nordic Council of Ministers adopted a risk fund ("NEFCO") of 30 million ECU, to be used in joint venture projects to stimulate environmental techniques.

Arrangements to finance and implement projects identified in the program will require more detailed feasibility studies. These studies will be undertaken where there is a high probability of implementation. It is expected that implementation of the program at twenty-nine priority municipal and industrial hot spots will annually reduce BOD_5 by 300,000 tons, nitrogen by 33,500 tons, and phosphorus by 8200 tons. Nineteen of the hot spots are situated in Poland, and ten are in the St. Petersburg area. The immediate actions to implement the program cover the period 1993–1998, or phase 1. Activities for phase 1 were selected on the basis of their potential impact on water quality and the ecological systems of the Baltic Sea, its coastal waters, and its catchment area, in conjunction with present institutional capacity for project preparation and financing. The cost of funding feasibility studies for priority projects is estimated at $30 million ECU. Four principal investment priorities have been identified:

- Emergency support and warning systems to avoid potential public health risks.
- Improvements in combined municipal and industrial wastewater treatment systems. Completion of unfinished treatment facilities and repair of inoperable ones (such as those at St. Petersburg, Tallinn, Riga, and so on); installation or upgrading of industrial pretreatment projects; elimination of the uncontrolled discharge of wastewater into the environment; and expansion and improvement of the safe disposal of sludge).
- Rational industrial pollution control by sector.
- Control of pollution loads from the agricultural sector (nutrients and organic substances), and management of coastal lagoons and wetlands.

As mentioned earlier, during the April 1992 diplomatic conference in Helsinki, the contracting parties adopted and signed a new convention on the protection of the marine environment of the Baltic Sea area in which political and socioeconomic changes of the last decade are reflected. The new convention will apply to the whole Baltic Sea Area, including the national waters. Moreover, the recommendations made by the commission will be legally binding to all contracting parties. Most important, the European Economic Community is a new contracting party of the convention. Cooperation between the old East and West is entering a new era.

Looking into the Future

The JCP adopted by the diplomatic conference in April 1992 is a significant step toward protecting the Baltic Sea. On the other hand, the actual time of evaluation (less than a year) is too short to reveal any marked improvement in the marine system. The suggested measures are directed primarily at large industrial and municipal point sources, but no clear evaluation has been made of the extent to which these measures will reduce total loads. It is evident that for many pollutants the nonpoint sources (diffuse), such as nitrogen emissions through rivers and via atmospheric fallout, are overwhelmingly important. Also crucial for the evaluation is the determination of the spatial (local or the entire Baltic) and temporal (short term or long term) results of these measures.

An informal, interdisciplinary group of Baltic scientists concerned about the future of the Baltic environment met at Tvärminne Station, University of Helsinki, to discuss priorities for restoration of the Baltic Sea based on current scientific knowledge (Wulff and Niemi 1992). The principles discussed at those meetings are summarized in the rough graphs of figure 7.7. The upper graph illustrates a possible scenario in terms of future costs and efforts. The lower graph pictures how the Baltic could change as a result of implementing the plans. It will take several years to reduce the concentrations of nutrients by 50%, even if the load could be halved immediately, given natural events and the long retention time of the water and pollutants already in the system. In open coastal regions, no improvement will occur until large-scale conditions change. From a local perspective, investments in municipal sewage plants situated in enclosed regions and archipelagoes will be most beneficial in the short run. From an international, financing perspective, coastal sewage treatment plants should be given the highest priority. Industrial investments will rise rapidly once the old technology in the former Eastern countries is replaced to produce goods that can be sold on the world market. Modern technology and the public demand for clean production could reduce the load of pollution. A cleanup of existing industry will only result in a small reduction of industrial effluents at very high costs. When we propose a cleanup of existing industries, it is a dangerous illusion to believe that the end-of-pipe technologies will have more than marginal effects on the Baltic environment.

How can we effectively stop the eutrophication of the Baltic Sea? Both nitrogen and phosphorus must be removed to reverse eutrophication. It is evident that the essential sources of nutrients are diffuse sources entering the marine environment through rivers and the atmosphere. It is often very difficult to trace the sources. For instance, nitrogen entering the Baltic Sea via Polish or German rivers may have its origin in atmospheric emissions from outside the Baltic drainage basin. Local pollution levels are severe in many regions of Baltic Europe, and the cost of reducing these levels is enormous. The mistakes made in highly industrialized Western countries should not be repeated. In these countries losses of nutrients have increased dramatically due to the drainage of highly fertilized agricultural land. The belief that we can replace natural processes by technology is a social trap. What is needed is a more rapid increase in our efforts to reduce diffuse pollution sources by ecotechnological measures. Examples are the restoration and conservation of wetlands, and changing agricultural practices in the entire drainage area. Such methods require low investments and have low

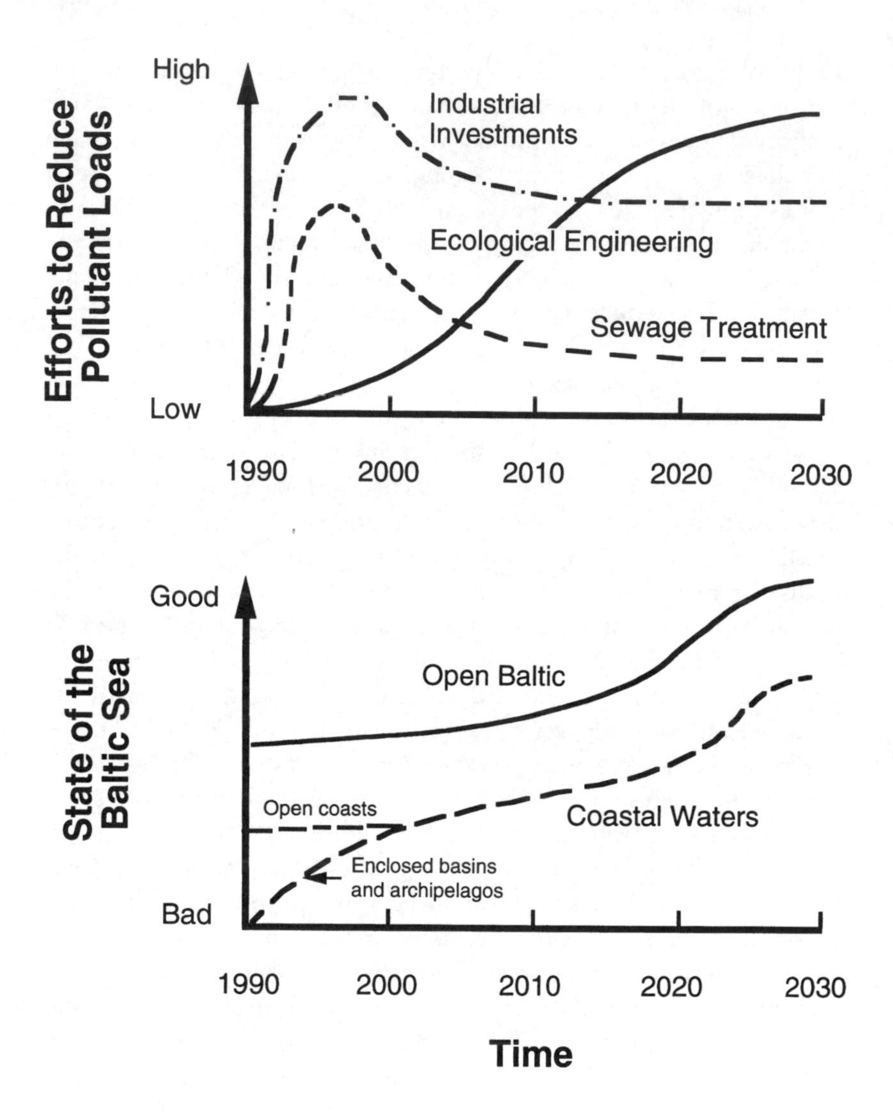

FIGURE 7.7
A possible scenario of the redevelopment of the Baltic (Wulff and Niemi 1992). The upper graph pictures the construction of classical sewage treatment plants starting with the bigger cities, preferably at the coast. The industrial investment also has to be high in forms of maintenance. Developing "green technology," letting nature pay for the maintenance would help lower the costs. The lower graph shows the accompanying improvement of the Baltic water quality—faster in coastal areas with their smaller turnover times, longer for the large open Baltic.

energy demands. Geographical Information Systems (GIS) describing land use, coupled to hydrological and ecological/economic interactions in different parts of the drainage area should be supported to model the effects of different future scenarios.

As a final result of the informal discussions there was a strong recommendation to the HELCOM Task Force Implementation Committee to formulate a common international and national strategy, with clearly defined priorities and responsibilities that could be applied during phase 2 of the JCP Programme implementation.

Interpretation of System Dynamics

When we look into the tangled mix of processes and events in the coupled systems of man and nature, a number of questions arise. Has there been a consistent pattern in time and space, going back to the basic concepts of energy and material flows? To what extent is Holling's infinity loop applicable to Baltic Europe? What can be said of the potential stability of the total system? We attempt to answer these questions, starting with a conceptual model of the system dynamics of the Baltic.

Order and Disorder—Exploitation and Degradation

Holling's infinity loop is a theoretical concept, which, applied to systems with well-defined boundary conditions, can be demonstrated in fair detail (figure 7.8). Because of the hierarchical structure of nature, based on scales of time and space, the loops are run simultaneously within and among the different levels. As a first phase in tracking the network of connections between the conceptual loops, it is tempting to speculate with our present knowledge of the system dynamics as background.

In the natural systems light and temperature are probably important forcing functions in synchronizing separate loops of the cycle. The coupling between the pelagic and soft-bottom systems in the Baltic can act as an example. In the surface water the microscopic plants, the phytoplankton, start to grow rapidly when light increases in early spring and the amounts of nutrients in the water are high after the dark winter period. This "exploitation phase" is very short and, as mentioned before, a large amount of the organic matter produced is exported to the soft bottoms through sedimentation of especially heavy diatoms.

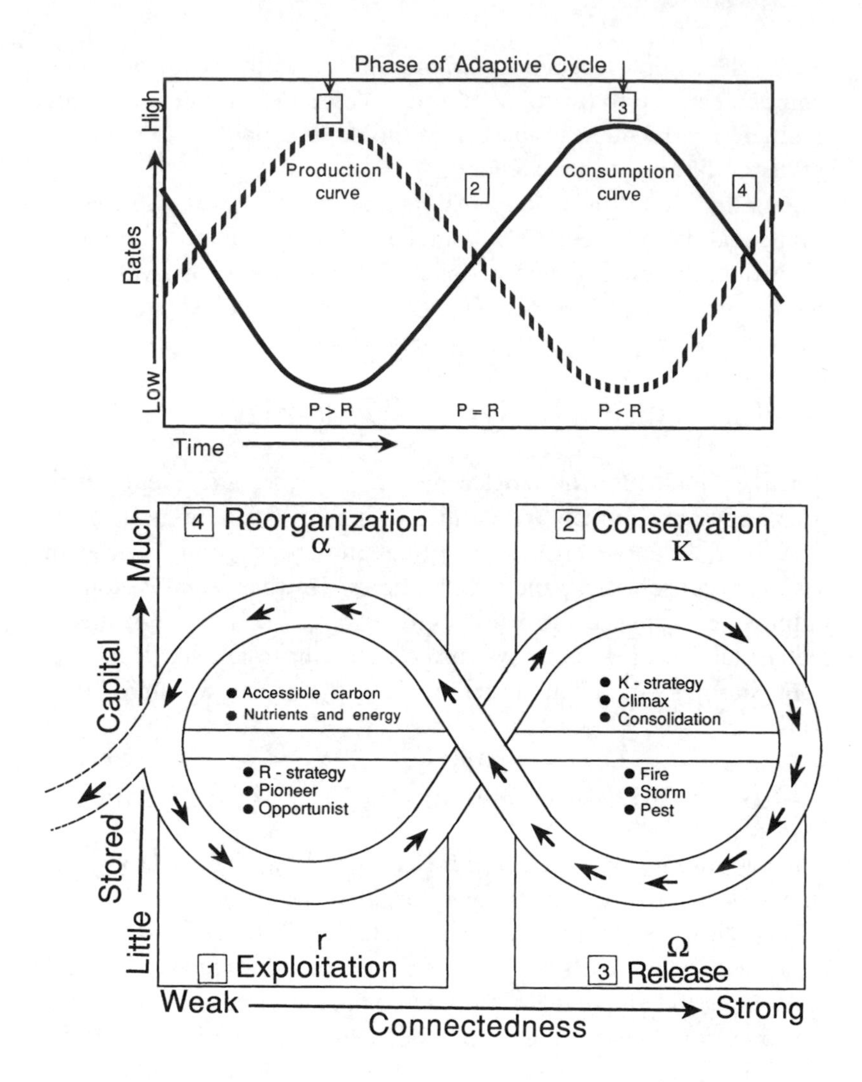

FIGURE 7.8

Comparison between the Holling infinity loop and an idealized production/consumption module. Both sets of curves track the same phases from reorganization (consumption/respiration) to conservation (or production). The difference is mainly in the tracking of the P/R curves along a time axis, in the Holling diagram replaced by degree of connectedness.

Whether the pelagic system ever reaches the conservation phase other than for short periods because of the stochastic disturbances through turbulence and advection is a matter of considerable interest. The soft-bottom system is usually in an exploitation phase during the spring bloom, with new generations of crustaceans, for example "waiting" for the injection of high-potential food from above and then reacting to it with increased growth. The decomposition and remineralization of accumulating organic matter reach their peak in the autumn. This "release phase" causes a buildup of nutrient storage in the bottom water during winter and "reorganization" during early spring, when accessible carbon, nutrients, and more light prepare for the next spring bloom, which ties in with the soft-bottom system again.

In the offshore areas the anoxic periods resulting from stagnation of the deep water can be translated as reorganization phases in which the formation of hydrogen sulfide triggers the release of phosphorus from the bottom sediments. An inflow of oxygenated North Sea water with accompanying faunal elements starts a reorganization phase, the length and fate of which are determined by the depletion rate of oxygen in the bottom water. Denitrifying bacteria release nitrogen to the atmosphere during periods of low oxygen. On average, 50% of the nitrogen release from the Baltic proper occurs during stagnant periods (Shaffer and Rönner 1984). Heavy metals are immobilized (bound in water-insoluble compounds) in anoxic sediments. The immobilization of metals is beneficial to other parts of the total system but poses a dilemma for management in the deeper parts of the Baltic: We might, for example, buy living macrofauna at the price of releasing hazardous substances (Jonsson 1992).

The socioeconomic system and its historical development are so complex that the brief efforts here to test the validity of the infinity loop are mere speculation. Some examples from the preceding historical description might stimulate future and more penetrating analyses, however.

Early in Scandinavian history the human population was small and the inability to use fossil fuels resulted in a scarcity of food. The exploits of the Vikings, who frequently ravaged regions outside their home area, constituted an effort to widen the resource base. For most people the Viking era has a strong flavor of destruction. Nevertheless historians say that they had a constructive effect on northern Europe. Is this a period that provides an example of creative destruction? During the prefossil fuel period the resource use was dominated by agrarian products. Arable land still dominates the land use pattern in four of the Baltic countries and exceeds 30% in six (table 7.3). Because inhabitants were restricted to renewable, "natural" energy sources, overexploitation of the natural resources was difficult. The limiting factors for production were access to arable land and human labor, including those occupied in the transportation sector. Overfishing in the open sea was not possible because of technical inefficiency.

The Hanseatic League was able to couple these resources and exploit resources from outside the usual system boundaries, enabling growth in human population and infrastructure during a long conservation phase. The growth was checked by the crash of the herring population. The crash was one example of the mystery of fluctuating fish numbers and happened again on the Swedish west coast during the 1700s (Höglund 1972). The league disintegrated, losing its niche to the Dutch, who by extending their network began into a new cycle of exploitation. Now, 400 years later, the New Hanseatic League (Andersson 1991) emerges after the creative destruction of the Soviet Union, as a Venus born in the Baltic Sea and delivered by *glasnost!*

The introduction of fossil fuel into the socioeconomic system totally changes the pattern, generating new cycles almost everywhere in society and nature. The rapid increase in availability of fossil fuel, the most revolutionary and powerful surprise in the history of mankind, theoretically meant that the operational time and space scale for the human system suddenly increased. New means of transportation allowed transfers over long distances in shorter times. Machines increased efficiencies in processes, allowing intensified use of natural resources within agriculture, forestry, and fishing, and thereby making overexploitation possible. Fertilization of farm fields no longer was based on manure or other types of natural fertilizers but could be industrially synthesized using chemicals. This strongly affected not only the socioeconomic system but also the natural marine system of the Baltic Sea, where the excess nutrients in the land runoff successively affected many cycles.

The accelerating fossil fuel use also meant an increase in complexity over the centuries. European society grew toward a climax. The release came with World War I, a cruel, inhuman affair. Coldly speculating, one could call it a "constructive destruction" from the viewpoint of politics. It created an independent Finland, Estonia, Latvia, Lithuania, and Poland, which were able to develop specialized industrial activities. World War II checked this consolidation. The small states became incorporated into a larger political system, and their resources were overexploited as the consumer base was greatly extended and their identity depressed. *Perestroika* released another "creative destruction" of the old system. At present the reorganization phase is in full swing, and money is its chief limiting factor.

Diversity and Resilience

Complex, large systems usually have several areas of equilibrium or stability regimes within which they move, depending upon the magnitude of the outside forces or boundary conditions. Stability of a system has to be defined for a specific time and space scale. Neglect to do so and a wide diversity of both species functions have led to seemingly different views in the past (McArthur 1955; May 1973; Orians 1975; Jernelöw and Rosenberg, 1976; van Voris et al. 1980; O'Neill et al. 1986). An important property of a system is its resilience, which means the amount of disturbance it can absorb without changing structure or behavior (Holling 1973; Holling et al. 1994). A highly resilient system can lose a fair number of species but still maintain its characteristic functions. The functional diversity—the different processes that enable the system to fix, process, and recycle energy and matter—is therefore more critical than the diversity of species. The Baltic Sea, with a small number of species carrying out the different basic processes, is less resilient than, for example, the North Sea, which has at least ten times more species. The absence of clams and mussels in the soft-bottom macrofauna of the far north of the Baltic Sea and therefore the lack of the filter-feeding function have resulted in an impairment of the decomposition of the organic material (Elmgren et al. 1984). The small redundancy of the Baltic Sea system is a strong barrier to unlimited exploitation of its natural resources.

If it receives 1 million tons of nitrogen per year from air and land, how long will the present marine system last? The increasing sedimentation of organic material from expanding filamentous algal belts and plankton blooms means increasing anoxia in the sediments and increasing denitrification, but it also implies increasing immobilization of heavy metals and toxics. The levels of phosphorus and nitrogen in the open waters have been more or less constant since the mid-1980s. Is there a breakpoint leading to another type of system, and if so where is it and what will this new system look like? Will it be a system with hydrogen sulfide up to the halocline, without cod populations and therefore changing populations of its food items—herring, sprat, and eelpout—and of its food competitors, salmon and seals? How near is another area of equilibrium?

How is the diversity concept applied to the socioeconomic system? There is considerable linguistic diversity in Baltic Europe. The pattern of coupled languages, customs, and temperaments indicates that the present nine-nation system may be stable. The terminated 50-year "experiment" of incorporating four and a half states into a larger and different system resulted in strong repellent reactions and emphasis on individual uniqueness. The functional diversity of the eastern Baltic states, expressed, for example, by the developed industrial sector in Estonia and Latvia, might be a sign of potential stability of the larger system. As stated by A. M. Jansson (1991), sustainability is reached only if the boundaries between the different countries are opened to a free flow of capital, people, and ideas. An industrial society can only be maintained, however, if it is able to develop feedbacks that increase the life-supporting capacity of the total region.

Spatial Diversity

As stated by a systems ecologist, the main respiration or consumption of Baltic Europe takes place in the populated areas of the drainage basin. The system of man and nature, now heavily subsidized by auxiliary energy, shows considerable patchiness, often with practically pure primary production centers (natural land) far from areas of heavy consumption (urban centers). Marshes, swamps, wetlands, water, and forest occupy more than half the land area of Sweden, Russia, Estonia, Latvia, and Finland (table 7.3). The urban areas represent centers of intense consumption, dissipation of energy, and generation of waste. Only in Denmark do they occupy more than 10% of the land area. The marine system as a whole is a consumer system, successively driven further into heterotrophy because of increasing inputs of nutrients and organic material. Man's part in the heterotrophic activity consists mainly of fishing and transportation; both activities continue to increase. The preindustrial transportation based on wind energy had little effect on the system. Transportation has increased enormously after the industrial revolution. Compared to nature's agencies of transportation, such as wind and water, vehicles of human transportation are consumptive and create considerable waste products, most of which end up in the Baltic Sea. Nearly 60% of the emissions of oxidized nitrogen in Northern Europe are due to the car (Pacyna et al. 1991).

Usually, the production (incorporation) area and the respiration (release) area of nutrients are spaced apart, and the regenerated nutrients are only to a limited extent brought back to the incorporation centers through feedback. The originally oligotrophic, or nutrient-poor, Baltic marine system is gradually aging and being transformed into later stages of eutrophy, nutrient-rich stages where decomposition dominates. Is the Baltic doing most of the denitrifying service for the Northern Europe system?

Validity of the Infinity Loop

The strength of the infinity loop is a seductive visualization of the basic perpetual processes of transformation of matter. Based on the fundamental order/disorder concept, it also ties in with the thermodynamics of living systems. Because it's a synthesis of innumerable successive processes, it is interesting to compare, with the classical pulses of production and consumption tracking each other in a living system. Idealizing the curves (figure 7.8) and inspecting one cycle provide a configuration of activity that resembles the infinity loop. Starting with the production process, the first phase shows a successively increasing gross production, reaching a maximum that corresponds to the exploitation phase of the infinity cycle. Production then declines and crosses the consumption curve, indicating the steady-state phase of the mature system when production balances consumption (one type of climax system). The system then becomes senescent, brittle, and sensitive to a catastrophe, declining to a minimum stage that corresponds to the release phase, when decomposition dominates. During the next phase, production increases and consumption decreases, as they outbalance each other in the reorganization phase. The sum of the production and decomposition processes of a system shows the same phases as the infinity loop, the difference being that degree of connectedness is the abscissa in the first case and that time is in the other. Both diagrams indicate the basic shifts between order and disorder in systems.

Evolution of Adaptive Management

Invalidation of Postulates

Does the description of the natural events and the human responses in Baltic Europe fit the presented postulates? Can we recognize the different phases of action at the scientific, institutional, public, and political levels? Or does the Baltic present a totally different dynamic? We attempt responses in the following sections, which should be regarded as biased because our primary knowledge is that of our own countries.

The chronology of resource management has been a history of crises and unitary responses, generating polarization and conflict. Numerous examples of this pattern are found in the history of the Baltic. We have described many triggering events, ranging from those affecting the whole Baltic to those affecting more local areas. When the mercury crisis was recognized for the Baltic, fish and birds had been exposed for a long period. Action at the management level was fast and successful. Now, nearly three decades later, a second mercury crisis is feared, this time due to acid rain, which mobilizes mercury in the soils. The time and space scales for this mobilizing process are much longer and larger, involving atmospheric fallout from northern Europe. Eutrophication was regarded as a local phenomenon in the 1950s until Fonselius's publications in the late 1960s suddenly transformed the problem to a regional or larger scale. Polarization, both at the scientific level and at the management level, certainly did occur. One analysis by a university scientist (Odén) of the situation with independent researchers was met with skepticism by governmental scientists and the management institutions. One of the main criticisms was the brief time scales of the university projects. That this was mainly a function of money was not considered. The managing institutions reacted quickly, however, perhaps because of competition with the universities. The classical monitoring of the fishery boards in the open Baltic was combined with less institutionally secure studies of the coastal areas by the universities, a combination that still exists and that has produced valuable insights into system dynamics. The myth of the "endless sea" existed for many years, among both scientists and managers, who blamed the natural stratification and intermittent water exchange of the Baltic for the anoxia. The development of national and international monitoring programs succeeded in precipitating a dialogue between the scientists and the managers who had to allocate the money. As always, the scientists thought the coverage was too restricted both in space and time and that the limited funds could be better used. A question that was always in the background and sometimes was expressed was, "How much money do we allocate to stop the pollution flows, and how much do we need to spend on measuring the effects?"

A foundation of understanding began to emerge following periods of crises and polarization. Understanding the sources of the surprise usually came gradually. In the case of toxic substances, the point sources were identified fairly soon. It was more difficult to learn how they acted on targets in the system, the population, and the individual. In the gray seals the high levels of DDT and PCB were registered in 1969; sterility was suspected, and aborted pups were found in 1970–1972; occlusions of the uteri were noted in 1975; in 1977 PCB was suggested as the likely main cause following feeding experiments with mink. A beautiful job by the scientists! The large-scale eutrophication of the Baltic was a less clear-cut affair, because, according to geological evidence, "dead bottoms" had existed before industrialization. It took some 10 years before physical oceanographic data (Shaffer 1979; Stigebrandt 1983) showed that decreased water exchange could not be the ultimate reason for the lack of bottom oxygen over large areas. The informal compilation by Larsson and colleagues (1985) of HELCOM data, not yet authorized by the commission, also definitely buried the previous speculations that leakage from the sediments of accumulated nutrients was responsible for the increasing concentrations in the water column. Scientists gradually were represented on relevant boards, and the barriers were more economic than personal or institutional. One example of new scientific ideas stimulated by crisis occurred when Walin (1977) used a coordinate system of isohalines (lines drawn through stations of the same salinity) that greatly facilitated water transport calculations.

In conjunction with the emerging foundation of understanding, an informal collaboration developed among scientists in various institutions. The importance of the CBO and BMB as informal organizations of scientists has already been stressed. Although the ecosystem approach was readily adopted, cooperation across scientific disciplines was slower. Although cooperation was facilitated through the administrative structure at the institutes for marine research in Kiel, Warnemünde, and Helsinki, the process was more difficult at the universities, where the marine activities are split up among institutions much broader in scope. International projects like SYSEX, PLEX, and Large-Scale Processes in the Baltic have bridged the information gaps. Modeling has concentrated and tested the present information, and strong groups have emerged in Copenhagen, Kiel, Rostock, Tallinn, St. Petersburg, Helsinki, Stockholm, and Gothenburgh. The collected knowledge has been presented and disseminated in frequent seminars, working groups, and symposia, both in the Baltic region and in the wider international scientific community.

An informal consortium of concerned citizens, scientists, and business leaders was established. The magnitude of the Baltic crises covered a range of hierarchical scales, from loss of species (white-tailed eagle, porpoise, ringed seal, gray seal) and changed subsystems (e.g., hard bottoms transferred to soft bottoms because of increased sedimentation through algal growth), to large soft-bottom communities changed to enormous cultures of sulfur bacteria. This and restrictions for human consumption of some fish species aroused everyone to action. In several cases (e.g., Sweden), competing research groups now cooperate closely and have learned to put aside personal interests for the benefit of the whole. There are certainly differences among countries concerning the relation between the controlling institutions and the informal networks. In Sweden the small population and the efficiency of communication among politicians, administrators, and citizens speeded up official reaction to the surprises. NGOs and individual concerned citizens usually appeared at the initial negotiations of major environmental suits. One example of an ecological concept that has gained interest is the use of wetlands as nitrogen traps. Another is based on the large filtering capacity of blue mussels and suggestions to use them as a biological filter in coastal receiving areas (Folke and Kautsky 1989).

It is hoped that these examples of ecotechnology will induce an expansion of that field also in the Eastern countries, which are now struggling to build a new structure for a functioning society. The present shutdown of many industries means a decrease in pollutants. In the turmoil of this reorganization phase, with a mixture of foreign institutions, private consultants, and modern brigands, it is most important that the reorganization of a sustainable society be carried out within the dynamics and carrying capacity of the natural life support system.

Catalysts triggered political actions that resulted in system reconfiguration and renewed commitment to broader-scale ecosystem management. In some countries, single surprises added up to large-scale political actions. One example is the formation of the Action Group North, South, and West in Sweden following the Laholm crisis. The groups quantified total sources and emissions in the watersheds and suggested means of reduction. A task force created by the Baltic Sea prime ministers conference in 1990, within HELCOM, took a systems view and included economics in the resulting Joint Comprehensive Program. A more concrete example is that of the foundation of the three Swedish marine research centers in 1989, with responsibilities for multidisciplinary research and education and basinwide monitoring and information, constituting a substantial economic input to Swedish marine science.

The preceding events resulted in a new institution characterized by systemwide change, adaptation, and adequate funding. Nobody in the Baltic region who interacts with the mass media in any way has been able to avoid knowing of the present condition of the Baltic Sea and the necessity of restoring it. A clear example that this has cut through the whole society, even through the most urbanized sector, is provided by the Association of the Baltic Sea Towns, which will optimize the socioeconomic exchange and minimize the effects on the environment. Another is the New Hansa, which is a metaphor for a regional collaboration around the Baltic aiming at an expansion of networks between the creative regions of the north. "This will undoubtedly favour the cultural and economic development of the regions around the Baltic Sea, and the evolution of an integrated ecological strategy for the region" (Andersson 1991).

From the preceding examples it should be obvious that the postulates are valid for the Baltic case, at least as far as we know. *Perestroika* has precipitated a new era, especially in the Eastern countries. Future development offers opportunities to test the postulates further and new turns of the infinity cycle.

Barriers and Bridges

One of the classical myths is that nature is constant, endless, and totally resilient. Our case study has shown that this is not true. The ecosystem is developed and maintained from pulses and feedbacks, many of which we call surprises, as we are unable to step out of our hierarchical level and apply a longer predictive time scale. We are aiming toward a sustainable society. How can we adapt to a dynamic life support system? What are the barriers and the bridges to adaptive management? In the complex picture we have described they may stand out more clearly if classified into five problem areas: limits to adaptive capacity, myths, the relationship of science and policy, institutional capacity, and the level and degree of collaboration.

Limits to Adaptive Capacity

General understanding of the system is a necessary base for any type of constructive interaction with the system. The past is full of examples of misunderstandings because of ignorance of a system's behavior. Every exceptionally sunny spring we are approached by worried citizens who have been looking at the floating algal mats after the spring bloom and believe that the Baltic has suddenly become even more polluted. Algal growth, however, needs not only nutrients—which have been fairly stable since the mid-1980s—but sunshine. Growth will be faster during sunny conditions. The media's sensational coverage of toxic plankton blooms has led the public to feel all plankton blooms are negative, when in fact most are the basis for marine life. The importance of the spring bloom in the Baltic for the soft-bottom system has already been stressed. As we write, a big storm has wrecked a large Polish ferryboat, with many casualties—a bad surprise. At the same time, this storm is bringing in large amounts of salty, oxygenated North Sea water, which will improve conditions for a successful cod spawning this spring. Is this an example of creative destruction?

The importance of adequate time and space scales is exemplified in the residence time of the water in the Baltic. Even if we stopped all inputs of nutrients, the effects would only be measurable in the open sea after many years. A point source in the immediate receiving area, of course, has more rapid consequences than one more distant. Changing the water quality offshore means changing activity and discharges in the whole drainage area, an area four times that of the Baltic. Therefore setting the systems boundary is a crucial bridge.

Discussion of the potential effects of the planned bridge between Sweden and Denmark, the Öresund bridge, has focused on the possible negative effects on the water exchange through this critical strait. The official environment impact assessment totally ignored the possible secondary effects on the water quality in the Baltic. The expected increase in car traffic will likely increase acid rain, thereby releasing nutrients and hazardous substances from the already saturated soil to the coastal waters. Likewise, an expected increase in tourism means increased exploitation of the coastal area, with the well-known effects of eutrophication of the water, anoxia, and changing fish fauna (figure 7.9). This is an example of using inadequate time and space scales and inability to see feedback effects.

Time lags between surprises and responses are longer for larger systems. The time span for solving a single environmental hazard, such as mercury, is much shorter than that for averting the threat of toxic substances, as we are continuously emitting new, untested chemicals. A specific substance produced by a certain type of industry is easier to check than, for example, the flow of nutrients generated (over much larger time and space scales) by much larger systems: agriculture, forestry, and traffic.

Synthesizing the information is most important. One useful way of integrating scientific information is to build models. The mere construc-

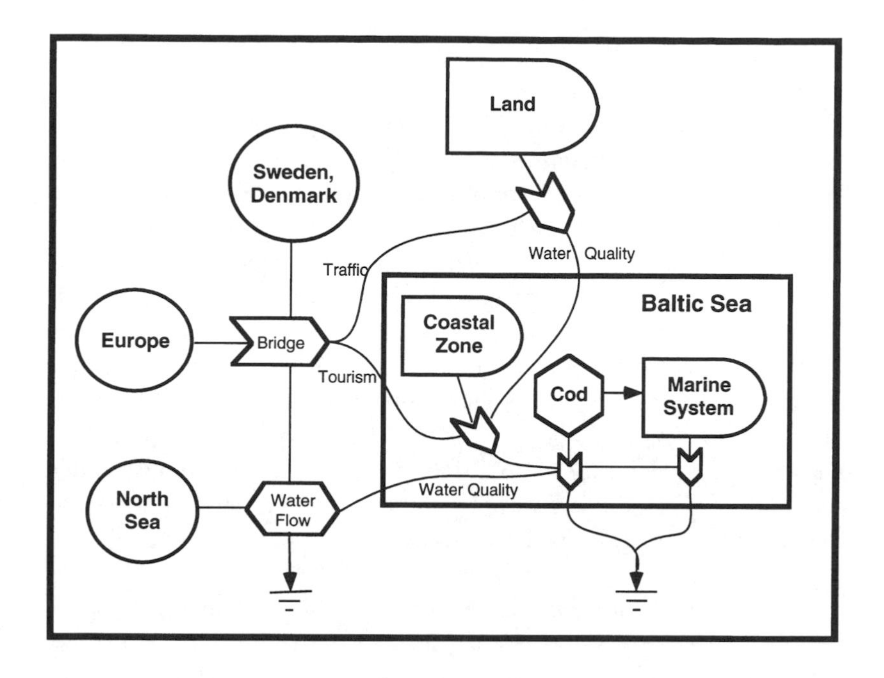

FIGURE 7.9
Conceptual model of the main processes connected to the planned bridge between Sweden and Finland—the Öresund Bridge. Discussing the effect of the bridge on the water quality in the Baltic, the planners and the authorities have concentrated on the potential check on the water flow. The "indirect" effect from increasing car traffic, acid rain, leakage of the soil, and increased exploitation of the coast might be more serious.

tion of a box model is an exercise that forces decisions concerning boundaries—time and space scales, the main processes, feedback loops, and so on—and this is good insurance against future surprises. Models can be used to show the coupling between subsystems (Jansson et al. 1984), prediction (Stigebrandt and Wulff 1987), and blueprints for multidisciplinary and multinational research teams (Wulff 1990). Informing the public is a matter of highest priority, and one that is shamelessly neglected. Providing correct, balanced, interesting, and understandable information to laymen is very difficult, and something of an art. Disregarding the fact that taxpayers have the right to know what scientists are doing with their money, understanding of environmental problems has a tremendous information value to society.

Myths

Myths are concepts held to be true, whether tested or not, and can be a natural, although dangerous, aspect of society. Here are some myths we have encountered while working with problems of the Baltic.

Myth 1: The endless sea. Leading among the old myths, both in toughness and in influence, is the already mentioned idea of the "endless sea." The Baltic fits that description very poorly, being semienclosed, tideless, and cold. Still, in the 1950s, there were serious plans to transport the untreated sewage of the city of Stockholm to the coast and dump it in the Landsort Deep, illustrating as well the myth "If you do not see it, you've safely disposed of it." This also illustrates the well-known slogan "Solution to pollution is dilution," which holds under certain conditions, but is seldom true in the Baltic. In fact, it is only the shallow coastal systems that have more efficient built-in mechanisms of degrading and recycling matter.

Myth 2: The natural boundary of the Baltic Sea ecosystem is the water line. Although many processes are restricted by that border, coastal seas (and especially enclosed ones) are greatly affected by processes in the drainage areas, processes that have to be included within the boundary conditions.

Myth 3: Increasing the water exchange by dredging the shallow Öresund would improve oxygen conditions in the Baltic. This is false. It would certainly mean an increased inflow of salty, oxygenated water below the outflowing Baltic surface water, but the stratification would increase and the deeper bottoms would become even more loaded with phosphorus, released to water during stagnant periods.

Myth 4: The pollution problems of the Baltic can be solved by building and refining more sewage plants. This means a continuation of the output-management, end-of-pipe strategy, a strategy that should be replaced by an input management. First, land runoff and atmospheric fallout can only be stopped at the source; second, the maintenance costs of the plants are very high. Sewage plants are probably needed for the larger urban centers, but they should be replaced whenever possible by ecotechnological solutions that make better use of the services of nature and have cheaper, long-term maintenance costs.

Myth 5: Once an environmental problem is institutionalized, we can relax. Experience shows that every institution—from small local boards, sewage plants, and monitoring programs to national boards—needs evaluations, stimulations, and renewal of methods and ideas. Fresh insight from the public and a strengthening of the personal responsibility are often needed.

Myth 6: Information to all concerned categories is one of the most important issues. This is true, but it has to be balanced, understandable information. It is difficult to get a diversified statement printed, but it is important always to try. Scientists should participate in the public debate, because they are the experts who should have the facts and because they are also citizens and part of the same problem.

Myth 7: Growth of society is necessary for the sustainability of man. This is a worldwide myth, which is impossible to discuss if growth is not explicitly defined. Growth should be possible even while using less fossil fuel. Natural systems are creating growth by increasing functional diversity and speeding up recycling. This is an urgent, basic, and fascinating area for intensified research.

The Relationship of Science and Policy

One good example of events that showed the managers and policy makers that science is an important means for understanding our lives is the mercury problem. Dr. Karls Borg's findings in the late 1950s of the decline of Swedish yellow sparrows, supported by the observation of the famed Swedish bird-watcher Erik Rosenberg and Rachel Carson's book *Silent Spring* (1962) were warnings that the environment was not a forever stable, all-forgiving panorama for our different performances continuously renewing itself, like the still water surface after the diminishing impact of a thrown stone. In Sweden scientists were listened to by both politicians and managing institutions. In Finland, the scientists' warning of the state of the Baltic inspired the politicians to design bilateral cooperative agreements with the U.S.S.R. on the Gulf of Finland in 1968 and with Sweden on the Gulf of Bothnia in 1970.

Scientists in Poland had greater difficulties transferring practical reactions into policy. Part of the public looked upon the scientific work as unproductive and postponed action on the management level. There was practically no dialog between leading scientists and the regional coastal management offices. The fact that most of the government ministers were also professors was not viewed positively. At the same time, public pressure was necessary for getting the politicians to act on the scientist's advice. Poland's active fishery biologists claimed pollution, not overfishing, as the main reason for the decline in fisheries, stressing that sturgeon and shad had vanished from their coastal waters and that bottom fish showed more skin diseases.

The DDR was very active in Baltic research, both with extensive offshore monitoring as a basis for following environmental changes and with long-term studies of the coastal area. Political conditions made follow-up at the management level difficult. After incorporation with the FRG, the former Institut fur Meereskunde in Warnemünde was reorganized into the Institute of Baltic Research, with a multidisciplinary structure and a section for monitoring studies.

The FRG came fairly late to the managing sector, partly because of its long and famous tradition of fundamental marine science and partly because of a certain skepticism about the extent to which human-caused nutrient flows have affected the conditions of the sea (Gerlach 1985). Nevertheless, the FRG played an important role in scientific collaboration through its large resources and generous attitude about collaboration. The critical oxygen conditions in German coastal waters, both in the North Sea and in the Baltic in the early 1980s, prompted the minister of the interior to establish a working group on eutrophication of both seas in 1984.

Scientific information has proved important for policy actions both in Finland and in Sweden. A decrease in the seaweed communities because of the large-scale increase in turbidity is one example of science intersecting with management actions. The series of action plans and action groups set up in response to the eutrophication events and the high levels of hazardous substances in Swedish coastal waters resulted in representatives of the scientific community being involved. They had a strong influence on the formulation of action plans, although their full recommendations were not taken in tailoring the limits of discharges. Swedish industry spokesmen felt that the authorities listened too much to advice from the scientists. In fact, the close collaboration between the Swedish Environmental Protection Agency (SEPA) and the scientific community has been most important for the growing success of the redevelopment of the Baltic ecosystem. A good number of SEPA administrators are scientifically educated, creative, very sensitive to new scientific findings, and quick to sponsor environmental research even of a basic nature to broaden the necessary base of knowledge. On the basis of this well-functioning integration of science and policy, including the transdisciplinary coalescing of scattered information, Sweden should follow the proposed action of the US Congress and elevate the Environmental Protection Agency to cabinet status (Koshland 1990).

International NGOs

The informal scientific organizations CBO and BMB have played a significant role in discussing environmental changes from a scientific point of view, spreading new ideas, and pushing the policy organizations in the different countries to support the advice of foreign colleagues. In addition, they played a basic role in educating graduate students and young scientists, who often made their first public appearance at the alternating biannual symposia of the organizations and made friends with the "big shots" and colleagues in other countries. The resulting large family of scientists around the Baltic has been an important force in managing Baltic Europe.

The creation of the Swedish Marine Research Centers was based to a large extent on the joint scientific experience from decades of discussions within the scientific community. This should be seen as a result of increasing collaboration between scientists and policy makers since the 1960s in a small country where informal contacts have been easy and important. It takes both scientists and administrators to build bridges over environmental abysses.

Institutional Capacity

National Portraits

In Finland and Sweden the environment is an extremely important political issue. Some 20% of Sweden's population like to fish in their spare time, and fishing along the Baltic coast is free. Comparatively large areas of land belong to the state and *allemansrätten* ("all man's right") expresses the feeling that all land, including privately held lands in Sweden and Finland, belongs in certain respects to everybody. Thus the public freely strolls around in the fields and along the coast. In Finland, land ownership is "holy," and the quality of the environment in some cases has become more important than economic considerations. This has been the main force behind the creation of institutions for the environmental policy process in the two countries. Finland was the first nation in Baltic Europe to have a ministry of the environment, and in 1971 an advisory board was established to inform politicians on environmental matters. The National Board of Waters and Environment is responsible for what happens with the more than 5000 lakes and the Baltic Sea. Current policy gives a certain feeling of controlling the situation, together with the numerous projects of the Finnish Water Research Institute. Improved treatment procedures applied during the 1970s have borne fruit; water quality has improved along the coasts around urban and industrial areas, and bottom communities are recovering. Finland has a more stoic attitude toward the Baltic problem. Changes in such a large body of water require a long time for definite proof, and who knows if they are caused by man? The Baltic comes after problems like acidification, decrease of natural resources (including rare species), regulation of hydroelectric power, and spreading urbanization.

In Sweden increasing public concern for the environment during the 1950s and 1960s could be measured in the increasing number of bills on conservation issues in the Riksdagen, the House of Commons (Söderqvist 1986). These bills had their roots in the increasing number of scientists and members of NGOs (such as the Swedish Society for Conservation of Nature) who had seen their beautiful country change during the process of industrialization. One of the effects was the creation of the National Board of Nature Conservation in 1963, which 4 years later received an increased mandate as the National Board of Environmental Protection, previously discussed as the Swedish Environmental Protection Agency (SEPA). These and similar events started an "ecologization" not only in academia but in the whole society. In the environmental sciences this process was natural within the water pollution sector but slower in the area of toxicology. The early mercury problem (1962–1967) was strictly a chemical/physiological problem until Professor Alf Johnels in 1966 coined the term *ecological biocide* research (Söderqvist 1986). Our description of the evolution of an ecosystem approach to the Baltic could be seen as part of this process. The shift of mandate of the relevant minister, first to the Ministry of Environment and Energy in 1987 and then to the Ministry of Environment and Natural Resources in 1992 could be interpreted as the gradual political acceptance of the concept of nature as the endangered life support system of man.

Denmark has always been strong in agriculture, fisheries, and related sciences. This has led to active utilization of its natural resources, affecting both living marine resources and water quality through excessive use of chemical fertilizers. The short coastline along the Baltic and better ventilation in the Belt Sea have probably contributed to Denmark's low profile in the management of the Baltic Sea, although Denmark plays an important role in Baltic science, with its outstanding research laboratories and its long experience of water management through the Water Quality Institute and as host of the headquarters of ICES. The Laholm event put pressure on the Danish authorities, forcing politicians to decide on strong goals for cutbacks of the nutrient flows from land in 1982.

Poland is an example of the special conditions that exist in the former states of the Eastern bloc. The situation has been different, mostly because of the political regime and limited economic resources allocated to management of the environment. Poland's scientists could not move the government to act without strong support from the public. A new monitoring program for the Polish coast was established, thanks to public action. The Catholic church has played an important role locally in checking the location and expansion of nuclear power plants. This contrasts strongly with the picture in Finland, where management institutions were active without strong public pressure.

HELCOM

In this diverse Baltic family of national managing organizations, HELCOM has played an important role. According to Polish opinion, this type of organization is the only way of generating international cooperation and accomplishing such things as the standardization of monitoring the environment. HELCOM's drawback is a result of having to function in a difficult political climate: It can only make recommendations and has authority over only the international waters—that is, the open Baltic. For the important coastal areas it must appeal to the goodwill of the separate states. This is now being improved, as the catchment area will be included in its jurisdiction. Certain Finnish and Swedish voices, complaining of the information gap between HELCOM and the national scientists and managers, signal inefficiencies in the national information network. HELCOM has been a study object not only for the Mediterranean countries looking for a common organization but also for China and India. International groups have been impressed to see how the commission, with such a limited budget (in 1986 about $800,000 U.S.), has been able to act despite such large international differences. The status of HELCOM must be improved, however. The agreements must be made mandatory. The ministerial agreement in 1988 for reducing the emissions of nutrients, heavy metals, and persistent organic substances by 50% by 1995 is unlikely to be achieved. The new political era is evident in the meeting of the Environmental Committee in October 1992, in which were discussed Russia's establishment of a working group for mapping dumped munitions and HEL-

COM's proposal for a seminar through which the eastern Baltic states could learn working procedures. Scientific cooperation must be greatly improved if the Baltic ecosystem is to be restored. Political and economic restrictions still make work from research vessels very difficult, and sometimes impossible.

International Baltic Sea Fishery Commission

Few problems are as delicate and challenging as fishery management. The object, the fish, is a moving resource, affected by changing environmental conditions, including man-made ones, and subject to strong economic pressures. The Baltic was divided into seven national zones in 1978. As a result of increasing fishing pressure, the total biomass of fish in the Baltic declined by 30% between 1970 and 1985. The total landings have at the same time increased to a maximum of nearly 1 million tons in 1985, and then they declined. Reliable statistics on fish stocks date only to 1970, with the state of each fish stock having been collected by the ICES, which also has an Advisory Committee for Fisheries Management. It reports each spring to the International Baltic Sea Fishery Commission (Warsaw), created in 1974 by the seven littoral states. This information is not processed until September, when the total allowable catch of each species for the following year is determined and then split into national quotas. This means a 2-year delay between gathering the basic data and establishing the quota—a long time for a dynamic natural resource.

There are several stocks of herring, sprat, and cod in the Baltic, and they cannot be dealt with as one unit. Since 1982 the commission has not been able to agree on a quota for cod, and there is a tendency to surpass the recommendations of the ICES when it comes to the national quotas. When the catch of a species decreases, this should be a sign to decrease the fishing pressure, but the rule has been precisely the opposite! The Baltic fishery therefore is seriously overexploiting some of its fish stocks.

Recent discussions have recommended that fish quota decisions be made in June instead of September, which would make successive revisions during the year possible. Second, the fishing capacity in coastal waters, where trap nets and set nets are mostly used, should be controlled. The trawling in the offshore waters should be checked by limiting the potential fleet power consumption, as this has been shown to relate to the fleet fishing pressure (Thurow 1989). Theoretically, this is an interesting proposition. This approach suggests the system be regulated by changing the input not of energy but of fossil energy and by allowing the system to reorganize itself. As systems ecologists we are very curious, eager, and optimistic to see the results!

Nongovernmental organizations (NGOs) play an important role in management of the Baltic. Important organizations are appointed observers in HELCOM. NGOs present their own detailed action plans, such as the Swedish Society for the Conservation of Nature, which, together with its Finnish counterpart, has taken the initiative of founding an umbrella organization for the Baltic environmental organizations, Coalition Clean Baltic (CCB). WWF has a Baltic program and a bulletin that presents diverse views and information, mixing articles on eastern Baltic landscapes with reports on fishing quotas, environmental courses, and cruises. Greenpeace has a Baltic campaign and participated actively in the Baltic Sea Prime Ministers Conference in Ronneby. The NGOs are playing larger roles on governmental boards and in management institutions, providing an important bridge between the public and policy makers by increasing the spatial extent of interactions. Their urge to redevelop the Baltic rapidly contains one negative side. Caught up in their own institutionalization and growing militant behavior, they forget to advertise the fascinating and pleasant sides of nature that catch people's interests and create positive change.

The public has exerted strong pressure on the managing institutions and politicians to stop the degradation of nature. The love of outdoor life in the Nordic countries, where many families have summer houses, results in personally witnessing the environmental changes. Ownership or sense of place seems also to be a strong force. When it is supported by interesting information of the fascinating ways nature functions, it supports the redevelopment of the ecosystem.

The inherent inadequacies in the Baltic management system can be summarized as a lack of common systems knowledge and a lack of institutions with adequate political status, mandate, competence, and economic resources. Within Baltic Europe there is systems knowledge adequate for a successful adaptive management, but it is mostly confined to scientific groups when it should be common knowledge of the managers. Basic knowledge of time and space scales and of the interrelatedness of industry, agriculture, and forestry are critical to adaptive management. Adaptive management of the Baltic Sea must be international. Scientific competence is fairly widespread, and an institution at the proper scale (HELCOM) exists. However, political stability and money are still lacking. The European Community could fill these needs, but it must acknowledge that nature is the life support system of man and that society must be governed according to nature's rules. Governing Baltic Europe as a sustainable society; keeping the fascinating diversity in the natural and urban systems; and following the pulses of nature are not only challenging but vital.

Research Recommendations

We may ask to what extent lack of scientific knowledge is a barrier to creating a sustainable Baltic Europe. The scale and diversity of the total concept exclude a thorough analysis of the question, but some main challenges within different domains of knowledge should be pointed out.

The present development within basic natural science appears satisfactory, and our understanding of large-scale dynamics of limiting nutrients in the marine system is rapidly increasing. The unification of large-scale processes to form a coherent picture of the total system is also under way. Moreover, the present collected data base is probably unique in its completeness for such a large system. Little has been done, however, on the theoretical level. Therefore, research on the dynamic behavior of large systems, such as the Baltic, must address the concept of multiple stability domains.

Within the area of complex systems of man and nature there is a rich field of applied problems connected to the sustainable use of natural resources. Natural resources management, including the crucial area of ecological economics, is central to bridging the gap between practical use and theoretical knowledge. A number of challenging tasks remain. The following are some examples:

Ecotechnology (nature's services to man, such as producing clean water, creating clean air, and recycling nutrients). This should partly replace human energy-consuming substitutes.

Land use/water quality. This is a crucial problem for all semienclosed seas and coastal areas, where new, powerful techniques have opened up systems ecological approaches.

Long-term exploitation of pulsing natural resources. Fisheries constitute a typical example through the often periodic appearance of fish schools. A more flexible way of harvesting than the present continuous pressure from large, static fishing fleets with huge maintenance costs has to be explored.

Transportation systems networks. Natural systems are structured with flows. Transports of goods and people of Baltic Europe have to be tailored from the aspects of time, energy costs, environmental disturbance, including fragmentation of landscape.

One prerequisite for a successful research of such a diversified system as the Baltic Europe is the open and respectful acceptance of the natural scientists that the soft sciences are equal partners in the formidable task of exploring the combined dynamics of society and nature.

Epilogue

January 11–15, 1993. The Öresund bridge project, the first bridge to be built between Sweden and the continent and by some called a bridge, by others a barrier, is tried for sanction according to the law on protection of the environment. The primary background information consists of investigations by experts from the consortium of the construction project and an Environmental Impact Statement made by consultants at the request of the consortium. Both have focused the interest on the water flow through the Sound, claiming small effects. Greenpeace, World Wildlife Federation, the Swedish and Danish Society for Nature Conservation, and other NGOs plead for rejection because of considerable negative effects on the aquatic environment. University systems ecologists argue that in addition to the effects from dredging on the fish spawning grounds in the Öresund the crucial effects on the Baltic will be those emanating from car traffic and the increased tourism (figure 7.9)

February 22–26, 1993. An international expert panel consisting of biologists and physical oceanographers appointed by the Swedish and Danish governments meet to discuss the project. They point out that their mandate and expertise are restricted to water flow and marine biology but that the socioeconomic aspects of the bridge are "exceedingly important and must be carefully considered in an overall evaluation of the environmental effect of the proposed bridge-tunnel." The hydrographers are not convinced of the two-dimensional models used for simulating the effect of the bridge on the Baltic. A three-dimensional model should be tried together with field measurements. These tests might be ready in 1994.

February 25, 1993. The Commission of Environment declines the project, referring to the precautionary principle. The beneficial role to Swedish society and the small impact on the Baltic have not been convincingly proven, and the effects of increased traffic will be negative, according to the commission. The Swedish prime minister's main comment was, "We shall get a bridge sooner or later."

September 27–October 1, 1993. The Swedish Water Court tries the project. The main actors from the January meeting play again and present the same arguments with little additional information. The court will decide on November 15, 1993.

November 15, 1993. The Water Court approves the project provided the water flow from Kattegat through Öresund to the Baltic will not be checked and sediment spill from dredging is minimized in amount, space, and time.

June 15, 1994. After two additional rounds through the Water Court, checking changes and results of two proposed two-dimensional models tests, the Swedish Government issues the permission for construction of a permanent link over Öresund.

Following the decision, the Minister of Environment in Sweden and the Minister of Finance in Denmark resigned.

This vignette is our main thesis in a nutshell: an actual bridge that may turn out to be a barrier in the long run: scientists blaming the politicians for immature decisions; the managing institutions pushing their special interests; the Environmental Protection Board with sound views but bound by clumsy legislation; the Greens; the public. What is the long-term solution? Who has the answer? Who has the answer for Baltic Europe? We leave these questions for future historians to retell in the the next chapter of the "Sea of Surprises."

Acknowledgments

We dedicate our chapter to the late Professor Ian Morris, former director of the Center for Environmental and Estuarine Studies, University of Maryland, who not only incorporated us in his early Governance of Coastal Seas Project but gave so generously of his vast knowledge and warm friendship. We sincerely thank Drs. Ạke Niemi, University of Helsinki, Finland; Alf Johnels, Swedish Museum of Natural History, Stockholm; and Ragnar Elmgren and Fredrik Wulff, University of Stockholm, for their constructive criticism of the paper. We admiringly acknowledge the positive patience of the editors.

Part 3

Essays

8

Ten Theses on the Promise and Problems of Creative Ecosystem Management in Developing Countries

Steven E. Sanderson

The cases in this volume are overwhelmingly from big, powerful social systems in the north. They are relatively closed to exogenous international political change, though on a gradient from the closed, single-state Everglades system to the multinational Baltic Sea, which is a much more open management system. The cases operate in a milieux of long-standing administrative rationality (in the Weberian sense and not an implication of "rational use" of resources). Policy decisions and daily management are organized on formal administrative lines. The political systems that host our cases are blessed with enviable internal political stability, again excepting the greater Baltic area. Even there, macropolitical instability struck only in the 1990s. And none of the countries in which the cases occur is buffeted by the international system in the same way as are developing countries.

This is but a beginning litany of differences among the cases in this volume and the characteristic difficulties facing southern political systems, where colonialism, vulnerability to external shock, internal administrative weakness, and political instability cloud the promise of transporting conclusions from north to south. On the other hand, southern society looks more and more like northern society, as the processes of industrialization, agricultural modernization, bureaucratic rationalization, and social differentiation become universal attributes of twentieth-century society. So what are the implications of this volume for developing countries?

Organizing some principles behind an answer to this question is my charge in this essay. To begin to answer such a big question, we must ask whether ecological conditions and policy communities are the same north to south, or whether there may be some qualitative difference in developing countries that must be taken into account—some special barriers or bridges. As development economists asked a generation ago, "Is there a valid set of rules that govern all economies (the monoeconomics claim), or is there a case for developing country exceptionalism" (Hirschman 1981)? Are countries the same but for their wealth, or do the conditions of wealth and poverty signify something more profound that separates some states from others? And as global-change ecologists might ask, "Do life zone, potential evapotranspiration, or other biophysical features of north and south compromise the comparability of landscape-scale ecosystems?"

I am not prepared to address the latter question, though it would be intriguing to stratify ecosystems according to biome or biogeographical region, as was done in the recent global biodiversity survey (WCMC 1992). Such a stratification, taken to the appropriate scale for ecosystem management, might suggest important differences of scale or "connectedness" of ecological variables that affect management strategies. Conversely, social organizations might show scale-dependent degrees of "embeddedness" in their environments, which Westley suggests elsewhere in this volume might relate to the harmonics of managing ecosystems.

As to the matter of "developing-country exceptionalism," part of the answer must be found in what the case study authors have identified as (albeit in other words) the relationships among structure (the architecture of the social system), agency (the politically relevant actors in the system), and choice (the options chosen by actors and their societies), and the changes in those relationships within historical periods and across macrohistorical changes. Policy choices are never made in unconstrained environments, nor do agents behave as if history and structure were absent or irrelevant. History is relevant in many aspects: as a discontinuous, multiscale structuring process recorded over time, as a set of shared meanings in a given cultural or social setting, and as an ongoing social dynamic between agents endowed with vastly different resource capacities, including power. It goes without saying that our understanding and portrayal of history are loaded with subjective meaning, which disposes institutions and actors to view the future differently. Donald Michael (chapter 11) points to a need to be aware of the "forward meanings" and "backward meanings" of concepts, which evokes both the burden of historical meaning and the need for a prospective turn of mind in order to make constructive policy.

This reminds us that overcoming barriers and constructing problem-solving bridges involve nothing less than the creation of viable solutions for an unknown future, usually on the basis of some limited and dangerous assumptions about what has gone on in the past and what is going on now. It is difficult to argue that such assumptions about human nature are objective, rather than seeing them as a loose understanding about the world, a world view produced by continual historiographic struggle. Likewise, our projection of that understanding into the future is based on assessments of the state and viability of the social order. This is not new. Marx's famous dictum "Men make their own history, but they do not make it just as they please" (Marx 1972) is an appropriate source for such thinking. To bring that message home, Holling suggests in his introduction to this volume that the historicity of a "nested cycles" approach to social change contrasts with the immutability of Cartesian science (Harrison 1992).

These (over)general observations focus this essay's attention on whether we can draw lessons from the cases in this volume for developing countries. I will try in these few pages to begin to address the "special" quality of developing countries, the resulting difficulties in managing ecosystems flexibly in developing-country settings, and the importance of a prospective turn of mind for policy innovation. Throughout the essay are sprinkled likely propositions that might guide future research on the appropriateness of northern studies for developing countries.

The Case for Developing-Country Exceptionalism

Machiavelli thought of the purpose of politics partly in terms of an economy of violence, political virtue being defined by the prince's ability to devise a science of coercion, in which power is measured in strategic fashion (Wolin 1970). Alone, every prince thinks of power in a positive way, but among their rivals only the great powers enjoy such luxury. From below, the weak view power as a negative force, limiting freedom of action and constraining the political actor's path to optimality, or at least to success.

In some measure, power inequalities define the importance of structure in the international system to political organization and outcomes. Powerful states are more able to resist exogenous shocks (who would think of the United States submitting to an external stabilization scheme mandated by the International Monetary Fund?), whereas weak states are not. Large states can maintain closed economic borders, whereas small states cannot. Independent states are makers of alliances; dependent states are followers of such accords.

But these generalizations are about large states and small, not necessarily developed and developing states. Are exogenous variables qualitatively different in developing countries, or could we just consider external vulnerability to be measured across north and south equally, according to such traditional indicators as economic openness? Economic openness (traditionally measured by the ratio of exports to GDP) crosses the north/south divide easily, commingling open systems such as Hong Kong with Luxembourg, Belgium with Gambia, Botswana with Ireland. The most closed systems include many developing countries, but prominent among the most closed are the United States and Japan—united in their openness and at completely different ends of the scale in their geographic size. These matches are confused further by interannual variation in the rankings.

Openness itself is a function of historical patterns of economic development and a country's connection to the exterior, so simple ratios of trade to economic output are wholly inadequate measures. Nevertheless, it does seem reasonable to conclude that a country's ability to manage anthropogenic influences on landscape-scale ecosystems is affected by the role of the exogenous in the behavior of the ecosystem. Openness is one rough guide to the role of extra-system variables, but it must be tempered by historical influences on structure, agency, and choice.

The analysts in this volume stress system dynamics and organization, in a manner that roughly parallels the concerns of "historical institutionalists" (Steinmo, Thelen, and Longstreth 1992). Historical institutionalists consider variations in institutional and organizational behavior to be relevant to agency and choice; that is, they would be interested in institutional or organizational flexibility, learning, and adaptation.

But they do not address the question of whether different systems have qualitatively different places in the global political organization of social life. That question is not addressed by case analysts in this volume either. But their systems approaches and narrative voice suggest a rough fit with historical institutionalists.

The two alternatives to institutionalism are structuralist and rational-choice approaches. Structuralists are more disposed to look at the exogenous variables that frame developing countries' choices. Structuralists argue that developing countries are, indeed, different. They are constrained to behave in ways acceptable to their more powerful "structuring agents," whether those be the imperial powers of the recent past, current international regimes built on a system of independent nation-states, or the supposedly neutral organizational force of a market-based hierarchy. Rational-choice theorists, on the other hand, emphasize agency overall, by framing politics in terms of identifiable choices and logics among actors in the system.

This is not just a political scientist's digression. The analysis of landscape-scale ecological change under human management is only susceptible to comparative analysis of the kind proposed in this volume if larger structural issues are taken into account. It is not necessary that all cases become homologous, but that comparative analysis at the mesoscale allow for governing differences at the macroscale. But what does this mean to our cases and their implications for developing countries? And how does it relate to the agency so apparent in each of the case studies described in this volume?

Proposition 1: Politically, developing countries are more structurally constrained than developed countries. They must craft their internal strategies according to relatively fixed macropolitical structures, but against a large array of actors with superior power capabilities and greater strategic flexibility. At the other extreme, great powers can consider their external environment *and* the macropolitical framework more strategically, making policy moves according to complex, interactive, and ultimately more flexible dynamics of the interstate system. Poor countries, in regard to structural questions, engage the international system via rather rigid parametric behavior (Elster 1979), defined in their range of motion by the rules of the international system, over which they have little control. So Peru, in managing the guano boom of the 1860s and 1870s, reacted to the external demand for phosphate by mining fossilized guano (Gootenberg 1998; Levin 1960). The case describes the classic export enclave, in which a poor country exploits its natural resource base strictly according to the demands of the external sector. European agriculture is the driving force; the main question is the degree to which benefits return to the domestic economy (Bertram and Thorp 1978). How the state of poor countries differs from the general condition of small states is a matter of considerable controversy (Thomas 1988), but a broad range of historical literature indicates that wealthier northern small states have considerably more flexibility in regard to their surrounding system (Fox 1959; Katzenstein 1985).

This might lead us to hypothesize that developing countries are more likely to be dominated by "outside" management organizations, with less "embeddedness" in the environment being managed (Spooner 1987). The place to test such a proposition is probably at the margin of some agroecological zone, where the propensity is high for management to change environmental outcomes and where the condition for previous human survival was the careful endogenous management of institutional arrangements (Blaikie and Brookfield 1987). A likely case would be the Sahel, or other semiarid pastoralist systems. External agents, in the service of development, have often overcome local knowledge to force development alternatives on vulnerable populations, only to see them wither in the face of the next environmental pulse. Likewise, technical solutions and expertise will dominate, even in the face of greater local heterogeneity than is found in the north. This is discussed in more detail later.

Proposition 2: Developing-country politics are structured by global-scale economics to a greater degree than in the developed world. Although one might argue that north and south are on a single gradient in this regard, perhaps measured by such indicators as openness, a more reasonable hypothesis would cluster countries around a series of characteristics, including the history of colonialism, aggregate economic output, degree of industrialization, or energy allocation, as well as economic openness. So to Holling's allusion to cycles of crisis in management, we must add powerful exogenous cycles of crisis, in which developing countries are adrift in the turbulent currents of international economic change. That is, the nested cycles he refers to extend beyond the region to hemispheric and global scales. In their vulnerability to external cycles, developing-country ecosystems are more akin to New Brunswick in the nineteenth century, when it was considered a timber colony of Great Britain (Baskerville this volume; Richards 1990).

Proposition 3: Those countries facing the international system with a power deficit focus more attention on the historical processes that trap them. I have argued elsewhere (Sanderson 1992) that Latin American countries view trade through a much more jaundiced lens because of its historically heavy meaning for their national independence and because of the difficulty of seeing a progressive future without undoing that history in some fundamental sense. History defines the turn of mind in poor states' public policy in a manner evoked by Walter Benjamin earlier in this century:

> A Klee painting named "Angelus Novus" shows an angel looking as though he is about to move away from something he is fixedly contemplating. His eyes are staring, his mouth is open, his wings are spread. This is how one pictures the angel of history. His face is turned toward the past. Where we perceive a chain of events, he sees one single catastrophe which keeps piling wreckage upon wreckage and hurls it in front of his feet. The angel would like to stay, awaken the dead, and make whole what has been smashed. But a storm is blowing from Paradise; it has got caught in his wings with such violence that the angel can no longer close them. This storm irresistibly propels him into the future to which his back is turned, while the pile of debris before him grows skyward. This storm is what we call progress [Benjamin 1969].

Proposition 4: Developing countries, as they are hurled forward by the storm of progress, are linked inexorably to a development design not of their making and beyond their control, which affects their management of social and ecological systems and threatens to compromise their natural resource base. Yet the imperatives of economic growth drive them to manage natural resources intensively, in a manner that this volume leads us to believe will lead to brittleness, surprise, and catastrophe. This constitutes an additional paradox to those that Holling names in this volume's introduction. As if this were not enough, developed countries, having confronted their own "piles of debris" littering the path to progress, are changing the international system to increase conservation pressures in developing countries, without changing the fundamental forcing dynamics of the system (trade, direct investment, etc.) at all. The World Bank's imperatives now include "sustainable development" (World Bank 1992), which insists that developing countries manage their resources in the general context of development, but without changing the logic of development itself. The contradictory pressures of growth vs. preservation are played out on the field of trade liberalization as a guiding principle vs. trade restraints on scarce commodities, such as tropical timber, and of technological modernization for the sake of efficiency vs. preservation of traditional technologies as a rate constraint on environmental degradation.

Proposition 5: If the persistence of the current international structure forces an unhappy paradox on developing countries, their ability to break out of its destructive consequences may depend on their forcefulness in changing the rules of the international system itself. That calls on developing-country resource management strategies to focus not only on the point of impact, but also on the overarching macrostructure. That strategic focus is undermined by the north/south power gap.

Jon Elster (1979) separates the actions of humans from other organisms by distinguishing human intentionality from what he calls evolutionary gradient climbing. Despite its unfortunately linear connotation, this distinction is an apt metaphor for the paradox of developing countries vis-à-vis the international system. The model of development proposed by the north involves slow gradient-climbing for the south, without conscious amendment of the rules of the development game. Yet that model yields poor environmental results and forces more intensive and less adaptive management of natural resources. For developing countries to improve qualitatively in their environmental management requires the improvement of internal organizational design, or other kinds of tinkering with the system, but it also demands that the south amend the development model to suit its special needs, or that it insulate itself more from the pressures of external change.

But developing countries would presumably engage in such radical politics in the service of some identifiable set of values consistent with the recommendations in this volume: enhancing local, or "inside" management, abjuring single targets in favor of managing for ecosystem architecture, designing management systems built around dynamic and multiple equilibria, and so on. But such broad-gauged objectives do not thrive in a political vacuum, nor do they prosper in a politically heterogeneous, conflict-beset environment.

Proposition 6: Developing countries are institutionally more heterogeneous than their developed counterparts. In the language of the common property literature (Bromley et al. 1992; Ostrom 1990), the governance of common pool resources varies enormously, from reliance on the state or the market or local institutions. The durability of indigenous institutions and property regimes in multiethnic societies also affects the relationship between human communities at the local level and national or supranational efforts to manage fragile ecosystems. For example, throughout Latin America, differences in use rights among indigenous, nonindigenous local, and outside claimants affect state as well as private strategies toward conservation. Debates over differential rights to resources in developing countries mirror conflicts in Canada and parts of the United States, in ways not represented in the more homogeneous environments portrayed in this volume.

Defaulting to Developed-Country Models

Developing countries, in sum, are overburdened by history, underendowed by administration, and forced to face persistent demands for natural resource management from all sides. But short of revolutionary breakouts by the developing world, the temptation is to adopt models of environmental management from the developed world. Certainly the overriding influence of the developed countries and the multilateral development assistance community has been for developing countries to abandon state-based management strategies in favor of market incentives, private property regimes, and even local participation (Ascher and Healy 1990; National Research Council 1993)—as if these strategies had the same meaning in different cultural and ecological contexts. And the preeminent management target—sustainable development notwithstanding—is maximizing output.

The developing world suffers these extraordinary pressures to manage, disposing them to the inflexibility and brittleness Holling has predicted. The connection of that pressure to studying, counting, monitoring, and lifting "technological packages" from outside experts is profound. The fragile *cerrado* of Brazil or the semiarid north of Mexico is converted to agriculture as part of internationally assisted agricultural modernization drives. Ecuador converts its coastal marine resources (and loses its mangroves) to shrimp mariculture for the international market, as part of an export promotion drive. And pastoralists in the Sahel submit to settlement and herd-building programs that undo their traditional management institutions. The north is arguing for much more ecosystem management in the south, but driven by northern ideological templates that fit poorly with southern political and social circumstances. This creates three important and potentially destructive dynamics related to the lessons of this volume:

- A forced predilection toward rigid bureaucratic behavior.
- The uncritical acceptance of northern policy packages.
- The attenuation of local learning and adaptation.

However, it also holds the possibility of creating cross-cutting "epistemic communities" (Haas 1990) that build new consensus on the desired relationship between nature and society, and the implementation of ecosystem management policies that seek new values. Such epistemic communities in the case studies discussed in this volume have emerged as transient, relatively flexible, informal organizations.

The common dangers of a bureaucratic approach are obvious from the cases in this volume. Bureaucratically rigid behavior stands in opposition to leadership, vision, and institutional flexibility. The difficulties in making inferential leaps about who learns and in what institutional circumstances should make readers of this volume very shy about recommending portable management packages for export to the south. The role of consensual knowledge and the limits of charismatic leadership suggest that successful management is not very portable.

This lesson is not new to students of north/south relations. More than 20 years ago, Albert Hirschman (1970) wrote a stinging review of northern optics on southern development, in which he described the developing countries as "objects of history" upon whom were visited "deluges of paradigms and models" to describe the complex realities of their history. Ethnocentric models of social change and economic modernization became targets of Hirschman's critique, even as southern countries were adopting them without historical reference.

Yet external policy influences on developing countries push them to adopt such models via multilateral development bank conditionality, bilateral development assistance, and the formulaic approach of international nongovernmental organizations. Twenty years ago, the emphasis was on developed-country models of infrastructure development or agricultural modernization; today those models continue, despite widespread criticism over their environmental impacts. More important to our cases, though, are the current models of conservation or ecosystem management being forwarded from outside by international nongovernmental organizations. From species and habitat protection to national conservation areas, northern models—with their very different social histories, resident populations, and ecological dynamics—grind away in the south, but with generally modest results.

The second destructive dynamic, the uncritical acceptance of northern management packages, stems from this external manipulation and from the fashion of "sustainable development." The suggestion that ecosystem imperatives are compatible with human use is attractive; burgeoning human populations in a growth-oriented world demand such a notion. So the World Conservation Strategy and its successor, Caring for the Earth, stipulate without systematic empirical examination that conservation and use not only are compatible or necessary partners (Redford and Sanderson 1992), but are, in fact, practically identical (Robinson 1993).

But after 20-odd years in which the notion of sustainable development has been around, one finds precious few politicians, bureaucrats, or other leaders who manage multiple-use ecosystems for the elegant values described in the four-box heuristic. In resource-based developing societies, where are the cases where intergenerational equity drives development policy? With notable exceptions in the Arab OPEC countries, what countries are leaving oil in the ground against some future oil scarcity or in favor of relatively untouched forest? Where in the south are fallows being lengthened or agricultural use deintensified? When intense conservation pressure emanates from the dominant structures of power, it should be viewed skeptically. Before privatizing property rights for the sake of better environmental outcomes or using hotspots to define biodiversity priorities (Prendergast et al. 1993)—both cutting-edge approaches pushed by northern organizations—a proper evidentiary base must be built *in situ.* Recognizing the difficulty of transporting northern rules of thumb to southern situations, it may be in the best interest of the dominated to defend the status quo rather than to push rule changes further.

Third, who learns and who adapts in these borrowed frameworks? The adoption of developed-country property ideologies for developing-country ecosystems, for example, suggests that the meaning of the key management values is brought in from outside the management domain. Can this provide a proper environment for flexibility, or does flexibility—and, for that matter, adaptation—come to mean tweaking a foreign management system to fit somehow with local realities? This returns to the question of what scales and what institutions are most appropriate to ecosystem management, partly from an ecological perspective, but also from the standpoint of the political process of delineating, negotiating, allocating, and adjudicating use rights (Naughton and Sanderson 1993).

The overall sense of the propositions I have offered indicates that to derive policy packages for developing-country ecosystem management from a set of developed-country cases creates a multilevel cascade of errors. Developing countries are different in their structural constraints, in the composition and power of their political agents, and in the choices they make. These terms are all framed in a different context according to individual country and institutional histories, which in turn predispose organizations and individuals toward different adaptive strategies and preferences.

Does this mean that this volume offers no lessons for developing countries? On the contrary, the cases here and the overview essays that so ably accompany them combine with the preceding propositions to yield the following developing-country implications:

Proposition 7: Policies for ecosystem management must reflect a clear (or at least explicit) understanding of the natural and social history of the region in question. This is crucial not only for establishing baseline values for ecosystem management, but for determining the legitimacy of human claims to resources in the region. This militates against formulaic consumption of sustainable development ideas, or of technological packages for energy-intensive agriculture, for example. A good example of the importance of natural history and ecosystem knowledge for development policy is found in Lake Victoria, where a once diverse system now relies on three species, two of which have been introduced in the past three decades (Kaufman 1992; Lowe-McConnell 1993). Lake Victoria is interesting not simply for its size and transborder institutional complexity (it encompasses parts of Kenya, Tanzania, and Uganda, and a number of different fishing communities and technologies). It also provides an important illustration of the need to concatenate fisheries policies with natural system dynamics—not only according to fishery abundance in a given span of time (harvest cycle), but according to total ecosystem productivity, ecological function and system organization, and large-scale ecological processes.

Proposition 8: The openness of the political system and the commonly understood urgency of the management problem are two key indicators of the portability of institutional lessons learned. Simply put, this means that systems that are more vulnerable to external influence will be more likely to import answers. Those systems that are more transparent in their political processes domestically will be more conducive to the creation of epistemic communities. And those that are concerned with a commonly understood set of short- and medium-term policy problems are more likely to address them pragmatically.

These are little more than initial hypotheses that follow from the preceding arguments. Supposing that they are generally close to the mark, we derive a bias toward a more open international system and toward more democratic politics. These are important leads to follow, because they may run against the grain of some of this essay's argument (about the dependency of the south on imported solutions) and against the "authoritarian environmentalism" that has been such a strong element in the literature in recent years (Hardin 1968; Dryzek 1987; Paehlke 1989). If democracy and openness do run together, it must be in the context of a bidirectional openness (i.e., the openness of larger-scale drivers to local-scale technologies for appropriate management).

Proposition 9: The efficacy of using political mechanisms to manage ecosystem dynamics depends on whether, in Westley's terms (this volume), a single design cuts across the workings of nature and the workings of humans. I find such a notion implausible. This relates to the problems in arguing the parallelism of natural system performance and human ecosystem management. In the worthy effort to marry natural and social dynamics at the regional ecosystem level, a couple of key linkages are missing.

First, to argue that Schumpeterian interpretations of economic history (or for that matter, some others, including Marxist) parallel evolutionary models is attractive. But the cycles of capitalism—or more narrowly, capital accumulation—are at a different scale than the management cycles. It is extremely difficult to argue that capitalism is *managed* for the dynamics Holling and Schumpeter describe; and there's the rub. The intentionality of management and its uniform modern obsession with sustaining maximum output from a given social system with a given production function separate it from the natural dynamics of evolution.

Second, the lesson I draw from the case studies in this volume and from the record of global biodiversity conservation is that the architectonic concern for overall ecosystem structure is virtually unknown in developing countries' policy-making and only occurs sporadically in response to extreme crises in the developed world. Management can only partially defend what a renewed ecosystem might look like in relation to its previous historical states, and policy almost always points at much smaller targets: storm water treatment in the Everglades; cod production in the Baltic; forest output in New Brunswick; species richness in the tropics. One might suggest that the management world is built upside down, at least in terms of the hierarchy suggested by Holling. The role of epistemic communities can be to force a periodic restructuring of strategic vision in the management domain, forcing a new survey of the boundaries of that domain (see proposition 10) and unlocking the torpor of conservative bureaucracy.

Finally, we return to history, or at least to its expression in terms of decision making burdened by memory. Knowledge influences decision making but always makes its way to policy through such mediations as memory and ideology. In two aspects this burdens developing countries in special ways: The prospect of a public sector policy toward ecosystem management is at once historically attractive and repulsive. The public sector is attractive because it has always complemented (some would say supplanted) private-sector initiatives in unevenly developed markets and has often been the standard-bearer of progressive social policy. On the other hand, as recent years have shown, the state has been a failure in terms of its ability to control its macroeconomic and macropolitical instruments of management. Since the advent of the debt crisis in the early 1980s, the state in many developing countries is broken, at least in terms of its fiscal and monetary power (Fishlow 1990). Nevertheless, it is extremely difficult to conceive of the private sector's undertaking a responsible social role. The state is the default management medium, and it is in default, politically and economically over large parts of the developing world.

These propositions lead to a more general thesis:

Proposition 10: Greater size and complexity in an ecosystem may disengage the management task from specific, often crippling historical burdens or from local incentive systems, to offer the possibility of flexible institutional development and effective management. This possibility is offset, of course, by the increasing difficulty of creating consensus across political boundaries, management values, and ecological dynamics. Optimism is also tempered by the difficulty of pitching management interventions at the appropriate scale for the ecosystem values being sought.

There may be lurking behind this proposition a complicating issue of developing-country spatial scale when compared with cases in this volume. The Great Lakes Basin Ecosystem, for example, would be the thirty-sixth largest country in the world by area, substantially larger than France, Kenya, Thailand, Spain, and Chile. The largest tropical lake in the world, Lake Victoria, would rank 112th, slightly smaller than Ireland and about three times the size of Rwanda and Burundi, or Haiti, countries with extraordinarily complicated and important ecosystem management issues; they are more the size of the south Florida watershed. And Costa Rica, with its complex of parks and watershed management areas, is only about four times the size of the Everglades.

The largest protected areas in developing countries combine high complexity with small spatial scale. Rarely does a single protected area constitute a significant percentage of a biogeographic province (WCMC 1992). Only for a limited suite of species are protected areas larger than home range, even more rarely for taxon range (Harris 1993). Yet many developing-country protected areas or landscape-scale systems are extremely diverse in terms of governance. Lake Titicaca alone, for example, has some 150 communities, each acting as a system manager at one (or more) scale. Lake Victoria, another example cited earlier, includes three countries and manifold fisher communities.

What does this mean? If one believes in the virtues of comanagement, bottom-up, participatory, user-group-oriented decision making might be enhanced by small spatial scales; government might be kept at bay; and local prerogatives might prevail (Gadgil, Berkes, and Folke 1993; Levieil and Orlove 1990). But viewed from another perspective, local interests stand in the way of effective management, which requires superimposing global goals on individual self-interest (Costanza, Kemp, and Boynton 1993). This is a new rendition of the particularism vs. universalism debate from the literature on social modernization.

Taken together, the literature suggests that management virtues are scale dependent. Local self-rule via community institutions may effectively police resources at the microscale, where larger forces (e.g., government) would err. Regional institutions, to the extent they reflect a resource-driven evolution of management goals over time, offer a larger scale on which to militate on behalf of a given ecosystem. And in systems vulnerable to external shocks, national-scale management is required for system maintenance and preservation.

If there is some direct and general relationship between size and management flexibility, though, it is counterintuitive. Perhaps the argument in this volume with the greatest implication for developing countries is that the key to regional ecosystem management resides in a nested set of dynamic policy settings, in which the natural variability and diversity of the ecosystem itself should guide policy targets over management domains that are periodically adjusted. It might be argued that the hierarchy of policy should be reversed, with regional agents given free rein to monitor and adapt to the pulse of the region, but only within the nested hierarchy of global system values. Now, to find a bureaucratic organization (for there is nothing else on the horizon, except at the local scale and in the periodic pulses of epistemic communities), that permits such flexibility would really add to the arsenal of progressive environmental policy, both north and south.

Conclusion

The developing world certainly offers a different set of challenges from those posed by the cases in this volume, but they are different because of the defining characteristics of poor, vulnerable states and institutional heterogeneity, entrapped by the web of development. The principles evoked by the cases from the developed world are certainly appropriate to the consideration of management tasks in the developing world. Clearly, there is a mandate for better, sharper images of the natural dynamics of landscape-scale ecosystems in developing countries, and a need for robust local management institutions among small producer communities.

Larger challenges lurk behind this more obvious agenda. The first is a quest for management flexibility; the second is to place that flexibility at the service of a new development design, based on different system values. The developing world needs to unwrap that web of externally induced development in favor of its own economic and social agendas. Finally, for local initiatives to marry with macroscale driving forces requires a much deeper understanding of cross-scale human system dynamics, so that the interaction of global and local systems does not lead to our expected outcomes of brittleness, surprise, and ultimate catastrophe.

9

Governing Design: The Management of Social Systems and Ecosystems Management

Frances Westley

What had that flower to do with being white,
The wayside blue and innocent heal-all?
What brought the kindred spider to that height,
Then steered the white moth thither in the night?
What but design of darkness to appall?-
If design govern in a thing so small

ROBERT FROST, FROM "DESIGN"

The preceding lines were penned by the poet Robert Frost, surely the greatest spokesman in the artistic world for the strange and haunting similarities between the processes of nature and the symbolic constructions of man. These lines were inspired by the discovery by the poet, in an early morning walk, of a white spider poised on a white albino flower, holding up the wings of a dead, white moth—elements in nature of death and life, camouflage and discovery. For Frost the whiteness also symbolized good against evil, light against dark. And so, in a typical Frostian twist, he sees "design," an order that joins man and nature, defying the forces of disorder and darkness and then almost self-mockingly questions his own perception: Does design operate in things so small?

This rich book of essays is about this question in its most profound sense. Is there a design that cuts across the workings of nature and the workings of humans? Can we grasp its outlines through the careful observation of detail? Can we learn wisdom from the patterns of change and stability that we observe over time in the intertwining of ecological and social systems? In his conception of "creative destruction" Holling (chapter 1) proposes a model that he argues convincingly is a powerful explanation of processes observed across a variety of ecosystems; the material in these cases pertaining to human system suggests similar forces may be at work there. In employing this model and gathering together these case studies, this book is exemplary in its ambition to marry macrolevel theory to a level of detail, as rich as that which Frost observed. But unlike Frost, this book not only contemplates the possibility of design, but seeks to offer a ground for practical action, a stimulus for pro-active change on the part of scientists and managers responsible for the future of fragile ecosystems.

As a social scientist and researcher in the area of management of contemporary organizations, I come to this ambitious project from the opposite direction of most of the case writers. However, my aim is the same, to contribute to helping to build the bridge between the biological and social sciences in an effort to understand and perhaps better manage this interface. More specifically, I will set out in this chapter to address three themes raised across these cases, the theme of the relation between research, policy, and effective action; the theme of collaboration/consensus building; and the theme of system change and learning. My objective is to explore each of these issues from the point of view of current theory and practice in management.

Management itself is a discipline born from the coming together of old knowledge, from diverse sources, into new perspectives. It is related to and fuelled by many of the disciplines in the social sciences, but it differs from the social sciences in its overarching drive toward practice and its concern with technical knowledge in the widest sense of that word (technology as a set of knowledge and beliefs on causal relations; a logic that is complete when the system is closed [Thompson 1967]). Hence it is a discipline that, at its best, offers practical guidance as well as theoretical reflection about the challenges raised in these cases.

This chapter will begin with an exploration of the nature of planning, the ways in which conventional planning can be an impediment to organizational responsiveness, and the means researchers and practitioners have evolved to circumvent these difficulties. In particular, in this section I will deal with the following subpoints: (1) the role of planning as an intervening variable between knowledge and action, (2) under what circumstances planning is receptive to scientific knowledge,

(3) why action is so seldom an outcome of planning, (4) how vision acts as an action generator, and (5) how learning may be an effective bridge between knowledge and action.

Second, I will look at the literature in management on interorganizational collaboration and networks. As all the cases in this volume have underlined the importance of consensus building and collaboration in solving problems, it is useful to reflect on what social scientists have learned about which kind of dynamics result in successful collaborations, how interorganizational networks originate and what are the effects of these different origins, and what are the limits of consensus building and collaboration (e.g., the point at which networks of human organizations, like populations of other organisms, become overconnected and hence vulnerable).

In the third and last section I will look at the recent literature on managing organizational change, as it pertains to organizational revitalization. If organizations and social systems go through the same cycles of creative destruction that Holling and his associates have discovered in ecosystems, is there any way to short-circuit the process, to avoid or at least shorten the periods when the system is rigid and unresponsive, maximize the periods in which the system is tuned to its environment and responding creatively? As Holling (this volume) has pointed out in his essay, the ideal may be the social equivalent of the endotherm: some exchange of loss of internal variability (as long as it is associated with specific kinds of regulation), for heightened ability to explore, sense, and respond to a variety of external environments. How would these principles translate into management of change?

We shall therefore begin this chapter with a description of how responsive action is created in individual organizations, continue with how collaborations emerge, and conclude with a discussion of how a continual change in organizations as well as ecosystems can be managed. In other words, does the field of management offer any clues as to whether creative destruction of social systems as well as ecosystems can be managed, and if so, how?

Strategies for Responsive Action: Managing Small Details

One of the rewarding aspects of reading the cases and essays in this book is that they are integrated around several simple, elegant models or images. The first of course is the four-box model of creative destruction that offers an integrated explanation of the dynamics of social and ecological systems. The other is a model of adaptive management and the image of organization/environment relationships implied by that model. As I read it, adaptive management is a way of managing in order to ensure that the organizations responsible for ecosystems are responsive to the variations, rhythms, and cycles of change natural in that system and are able to react quickly with appropriate management techniques. The image suggests that certain guidelines, based on a scientific understanding of that ecosystem and a related definition of ecosystem health, act as umbrella principles to integrate action.

It is a compelling image, with an appeal that transcends the management of ecosystems. In the past decade—in response to radical shifts in world economies, resource bases, population dynamics, and competitive structures—private- and public-sector organizations in all domains have wrestled with similar challenges. The field of studies in strategic management has struggled with the problem of how to position an organization, in an ever-changing environment, to ensure system health and survival. As in many of the government organizations described in the cases, the dominant focus for many years has been on control. The organization has been viewed as "a lone gunman in the wild west . . . at war with its environment. Its tools were analysis and planning . . . careful evaluation of the opportunities and threats, strengths and weaknesses . . . in an attempt to forecast the future, master and tame the environment, and use it for organizational ends" (Astley 1984). The process was viewed as deliberate: "most strategists (strategic management researchers) adhere strongly to a belief in systematic, definable strategy procedures and structures that can be measured, analyzed and compared" (Daft and Buenger 1990). The effective organization was viewed as a machine; strategic planning was the engine that ran it, rational, mechanical, analytic, and programmatic (Bowman 1990).

Recently, however, there is evidence that this paradigm is eroding in favor of an image of organizations that are much more "embedded" in their environments (Granovettor 1985) and strategy as a more natural, emergent process, the end result of the creation of meaning within the system and of ongoing learning linked to these meanings.

This fundamental shift in paradigm has been led as much by experience of practitioners as by theory in management science. North American companies have become disenchanted with strategic planning in recent years because of the low success rate in implementing them (a review of implementation rates of major corporations, conducted by the leading business publication in the United States in the late 1980s, suggested that only about 3% of strategic plans were actually implemented), and the relatively poor showing of North American companies in a variety of industries, when compared with their counterparts in Europe and Japan (who had never practiced strategic planning with the same fervor) (Hayes,1985). Disillusioned, American corporations have abandoned their "love affair with corporate grand strategy" (Pascale 1987).

The implications of this paradigm shift are far-reaching for both theory and practice in management. The shift of focus from control to responsiveness has meant a reevaluation of the function of planning and a search for alternate processes better at generating learning and meaning (all considered key criteria of responsive action). In this section, therefore, I will begin by exploring in greater depth the problems with planning in relation to responsive action. I will continue by examining processes that (1) contribute to the generation of meaning and (2) contribute to the generation of learning. I will conclude with some thoughts of the practical applications of these processes.

Planning as Intervening Variable: The Knowledge/Planning/Action Connection

When one thinks of adaptive management, one thinks of responsive action, action that is triggered by change in the environment. The smaller or more subtle the trigger, the more "responsive" the action. But the link between stimulus and response at the level of the organization is less direct than that in single organisms (and it is far from direct even there). For in organizations, numerous actors must be coordinated to interpret the stimuli and to integrate their response. Enter the need for planning.

Modern theorists in social science argue that for conditions for social action to be optimal, three aspects of social structure must reinforce each other (Giddens 1984; Collins 1981). These three are "structures of signification" (the interpretative schemas that give meaning to our activities, sometimes identified as the myths, paradigms, mind-sets, or ideologies that "frame" our activities), "structures of legitimation" (the rules and norms that organize our activity, and that govern the routines that make up our daily life) and "structures of domination" (the allocation of resources and decision-making power that governs our ability to take effective action).

Planning, in all its forms, is a structure of signification, functioning primarily as a means of organizational sense-making. Ideally, the planning process reduces equivocality of information so that choice is possible. Planning is not in itself a paradigm. But as a technology for sense-making and choice generation, its form is fundamentally determined by the myths or paradigms that dominate a given organization, determining the perceptions of the environment and of the organization's role in that environment. So planning acts as an intervening variable between knowledge and action in large, complex systems. But under which circumstances is it a barrier and under which is it a bridge?

When Is Planning Receptive to Scientific Knowledge?

One of the issues raised in these cases is under what circumstances is policy-making receptive to knowledge generated in scientific studies. Integrating such knowledge into the planning process is only the first step in creating responsive action, but it is an important one. The answer, from both the literature on planning and that on evidence in these cases, is that utilization depends on (1) the *form* of the scientific knowledge and (2) the strength and dominance of the organization paradigm informing the planning process.

Formal planning processes, as we have noted earlier, are based on myths of the relationship between organization and environment as one of "instrumental rationality." Such exercises are highly linear processes, involving a series of preformulated steps and systematic scanning processes. Scientific information can be useful, but only if it is packaged in such a way that it is easy to plug into such formulas. It cannot be ambiguous, excessively complex, or subject to multiple interpretations. It must fit with the "mental maps" (models of reality based on past experience, assumptions, industry recipes that inform the planning processes); otherwise the information will be filtered out as not being pertinent (Aguilar 1967; Spender 1989).

This may explain the fact (mentioned in the Great Lakes case in chapter 6) that when the government commissions specific studies, it is more likely to incorporate the findings in planning processes, as these are likely structured in ways that fit with those processes. Indeed, under such circumstances, the knowledge/planning link becomes so well established as to be well worn, almost routinized. Francis and Regier (chapter 6) point out that in the Great Lakes case "governments commissioned the IJC to organize increasingly convoluted, drawn-out studies (almost continuously over the last 25 years)." However, new theories of science, based on a view of natural systems as "emergent, evolutionary, and open" represent too great a challenge to the assumptions underlying formal planning processes to be easily entertained by these processes.

So formal planning processes are able to incorporate stimuli from the environment, in the form of scientific information, as long as that information does not challenge the paradigms upon which the planning processes are based. The stronger (strength = closed, focused, monolithic, and orthodox) that paradigm, the more unreceptive to stimuli the organization becomes. Studies of highly successful firms that create intensive focus and unified cultures indicate they do so at the expense of responsiveness. The singlemindedness that initially gives them an edge over competition and results in success, over time reduces internal diversity. Certain functions are cut, as they are not seen as core or central; disconfirming information is neither sought nor fully entertained. Deviants are expelled as extraneous, and successful routines are rigidly maintained (Miller 1992). The result is that the highly focused organization over time ceases to pick up stimuli signaling fundamental changes in the environment and gradually reduces internal diversity until it is insufficient to respond to new demands from the environment.

Fortunately for the link between knowledge and planning, the large bureaucracies, such as governments, most likely to engage in formal planning processes are also least likely to have strong and unified ideologies (Gouldner 1976). Unfortunately for responsive action, however, that same absence of strong, overt ideologies represents a positive barrier to action.

The Failure of Planning to Produce Action

Studies in strategy process have indicated that action is a fundamentally irrational process. Action is made up of two components, the motivation to act and the availability of resources to support action. The first, motivation, is grounded in emotion, as it is through emotion, not logic, that energy is mobilized (Hochschild 1983). Availability of resources depends on how physical and human resources (money, time, space, technology, authority) have been organized. Resources do not flow equally to all parts of an organization, but tend to lump in certain functional and hierarchical pockets, according to the ability of groups within the organization to claim and control the distribution of resources (Ransom, Hinings, and Greenwood 1980). The link between the two is the degree of authority of dominant paradigms or structures of signification.

Contemporary literature on organizations uses such terms as *organizational cultures* (Schein 1985), *ideologies* (Brunsson 1985), or *myths* (Jonsson and Lundin 1977) to refer to these interpretative schemas that provide a unified conceptual field and a shortcut to action. Common interpretative schemas "substitute for decisions. Many organizational actions arise without decision making, because the actors perceive situations similarly and share expectations and general values" (Brunsson 1982). When all organizational actors share the same paradigm and it is strong and overt, the problem of *what* to do (choice) is greatly diminished. For example, a university department with a strong emphasis on research and a belief that certain methodological approaches are superior to others in the execution of research will find considerably fewer candidates to choose between, and the choice will be simpler. The emphasis can therefore be in "creating expectations, motivations, and commitment"—energizing organizational actors to engage in activities that will secure the chosen candidate (Brunsson 1985). Such organizational contexts are often described as strong cultures and are excellent action generators (Miller 1992; Peters and Waterman 1982).

If, however, there is much dispute among department members as to what kind of colleague is best, then the ideology is "nonconclusive," and considerable effort has to be expended on making the choice. Formal planning procedures as rationalistic decision modes are good tools for choice. Planning, as noted earlier, is a linear, rational process. At its best it scrupulously avoids the irrationalities that make for strong commitments, the synthesis that motivates social action (Brunsson 1985). The logics of choice and action are fundamentally different.

This may explain why in the Great Lakes case scientific study seemed to be carried on *in lieu of action.* Although in this case inaction was probably beneficial, since action might have resulted in the kind of engineering projects that caused problems in the Florida Everglades, the pattern of scientific study as alternative to action is similar to that which Jansson and Velner found in the Baltic Sea and Baskerville found in the forestry case. In such cases, the planning process has taken on a magical, tension-reducing function, much like witchcraft in traditional societies (Gimpl and Dakin 1984). It cycles repeatedly between information and choice; action is almost irrelevant.

Organizations that engage in formal planning processes and rely on these, implicitly or explicitly, split themselves into formulators (thinkers) and implementers (doers). However, they assume that their strategies in themselves will act as a motivator for the doers. This does not generally prove to be true; instead the recipients of the strategic plans, from middle management down, often find the plans bewildering, demotivating, and alienating (Westley 1990). Evidence suggests that formal planning processes are already devoid of the richness of information necessary to generate meaning. In addition, as plans filter down through hierarchical levels they lose more nuance and the process rarely allows for face-to-face communication complex enough to generate understanding at lower levels (Daft and Lengel 1984). Unless the poverty of understanding inherent in the planning process is compensated for by clear values and ideologies, middle managers are poorly motivated to act. Powerless and alienated, they are likely to engage either in increased political activity or in increasingly bureaucratic behavior (Izreali 1975).

The Role of Vision in Generating Action

In sharp contrast to the formal planning process described earlier is the kind of sense-making generated by visionary leaders. Whereas planning is a technology for institutionalizing vision, visionary leaders shape and reshape the myths themselves. These are sense-making processes akin to second-order learning.

Studies of visionary leaders indicates a strong facility with creating and manipulating emotionally evocative symbols (Weber 1922; Conger 1989; Westley 1991, 1992). Again, symbolic language is qualitatively different from the language of science and planning. It is colorful, emotional, heavily dependent on literary devices that build a bridge between the communicator and the audience (Burke 1950). It is inspired, however, by the material at hand: The followers "lead" the visionary as much as they are led by him/her (Westley and Mintzberg 1989). Visionary leaders throughout history are brilliant *bricoleurs*; they fabricate new and vital meanings out of the fragments available (Wallace 1961). In so doing, they overcome contradictions and create new synthesis. Myths are powerful devices for reconciling seemingly paradoxical elements in cultures, for making sense of the nonsensical (Levi-Strauss 1955; Leach 1964). In addition, visionary leaders rely heavily on face-to-face exchanges and on generating intensive communication exchanges within their organizations. They appeal directly to the middle and lower "action" levels of organizations alienated by planning processes (Mintzberg and Westley 1992; Vredenburg and Westley 1993). In visionary models of strategy the system remains flexible and responsive not because of a nested system of decision rules, but because of nested authority and meanings. Those closest to the action are empowered to act, and they do so in the interests of a common purpose and mission.

Visionary leaders play a key role in all the cases in this volume. Whether it is an Art Marshall in the Everglades or an Odén in Sweden, they have proved critical to the evolution of the social system and its relationship to the natural system. They have appeared at times of crisis, to forge new alliances between knowledge and action when the paradigms that forged old bridges had proved bankrupt as a platform for effective management of ecosystems. With time, though, these intense visionary perspectives must be routinized into less focused structures if the organization is to remain adaptive and responsive to stimuli from the environment.

In sum, then, responsiveness, if defined as the organization's ability to detect and understand changes in stimuli coming from the environment, is helped by nonconclusive ideologies and rational choice processes, such as planning. On the other hand, responsive action, defined as the organization's ability to translate the perception of changed stimuli into appropriate action, is inhibited by nonconclusive ideologies and rational-choice processes. Or put another way, strong ideologies, myths, and paradigms are important to action but potentially detrimental to interpreting and incorporating new information about the environment. It may be assumed (although we do not have enough detail to be sure) that some instances of the failure to use scientific information (New Brunswick Forestry, chapter 2) were due to the presence of strong ideologies and action tendencies in the government at the time. Conversely, the failure to act in the "Turning Green Lines into Red" era (Florida Everglades, chapter 3), despite outcry from scientists and "landmark legislation," signifies the absence of a strong ideology (perhaps due to Bill Storch's death) and a tendency for the government to engage in rational choice processes as opposed to action.

This discontinuity between knowledge and action and the role played by structures of signification presents an important challenge for those wishing to make management of ecosystems truly adaptive. It is clear that for adaptive management to succeed, organizations must find sense-making processes that *simultaneously* open the organization to new stimuli and provide strong action generation.

How Can Learning Be an Effective Bridge Between Knowledge and Action?

One of the reasons that dominant ideologies or paradigms are so resistant to change is that the dominance is taken for granted by organizational members. Normally, in most large organizations, a variety of different myths/paradigms and ideologies abound (Jonsson and Lundin 1977), each representing the viewpoint of different groups or "communities of practice" within the organization (Brown and Duguid 1991). When a single paradigm dominates an organization, this is generally because a powerful visionary or coalition has also dominated, controlling the flow of information and resources in a way that is unquestioned and unchallenged by others (Ransom, Hinings, and Greenwood 1980). Securing access to strategic conversations in order to influence interpretive schemas or to secure resources in order to fund divergent action is rarely even attempted (Westley 1990).

The organization so dominated does, as I have noted, become increasingly resistant to new sources of information, either that coming in from the environment or that coming from inside the organization. The result, as illustrated by examples, particularly from the Florida Everglades case, is management systems disconnected from the environment they seek to manage. Crisis is needed to shake such conclusive ideologies, and organizations in this state are prone either to crisis or to demise.

However, it is possible to design "changeful" organizations (as opposed to the "change-prone" organizations described earlier) (Brunsson 1985). Studies of highly adaptive systems suggest that the design need only provide mechanisms that facilitate the learning processes inherent in all human activity and that ensure the dissemination of that learning throughout the organization. Learning provides an alternative to crisis, as it introduces redundancies and inconsistencies into the structures of organizations that may serve to modify the conclusive nature of existing ideologies. Like planning processes, learning designs ensure greater receptivity to environmental stimuli. The same processes, happily, also seem to act as a functional equivalent of ideologies in generating action.

Learning and innovation as mental processes are "an almost instinctive propensity of the human organism, activated under the merest provocation of desire for a richer or more orderly experience" (Wallace 1961). The challenge is not, therefore, so much to structure organizations to learn as to structure them to take advantage of and incorporate the ongoing learning that is occurring, what is called the "tacit" knowledge of all organizations.

Several years ago Xerox vice-president John Seely Brown commissioned a qualitative study of how "learning" occurred at Xerox. An anthropologist was hired to follow around technicians as they made service calls and product development people as they worked on new products and product modifications. What the study revealed was that (1) formal routines and procedures were used not so much to guide action but to compare end states (i.e., as justification), (2) learning was socially constructed through exchange of stories ("war stories") based on improvisations in a problem-solving context, and (3) innovators let the world do some of the work (i.e., they took their solutions from those suggested/provided by the environment, not from analytic or abstract reasoning processes) (Brown and Duguid 1991).

What these findings suggest is that at the grass-roots or "local" level, individuals in organizations constantly respond to subtle changes in their environments and that these responses represent sources of innovation and learning. Second, they suggest that informal, face-to-face conversations are the best way to transmit learning. Finally, they suggest that the rules, procedures, and routines in most organizations act as barriers to learning unless they are treated as purely heuristic. The study concluded that much innovation directly contradicts the officially sanctioned formal operating procedures of organizations, as well as the rational decision rules. Learning, like action, is an irrational, highly social activity more connected to the construction of meaning (structures of signification) than to rules or authority (Weick 1991). If organizations, particularly large bureaucratic organizations, wish to increase responsiveness and adaptability, they must harness the instinctive learning of the front lines, as opposed to actively inhibiting it (Hamel and Prahalad 1989).

Practically, implementing the preceding conclusions involves providing opportunity for face-to-face exchanges horizontally, between functions. Such horizontal, cross-functional contacts provide synergy—integration without loss of individuality. Unlike conclusive ideologies that act as "restricted codes" to limit what data are made available for interpretation, cross-functional discussions provide a forum for "elaborate codes"-discourse linked to problem solving (Bernstein,1971). Such exchanges put learnings generated at the functional and technical level into the context of the whole organization, a context in which strategic implications are clearly recognizable (Collins 1981; Westley 1990). Such exchanges have been correlated with organizations capable of ongoing innovative action in response to their environments (Kanter 1983; Quinn 1985; Mintzberg and McHugh 1985).

Second, implementing learning systems involves structuring a vertical flow of strategic information. In an interesting study of Honda's entry into the North American motorcycle market, Pascale (1984) showed that the insight that led to the introduction of the small motorbikes that revolutionized the motor cycle industry came without any planning at all. Honda had *planned* to introduce its large machines and had sent over two salesmen to North America to set up distribution channels. After months of discouragement and on the brink of abandoning the project, one of the salesmen observed that many people had been stopping him on the street to inquire about where he got the small motorcycle he was riding (which he had brought for his own transportation). The salesmen thought that because of this expressed interest, there might be a market for the small motorbikes. On the strength of that hunch the top management of Honda decided to radically switch their "strategy" of entering the North American market. Small motorbikes were a huge success, and the motorcycle market was revolutionized.

What is useful here for the management of ecosystems is the recognition not only that the strategic innovation emerged from the lower levels of the organization (where stimulus response times are generally and necessarily shorter than those at higher levels of the organization), but that the strategic apex was so responsive to the hunches of two people close to the market but far from the top of the organization. In terms of ecosystem management, the equivalent might be the readiness of policy makers to be responsive to the input and recommendations of practitioners in the field (such as field biologists and wildlife managers) as well as to those of scientists and policy specialists.

The ability to integrate the highest and lowest levels of the organization is critical in a political system where, as is pointed out in numerous cases, the policy level is often very removed physically, conceptually, and technically from those individuals at lower levels in the system who are in touch in an immediate, day-to-day way with the environment and are hence in a position to detect changes most easily in that environment. Unfortunately, such managers are often low in status in the overall system and are disregarded as poorly trained "scientifically." Similarly, in the North American corporation there has been a tendency to ignore "intelligence" collected by those in the front lines (Albaum 1964) not only because strategy is viewed as the purview of the strategic apex but because reliance on statistical analysis of market surveys, for example, is seen as more valuable intelligence than the "hunches" or insights of those on the front lines. Many middle- and lower-level managers are deliberately excluded from the rich, face-to-face discussions that forge the backdrop, the meanings, and the frames for policy decisions. Participation in such activity is emblematic of elite status. Instead they are presented with formal policies, sets of statistics and planning documents stripped of the context and divorced from the stimuli that occasioned them (Westley 1990).

So why was top management at Honda prepared to listen and to entertain the hunches of their salesmen? Pascale indicates two important structuring devices that the company uses to make the communication flow from the bottom up as smooth and effective in terms of action as that flowing from the top down. The first is the fact that Honda does not isolate top managers from other levels of the organization. Top managers are not even assigned offices. Instead they are given desks in the corner of workstations. Their time is seen as valuably involved in staying in touch with the ongoing activities of their workers. Second, there is a careful effort to create a balance of elites in the organization, so that no single function becomes the star function in terms of power and influence over the strategy-making process. Again in many North American corporations, as well as government organizations, finance holds sway as the important function and most decisions are made with the bottom line in view. Such dominance breeds conclusive ideologies. It also blocks learning emerging from other functional areas that have strategic significance for the organization as a whole (Kanter 1983). Finally, it prevents the cross-functional integration vital for generating innovation.

In addition to cross—functional discussions and involved top management, studies of large organizations suggest that the role of middle-level management may be crucial in ensuring responsive action (Burgelman 1983a,b,c; Nonaka 1988). Middle managers are in critical positions to act as information brokers and shock absorbers between the strategic apex and the technical core. If they are allowed to question the strategic apex about the rationale behind strategic decisions, they can understand the significance of particular innovations for overall strategic directions and can act as interpreters of strategic directions to the technical core and champions of innovations and intelligence emerging from that group. This can be facilitated if managers are better trained in symbolic as well as rational forms of discourse, a practice common in classical training of leaders (Burke 1950) but no longer followed in modern, technologically driven organizations (Gouldner 1976).

However, no matter how skillful middle managers are as change agents, their role is restricted by the rules and norms that govern the organization. The day-to-day interactions of superiors and subordinates must be structured so that the subordinate is allowed to challenge decisions made by the superior without necessarily dominating the outcome and, equally important, be respected for the concerns and innovative capacity that such a challenge implies. Superiors who allow such challenges are more likely to have more innovative and responsive subordinates. Such conversations act to "nest" not only decision rules but meaning and authority structures (Westley 1990). Although the accusation is often made that managers and policy makers focus on the short term and therefore fall short of a vision of a management system that could encompass the scale of changes such as those in ecosystems, in fact humans tend to focus on the mesoscale, thinking in terms of structures and organizational systems. They fail equally in focusing on the microdynamics of interaction, the level at which these structural elements are produced and reproduced and where change, if it is to occur, must begin and end (Giddens 1984; Collins 1981).

Of course, in more decentralized, less hierarchical organizations such as those Mintzberg described as "adhocracies," the learning model of strategy is epitomized. Here, as in many Japanese and European organizations, strategy is never formally announced; it emerges and evolves from the collective activity. As such, it is only partially under the control of conscious thought and formulation, and implementations are virtually indistinguishable. Thus, it is the ultimately adaptive form (Mintzberg and McHugh 1985), but very remote from the bureaucratic systems that govern the management of many ecosystems. In the latter systems, discontinuities between knowledge and planning or between planning and action can best be bridged by a "learning" design-one that encourages strategic conversations across functions and levels, simultaneously regenerating meanings and transmitting learnings.

Summary

In the preceding discussion I have reviewed five themes that have implications for the cases in this book as well as for our purposes of finding practical perspectives on the challenges involved in ecosystem management. Our discussion so far has stayed at the level of single organizations and the issue of how to link knowledge to policy and policy to action. I have suggested that for successful social action rules, resources and meanings need not only to be nested at each hierarchical level, but also to be integrated across functions with each other. In practical terms it suggests that managers wishing to ensure a more adaptive management system should be sensitive to the following process issues:

1. For management systems to be adaptive to ecosystem dynamics, formal planning procedures should be minimized, or at least treated experimentally.
2. Strong ideologies should also be treated with caution. Although forging meanings, which are nested and coupled with a more even distribution of authority across the organization, is necessary for action, these meanings should not be maintained at the expense of diversity.
3. Middle managers should be encouraged to develop symbolic skills and to act as integrators between the strategic apex and the operating core. Mechanisms should be designed to ensure strategic conversations across functions and between levels. The more the strategies of the top can be influenced by the learning of the bottom, the more responsive the organization is likely to become.

I shall now turn to the second major theme of this chapter, that of the conditions facilitating and limiting interorganizational collaboration.

Interorganizational Collaboration and Consensus: Weaving Webs

Throughout these cases the need for interorganizational collaboration is stressed again and again. This is not surprising considering the complexity of organizations and jurisdictions represented by most of these cases. We rarely find human management systems patterned in terms of an ecosystem. Instead we find a number of "stakeholders" who have a vested interest in the ecosystem (Vredenburg and Westley 1993). In the cases in this book these range from government organizations to university consortiums, international commissions, citizen organizations, public and private companies, and native peoples. Some of these stakeholders have a central concern with managing the ecosystem (such as the Corps of Engineers in the Everglades). Others are concerned with the ecosystem as a source of raw materials (as in the New Brunswick Forestry) or as a disposal site for waste (as in the Chesapeake Bay), a recreation site, or a primary dwelling place (as in the Columbia River). Overall stakeholders represent a highly fragmented group whose interests concerning the ecosystem are very diverse and differ in intensity. It is not surprising therefore that the result, as described in these cases, is one of disconnected initiatives, sometimes conflictual, rarely cooperative, that bear little resemblance to a managed entity.

Yet the need to forge collaboration and consensus is a critical one. For one thing, no one organization, even in the case of the least complex (jurisdictionally) ecosystem, can solve the problems of ecosystem management unilaterally. These are metaproblems, or problem domains that demand "cultivation of domain-based, interorganizational competence" (Trist 1983). Yet society in general is weak in these capabilities.

Although the cases here describe numerous unsuccessful collaborations and some few successful ones, little attention is given to the microdynamics of what makes a successful collaboration. We are aware, on reading the cases, that the collaborations differ in kind, particularly that some were government initiatives and others (the "shadow networks") stemmed from efforts of citizen groups or university-based scientists, but there is little discussion of how these collaborations originated or of whether these differences in origin are related to the success/failure of the collaboration. Finally, there is a general recognition that collaboration is necessary there and an implicit assumption more collaboration/consensus will make the successful implementation of adaptive management more likely. But is there a limit to collaboration? Can there be too much consensus at the interorganizational level? These are questions prompted by the descriptions of collaboration in the cases, for which research in management has potential answers. I will deal with each in turn.

What Makes a Successful Collaboration?

Considerable work on the dynamics of interorganizational collaborations has established a number of features that make for successful collaborations between organizations. In the early stages the need to define the problem is paramount (Gray 1989). Therefore the extent to which all stakeholders can be brought to the table simultaneously will impact on whether the problem is defined with a sufficient degree of complexity. The process is iterative. Problem definitions will result in more stakeholders surfacing, who will then enrich the problem definition further (Westley and Vredenburg 1991; Gray 1985). If major stakeholders are left out at the problem definition stage, however, the chances of successful problem definition are reduced (Gray and Hay 1986).

Willingness to come to the table may also be an issue, affected by such things as mutual recognition of the need for collaboration, perception of legitimacy of all involved stakeholders, and the presence of a legitimate convener (Gray 1985).

Once they are brought to the table, stakeholders will attempt to set the direction for collaboration. Whether they are successful or not will be determined by such factors as the coincidence of values and the dispersal of power. The representatives of the stakeholder organizations who come to the table are limited by such factors as the level of commitment of their home organizations, the amount of resources at their disposal with which to negotiate, and their own conflicting loyalties (Kanter 1989). Inequalities in any of these areas, a common occurrence in complex collaborations, will weaken the collaboration, signaling to the collaborators that they are not equally needy and therefore will not benefit equally from the collaboration. For example, in describing the collaborations between business leaders, government, and environmental groups on Canada's National Roundtable, one observer described said,

> It is a strange sensation. You have the provincial treasurer sitting there with his flunkies behind him, the environmental minister sitting there with his people behind him, the man from the premier's office with his people behind him, and the chairman of this or that large company with his resources all lined up on one side, and cowering at the other end of the table are the few lone environmentalists with their bits of paper. This forces the environmentalists into strange stances. They either drive toward a purely ideological stance or begin to get coopted. They tend either to get extremely obstreperous or to say, "I can't argue with this; I'll give in" or to just walk away from the table.

From the preceding it is evident that it is not only shared understandings, myths, paradigms, and values that make for successful collaborations. Power is a central and underaddressed issue in these cases. Much of the aggravation attributed to the special interest groups in the

New Brunswick Forestry case, who used the media as a tool to gain access to the public arena and lobby for single-issue reforms, is characteristic of a situation in which there is a perceived power imbalance. In such cases appeals to the media can be an effective means of righting such imbalances. However, once a problem needing collaboration moves into the public arena, stakeholders tend to become frozen in polarized positions, and any real negotiation becomes difficult (Hilgartner and Bosk 1988).

If the public arena is to be avoided, however, great sensitivity about redistribution of power on the part of those with the most resources is critical, but not easily accomplished. The tendency is toward strong demands for equality from those less powerful and little concern for equality on the part of the powerful. Members of one public service organization that I recently advised were quick to point out that other service organizations got special treatment from the municipal government because their results were easier to measure. They felt this should be changed. They greeted with alarm, however, the idea that they should take care to give credit to these same service organizations for their less visible part in emergency rescue operations. They were used to receiving the lion's share of media and public attention and were reluctant to share it. Both inequalities (distribution of government resources and distribution of media attention) were detrimental to building the ongoing collaborative structures necessary for dealing with the increasingly large scale of public disasters with which all had to contend. But most organizations hoard power.

If the organizations are successful in setting directions, there remains the task of maintaining ongoing relationships, dependent on achieving a balance of power among participants and voluntary alignment of directions (Trist 1983). In this stage problems often result from the inability of the collaborators to maintain commitment "back home." This may be particularly true of organizations in pivotal positions of bridging two previously polarized camps. Such bridging attempts are inherently fragile because the bridging organization will have more invested than other organizations and must struggle with the same need to secure the commitment of their home constituencies (Westley and Vredenburg 1991a).

In sum, most individuals socialized in hierarchical organizations are not prepared for the kind of adaptive, interactive negotiation under relatively unstructured conditions (in terms of clear authority, rules, and meanings) of the successful collaboration (Kanter 1989). Individuals who work in adhocracies (and here I would include many research institutes and university settings) are used to a freer, more egalitarian setting and have less need to "represent" their organizations when they engage in collaborations outside their "home" institutions. This may account for the vitality and resiliency of the "shadow networks" described, particularly in the Everglades and Great Lakes cases. This resiliency may also be due, however, to the way in which these shadow networks originated, as compared to those collaborations initiated or mandated by the government. It would appear that the origin of the collaboration does have an impact on its trajectory.

The Origins of Interorganizing

In these cases three types of collaborations may be distinguished: those that originated by the organizing and inspirational leadership of a visionary, those that originated by being mandated by government, and those that seemed to spring up spontaneously, such as citizens movements and interuniversity networks. This is consistent with the pattern of interorganizing that we have discovered in a variety of collaborative domains. In line with the models of strategy making within organizations discussed earlier, we have termed these three, respectively, "vision-led," "planning-led," and "learning-led" forms of organizing (Westley and Vredenburg 1991b).

Each of these three forms, it would appear, has particular strengths and vulnerabilities, in addition to those described earlier, that are evident across all forms of collaboration. These become apparent when the forms are compared on the basis of the fundamental tasks that need to be accomplished for ongoing collaboration. Much has been written about the early stages of collaboration, the need for issue definition, stakeholder convening, and direction setting. However, consistent with what we know about strategic action within an organization, it is evident that in addition to the necessary structures of signification, realized collaborations must also create structures of domination (e.g., a mobilization of resources and empowerment for action) and structures of legitimization (e.g., institutionalization of the collaboration, development of norms for interaction, terminology for expectations, rules for balanced and productive participation).

The three forms of interorganizing may be said to vary in their ability to complete these four tasks (issue definition, resource mobilization, action mobilization, and institutionalization; table 9.1). The planning

TABLE 9.1
Issue Definition, Mobilization of Actions and Resources by Mode of Organizational Change

	Issue Definition	**Action Mobilization**	**Resource Mobilization**	**Structuring**
Planning Mode	"Public arena" dynamics may force early closure of issue definition without sufficient data	"Public area" dynamics immobilze stakeholders, making coalition cooperation difficult	Resource channels secured often in advance of issue definition	Procedures/ normal task allocations often limited to preexisting structures
Learning Mode	Incremental issue definition through individual initiative or negotiations	Commitment in advance of issue definition	Need to "piggy-back" on other institutions to mobilize. Resources may be coopted in process	Lack of resources may make structuring difficult
Vision Mode	Visionary particularly skilled in issue definition process	Link between affect and action fully utilized	Creative resource mobilization	Overdependence on visionary leader. Failure to institutionalize process or assure resource flow

mode—exemplified in this case by the various joint commissions, government tasks forces, and think tanks mandated by government—helps in the mobilization of resources and structuring; it is weaker at issue definition and action mobilization.

Issue definition and action mobilization are hampered in the planning mode by the fact that it is difficult and rare for politicians to operate out of the public arena. As noted earlier, however, decision making and negotiation in the public arena are often constrained by media coverage, which has a tendency to distort and polarize issues, and by the pressure from the public and back-home organizations of the collaborators to become paranoid about whether or not their issues are being adequately defended. Hence the issue definition may be prematurely closed to avoid undue controversy, losing in the process sufficient complexity to represent all stakeholders adequately on an ongoing basis.

This has a disempowering impact. Stakeholders may withdraw and feel slighted. Also, as action mobilization is dependent not only on the representatives but also on the willingness of back-home constituencies to commit energy to the ongoing collaboration, the public arena model may severely hamper implementation of decisions arrived at by the collaborators.

On the other hand, planning-led collaborations are particularly strong at resource mobilization, for government agencies can mandate the resources necessary for setting up the collaboration, and often do so before the issue definition stage begins. As far as institutionalizing is concerned, planning-mandated collaborations often use procedures set up by the government for its ongoing operations. The institutionalized collaboration is then subsumed into the bureaucracy of government, which may help perpetuate it but does not strengthen its action-generating capabilities.

In contrast, the vision-led collaborations are often strong at issue definition and action mobilization and are creative at resource mobilization. As noted earlier, visionaries excel at using symbols to capture the complex and various interests that must be integrated and forging them into a compelling scenario. Their use of symbols and emotionally evocative language and the intensity of their personal commitment create a "hunger for enactment," a powerful motivation to action, in those whom they reach (Westley 1992).

Visionaries are often also adept at creative resource mobilization. Many visionaries operating at the interorganizational level avoid more conventional channels of resource mobilization, such as government grants, out of fear of the restrictions that may accompany them (Westley and Vredenburg 1992). Instead they often mobilize sufficient commitment on the part of followers so that they are willing to divert personal and "home organizational" resources to making the vision real. This may have the negative side effect of network burnout as network participants become exhausted by the extra demands for resources that the leader inspires (Westley 1992).

On the other hand, visionaries are notoriously bad at the institutionalization tasks necessary for ongoing collaboration. As visionaries tend to resist structuring and place a high value on creativity, "routinization" processes often do not begin until after the visionary leaves the domain, that is, if the collaboration survives his or her departure. The death of Bill Storch in the No Easy Answers period of the Everglades left a void that was only filled by another visionary, Art Marshall. Marshall "carried the mantle of high priest of the environmental community in Florida," and after 10 years his vision was incorporated into public policy. It would be interesting to know, however, to what extent it was Marshall himself who was responsible for the institutionalization or whether others handled the structuring issues, as is often the case with visionaries (Westley 1992). Key to the continuity over time of vision-led collaborations is the development of a stable team, capable of turning visions into structures.

Perhaps the most interesting of the three modes of interorganizing are the learning-led collaborations, partly because their origins are harder to grasp from a research point of view. Learning-led collaborations seem to emerge from a groundswell of concern, the composite of experiences, and reactions of many individuals simultaneously to certain stimuli. Under certain conditions, concern, usually stimulated by the media or by a visible crisis, will snowball into action. Clearly, the action is triggered by individuals, but they are not necessarily visionaries, and they do not use symbols to motivate action. Rather, like setting a match to dry kindling, a small initiative will result in a conflagration of like initiatives, and a direction will emerge without ever having been planned. (See Mintzberg and McHugh [1985] for an excellent case description of such emergent strategies.)

Clearly, in learning-led strategies there is no difficulty in issue definition, for by the time the issue emerges, consensus (achieved through much discussion) has already been reached. This may be helped by the fact that participants in such collaborations often consider themselves to be representing only themselves. Hence a shadow network of interested scientists does not necessarily mean that a network has been forged between institutions (X, Y, and Z university, for example). It is individuals who, as professionals, consider themselves independent agents with regard to their activity in the problem domain who unite to collaborate. The challenge of having to "sell" the collaborations to back-home organizations is therefore minimized.

On the other hand, when the coalition starts to enter into negotiations with other organizations, it may be at a disadvantage because it will be relatively resource poor. Learning-led networks or coalitions are often "thinly institutionalized" (Zald and McCarthy 1987), which means that they do not have a rich resource base, a foundation of action routines, or established structures of significance on which to draw. This means that they may be at a disadvantage when attempting to collaborate with more established organizations. Sometimes this disparity can be outweighed by such factors as the scientific reputation of members, but with citizen groups such as those described in the New Brunswick Forestry case and the Great Lakes case, even this resource may be unavailable.

Therefore although action mobilization may precede issue definition (as emergent strategy is action followed by justification), learning-led coalitions often founder on the lack of availability of resources to fund their activities. They are forced to "piggy-back" on larger institutions (e.g., the media, the church, private-sector organizations) in order to distribute their message or underwrite their activities. In attempts to enter larger collaborations, such as those surrounding management of an ecosystem, they may find their position distorted by their associations. For example, an NGO with a government grant may lose some independence of position, for using the media as an outlet means taking on whatever distortions the media introduce. Without such associations the power imbalances described in the roundtable example earlier may result. Scarcity of resources may also hamper the development of an infrastructure to sustain the collaboration. Hence many learning-led collaborations are temporary coalitions that enter into negotiations within the problem domain for a period of time and then disappear.

In summary, then, the three different kinds of collaborations that are found throughout these cases may be characterized as planning led, vision led, and learning led. Planning-led collaborations tend to be long on resources and structuring potential; vision-led collaborations are powerful instruments for issue definition, action mobilization, and creative resource mobilization; and learning-led collaborations are strong at issue definition and action mobilization. As an actor in any one of the three types of collaborations it is helpful to recognize both these strengths and the vulnerabilities. All three types have their successes, but collaborations that are planning led must be particularly careful at the issue definition stage, those that are learning led must be careful in mobilizing resources, and ones that are vision led must be careful in institutionalizing processes.

But these cases, with their sweeping overview, offer food for thought at another level of analysis. If attempts at managing ecosystems such as these exhibit such a rich array of different forms of collaboration, all existing at the same time, what is the relationship among the social forms? Should one form supersede all others? Should there be a kind of rationalization process in the domain, with these apparently "redundant" collaborative forms merging into one central referent organization, as has been suggested by some domain theorists (Trist 1983)? How much organization is good for a problem domain, in terms of the ultimate goal of problem resolution (or adaptive ecosystem management, as is the case in this book)?

The Limits to Interorganizing

One fascinating fact about the multiple collaborations that seem to occur around attempts to manage ecosystems has to do with the relationship among the different collaborations. Is the presence of numerous networks, coalitions, and task forces at the interorganizational level a positive or a negative factor?

In the New Brunswick Forestry case it would seem evident that the presence of citizen's lobby groups was counterproductive in terms of the government's efforts to solve the problem in a sustainable manner. In fact, Baskerville states baldly, "In general, special interest groups stop things" (chapter 2). In his view the presence of multiple groups with multiple viewpoints heightens conflict. On the other hand, certainly the most "successful" case of ecosystem management would appear to be the Chesapeake Bay study, which also demonstrated the most unified action among the groups, and the highest intergroup consensus around issue definition. In fact, Constanza and Greer (chapter 4) isolate this creation of broad consensus as the primary factor in the effective action that occurred in the problem domain.

Scholars in the area of collaborative theory have long viewed networks in particular as an inherently unstable form, a stage in the development from an underorganized to an organized domain. Although such networks serve to link organizations, theorists such as Trist (1983) and Gray (1989) have argued that an attribute of successful collaborations is the establishment of more permanent, less fluid, and more centralized organizations. Ideally, such "referent" organizations are democratic, are nonhierarchical, and do not constrain members of collaborations, but they help to focus all activity on a unitary purpose: to achieve effective action within the problem domain. Networks, unless they become "centered" in such a referent organization, are "not in themselves purposeful" (Trist 1983).

Hence theory would argue with the evidence in New Brunswick (chapter 2) and the Chesapeake (chapter 4) cases: a redundancy of collaborations—planning led, vision led, learning led—merely introduce conflict and confusion into the management of ecosystems. What is needed for effective action is greater organization among such collaborations, to avoid redundancies and create a more ordered focus.

Yet such a conclusion sets off alarm bells. For what then do we make of the Great Lakes case, where a multitude of different types of collaboration coexisted over time, often at cross purposes and with little consensus, and yet appeared to offer a rich field for action? Although the redundancy might seem dysfunctional from a system perspective, it appeared that individuals used it for their own ends. If they were blocked in one interorganizational forum, they would try another avenue. For example, individuals from the IJC who had been instrumental in innovative policies in this planning-led context, then used a coalition of senior officials of fisheries industries to introduce a similar ecosystem approach to the fisheries. Some of these same individuals also participated in the university-initiated networks, making recommendations to the GLFC (Great Lakes Fisheries Commission).

As Regier noted (personal conversation), individuals who have made their career in attempting to save and manage the Great Lakes ecosystem, might, over 20 years, move through a variety of positions in different kinds of networks, carrying with them concepts and creating delicate linkages between different types of collaborations. Activists in environmental lobby groups who propose changes go on to be the commissioners who implement those changes 20 years later. A number of the case writers in this book have themselves played different roles in initiating, supporting, and using a variety of collaborative contexts to realize their aim. In the opening paragraphs of his text, Baskerville (chapter 2) admits to being student, researcher, professor of ecology, assistant deputy minister, policy consultant to the provincial government, and chairman of the board that heard disputes with respect to Crown forest management. Cutting across the seemingly redundant collaborative initiatives are the traces of individual career paths; people use the redundancy to achieve purposes that might not be achieved in simpler, more centralized structures.

From the perspective, therefore, of the individual, redundancy and variety in types and forms of collaborative initiatives within a problem domain may provide room to maneuver. Human beings find considerable freedom in the ability to change roles and move from one context to another (Rubenstein and Woodman 1984). Such freedom is also associated with the creation of energy, energy that can be contributed to problem resolution (Marks 1977). At the micro level of human days and ways, therefore, redundancy may be important for the creativity and energy required to solve metaproblems.

At the macrolevel, on the other hand, one is tempted to extend Holling's endotherm analogy (chapter 1) to the interorganizational domain. The larger social system represented by all the stakeholders within the domain may be more resilient if, in fact, there is a redundancy of mechanisms for sensing and responding to change. To the degree that these become rationalized and focused, the system may seem more efficient in the short run, but it may actually become more vulnerable.

Another way of looking at it is to think of the stakeholder organizations as patches in a metapopulation. Metapopulation theory suggests that it is not the survival of single patches that is important. Rather, it is the nature of the linkages between the patches that determines the survival of the metapopulation. If the contact is too restricted, the isolated patch may be short-lived; but if the contact is too dense, the metapopulation is also at risk (Shaffer 1985; Oliveri et al. 1990).

In short, there may be limits to the amount and kind of consensus and focused collaboration that is desirable. It is preferable to have redundancy if that allows creative individuals multiple avenues to reach their ends and increases the amount of energy in the system. It is better to have imperfect consensus, even conflict, if that maintains a situation of some diversity of approach and hence flexibility in the system to explore radical options. As in the case of the excellent organization that fails by the same means that it succeeded (Miller 1992), an overly organized problem domain with a high level of consensus may be less responsive to ongoing change than one that is less organized.

This leads us to consider the interorganizational domain as a system in its own right and to wonder about long-term system dynamics in terms of Holling's four-box model. Perhaps at their optimum level, many learning-led networks, such as informal collegia of scientists, act as a resource and a fertile redundancy, "advancing fundamental reframing and innovation" (Everglades case). Systems in which the variety of coalitions and collaborative forums offer this stimulating and healthy kind of exchange may well be in the box 4 "reconfiguration" phase.

However, as such groups continue to proliferate, as the domain becomes more organized, issues become more clearly defined and viewpoints become more entrenched (Westley and Vredenburg 1991). As in the exploitation stage (box 1), either the appearance of a visionary provides a centralizing and integrating momentum (as was the case with an Art Marshall), a strong, planning-led initiative demands integration across initiatives, or the groups themselves will begin to act competitively and opportunistically to secure resources (niches), with manifest conflict (as in the Forestry case). Ultimately, there will be a shakeout, as the stronger coalitions survive and the weaker ones are eliminated.

If the first or second form of exploitation occurs, the result may be highly productive consolidation (box 2), of the kind seen in the Chesapeake Bay case. However, the dynamics of Holling's model would suggest that this consolidation, although seemingly effective, may not in fact be as resilient as the system in a box 4 or in box 1. As Baskerville notes in chapter 2:

> Attention to structures that facilitate all party learning is urgent. There is a need to learn how to embrace error and to break from the "priesthood" approach where only a single group or agency holds wisdom. Clearly the forest will continue to change *whatever policy is in place,* and industry will change in response to the forest and in response to international market pressures, as well as in response to public policy. The key issue will be the degree to which these policy and forest management changes are reasoned (adaptive).

It is probable that a consensus of all parties around an adaptive management approach is likely to be more resilient than a consensus among all stakeholders around, say, a control and harvest approach, as the former puts emphasis on scanning, testing, and responding to changes in the ecosystem and the latter is concerned with controlling the variability within the ecosystem to ensure a steady supply for human consumption. However, it is important to realize that consensus on *any* policy if it includes the whole interorganizational domain represents a state of consolidation and conservation and will tend to become increasingly centralized, routinized, fixed in single-loop learning, and committed to fixed allocation of authority and resources. In short, consensus and consolidation go hand in hand. And as we discussed earlier, such systems, be they single organizations or interorganizational systems spanning problem domains, are more vulnerable to crisis. A boundary created around an interorganizational system, by the simple mechanism of strong ties within the system (ties of meaning, resource exchange, or norms of action) is still a boundary. The bounded system is vulnerable to crisis coming from outside its boundaries.

In a landmark study Granovettor (1973) compared two communities in the Boston area, both threatened with urban renewal plans that would destroy their integrity. One was a close-knit ethnic community with many bonds of kinship, shared culture, and clear social norms. The other was a middle-class "bedroom" community with very little internal organization. Although the shared value systems and tight organization of the ethnic communities would suggest that they would be better able to mobilize to react to the crisis, it was the middle-class community that survived. Granovettor attributes this to the "looser ties" that connected the middle-class residents not only to others within the community *but also to those outside the community,* whose influence and resources they were able to tap to stop the project. There seems to be an inverse correlation between how dense the ties or connections are within a bounded system and how many "loose" ties exist between members of that system and others outside it. If change therefore is coming from outside the social system (influences from other social systems, for example) or crisis to the natural system originating outside the ecosystem, the interorganizational domain in a box 2 "consolidation" state will be more vulnerable than those in a box 4 or box 1. Redundancies, as long as they represent multiple sensitivities and the ability to explore other "environments" for opportunities, may be as important for social systems as for organisms. Disorder in problem domains may be as valuable as order, diversity as important as consensus.

Summary

In this section I have again explored a literature that has developed separately but in parallel with the work on adaptive management: that concerning the microdynamics of successful collaboration, the origins and types of collaborations, and the relationship between the pattern of collaborations and system health. These cases offer a rich description of the multitude of collaborative initiatives, networks, and coalitions that spring up in attempting to negotiate problem domains as complicated as the management of ecosystems. The following are some practical lessons that can be drawn from analyzing the social system dynamics from a management point of view:

1. Although consensus building is clearly a critical issue in the management of successful collaboration, power dispersal is equally important. Actors involved in collaborative efforts must

ensure that some equal access to resources is provided, even if this involves designing processes that give a higher profile to stakeholders who are weak but important to problem resolution or sensitizing powerful stakeholders to the need to share resources.

2. A variety of different kinds of interorganizational collaborations can be recognized in these cases. We have identified three generic types: *planning-led* collaborations, which tend to take the shape of task forces, roundtables, committees; *vision-led networks,* associated with the activities of single visionaries and their supporters; and *learning-led networks,* which take the form of social movements, scientific consortia, community forums. Over time all three, if successful, have a tendency to crystallize into formal organizations. However, each of the three has different vulnerabilities. Actors involved in planning-led initiatives should be aware that such initiatives need careful attention to issue definition, to avoid premature closure and alienation of important stakeholders. Those involved in vision-led initiatives should be aware that the demands for resources of time and money may exhaust network members. Burn-out is common, and the visionary is unlikely to be concerned with evolving institutional structures to support his or her ideas. Finally, those involved in learning-led initiatives must recognize that the crucial challenge is to secure enough resources to survive; "piggy-back" arrangements may be necessary and should be pursued as far as possible without compromising the issue at hand.
3. Looked at from the macroperspective, the interorganizational domain is a larger social system and therefore may be affected by the same system dynamics that affect single organisms or ecosystems. Although consensus and organization at the domain level are clearly desirable, too much consensus and organization may make the interorganizational system vulnerable. For the actor in such systems, therefore, it is important to resist too much organization and centralization. Shadow networks should perhaps resist being turned into task forces; task forces should perhaps resist being turned into intergovernmental committees. Although the idea that policy should be treated as experiment is a brilliant one, it is difficult to achieve. Scientists should continue to be wary of politicians prepared

> to turn theory into policy. A healthy tension between the two, a redundancy of efforts and activities may offer the most fertile ground for individuals to manage the process of change.

But how manageable is this process? In this final section I shall address my third and last theme: Can (and should) human systems be managed to avoid the cycles of creative destruction?

Managing Cycles of Change: The Grand Design

Anthropologists, economists, and sociologists interested in change at the macrolevel have long recognized cycles of change similar to those that Holling's model captures at the ecosystem level. In addition to Schumpeter (1934), scholars such as Weber (1922) and Wallace (1966) have postulated that social systems go through a round involving (1) the creation of a new social order, often associated with a visionary leader, that is then encoded and institutionalized, a pattern Weber refers to as the "routinization of charisma." Eventually, it becomes (2) consolidated into an organized set of structures containing action routines, taken-for-granted assumptions about the meaning of such routines, and patterned flows of resources and authority. Such highly structured systems are not devised for learning; rather, they are devised for efficiency and routine (Weick 1991). It seems that as the consolidation progresses, adaptability decreases. However, change continues outside the system, and if the organization fails to adapt, it will become increasingly cut off or closed and subject to entropy. Eventually, (3) a crisis is produced both at the individual and at the organizational level. At the level of the organization this is often experienced as a sudden dropoff in performance and accounted for by such phrases as "the market suddenly dried up" or "the market moved out from under us." At the level of the individual, this is often experienced as a crisis of meaning. Rituals and routines that sustained daily life and guided daily contact seem suddenly meaningless or to break down altogether, occasioning acute psychological distress (Geertz 1976). In some cases the organization or social system goes into slow or rapid decline. In others, however, a period of (4) revitalization or reorganization occurs. This is characterized by highly individualistic, apparently chaotic behavior. A plethora of new ideas, initiatives (say, in the area of start-up ventures), and myths seem to circulate, some of which are imported from other systems and others of which are forged by individual actors within the system. Many of these initiatives seem contradictory. The possibility of conflict and generalized disorder is high, and there is an absence of overall direction in the system. Despite the learning going on, for new order to emerge, these learnings must be reintegrated into a new vision or myth and encoded in a new organizational structure. This often occurs because of the intervention of a visionary, and the cycle starts again (figure 9.1).

It is easy enough to overlay this cycle of social change onto the four-box model of ecosystem change. It has in it the same sense of inevitability, of natural rhythm that the ecosystem cycles contain. But from the point of view of the actor, of the manager existing within a system and attempting to solve such problems as how to manage an ecosystem adaptively, this similarity is not comforting. For we seek from our social systems the same overall stability that natural resource managers have sought for in ecosystems: change, yes; growth, yes; learning, yes—but against a stable backdrop, a structured order.

Perhaps this is a mistake. Perhaps the crisis and renewals that occurred in the Florida Everglades management systems are as natural

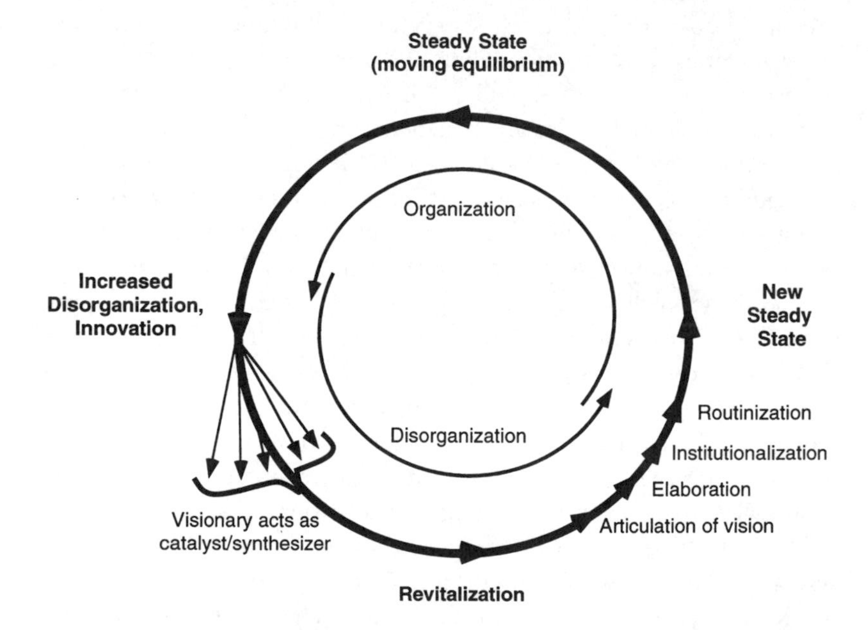

FIGURE 9.1
Cycle of revitalization (Wallace 1966).

and as *healthy* as those occurring in the Florida Everglades. Perhaps, as we have suggested earlier, no one-policy approach, even an adaptive one, should dominate for too long a time. Maybe *should* is irrelevant because the cycles are inevitable.

On the other hand, longitudinal studies of organizations have suggested that those that survive over long periods of time *do* find ways of managing (rather than avoiding) such cycles. The key seems to be in avoiding the extremes of order and structuring, on the one hand, and of disorder and confusion, on the other.

Earlier in this paper we noted that strategy within organizations can be characterized as vision, learning, and planning. These three tendencies can also be recognized at the interorganizational level. Overlaying these tendencies on the cycle of change we can see that the stage of reorganization is dominated by learning processes; that of consolidation, by planning processes; and that of renewal/exploitation, by visionary processes (figure 9.2). Clearly, each process has a role in the overall cycle. If the organization is to survive, however, the planning stage can never become so rigid as to prevent adaptation completely, and the learning stage can never dominate so completely that coordinated social action becomes impossible.

A series of longitudinal studies of strategy as a pattern in a stream of activity carried out at McGill University suggests a number of different configurations equated with survival. Three of these concern us here: the pattern of periodic bumps, the pattern of oscillating shifts, and the pattern of regular progress (Mintzberg and Westley 1992; figure 9.3).

In the pattern of *periodic bumps*, organizations go through long periods of stability and then experience a quantum shift, a sudden reorganization. Such organizations do not adapt easily. Change and adjustment are avoided in favor of routine and stability until, no longer avoidable, comprehensive change of the nature of a turnaround, or revolution is effected (Mintzberg 1978; Mintzberg and Waters 1982; Miller and Friesen 1984 for case descriptions of this kind of change). When viewed closely, however, it appears that when such sudden shifts occur, incremental adjustments through grass-roots learning have undoubtedly been going on for some time. As in the case of individual conversion, months of internal reorientation lead to a moment of gestalt shift, experienced as a sudden and dramatic change (Westley 1977; Gerlach and Hines 1970). Like the carapace of the spider, however, which falls off in an instant, a new one hidden under the old is ready to take its place. Before a radical change new myths abound in organizations,

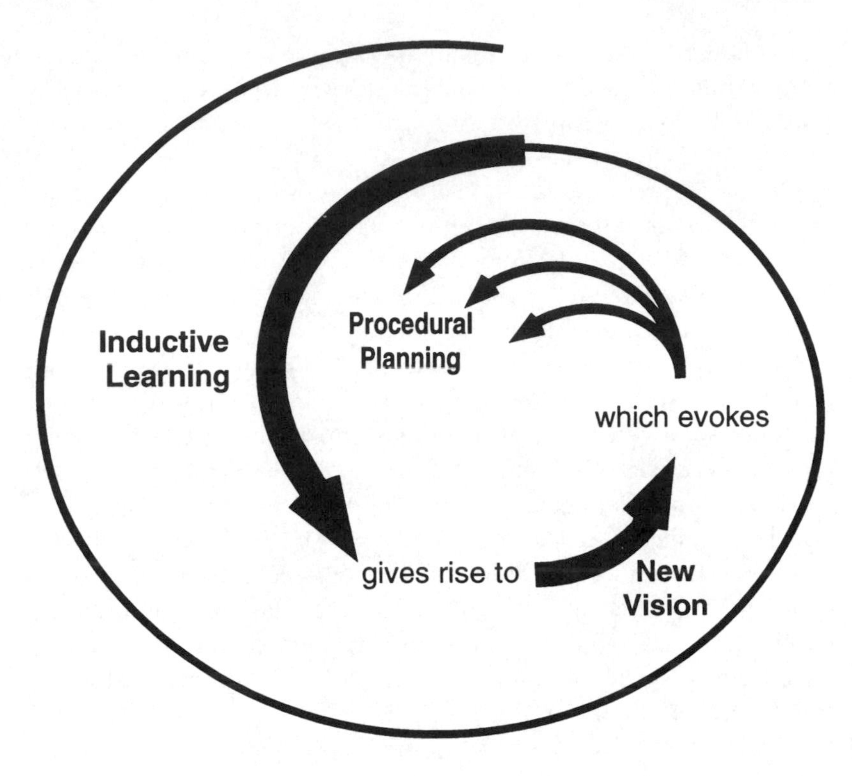

FIGURE 9.2
Sequences of the means of change (Mintzberg and Westley 1992).

each one a candidate for dominance (Jonsson and Lundin 1977). When crisis occurs, and the old myth disintegrates, another is ready to take its place. Still, when looked at over time, such periodic bumps appear as long periods of stability followed by sudden shifts. The moment of shift is perilous. Those organizations that survive are often dependent on leadership with the vision to make new sense out of the diversity.

The pattern of *oscillating shifts* represents a less radical or extreme form of change. An example of an organization that has demonstrated this pattern is the National Film Board of Canada. A publicly owned documentary film company, the film board is organized in terms of studios, each of which provides a basis for independent film makers. The film makers have complete latitude as to the films they make. The result of all this independent activity, however, when viewed over time, is a pattern of startlingly regular cycles with convergence around a particular theme occurring for about 6 years followed by 6 years of divergence and experimentation. None of the convergence was leader led. Rather, it happened spontaneously because of the tendency for film makers to interact and learn from each other (Mintzberg and McHugh 1985). The particular structure of the film board (an adhocracy) was loose enough to allow for divergence without falling apart, tight enough in terms of structures of signification, interactive norms to stimulate repeated convergence. The pattern was learning led, therefore, but with

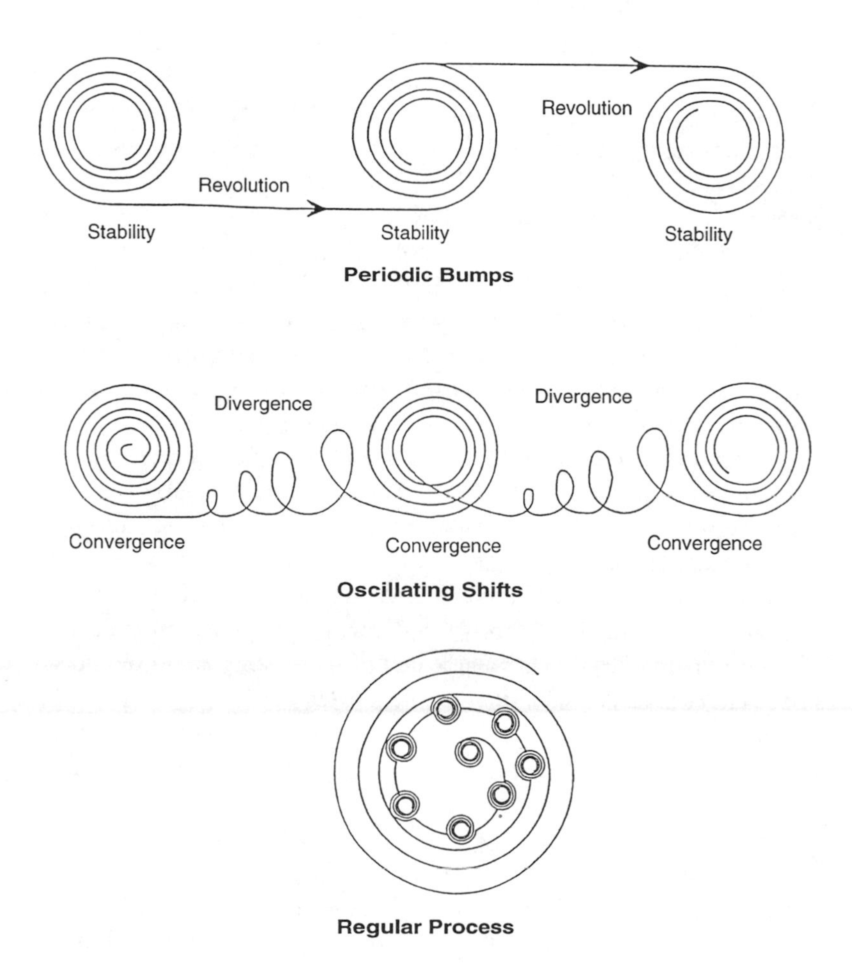

FIGURE 9.3
Patterns of organizational change (Mintzberg and Westley 1992).

enough planning on the part of management to ensure the maintenance of a stable resource base to support the experimentation.

The last pattern, one of *regular progress,* is one of continual and regular revitalization. One good example of such an organization is the Catholic church, which has survived for centuries, partly because of the (intermittent) ability of the popes to manage innovation in an enlightened way. At several important junctures, notably during the early thirteenth century and in the twentieth century, the church was headed by excellent administrators who carefully monitored new (potentially heretical) initiatives within the church and, instead of ignoring or expelling such movements, found a way to incorporate the new initiatives into the existing structures of the church, infusing the latter with new energy (Mintzberg and Westley 1992).

Another example is provided by McGill University, in which the activities at the grass roots are allowed to shape the direction of the university and the planners work not to control but to nourish and shape that activity. Such organizations change continually inside without changing position in a radical sense. Adjustment is continuous. The adaptation of the organization is led by the adaptation of responsive individuals connected to the environment (social) through a variety of professional networks (Mintzberg and Rose, in progress).

In each of these different patterns the balance between change and continuity is achieved in a different way, through gradual, hidden diversity and sudden shifts, through a rhythm of convergence and divergence resulting from patterns of interaction and patterns of creativity, through planned cultivation of creativity. In all three, however, there is exhibited a continual tolerance for diversity coupled with a positive response to orchestration. As Clark said of ecosystems, so one could say of social systems: The metaphor of the garden seems to work best. Resilient social systems, like resilient ecosystems, seem to be managed by people who ask, "What kinds of gardens do we want, and what kind can we get?"

It appears, then, that despite the fact that social systems, like ecosystems, evolve through a four-box cycle of creative destruction and that these are to some extent inevitable, it is possible for actors within these systems to manage in such a way that the crises are minimally destructive and the rigidity is not excessive, while the regenerative learning and sense of direction remain strong. Finding the balance point is a continual process of adjustment; however, the job is never finished.

And now I have come full circle in my discussion as well. For the key lies in the way in which process is designed in such organizations. Some organizations are more highly structured and hierarchical: For them the challenge is to create and maintain mechanisms for influence and communication from those levels where change and adjustment are a part of everyday life. Other organizations, such as adhocracies, seem to create themselves new everyday. For such organizations the challenge lies in creating enough structures so that convergence and direction are possible. The same processes that will allow the social system to remain resilient will also allow it to respond to the ecosystems it seeks to manage: tolerance for diversity, openness to new ideas and information, balance between efficiency and redundancy, willingness to move in new directions while maintaining internal stability. As one commentator on business and the environment noted:

> Companies that take the environment seriously change not only their processes and products, but also the way they run themselves. Often these changes go hand in hand with improvements in the general quality of management. Badly managed companies are rarely kind to the environment; conversely,companies that try the hardest to reduce the damage they do to the environment usually manage well. Why the link? . . . (such) goals demand a tolerance for ambiguity that irritates most managers . . . state-of-the art management tools to handle complexity . . . the skills to deal with multiple stakeholders and to think in networks, not hierarchies [Cairncross 1992].

It would appear that not only do dynamics in social systems create crisis in ecosystems and vice versa but also that the same means must be employed to create organizations capable of managing ecosystem resiliency and organizations capable of resiliency in their own right. Barren, overly structured organizations create barren, brittle ecosystems. Green, growing organizations may be able to manage green, living ecosystems. The clue to understanding this similarity lies in the construction of models, such as those in this volume, which charts this parallel on a grand scale. The key to managing this similarity lies in the design of the details within organizations and networks, where the grand order is founded in small processes.

10

Sustainable Development as Social Learning: Theoretical Perspectives and Practical Challenges for the Design of a Research Program

Edward A. Parson and William C. Clark

Meeting the challenges of sustainable development (development that meets the needs of the present without compromising the ability of future generations to meet their own needs [World Commission on Environment and Development 1987]) will require substantial advances in our understanding of how natural and social systems interact over long time periods and large spatial scales. Elements of the needed theory of ecological and earth system dynamics are beginning to emerge, as illustrated by many papers in this volume and the growing understanding being generated by research on global environmental change. Substantial progress has also been made in understanding the human side of the sustainable development equation. Most of that progress, however, has been from a static perspective that treats environmental change as either a consequence of, or an exogenous shock to, social systems. The bias of most research and policy programs has been toward institutions, processes, and practices that might maintain or restore some presumptive "equilibrium" with nature (Clark 1989). Generally lacking are theories of social dynamics that can complement the emerging theories of ecosystem dynamics to produce real understanding of the long-term, large-scale interactions of environment and development.

Among those social theories that are dynamic, the most striking common feature is reference to learning. Learning, as a manifestly dynamic process, provides a theoretical counterpoise to social theory based on power and interest, which characteristically yields a static formulation (Adler 1992). The tension between theories of learning and power, and their interpenetration in explanations of social change and stability, permeate the great social and political theories from the classics, through the Enlightenment, and into the present century. Our goal in this chapter is to explore what theories and metaphors of "social learning" might offer efforts to understand the barriers and bridges to sustainable development.[1]

The term *social learning* conceals great diversity. That many researchers describe the phenomena they are examining as "social learning" does not necessarily indicate a common theoretical perspective, disciplinary heritage, or even language. Rather, the contributions employ the language, concepts, and research methods of a half-dozen major disciplines; they focus on individuals, groups, formal organizations, professional communities, or entire societies; they use different definitions of learning, of what it means for learning to be "social," and of theory. The deepest difference is that for some, *social learning* means learning by individuals that takes place in social settings and/or is socially conditioned; for others it means learning by social aggregates.

This chapter is organized as follows. Some philosophical tensions that run through the concepts of learning and of social learning are discussed in the section after the introduction. Theories of individual learning and its social determinants are reviewed in the next section, followed by a section that discusses related work in which the perspective is shifted more strongly to the social level. After a general discussion of relationships between learning in collections and individual learning, the literature on learning in formal organizations is surveyed. The possibly unique situation of learning in science is considered in the next section, followed by reviews of theoretical and empirical studies of learning in politics, policy-making, and international relations. Evolutionary theories in social science and their relationship to learning theories are discussed in the next section. Tentative conclusions and assessments of how the literature of social learning might help to guide empirical research programs on sustainable development are presented in the final section.

Though our main focus is on theory, we have included in the review some applied and empirical work that uses the social learning concept and that shows parallels to this project. The overall approach is inclusive: a work's ostensible concern with social learning was a sufficient, but not a necessary, condition for its inclusion. Consequently, we have cast a broad net, and the treatment of many rich bodies of literature is highly schematic.

The long time scales relevant to sustainable development introduce one tension that we state but do not resolve: On such time scales, longer than many careers and many lives, substantial influence over the development of an issue may be beyond the capacity of any individual agent. Since concern for sustainable development is at least partly normative, however, it makes sense to apply evaluative criteria to decadal-scale social response. Some societies may do better than others in identifying significant risks associated with the interactions of environment and development, in marshaling research and information bearing on the risks, and in managing them. It is our hope that learning about long-term social learning processes may help some agents better understand how to play a constructive role or may suggest ways of structuring institutions, organizations, or negotiations to make effective learning more likely.

Philosophical Roots

Some of the conceptual differences among different students of social learning reflect deeper philosophical tensions that lie within the concept of learning itself, and in social learning in particular. On the confusing nature of learning in general, Polanyi (1966) cites the paradox first raised in the Meno—that to seek new knowledge and recognize that you have found it, you must have somehow already known it. Polanyi's resolution is that all knowledge, in addition to its explicit, conscious part, has a second part that is implicit, or "tacit," and that learning involves attending to different parts, different levels of a question or task in a way that moves bits of knowledge back and forth between the conscious and tacit realms. In anything that we really know, we know more than we can tell. Polanyi wrote only about individual knowledge, drawing on examples of language, such complex skills as music and chess, and scientific discovery; others have used his conscious/tacit dichotomy to refer to knowledge and learning of groups as well.

On social learning in particular, Friedmann (1987) argues that the term embodies tensions that were first articulated in John Dewey's epistemology and politics. Dewey argued that all knowledge comes from the interaction of people with the material environment and that collective learning from such practical experience, guided by the principles of scientific inquiry, could lead society progressively toward a golden age (Dewey 1927). Public decisions are viewed as a series of experiments with the world, which will yield progressive social improvement.

Dewey advanced several views on the difficult question of the validation of such learning and of knowledge in general. In "The Quest for Certainty" he proposed the pragmatic condition that knowledge is validated when it helps an actor settle a problem, granting that its "settledness" may be provisional or temporary. This condition, although reasonable for individual learning and consonant with the "learning-by-doing" approach of Dewey and modern students of innovation, poses problems for public decisions. It presumes that the settledness of an issue is not persistently opaque, as may be the case with certain issues of risk management. And it does not address the question, "Settled for whom?" or the possibility that whether a public issue is settled may be deeply contested.

Dewey later moved to the consensual theory that knowledge is validated by people's opinions (1980). He thus regarded the validation of knowledge, although not its production, as a social enterprise. But whose opinion validates? This view still presumes that no concepts are "essentially contested" and that there is no fundamental conflict between the primacy of public opinion and a scientific, experimental approach to public decision making.

This last tension, between democracy and reason, appears most sharply in "The Public and Its Problems." Here Dewey espoused an idealistic vision of neighborly discourse as the foundation of all political decision, but he also argued that most of the concerns of the modern state are technical matters, best left to the experts to resolve on the basis of the facts. He thus bequeathed to the modern debate on social learning its two most enduring tensions, one positive and one normative: the roles of the individual and of society in the creation and validation of knowledge; and the appropriate relationship between a learning, conversing public and political or expert authority in public decision making. We will see that these run through all the streams of literature surveyed here.

Individual Learning

The process of learning is so central to what it is to be human that any model of the person will address it. Theories of learning are consequently just as diverse as models of the person. Two extreme points for models of learning are marked by the rational-actor and the radical behaviorist models. In their pure forms these two share the characteristics of great cogency and parsimony, and strongly mixed empirical success; they differ in almost every other conceivable way. They embody venerable philosophical disputes on the nature of the mind, and pose strongly differing methodological agendas. As general models of the person, they are so well known as not to need summaries here, except to note that each in its way admits an extremely limited conception of learning.

Rational-actor theory—rooted in economics, political science, and philosophy—is a normative theory that also makes descriptive claims. It is first a theory of choice, but also implies a theory of rational belief, because to choose, individuals must predict the results of their choices. It is in forming and revising estimates of the consequences of choice that the rational actor can learn (Elster 1986).

The rational actor's learning is limited, though, because the actor already knows so much. Only two kinds of learning can occur: Bayesian updating of probability distributions of world states, and in interactive decision situations, reassessments of other agents' interests, available choices, and rationality. More substantial revision of cognitive structures, decision algorithms, or world views is excluded by the presumption of optimality. As for changes in interests or values, the theory does not exclude them, but it has nothing to say about them; it is purely instrumental.[2] The social context of choice and learning is reduced to its informational content—evidence to be used in rational reassessment of the decision situation.

The behaviorist actor's learning is also limited, because the actor knows so little. The methodological focus is strictly on behavior, viewed as a set of stimulus/response connections determined by the past reinforcement regime. The general principle is that behaviors that have been reinforced (rewarded) will occur more frequently in the future. Although the behavior exhibited can be versatile, complexly contingent on environmental conditions, even seemingly creative, "learning" is deemed identical to the observed changes in behavior. Concepts of retained information or image, or internal cognitive structures, are deemed unobservable and superfluous.[3] The social context of choice and learning is reduced to its reinforcement content—a set of socially delivered rewards and punishments.

If the rational actor has a nearly perfect internal model of the world and the behaviorist actor has none, any theory that can accommodate meaningful learning must lie between the two. Actors will hold incomplete, imperfect models of the world, subject to revision in the light of experience. There have been many branches of such theory over the last couple of decades, some growing from the behaviorist root and some from the rational- actor root. All these branches are informed by the empirical results of an increasingly powerful experimental technique. The discussion here briefly reviews a few of the major branches. At the same time, more theorists have looked at the social context of learning. In a somewhat arbitrary division reflecting historical lineages, one body of work stressing social factors in particular learning tasks is discussed here, whereas another that stresses social determinants of broader cognitive development is discussed in the section on codetermined individual and social learning.

One influential body of learning theory is known as Social Learning Theory (SLT).[4] SLT is strongly rooted in behaviorism, but it broadens the range of phenomena that are taken to affect individual behavior. In particular, it grants more standing to social determinants of individual learning.

SLT introduces three factors that determine behavior in addition to directly experienced reinforcement: observation and imitation of the behavior of others; symbolic representation of events and experience through language and other media; and self-generated rewards and punishments. Together these enable people to transform the stimuli impinging on them, and thereby partially to control their own behavior. This intermediate degree of individual autonomy, in Bandura's (1977) words, "neither casts people into the role of powerless objects controlled by environmental forces nor makes them free agents who can become whatever they choose."

The emphases on symbolic systems as reinforcement intermediaries and on self-generated reinforcements both suggest a role for internal representations that puts SLT at some distance from its behaviorist heritage. Different strains of SLT are more or less "representationalist," but the trend has been toward more representationalism (Bandura 1973; Bandura 1977; Aronfreed 1976; Mischel 1968; Kanfer 1971; Wren 1982; Rushton 1982).

A second major line of psychological theory, cognitive dissonance theory, focuses on changes in attitude and belief (Festinger 1957; McGuire 1966). This theory views belief systems as highly interconnected and people as seeking to maintain the coherence of these systems. Inconsistencies among cognitions, or between cognitions and behaviors, cause uncomfortable tension ("dissonance") that people seek to reduce by resolving the inconsistency. The standard experimental demonstration involves inducing people to act contrary to their beliefs and demonstrates that their beliefs then adjust. The less the externally imposed pressure or justification for behaving as they did, the greater the adjustment. The fundamental, highly counterintuitive result is that attitudes can adjust to behavior rather than the reverse, or in Bruner's (1979) words, "People act themselves into a way of believing as readily as they believe themselves into a way of acting."

There has also been development of new theory from the rational-actor pole. The rational-actor model has been softened by economists and students of artificial intelligence who realized that rational actors faced impossible computational and observational burdens. The goal is to relax assumptions of omniscience and reflect people's real perceptual and cognitive limitations while still retaining an evaluative framework.

Evidence of the need for such movement has been of two kinds: (1) empirical evidence of how people choose, learn, and solve problems and (2) attempts to simulate these processes on computers. Researchers in artificial intelligence have found that heuristic structures to filter and interpret information, and to short-cut lengthy optimizing procedures, are essential for computers to address decision problems of complexity that people handle routinely (Tamashiro 1984; Newell and Simon 1972). Studies of people's decision making, learning, and problem solving, on the other hand, show the working of strong ordering principles that are not describable as rational optimization. There are strong, systematic biases in the attribution of causes to other people's behavior (Heider 1958; Schneider et al. 1970; Jervis 1976), in the estimation of probabilities, and in inferences of causality (Tversky and Kahneman 1974; Hogarth and Reder 1986). Choices in uncertain situations depend strongly and systematically on the way the situation is described, or "framed" (Kahneman and Tversky 1981).

Early theoretical models moving away from the strict rational-actor approach were called *bounded-rationality* models (Simon 1955, 1959), stressing human computational limits that prevented optimization. The simplest such models originate in cybernetics and use a model of control (and learning, to the extent that it occurs) based on negative-feedback loops along a few simple information channels (Bateson 1967). More sophisticated versions explicitly substitute for optimization the search for an acceptable alternative, subject to information, calculation, and other constraints.

A more recent approach focuses on information and its processing, particularly on the need to select, filter, and impose structure on the vast excess of information coming in through the senses. People are viewed not as passive recipients of information but as active shapers of their experience, even at unconscious levels in the process of perception (Steinbruner 1974; Hochberg 1964). They construct powerful but fallible knowledge structures to make inferences about the environment and update them like "naive scientists" (Nisbett and Ross 1980). These structures reduce both information needs and ambiguity in the environment, to make it stable enough to allow people to choose and act.

Modern approaches largely share a focus on the internal cognitive structures used to represent the world. The convergence of these disparate bodies of research is called *cognitive science.* Gardner (1987), in a history of the cognitive science movement, summarized its distinguishing characteristics as follows: a focus on information processing; a reduction in the influence of context, affect, culture, and history; explanation using representations and internal cognitive structures; use of computers as analogies and research tools; and rootedness in old philosophical questions about the nature of mind (Hunt 1989).

Different kinds of mental representations are posited by researchers studying different mental phenomena—perception, categorization, imagery, problem solving, and memory. In one influential body of work, the constructs are called schemas (Minsky 1975; Schank and Abelson 1977; Rumelhart 1980). These are generic concepts stored in memory, collections of usual, representative knowledge. Schemas are not attitudes; they have no affective significance or evaluative component and are purely cognitive. They serve several functions: the selection of what is important from sense data; economical storage of information in memory (which they achieve by reducing redundancy); facilitation of "reasonable" inferences that go beyond the information available; and promotion of the envisioning and carrying out of sequences of actions. Understanding is viewed as a process of matching new information to existing schemas, with some analogic reasoning required, since every new stimulus has some unique features.

There are three types of schemas: scripts, metaphors, and personae

(Abelson 1976; Larson 1985). Scripts are characteristic sequences of events generalized from experience, such as "going to a restaurant."[5] Scripts mediate behavior in two ways: by the selection of a particular script as the best representation of a situation and by the choice of a role in the script. Metaphors are schemas selected as analogic models when there is no literal fit close enough to the present situation. A common example is the representation of complex situations in terms of sports or games (Schon 1979; Miller 1979). Personae are stock characters that permit inferences of someone's likely personality or behavior from their superficial resemblance to a persona (Nisbett and Ross 1980).

The three characteristic inferential biases observed by Tversky and Kahneman (1974)—availability, representativeness, and anchoring—fit well with a schema approach. Availability and representativeness describe efficient but fallible procedures for the selection of a particular schema to apply to a situation: the schema chosen usually is vivid, recent, or memorable (availability) and shows certain stereotypical, superficial similarities to the case at hand (representativeness). Anchoring describes the persistence of schemas once chosen and their resistance to change in the face of contrary information.

The economic storage of information that schemas offer, though, comes at the cost of a loss of flexibility in accommodating new situations or novel combinations. Holland and colleagues (1986) argue that a more general representation of human inference and decision making should be based on systems of mental rules, clustered in hierarchical structures. Rules can be synchronic or diachronic in their model and specific or general in their application. Several rules can be triggered simultaneously by any particular state of the environment; if simultaneously triggered rules conflict, other rules define precedence relations, and rules can gain or lose strength through competition over time. Induction processes include both parameter revision, such as changing rule strengths, and mechanisms to generate plausible new rules.

The list of characteristics that Gardner cited to define cognitive science did not relate explicitly to learning, and different cognitive theories address learning to different degrees. But a simple elaboration of Gardner's list would give the following characterization of cognitive theories of learning: Learning is an experience-driven change in the internal cognitive structures used to represent information. People respond to disparity between their cognitive structures and feedback from their behavior by revising their cognitions.

This description still subsumes much variation. The changing representations can be of very different kinds. Change can be incremental, quantized, or some combination of the two. And change may be principally mediated by motivational factors ("hot control," more connected to behavioral models) or informational ones ("cold control," more connected to rational-actor models; Nisbett and Ross 1980).

Codetermined Individual and Social Learning

Other streams of theory are substantially more social and less individualistic in their orientation than those reviewed earlier. They view social factors as completely dominant, or individual learning and the social environment as interacting so strongly that they are jointly determined. Either view calls for a less individualistic methodological approach.

One line of theory is descended from the sociology and anthropology of the early to mid-twentieth century, which argued that social constructs effectively constrain individual action and thought, and that these constraints are not themselves coherently reducible to individual phenomena (original statement of principles by Durkheim 1938). In its strongest form, as, for example, in the "thought style" shared by members of Fleck's (1979 tr.) "thought community," such a construct "sets the preconditions of any cognition, and it determines what can be counted as a reasonable question and a true or false answer" (Douglas 1986). The most fundamental organizing principles of thought—similarity and difference, classification, and causality—are seen as conditioned by the thought style. And the thought style is invisible to the individuals who participate in it. "The individual within the collective is never, or hardly ever, conscious of the prevailing thought style, which almost always exerts an absolutely compulsive force upon his thinking, and with which it is not possible to be at variance" (Fleck 1979).

For these theorists the focus of inquiry is the role rather than the person, and the norms or rules that constitute a role. The characteristic form of explanation is "functional," in which social forms are explained by the social benefits they bring about.[6] There has been substantial theoretical controversy over the possibility of sustaining coherent, autonomous functional explanations that do not collapse into either individual human intention or evolutionary selection (Dore 1961; Homans 1964). Moreover, functional explanations are static, and hence of limited value as theories of learning.[7] Granovetter (1985) calls such explanations "oversocialized," arguing that to assume rigid social determination of individual behavior is as limiting as "undersocialized" rational-actor explanations. He advocates a focus on ongoing, changing networks of social relations. Giddens (1984) presents a comprehensive theoretical attempt to reconcile networks of social control and individual human agency.

A more dynamic system of theory that put social factors as dominant causes appeared in the Soviet Union in the 1920s and 1930s, in the work of L. S. Vygotsky and his students (Wertsch 1985a, 1985b, 1991). The primary focus was on cognitive development and its interaction with cultural factors, particularly with tools and systems of signs. Vygotsky contended that culture affects not just the contents of thought, but the structure of its processes.

Wertsch states that three themes form the core of Vygotsky's theory: "A reliance on a genetic or developmental method; the claim that higher processes in the individual have their origin in social processes; and the claim that mental processes can be understood only if we understand the tools and signs that mediate them" (Wertsch 1985)—and that the focus on tools and signs is primary, for these are the socially determined elements that play an essential mediating role in all thought and action, and in the development of thought through instruction.

Vygotsky argued that the development of tools of thought depends on the child's entering into, and subsequently internalizing, a dialogue. The dialogue takes the form of guided participation in tasks slightly beyond their present ability, led by adults or more skilled children (Rogoff 1990; Bruner 1986). Vygotsky defined the zone of proximal development as that set of tasks that a child cannot complete unguided but can complete when guided or prompted with hints, questions, or examples. Vygotsky's contention that all learning takes place in the zone of proximal development has obvious parallels to Polanyi's "tacit knowledge," but on a social plane. Indeed, Rogoff (1990) broadens the view of the dialogue that guides learning to include tacit and nonverbal communication; she also stresses that the learning child is not passive in this process, but actively contributes to the interactive processes that in turn shape her own development.

In this view social causes and active individuals reach a dynamic equilibrium in which social/cultural forms and the cognitions of individuals are codetermined. Cole (1985) has argued that culture and cognition create each other in the zone of proximal development. Wertsch (1991) argues that individual and collective factors are so tightly interwoven that efforts to separate them and look at reciprocal effects in a research program are bound to be fruitless. He argues for a different level of analysis, in which "mediated action" is the primary unit of investigation. Such a research program would pose daunting methodological problems and would likely be cross-cultural.[8]

Organizational Learning

The next three sections discuss learning in various social aggregates: task-oriented groups, formal organizations, and professional communities. As we stated earlier, using the term *learning* to apply to a collection implies one of two forms of relationship between individual learning and changes taking place in the aggregate, which we call decomposition and analogy.

Decomposition treats group learning as the sum of learning by the group's constituent individuals (Udehn 1987; Elster 1989). It is not restricted to the obviously trivial case in which the individuals' group membership does not affect what they learn or how they learn it. What each individual learns may be complexly contingent on the choices and learning of other group members (e.g., in the pursuit of high-level coordinated performance by a group such as a basketball team or a string quartet). Or the means of individual learning might be through activities that depend on the participation of other group members, such as discourse, imitation, or shared activity (Bandura 1977; Argyris and Schon 1978; Habermas 1979).

Analogy treats group learning as autonomous, determined by group-level causal processes that correspond to the processes shaping individual learning. This view may simply represent a methodological convenience; even if all learning is reducible in principle to individuals, the most fruitful way to study groups may involve observations and theoretical constructs at the group level. The value of studying chemistry is not diminished by its being in principle reducible to physics. One may apply analogies between individual concepts—such as perception, reinforcement, memory, cognitive dissonance, or schematic change—and changes in group routines, stories, or behavior, without actually believing that the group sees, thinks, or remembers.

Alternatively, one may reason by analogy to individual phenomena and believe that the collective phenomena are autonomous, truly irreducible to individuals. This position originates in Durkheim's insistence that social facts be explained by social facts and his denunciation of dipping into the psychological level for explanations (Durkheim 1938).

Studies of learning in formal organizations include some instances of both decomposition and analogy, and some studies that combine the two approaches.[9] Although one body of theory, the neoclassical theory of the firm, analyzes organizations by analogy to the individual rational-actor model, the more frequent approach is to study organization-level variables, such as routines and procedures, with analogies to individual reinforcement or cognitive structures.

March and Olson (1976) presented a general model of interaction between an organization and its environment, based on a four-element feedback loop. The cognitions and preferences of individuals within the organization determine individual behavior, which determines organizational choices, which determine environmental consequences, which determine the preferences and cognitions of individuals in the organization. They describe this pattern as a complete learning cycle.

Each of these four causal connections can be broken, though, yielding a particular kind of incomplete learning cycle in each case. When individuals in an organization are not able to act out their preferences, the resultant incomplete learning is called *role-constrained learning.* When individual actions are not translated into organizational actions, the result is *audience learning.* When the causal relations between organizational choice and outcomes are not well understood, the result is *superstitious learning.* And when it is not clear to individuals what the consequences of organizational action were, the result is *learning under ambiguity.* Hedberg (1981) points out that when learning cycles are incomplete, direct empirical falsification does not occur and that consequently, organizations can persist in incorrect belief or ineffective action.

March and Olson's treatment is general and explicitly decomposes organizational learning into learning by the constituent individuals. A focus on organizational routines is also common (Cyert and March 1963). Heiner (1983), for example, argues that simple decision rules or routines are the appropriate response to an uncertain environment whose complexity exceeds the analytic capability of the agent. The larger the disparity is between environmental complexity and the agent's cognitive ability, the simpler and more predictable the decision rules are.

In a recent survey article, Levitt and March (1988) stress reliance on routines in their three basic observations on organizational learning: that behavior in organizations is based on routines, chosen more on a basis of legitimacy or appropriateness for the situation at hand than of outcome calculation; that organizational actions are determined strongly by interpretations of history and past outcomes and only adjusted incrementally in response to feedback (Lindblom 1959; Steinbruner 1974); and that organizations are oriented to targets, with success or failure assessed by the relation between observed outcomes and aspiration values (Simon 1955; Siegel 1957). Learning is more typically a response to shortfall or scarcity than it is to success, although affluent organizations may search out new opportunities (Hedberg 1981; Lewin and Wolf 1975).

Organizations learn by encoding inferences from history into routines that guide their behavior. In the simplest terms, a strict analogy to the behaviorist model of individuals applies; routines that are associated with successful attainment of targets will tend to be repeated (Cyert and March 1963). If the organization's environment and tasks are sufficiently stable, such learning will lead to long-term trends of improved performance. The clearest observed example of such learning is the "learning by doing" effect, through which production costs in a wide variety of goods and services decline with cumulative experience (Levitt and March 1988). A recent study has shown, though, that the learning effect of past production decays markedly with time (Argote et al. 1990).

Organizations and institutions can also be viewed from a rational-actor perspective as systems of rules and relations that rational agents construct to reduce transactions costs and defend themselves against fraud and exploitation (Williamson 1975, 1985). Organizational change can also be studied as rational response by organization members to a changing environment. In his study of institutional change, North studies the interaction of institutional constraints, organizational routines, and incentives of individuals within organizations to acquire new skills and information. Drawing on Polanyi (1966), he argues that organizational routines, like individual knowledge, embody both conscious, transmittable knowledge and tacit knowledge reinforced by practice and interaction (North 1990).

A more "cognitive" approach would recognize that organizations exert some control over what stimuli to notice; that the assessment of success and failure is problematic and related to internal organizational conflict; that the collective memory in which routines are recorded is imperfect; and that organizations share stories, assumptions, beliefs, and myths that guide action and give meaning to experience but retain important elements of ambiguity (Hedberg 1981; Levitt and March 1988).

Modeling organizational learning as a response to outcomes that are finally interpreted as either success or failure implies that learning is limited to means of achieving specified organizational goals. Argyris and Schon (1978) consider the deeper process of articulating and reconceptualizing organizational goals and perceived causal relationships, which they call "double-loop learning." Their model focuses on the difference between espoused theories and the "theories in use" that actually guide individuals' behavior in organizational contexts. Like Beer's (1972) "meta-logic," theories in use are typically inaccessible to the individuals who are guided by them. Changes in theories in use are inhibited by cognitive and behavioral feedback loops, which, among other functions, suppress recognition of the present theory. They propose a dialectical process of articulating, and thereby transforming, the unspeakable assumptions that guide organizational action. Though often treated as such, their theory is more than a simple distinction between learning means and learning ends; its connection to critical theories of social transformation (Habermas 1976) is direct.

Like individuals in Social Learning Theory, organizations do not just learn from direct experience; they can learn from observing others. The experience of other organizations can be acquired through the movement or imitation of technologies or procedures, contacts with or movement of personnel or consultants, or professional associations or public media. Organizations, like individuals, have some power to select or control their environments, and consequently to control their learning context. The movement of innovations between organizations has been modeled as broadcast and contagion processes,[10] sometimes with different degrees of match between the innovation and the organization being the most significant factor in explaining adoption (Mansfield 1968; Kay 1979).

Theories of organizational learning can be mapped onto the behavioral/cognitive/rational distinction presented earlier for individual learning. But theories of organizational learning also reflect differences in the fundamental metaphors used to characterize organizations. Morgan (1986) delineates different bodies of work that treat organizations as machines, organisms, cultures, political systems, "psychic prisons," and instruments of domination. Each of these metaphors implies a different view of the nature and determinants of organizational change, and of what it means for an organization to learn.

Although the goal of organizational learning theory has been to improve organizational performance, in 1981 Hedberg concluded that little progress had been made in the prior 20 years and that current knowledge did not permit much specific guidance. He offered a few general practitioner's suggestions: Promote experimentation by reducing penalties for failure, build in regular shocks to routines through such measures as time-limited management contracts, and don't filter the information reaching senior management too much (Hedberg 1981).

Learning in Science

Science may represent a special case of collective learning. Many social writers have granted a special place to scientific knowledge, believing that the superior verification of scientific method made the resultant knowledge more reliable and less liable to social explanation than other kinds of knowledge. Mulkay (1979) describes the "classical sociological view of science" as holding that scientific institutions and the social use of scientific knowledge were fair ground for sociological analysis, but that the conceptual content of scientific knowledge was not. The perception that science succeeds uniquely in generating new, reliable social knowledge lies behind the once common call for scientific principles to be applied to the solution of our social problems.[11] Contrary views have focused on the usefulness and legitimacy of transferring scientific thinking to the realm of public decision making, rather than on the special epistemological status of scientific knowledge (Lindblom 1959; Lindblom and Cohen 1979; Braybrooke and Lindblom 1970).

But this special status has increasingly come under question. It depends upon several presumed characteristics of scientific method: the generation of falsifiable hypotheses and their attempted empirical refutation; observable facts that are unproblematic and uncontaminated by theory; and gradual, cumulative progress (Popper 1959, 1972). But these have all been cast increasingly in doubt by new theoretical constructions of science and by observations of what scientists actually do.

Kuhn (1970) was the first to argue for shifting focus from particular theories to those larger macrotheories that temporarily bound particular scientific specialties, defining promising directions for research questions and criteria for correct answers. Kuhn argued that these macrotheories (which he called paradigms) are determined consensually and tend to persist even in the face of substantial empirical counterevidence. Lakatos (1970) refined the characterization of macrotheories (which he called *research programs*), distinguishing a hard center or "negative heuristic" of fundamental assumptions that are temporarily immune from criticism, from a softer periphery, or "positive heuristic," which provided general guidelines for the generation of particular testable hypotheses. Contrary to Kuhn, he argued that more than one research program can be active at one time and that rejection is not sudden but a gradual decay as other programs show more empirical success. Laudan (1977) proposed that macrotheories (which he called research traditions) can evolve over time, because although a core set of propositions is sacrosanct at any time, that set can change over time. He also argued that theories are tested by the balance between the empirical problems they solve and the conceptual problems and anomalies they generate.

These different views of how science works have increasingly been investigated through detailed historical study. Donovan and colleagues (1988) present sixteen detailed case studies of scientific change, examining hypotheses on the role of guiding assumptions and anomalies, and the character of innovations and revolutions.

Mulkay (1979) argues that the newer view, which regards science's epistemological status as more fallible, makes sociological study of science's conceptual content legitimate. Over the past decade there have been increasingly detailed sociological studies of the collective creation of scientific knowledge. Fleck's 1935 study of syphilis (1979 tr.), which argued the social determination of all cognition, was a remarkable early exception to the former deference. Recently, there have been many studies of routine operation of scientific laboratories and of scientific controversies (e.g., Latour and Woolgar 1986; Gilbert and Mulkay 1984).

Some studies have persuasively demonstrated social factors determining the acceptance of particular scientific theories or evidence,[12] but many others have yielded weak or ambiguous results. Laudan (1977) argues that the pursuit of social explanations of scientific knowledge has been too expansive. He accepts that the territory of scientific ideas can reasonably be partitioned into those parts determined rationally and those determined socially, but he contends that sociologists have too readily accepted the narrow, classical definition of rational explanation and thus have claimed an overly large remainder for themselves. He argues for a "history-of-ideas" approach to the study of science, following a methodological heuristic of first trying to account for scientific ideas on rational grounds of successful problem solving, coherence, and consistency with other bodies of theory, and only seeking social explanations when the results of this attempt are weak. The record of social studies of science would seem to support Laudan's agenda. It suggests that social factors play the largest role when a body of work is new, is incomplete, and has limited or unclear connections to more mature lines of inquiry; as a field matures, the rational pressures become more tightly constraining and the room for social determination of concepts becomes more limited.

Learning in Policy-making

A second special case of collective learning, one of particular relevance to the study of sustainable development, is the learning that goes into policy-making, policy change, and international relations. In contrast to the classical deference accorded to scientific learning, the traditional view of policy change is that it is better explained by power, interests, and coalition alignments than by learning.

The postwar rise of the disciplines of policy analysis and operations research has led to a contrary and excessive view of the role of technical and scientific knowledge, and learning, in policy-making. The following caricature captures the flavor: *Political leaders, in doing their job of making discrete policy decisions among well-defined alternatives,*[13] *realized that the increasing scientific and technical content of the decisions they were required to make called for outside advice. They turned to a professional cadre of analysts, who presented the relevant scientific and technical information. The decision makers, now knowing the relevant consequences of whatever decision they might make, then performed the appropriate balancing of values and interests to arrive at a decision.* (Note that the analysts operated only in a technical arena, and their only audience was the decision maker.)

This caricature is rooted in the reality of some specific policy and operational problems, many concerned with efficient military operations, that were solved in the early days of operations research. Were such separation of the technical and the political possible, the tension between democracy and reason described earlier in the discussion of Dewey would not arise. But for most policy and analysis, it is indeed a caricature. Even when policy makers know what technical or instrumental questions they want to ask, they are often ones whose answers science does not yet (or cannot) know (Weinberg 1972; Häfele 1974). And in an ambiguous, messy decision environment where goals are multiple or contested, they may well not know what questions they want to ask. In this environment, it is not surprising that so much policy analysis is bad, not listened to, or used for nonsubstantive purposes (Clark and Majone 1985; Sabatier and Jenkins-Smith 1988; Dunn 1980; Webber 1983).

But even with a more realistic view of the relationship between policy makers and analysts, there is much opportunity for learning in this system. One key result is that policy analyses often serve a longer-term "enlightenment function" (particularly relevant to a decadal time view), even if they have little immediate effect on particular government decisions (Weiss 1977, 1982). Clark and Majone (1985) point out that evaluating policy analysis using standards drawn from either pure science or pure democracy is liable to condemn too frequently, and they suggest a set of standards that mixes output, input, and procedural criteria. Increasing recognition of the role of deliberation and of the similarity of deliberative processes in policy and in the sciences has led several authors to call on analysts to acknowledge their role as debaters, persuaders, and rhetoricians and to jump into the fray (Majone 1989; Sabatier and Jenkins-Smith 1988; McCloskey 1985). As Rouse (1987) argues, even in the natural sciences there is a substantial collective, procedural element in determining who is right; the rightness of your view is determined by your ability, using the tools and results at your disposal, to persuade others of it. Majone (1980) extends the analogy between policy and science by arguing that policy change follows a process similar to Lakatosian competition among research programs in science.

An increasing number of case studies in policy change have focused on learning as a key and underaddressed issue (Keohane and Nye 1987; Breslauer and Tetlock 1991). The most common focus is on the learning and cognitive processes of top individual decision makers. Some studies apply models of learning from psychological research; others apply their own definitions and models. Jervis (1976) systematically studies the application of theories of individual cognition and perception, and related biases, to international relations. Larson (1985) applies psychological theories of attitude change to the changing views of four senior American officials at the origin of the cold war. Etheredge (1985) examines senior officials' decision making in recurring similar foreign-policy incidents and concludes that they do learn, but only under conditions of crisis. Richard Rose (1991) addresses the special case of "lesson drawing" across nations by senior governmental officials. His emphasis on the essentially political character of the officials' choice to seek rationally defensible "lessons" and to deploy them in the service of particular agendas is a valuable bridge between the political and scientific poles of the learning debate (see also Lee 1993; Majone 1991). Ernst Haas (1990) articulates two models of nonlearning adaptation and one of learning and applies them to a series of case studies of international organizations.

Steinbruner (1974) describes three characteristic biases, in cognitive processes in organizational decision making: grooved thinking, uncommitted thinking, and theoretical thinking. He asserts that these restrict learning and so characterizes organizational learning under a cognitive model as "constrained learning," but he provides no details of the structure or dynamics of learning under such a model.

Peter Haas (1990) takes more of a group focus in his study of the influence on Mediterranean pollution control agreements of a so-called community. He argues that this international group of like-minded officials, mostly scientists, reached a common understanding of the issue relatively early and then exploited their monopoly on relevant scientific knowledge and their positions of control in specialized government agencies to push their governments into strong international agreements.

Heclo's (1974) study of social policy-making in Britain and Sweden finds an important element of social learning among officials who develop policies. Hall (1990), building on Heclo's work, articulates a model of policy learning in the context of the theory of the state. Three central features of the model are that policy is strongly influenced by recent policy,[14] that a key role in advancing policy learning is played by experts in government or by advisers to it, and that states have substantial ability to act autonomously from societal pressures. Bennett (1990), in a study of learning among senior Soviet foreign-policy decision makers from 1973 to 1983, formulates a series of learning hypotheses on the individual, group, and governmental levels.

Two studies of long-term learning by governments are of particular relevance to a study of learning in sustainable development. Cooper (1989) studied the long-term development of knowledge, cooperation, and institutional capacity for combating international movement of diseases in the nineteenth and early twentieth centuries. Hall (1989) studied the development and movement of Keynesian ideas and policies in eight countries. He delineates three approaches to studying the political influence of economic ideas: economist-centered approaches, which turn on the ideas' intellectual acceptability to economists and presume that economists dominate economic policy-making[15]; state-centered approaches, which emphasize the institutional structure of the state apparatus and its recent experience with similar policies; and coalition-centered approaches in which a new idea's success turns on a politician's ability to construct novel winning coalitions around it. Hirschmann (1989), in his "comment" on Hall, stresses the intellectual component of even a coalition-centered approach: "Prior to Keynes, there was no respectable theoretical position between centralized planning, on the one hand, and, on the other, the traditional laissez-faire policies, with their denial of any governmental responsibility for economic stability and growth."

Hall concludes that ideas do affect policy, but the manner and extent of that influence depend on their economic, administrative, and political viability. He finds four central factors in determining the adoption of Keynesianism: orientation of the governing party, permeability of the civil service, concentration of power over macroeconomic management, and the power of the central bank over policy-making. (Central bankers did not like Keynesianism.)

In summary, these studies show two distinct approaches: studying learning among a few top political decision makers and studying it in the somewhat larger community of policy experts on a particular issue. With the former approach, the focus is clearly on individuals. Consequently, the large body of work on individual learning, attitude formation, and cognitive change applies directly. With the latter approach, the focus is necessarily on a larger and only roughly bounded group. Consequently, as in the preceding discussion of organizational learning, the relevant learning theory—theory of how the group learns—is not well developed. Several authors limit their comments on group learning to the observation that individual learning is necessary but not sufficient for group learning, for routines and conflict can prevent individual learning from being expressed in group activity.[16] The most frequent response is not to look inside the policy community but to ask how their learning affects policy change. On this question the theorizing is largely ad hoc, amounting to the assertion that learning by policy elites does affect policy change; change is not completely reducible to contending political interests and bureaucratic inertia. The theoretical tools for distinguishing the effects of learning from other forces of change, though, are still at an early stage of development.

Evolution and Learning: Complementary Models?

Aside from learning, evolution is the other manifestly dynamic process often employed as a model for change in social systems. An evolutionary approach can cut across the various categories of learning studies we have considered thus far. Evolutionary models have been applied to learning processes at the individual and organizational levels, in science and in policy-making, and to other processes of social change. This section discusses the biological basis of evolutionary models, their application to learning and other processes of social change, and the relationships between learning and evolution as theoretical models for social dynamics.

The basis of an evolutionary model is the selection among invariant forms leading to differential survival. In biology evolutionary models are based on stably propagating species defined by reproduction within but not across the species boundary; on variation among the individuals comprising a species, at least some of which is heritable; and on differential reproductive success of individuals leading to shifts over time in the genetic and somatic characteristics of the species.

The differential reproductive success of individuals is mediated by many factors, some of which are fixed (some variations are quickly lethal to the individual carrying them); some of which are contingent on particular characteristics of the environment the species presently occupies; and some of which are socially contingent, depending on the distribution of related characteristics or behaviors among other members of the same population. Overall adaptivity or fitness is not absolute, but contingent on many factors.

The shift in species characteristics over time and in response to changes in environment sometimes brings about the creation of new species. It is widely agreed that speciation requires that a population undergo substantial change during a period of effective physical reproductive isolation, which proceeds so far as to create effective biological reproductive isolation if and when the separated populations come back into contact.

The primary locus of selection, though, is in dispute. The standard view is that selection is among individuals, through processes of reproduction and death. However, one can construct scenarios under which marginal populations diverge quickly enough that internally stable, reproductively separate groups come into competition for the same environmental resources, with selection thereby operating among species through processes of speciation and extinction (Eldredge 1985; Curtis and Barnes 1989).

This basic structure of evolutionary thought—a mechanism of novelty generation and one of selective retention—has provided a richly provocative analogy, and sometimes formal models, for students of learning at the individual and group levels. The richest approach has been not to argue that learning shows particular characteristics because of adaptive advantage,[17] but to apply evolutionary arguments to nonbiological phenomena that show the same processes of (1) generally stable propagation, (2) some means of generating innovation and variation within the basic regime of stability, and (3) a separate mechanism of selection from among variants, such that some are more likely to survive and propagate than others.

Applying evolutionary thinking to ideas and cognitions themselves, Campbell (1960, 1974) has argued that human learning is at root an evolutionary process. Bateson (1972) proposes an ecological approach to broad classes of human cognition and behaviors. Popper's (1972) view of falsification and progress in science is expressed in explicitly evolutionary terms. Toulmin (1972) presented an evolutionary view of concepts in society at large, not just science, emphasizing that concepts that are dominated are not necessarily eliminated; like marginalized subspecies, they can remain available to exploit future opportunities.

At the organizational level, an evolutionary perspective can be particularly fruitful in view of the competitive relations among economic organizations. Nelson and Winter (1982) argue that competition among firms applies selection pressure to the processes of routine generation and innovation that operate within firms. Since the 1970s the emerging field of organizational ecology has sought to apply ecological and evolutionary principles to broader classes of organizations. Hannan and Freeman (1977, 1984) applied theoretical constructs from population ecology to organizations. They distinguish two forms of organizational change: learning from the environment and adapting to change in the environment. Organizations with too much inertia to change fast enough relative to change in the environment will be removed by selection pressure. The more recent literature has focused on the levels-of-analysis problem, arguing that the population is not the only or most useful level at which to study organizations, particularly the origin and diversification of organizations.[18] Young (1988) provides a forceful critique of biological analogies in the study of organizational change, arguing that such fundamental concepts as species, niche, and death are defined ambiguously or circularly.

In their work on mathematical modeling of cultural evolution, Cavalli-Sforza and Feldman (1981) argue that cultural artifacts, or "memes,"[19] can be modeled by the same processes as biological evolution, with two possible exceptions. First, innovations, unlike biological mutations, are typically created with a purpose or in response to a perceived problem, so the determinants of their adaptiveness may be structured differently, and the probability of innovations is higher than for biological mutations. Second, cultural selection may have complex interactions of either direction with Darwinian selection. Although most cultural patterns likely have minimal direct Darwinian effects, some clearly have negative effects (they cite ritual mutilations, particularly female circumcision, and risky sports), and some have clearly positive results (the adoption of agriculture, with resultant increases in local carrying capacity). Boyd and Richerson (1985) develop a "dual inheritance" model, in which formal Darwinian selection is applied independently to genetic and cultural inheritance.

In less formal treatments, Adler (1991) applied evolutionary concepts to government learning, explaining two major innovations in American foreign policy through a process of idea generation, domestic selection, and international diffusion. Schmid (1987) and Eder (1987) take an evolutionary approach to the explanation of social systems, norms, and rules. At the social level of analysis, theories focusing on the cognitive evolution of politically salient ideas have tended to become linked with normative discussions of progress in human affairs.[20]

An evolutionary model cannot be equivalent to a learning model at the same level of analysis. Evolution is based on selection among invariant forms through birth and death processes, whereas learning implies that the learning system survives its changes. Moreover, as Elster (1979) argues, evolutionary systems are restricted to changes through "local hill- climbing," selection based on the immediate adaptive advantage of each local increment, while human learners can proceed obliquely, take one step back for two forward, pause to reframe a problem, or otherwise circumvent the demands of continuous local improvement in pursuit of a superior global goal.

But an evolutionary model may be equivalent to a learning model at a different level of analysis. Selection among organizational routines may represent learning by the organization (Nelson and Winter 1982). Selection among particular organizations by birth and death may represent learning by the system in which the organizations operate. As Campbell (1974) and Popper (1972) have argued in different contexts, individual learning may be an evolutionary process at the level of cognitions, through which "our hypotheses die in our stead."

If evolutionary concepts are to be used in studying social learning, the selection must operate on the things learned—the concepts, names, slogans, images, acts, technologies, forms of social organization, political movements, opinions, attitudes, norms, and skills that people learn, individually or collectively. The evolutionary concepts of selection, ecological community, and species suggest many provocative analogies, hypotheses, and questions for the study of social learning. To determine how useful they are requires a careful look at the applicability of each of the constituent concepts: origination of variation, propagation, selection, and speciation and species stability.

First we need a description of origination. Although most would agree that particular new ideas are generated by individuals, there is more dispute over the relative effects of individual creativity and social ordering on the total set of new ideas generated. Social or cultural factors may interact with individual creativity to determine the range of novel concepts a member of a particular society is capable of inventing. Thereafter some social factors are likely involved in determining the intelligibility, sense of importance, and ease of communicating a new idea or image, and consequently its initial viability.

Next we need a mechanism of propagation. Different forms of propagation apply to different classes of things learned. Some things can be replicated cheaply and accurately by print or electronic media, others (e.g., behaviors to be learned by modeling) require more detailed information, and still others (e.g., skills and craft knowledge, forms of argument, persuasion to a new value) may resist explicit encoding completely and require direct communication from another person, guided participation, or practice.[21]

Models of information propagation alone may not be adequate to capture the full range of learning processes, though. In addition to the issue of tacit knowledge, there may be things learned that, contrary to the cognitive science model, bundle evaluation or affect with information.[22] The particular content of information and affect may change as a word or slogan is propagated and used, through the effects of both the channels of communication through which it passes and the people who hear and use it.

Third, there must be mechanisms to introduce variation in the propagation. Although physical processes of replicating information have low error rates, people's inclination to interpret and modify what they receive in view of their own experience, values, information, and cognitive processes will create pressure toward variability. One needs only to have played the party game in which a story is passed around the room in whispers, finally returning to its originator transformed beyond recognition, to realize how rapidly an idea can be transformed in interpersonal transmission.

Additional variation can also arise from the movement of ideas through various communication channels and to different audiences.[23] Certain kinds of ideas, such as technical concepts of specialized disciplines, may only be intelligible to a few people. When such ideas move through broader circles, they may be altered in ways that reflect the characteristics of the new, broader audience or of the channels of communication through which they pass (e.g., new scientific results reported on television or used in congressional debate). There is also something in the transmission of ideas akin to sexual reproduction in biology, with its mixing of existing genetic material to create new variations. Ideas, norms, and images come together in inter- or intrapersonal process—in introspection or in discourse—and generate new ones.

If there is to be a strong analogy to biological evolution, though, these forces for variability must be countered by forces to stabilize related collections of ideas. If the intrapersonal processes determining what is intelligible or worthy of consideration are not sufficient to effect this stabilization, then it must come from some interpersonal process—from discourse, from comparison to other cognitions and views held, or from argument and evaluation in view of a common shared body of knowledge.

This stabilizing force could operate as selection pressure on individual ideas to reinforce related systems of ideas. Bodies of ideas could be mutually reinforcing by being intelligible in terms of each other, reciprocally confirming, suggesting questions or actions whose results will further reinforce the collection. Disciplinary boundaries perform these functions in science, as do the various forms of macrotheory within disciplines (Lakatos 1970; Kuhn 1970; Laudan 1977). In political and social discourse, corresponding bodies of thought might be called ideologies or national characters (Geuss 1981; Bateson 1942a). The analogy to selection forces preserving the stability of species is evident.

There is abundant anecdotal evidence for the notion of bodies of related ideas defending their integrity and borders through repression of deviance and sniping at those who enter no-man's-land. This would parallel the argument that the relative paucity of smooth, large transitions in the fossil record shows the middle ground to be highly unstable. If you must change, you survive by evolving quickly across the boundary to find a new stable niche beyond. The often lamented difficulty of establishing and maintaining interdisciplinary scholarly careers and academic programs would fit this model. In this case the niches for which groups are competing are university appointments and research grants.

The competition will be intense. In the realm of political ideas, Chomsky's (1989) work on propaganda argues for the existence of social mechanisms that delimit the range of acceptable political debate in a society.

The adaptation of bodies of ideas, like adaptation of species, is relative. They compete not against objective criteria of value but against each other in view of the opportunities that the environment presently offers. This lends a provisional character to any presently ascendant ideology or thought system.[24]

Moreover, a great breadth of survival opportunities for idea systems will exist in a reasonably pluralistic society. The evolutionary analogy suggests that small populations of marginal ideas, like Toulmin's temporarily dominated concepts, may survive through long periods when the mainstream environment is unfavorable, flourishing anew when environmental conditions change in their favor. The persistence of neofascist movements among alienated young men in industrial countries and the rapid reappearance of such movements (and of ancient ethnic hatreds) in Eastern Europe and the former Soviet Union may be examples of this phenomenon. The analogy also suggests that there may exist many such small, marginalized subpopulations of ideas, collectively representing an important reservoir of cognitive variability that can increase society's resilience to extreme environmental change.

Conclusions: Guiding an Applied Research Program

The purpose of this chapter has been to gather together the major strands of theory related to social learning, with a view to complementing a research program of empirical case studies in the management of sustainable development. The best-developed theories of learning are clearly at the level of individual learners, and to a lesser extent at the level of small, face-to-face groups. Indeed, some writers limit their definition of social learning to processes that occur in such groups.[25] The studies reviewed in this essay, however, suggest that learning can be a meaningful concept to apply to processes that occur at all levels of society, from the individual to the international. A progressive research program on the role of learning in sustainable development must therefore address two sets of questions, one largely definitional and methodological, the other largely empirical.

On the definitional side, those studies are likely to be most useful that clearly specify their approach to four key questions: Who (or what) learns? What kinds of things are learned? What counts as learning? Why bother asking? These simple questions can be used in categorizing works in social learning. They concern what each body of work defines learning to be, what level of analysis they focus on, and how they decide whether or not learning has occurred. These questions are locators; they provide four dimensions on which to map the territory of theories of social learning.

Answers to these questions, perhaps tentative ones, define a research program. Once a focus is defined, a second class of empirical questions arises that the research program aspires to answer. What specific things did they learn? How did they learn them? Under what conditions and why? Did it make a difference? And how could the effectiveness of learning have been increased?

Who or what learns? Societies, governments, and organizations, or just individuals? Our basic conceptions of learning are formed from observing individuals, but collections of people large and small also exhibit clear changes in task performance, coordination, complexity of communication, and goals that look very like individual learning. Any writer who uses the term *learning* to apply to a collection speaks to this common perception, implying some form of relationship between what the collection is doing and individual learning. This implied relationship is typically of one of two kinds, which in our discussion of organizational learning we called decomposition and analogy. With specific reference to studies of sustainable development, at least five different groups of learners seem relevant: senior policy decision makers, scientific communities, industrial organizations making production and technology decisions, other nongovernmental organizations, and—to the extent that consumer market forces and broad political opinions and images condition or constrain elite decision making—the citizenry at large. The role of the mass media, as an agent of learning that may itself learn, is particularly problematic.

What kinds of things are learned? Individuals learn a vast array of different classes of things: behaviors, facts, concepts, words, skills, desires, opinions, attitudes, and values. There may also be a logical hierarchy of categories of things learned, analogous to the difference between improvement in one task and improvement in the rate of learning new, "similar" tasks, which depends on the second-order task of recognizing a context of "similarity."[26] Collections of people learn all the same sets of things, and such essentially collective things as languages, technologies, slogans, images, forms of social organization, and norms. It is reasonable to presume that different classes of things are learned in different ways. A useful learning theory must be clear about its breadth of application. Useful theory will also treat carefully the relation between "who learns" and "what is learned." Particular things learned will move differently through different communities. Some are only intelligible to particular subpopulations; others may rely on a prior favorable disposition. When taking a macroview of whole societies, particular learners (whether individuals or organizations) may become too fine to resolve, and the appropriate methodological focus may be strictly on the things learned—the ideas, facts, behaviors, and norms—and on their origination, propagation, and growth or decay. In studies of sustainable development, some of the most important things learned seem likely to include scientific facts and models, "policy theories," technologies, preferences, behavioral norms, images and names, or broad world views and conceptions of society.

What counts as learning? Any change in the phenomena studied, or only certain changes? If only certain changes, then what criteria distinguish learning change from nonlearning change? Short of calling any change learning, one might say that only change in response to identifiable stimuli or information is learning; or more restrictively still, that the change must follow rationally (defined somehow) from the stimulus. Some researchers propose narrower definitions still: increasing cognitive complexity, or increased effectiveness at attaining given ends. Others argue that definitionally, any cognitive change should be treated as learning, and questions of effectiveness or progress should emerge from the subsequent research (Nye 1987; Bennett 1990; Breslauer 1987). We have followed the latter advice in this review, since it is the more inclusive approach. Our findings lead us to believe that it will also be a more productive stance for the study of learning in sustainable development.

Why bother asking? Most students of social learning would answer, "In order to understand and explain the societal phenomena we observe." In this view a theory of social learning, like any theory, would be a set of general, causal statements whose purpose is to guide inquiry and explain observed phenomena. Others, working in the tradition of critical theory, take a more activist view of social theory. They would answer that we ask about social learning in order to bring about personal liberation and social transformation. In this view a theory is a set of propositions to guide communication between people, whose purpose is to bring about these transformations and which is validated by (1) its acceptance by the agent addressed in a noncoercive situation and (2) its efficacy in bringing about the desired transformations (Geuss 1981; Habermas 1979). Although critical theory does not address itself to learning as such, there are obvious points of contact with some of the strains of literature surveyed here—to individual Social Learning Theory as employed in therapeutic practice (Stuart 1989) and to some strains of the organization development literature (Argyris and Schon 1978; Friedmann 1987).

It is difficult to incorporate the critical perspective into a generally naturalistic inquiry into the characteristics and determinants of social learning, but this view cannot be ignored.[27] An inquiry directed to social intervention, as arguably any project in "policy research" is, may lie closer to the agenda of critical theorists than to that of positive theorists. In policy research, as in clinical practice, "what works" may indeed be the appropriate validation criterion.

Do we expect that research will provide simple or unambiguous answers to questions concerning the role of social learning in sustainable development? Of course not. The most relevant empirical studies that now exist (for example, Eder's [1987] work on the evolution of democracy, Hall's [1989] account of the spread of Keynesianism, Cooper's [1989] history of the development of public health practices, Peter Haas's [1990] work on the Mediterranean Action Plan, Adler's [1992] examination of the evolution of the idea of nuclear arms control, and Kai Lee's [1993] path-breaking work on the Columbia Basin) all point toward a fundamentally messy, contingent, and ambiguous intermingling of knowledge, power, interests, and chance in the workings of the world.[28] These studies nonetheless provide what we have found to be useful beginnings for efforts to understand and manipulate long-term improvements in the management of sustainable development policies. But to the extent that such improvements depend on the broad spread of images, norms, knowledge, and behavior through society at large, other models are also required. An evolutionary perspective that focuses on the population of things learned may be most suitable, but existing work in this area scarcely goes beyond provocative analogy. The immediate need is to increase the stock of empirically rich case studies informed but not constrained by today's body of theory relevant to social learning. The essays in this volume provide a welcome contribution to the task at hand.[29]

Acknowledgments

This chapter is a revision of contribution number I.6 (ver 2) to the project on Social Learning in the Management of Global Environmental Risks. It was partially supported by the National Science Foundation (through Grant number SES-9011503), the John D. and Catherine T. MacArthur Foundation, and the Social Science Research Council through its Committee for Research on Global Environmental Change. Members of this committee provided helpful feedback on an earlier version of the paper. For valuable discussions, criticism, and encouragement, we are indebted to David Ballon, Peter Cebon, Angela Liberatore, Vicki Norberg-Bohm, Douglass North, Barbara Rogoff, Thomas Schelling, and Nina Tannenwald.

NOTES

1. Our efforts thus complement those of Kai Lee (1993), who, in his extended case study of social learning in the sustainable development of the Columbia Basin, touches on many of the same themes we address here. Lee's work explores one particular view of how science and politics interact in the creation of policy. Although we are sympathetic with his approach, we attempt here to sketch a somewhat broader theoretical perspective within which his approach stands out.

2. Concise discussions of this model can be found in Elster (1986), Harsanyi (1986), and Allison (1971). Axelrod (1984), Taylor (1987), Rapoport and Chammah (1965), and Luce and Raiffa (1957) consider interactive decision problems from a rational-actor perspective.

3. Gardner (1987) provides a brief history of the behaviorist movement. Its furthest imperial expansion was Skinner (1957); the first rationalist counterattack was Chomsky (1959). Skinner (1971) is a clear popular treatment.

4. We will denote this body of theory as Social Learning Theory (capital letters), to distinguish it from the general topic of social learning.

5. Scripts can be of two kinds: episodic (a sequence of events described as a single experience [e.g., "Chamberlain at Munich"]), and categorical (generalization from common features [e.g., "appeasement only encourages aggressors to make more extreme demands"]) Abelson (1976).

6. Douglas (1986) argues that functional explanation requires that the benefits be unintended, indeed unperceived, by the individuals participating in the social form. Elster (1989) presents a skeptical view, arguing that even if norms are important determinants of individual behavior, they need not be "supraindividual entities."

7. There is, though, a recent body of empirical research on norms, based mostly on experimental studies in small groups, that focuses on norm creation and evolution and that is consequently not functionalist in orientation (e.g., Bettenhausen and Murnighan 1985; Feldman 1984; Handel 1979; Opp 1982).

8. Cole and Scribner (1974) provide a useful historical review of cross-cultural research on cultural determinants of cognition.

9. Hedberg (1981, p. 6) states that organizational learning studies mostly avoid the question of relationships between group and individual learning, using organization-level theoretical constructs but being based on observations of individuals.

10. Kimberley (1981); Rogers and Shoemaker (1971); Heclo (1974) use a contagion model to explain innovations in social policy. An early literature treating the spread of rumors with epidemic models was summarized by Dietz (1967).

11. Friedmann (1987) traces this "social engineering" view from Saint-Simon and Comte to modern muted strains in Simon and Dunn.

12. For example, Wynne's (1976) study of Barka's J-radiation demonstrating a consensus made and kept by highly selective use of evidence; Collins and Pinch's (1978) study of parapsychology demonstrating the force of unquestionable assumptions; and Frankel's (1976) demonstration that individuals' acceptance of particular claims can depend on their social position (e.g., outsiders take more risks).

13. This view, called *decisionism,* is from Shklar (1964).

14. Hall's presentation implies that he is distinguishing progressiveness in policy from policy that simply responds to present interest alignments, rather than asserting that policy learning is sluggish.

15. This approach has strong parallels to Haas's "epistemic community" work on environmental policy.

16. For example, Nye (1987); E. Haas (1990); Bennett (1990). This view fits neatly into the "incomplete learning cycle" of March and Olson (1976).

17.This is the approach of sociobiology, though, which asserts that some human behavior is genetic in origin. Wilson (1978).

18. Astley (1985), Carroll (1984), and Singh and Lumsden (1990) provide surveys of organizational ecology, distinguishing three levels of analysis: the organization, the population (of similar organizational forms), and the community.

19. Following Dawkins's usage (1976) for "units subject to imitation."

20. Emanuel Adler has been among the most careful and energetic proponents of this connection. See, for example, Adler and Crawford (1991) and Adler (1992).

21. These forms of propagation have been studied with various contagion models drawn from epidemiology (e.g., Cavalli-Sforza and Feldman [1981], and the studies reviewed by Levitt and March [1988]).

22. This applies not just to clearly evaluative phenomena such as attitudes, but also to names, definitions, and facts. For example, the concepts *racism* or *sustainable development*, or the fact that the "United States, with 5% of the world's people, consumes 25% of its energy" are learnable things in which cognition, evaluation, and affect are tightly bundled.

23. This argument is an inversion of Fleck's notion of a "thought community," with a core of initiates and a soft periphery of adherents who take the ideas literally and unquestioningly. In Fleck's view (1979), movement originates from the center and ossification occurs on the rim, but the reverse is also possible.

24. This has been most clearly recognized in the descriptions of science of Lakatos (1970) and Laudan (1977).

25. Friedmann (1987), for example, argued that significant social learning only occurs in small, task-oriented groups whose dynamics are not reducible to individuals' characteristics; that the learning is embodied in group relationships that are lost when the group dissolves; that learning occurs primarily through dialogue; and that such learning groups discover their objectives in the course of action.

26. Bateson (1942b) calls these classes proto and deutero-learning. Douglas (1986, p. 55), in a related comment, asserts, "Similarity is an institution," at least partly socially constructed.

27. Braybrooke (1987) provides a clear delineation of the naturalistic, interpretative, and critical approaches to social theory.

28. See also Kitschelt's (1986) policy history of the fast breeder reactor.

29. We are engaged in what we hope will be a complementary study focused on social response to global environmental risks.

11

Barriers and Bridges to Learning in a Turbulent Human Ecology

Donald N. Michael

How to Read This Chapter

In the introduction to this volume, Holling declares, "We want to discover if there are common features to all examples that identify critical barriers to and bridges for maintaining, renewing, or restoring the ecological attributes that underlie and provide services to the people and activities in a region." Among the major categories of barriers and bridges are those pertaining to learning: learning what needs to be done, how to do it, whether it worked, and how to apply the learning to the emerging consequences; learning what must be unlearned and learning what must be learned anew and by whom; learning about how to learn under the conditions that shape humans, on the one hand, and the environment, on the other. But these are no routine tasks! This is a world of increasing turbulence and uncertainties, of changing beliefs about rights and responsibilities, of changing ways of thinking and doing. Because change is so pervasive and profound, unlearning and new learning become imperative prerequisites for ecological management. But for the same reasons, becoming a learner is often a threatening and difficult challenge—hard to undertake and hard to maintain.

One of this book's aims is to provide leverage for doing better in the future. What can be derived from these case studies that will improve the future of ecological management so that attempts to manage ecological variables do not lead to "more brittle ecosystems, more rigid management institutions, and more dependent societies?" No doubt much can be learned, but, given the current turmoil over social priorities and appropriate means for thinking about and acting on them, we cannot presume a future like the past.

My intention here is to enhance appreciation of barriers and bridges to personal and organizational learning undertaken to improve ecological management. I hope to provide some useful insights about barriers and bridges to learning that the case studies experienced in their specific situations. But I will not use the case studies to verify or demonstrate my observations, since no single approach to dissecting human events can account for the unique properties of an event. Instead my emphasis will be on the emerging situation regarding society and the human psyche and the challenges posed to undertaking the difficult task of being a learner or a learning organization dedicated to ecological management.

To focus expectations about what we can expect from learning efforts and our insight into what might have engendered some of the behavior and events reported in the case studies, I will offer some observations about the emerging social context in which learning about ecological management may occur and unavoidable constraints that accompany efforts to learn. Finally, I will offer some suggestions about ways to enhance the likelihood of establishing and maintaining learning for ecological management. But before embarking on these discussions, some warnings are in order regarding the assumptions and premises that underlie my thinking about matters explored here.

Most of us, most of the time, take for granted our ways of thinking and using language: what we pay attention to, how we apply that attention to the definition of problems and opportunities, and what we can and should do about them. We assert that, "It's just a matter of common sense." But a deepening understanding about how we use language and how we construct our reality—our common sense—provides us with improved insights into the processes that encourage or inhibit learning. This is important! *Our conventional ways of thinking and speaking about language and social reality are inadequate for coping with our current circumstances.* Thus our understanding of the function of learning in coping with these circumstances, ecological management included, is distorted. Our semantic baggage from past experience is not matched to a reality of systemic interactions, circular feedback processes, nonlinearity, or multiple causation and outcomes. Implicitly, our conventional language relates us to a world of linear relationships, simple cause and effect, and separate circumstances, be they events, individuals, causes, or effects. But that is not the world we live in. Our world does not change in the same way that our language does. For example, Americans make much ado about where the buck should stop, but in a systemic world of proliferating interconnections, *should* the buck stop? Therefore a brief review of some language reframing will contribute to our learning about learning.

Given the reasons for this essay, it is important to note that a major effort to understand what is happening to the human condition takes the form of postmodern, constructivist analysis. Crudely put, the argument goes that we cannot discover social truth; instead we construct social reality. We create and choose among narratives—stories—that give motive and meaning to social action. We do not necessarily do this consciously—quite the contrary. Some stories consist of those that use the language and ideas of science, others are built around spiritual beliefs, and still others use an economics-based language. Many of these stories claim that theirs is the truth about social reality. Overall, the result is that, with regard to the social uses of physical science, economics, social sciences, politics, ethics, communication, the law, there is no agreed-upon place to stand, no "court of last resort." This current conceptual murkiness about the emerging human situation is unavoidably reflected in this essay.

A parallel intellectual turmoil accompanies the widespread, passionate, even violent arguments and actions about what should be preserved and changed in the conduct of societies and who is entitled to do so. Everywhere boundaries are challenged and defended. No explanation is generally accepted of how this society—this world—is working itself out or should. This is surely so with regard to all the issues that surround and pervade persons and organizations concerned with the goals, processes, and consequences of ecological management. We shall have to learn our way into the future and learn how to learn under the strange, uncertain, and contradictory circumstances confronting us.

I accept the well-documented premise that human behavior is overdetermined. There are always many ways to interpret what a human does, whether one is outside, observing another, or interpreting one's own behavior. Motives are always multiple, as are the causes engendering those motives. Some motives are current, whereas others are carried forward from the distant past. We construct our social reality out of an accumulated reality. I will revisit this point later.

I emphasize this at the beginning because what I shall attend to is unavoidably only a partial interpretation and explanation of the barriers and bridges to learning. Whether the complexity of human behavior can be quantized into a few primal variables, as Holling has done with ecological behavior, remains to be seen, although it surely cannot be done now. When it comes to causes and effects, human behavior is ineluctably a both/and relationship, even though we must act all too often as if it were an either/or relationship. This chapter, then, is a contemplation, an exploration, a learning occasion for me, and I am grateful for the opportunity to indulge in it.

It is imperative to free the idea of learning from its conventional semantic baggage. Learning used to mean (and for the most part still means) learning the answer—a static shift from one condition of knowledge and/or knowhow to another. This definition of learning leads to the organizational and stakeholder rigidification that Holling bemoans. But in the current and anticipated conditions of dramatic unpredictability, learning must be a continuous process involving: (1) learning to reperceive or reinterpret a situation, (2) learning how to apply that reperception to the formulation of policy and the specification of action (including evaluation of policy and action), (3) learning how to implement those policies and intended actions, and (4) learning how to keep these three earlier requirements alive and open to continual revision. These criteria suggest a thoroughly recursive way of being and doing. Learning, then, means learning content and, just as important, learning how to learn to attain these learning requirements. These are radical and upsetting processes that presage radical and upsetting consequences.

Be wary, then, of what you may presume about the forward meaning of words used here in contrast to the backward meaning we tend habitually to attach to them. With the context for reading this essay in mind we turn now to the context for learning.

The Societal Context for Learning

On the one hand, the world is becoming more interconnected. At all levels connections are becoming more systemic in that there are more shared signals that integrate and guide the feedback that regulates social processes. Among these integrating signals are the values, myths, and norms that regulate operating procedures pertaining to ecological management.

On the other hand, human groups (subsystems) everywhere are intent on going their own way. The increasing demand to do so is promoted under the banner of "rights." Often they are guided by internal feedback whose regulating signals have little in common with those of other groups. In this way, groups protect the integrity of their own regulative processes. (In less formal terms this is called *preserving ethnic diversity,* or *pluralism.*) Any survey of the current internal and external status of nations provides evidence that humans have much more experience (i.e., learning) in attaining and maintaining diversity than in creating systems integrated across boundaries.

As a result, democracies, certainly including the United States, face a profound and increasing tension between the claimed universalization of rights and the consequent claimed universal right to local self-government (Sandel 1982). Local preferences may conflict with claimed universal rights. Ethical and operational questions about who is responsible for the consequences of economic and time-related choices in ecological management result from those opposing views, and we are a long way from resolving this dilemma. We have learned how to behave in a world passionately pursuing the independence of diversity while mostly unaware of our accumulating *inter*dependences. Everywhere the boundaries delineating ideas, turf, relationship, gender, and status are challenged, defended, and changing. In such a setting of conflicting claims, generated and defended by different values and belief systems, learning becomes not only imperative but also very difficult.

The Function of Myth

To examine what is happening to boundaries we must first attend to the primal function of myth systems (otherwise called belief systems):

> Myth lies at the basis of human society. That is, because myths are general statements about the world and its parts, and in particular about nations and other in-groups, that are believed to be true and then acted upon whenever circumstances suggest or require common response. This is humankind's substitute for in-

stinct. It is the unique and characteristic way of acting together. A people without a full quiver of relevant agreed-upon statements, accepted in advance through education or less formalized acculturation, soon finds itself in deep trouble, for, in the absence of believable myths, coherent public action becomes very difficult to improvise and sustain [McNeill 1982].

Mythologies are (mostly tacit) social constructions of reality. The dominant mythology in the West, especially since the Enlightenment, includes beliefs that take the following as natural and right: scientific objectivity, efficiency, democracy, progress, competition, the earth as a resource, a "Greco-Judeo-Christian" god, moral superiority, technological knowhow, male dominance, and the inherent separation (through independence and isolation) of persons, objects, and ideas. These mythic attributes infuse our language, both in the meanings conveyed and the meanings we seek. (For example, in the United States it has been taken for granted that, as a separate individual, I am entitled to treat a "piece" of land I own as separate from other pieces of land, because it is mine. Indeed, in the United States a corporation is legally an individual, entitled thereby to freedom of speech.)

Challenges to and Changes in Boundaries

The preceding myth-engendered modes and expectations are reified, operationalized, expressed, and maintained by boundaries of many kinds—physical, temporal, ideological, territorial, factual, procedural, relational, and organizational. Boundaries are critically important for persons and organizations. They are supported by prevailing belief systems and in turn reinforce them. They keep threats outside and give support inside. They determine access, power, and legitimacy. In a word, explicitly and implicitly they define "me," whether the me is a person, group, organization, or state. As such, especially in our times, boundaries become the focus of action; many seek to shift or destroy boundaries, whereas others are intent on boundary preservation.

The boundary issue compounds and confounds the processes and purposes of governance. Externalities that were once outside the boundaries of organizations have become internalities. Environmental protection and child care services for working parents are two such examples. Matters once defined as internal to an organization, such as workplace safety and targets of "whistle blowing," now draw external attention. Such shifts in and challenges to *personal* perceptual and belief boundaries are also more frequent.

Many factors contribute to these boundary shifts. Information, especially in an open society, encourages a situation where the conventional mythology, and the boundaries sustained by that mythology, are under attack from many, often contending, sources. They are under attack because, for some, circumstances do not match a myth-engendered expectation, such as closing the gap between rich and poor. They are under attack because, for others, the myth-engendered expectations are deemed undesirable (e.g., "it's a man's world"). They are under attack because the promise that technology and the engineering approach would "fix" things has not been kept. They are under attack because the hardware and software of the information society have made it much easier for relevant information to be widely accessible and for like interests to aggregate around the information and to act on it. (This includes, especially, information that confirms doubts or invigorates aspirations, via protests, tracts, data, investigative reporting, leaks, monitoring organizations, monitoring technologies, judicial injunctions, and the like.)

Pushed hard enough, all dynamic processes can undermine their own stability, engendering a different phase, as described by Holling's' infinity loop for biological systems. For humans, as in certain ecological disasters, the system can unravel altogether and disintegrate. There are more dead civilizations than live ones. Short of that, the growing mess that characterizes so many human situations, including the environment, can provide sufficient disarray to make preservation of current boundaries seem less important than coping with the human–environment crises. Even then, the change sought is usually a change that reinforces the old system but in ways that work—or seem to work. New wine in old bottles.

But sometimes persons or groups seek to change profoundly unsatisfactory conditions in ways that change myths, resulting in corresponding changes in values, behavior, and boundaries. The long term and the larger interest become matters of here and now. Scientific data are among the realities that then can become relevant. Reason, shaped by these data, can play a significant role when people are ready to take the risk of reperceiving—of learning. But learning is, even with the best of will, difficult to undertake, to accomplish, and to sustain.

Constraints on Learning

Unavoidable constraints shape what a person or an organization wants to learn, can learn, and deems worth the risks and opportunities involved. These constraints are sociocultural, emotional, and cognitive.[1] To appreciate the implications for ecological management of these constraints it is necessary to discard some more semantic baggage.

Our conventional mythology and its semantic baggage view constraints negatively. We want to do and expect to do our own thing! However, for many cultures and for many activities in our society, constraints provide the discipline for focusing creativity, for reducing detours and dead ends, and for shaping reliable behavior. The controlled experiment, so essential in experimental science, exemplifies this virtue. In most cultures the societally stabilizing constraints of family, religious, and community obligation take priority. Even though we celebrate freedom from constraints, constraints persist, and they unavoidably shape learning for any purpose. Acknowledging constraints and working selectively with them will help learners learning be more realistic in their expectations and effective in their practice.

Contrary to popular belief, most people under most circumstances are not all that eager to learn. The belief that humans are eternal questers after new experiences, methods, and information derives from our Western (some call it our Faustian) mythology. It is a belief that is "naturally" held by those of us who have been rewarded for *learning* this belief and this kind of behavior–for example, the kind of people who write books such as this one and who read them, and, surely, those who espouse learning. Most humans have lived and still live content with believing and doing things as they have always been done, for the reasons they have been done, because, for them, that is the way things are and should be: Their myth system affirms this, and their ways of life confirm it.

When humans were evolving, the survival depended on attention to current conditions, familiarity with the local turf, and local group membership. Even now, this still is the preferred state of mind for most persons and groups most of the time. We are indeed creatures of habit—and habit, after all, can be seen as the residue of learned successful behavior for survival. So it is not surprising that it takes a special culture, a special mythology, to overcome these evolution-instilled tendencies (Robinson and Tiger, 1991; Hebb 1954; Barkow, Cosmides, and Tooby 1992). Schein's (1992) work documents that change in organizations requires disconfirmation of the adequacy of the current situation, anxiety about the consequences, and the psychological security needed to attempt change.

Sociocultural Constraints

Individuals, groups, organizations, and nations act according to the norms, history, and processes of their culture's myth systems. Implicitly and explicitly, these myths play a powerful part in defining desirable and undesirable ways of being. Generally, the boundaries that express and facilitate these definitions constrain doing and thinking to the "normal" ways. The constraints are all the more effective because, as with all viable myths, the mythic premises are tacit or unexamined and the boundaries seem natural, because they have worked. A culture is characterized by a shared set of tacit assumptions that have been learned through group experiences that have solved the internal and external problems of the group (Schein 1992).

Because learning is a deviation from the norm, the consequences of learning extend beyond the learner, and provides feedback to the learner. The learner must wonder and discover whether the shift in behavior, values, beliefs, and such will be compatible with those of the peer group, loved ones, or evaluators in other salient settings (Sarason 1972). In other words, will the learner be perceived as trustworthy? To the extent that a learner depends on peer group approval, and most of us do profoundly, learning new ways of doing or valuing may be threatening or rewarding to others and therefore to the learner. Our national culture asserts a belief that each has a right to one's opinion. But if learning another opinion or behavior threatens one's friendships and economic interdependencies, then one is less likely to undertake learning to learn or to practice such a change. One person's reluctance to risk rocking the boat can be obviated by group learning, but this, as we shall see, presents its own challenges.

Our belief that we are independent agents deters us from recognizing how very much our beliefs and behavior, our ways of evaluating persons and events, are shaped by our myths and our habits. Our prevailing beliefs about what is "naturally" worthy of aspiring to and doing—our myths—and reflexive dependence on what has worked well before—habit—are attractive ways to leave our minds uncluttered, our behavior reliable, and our anxiety levels low. Institutions and organizations depend on just those attractions to appeal to stakeholders and membership. Thus learning, which mostly upsets beliefs and habits in individuals and in organizations, is hardly likely to be embraced easily and enthusiastically, even though there is a growing, and sometimes powerful, recognition of the need for change.

One way or another, personal, community, or organizational myths tend to maintain themselves and to resist change—learning—unless they face extraordinary threats to or opportunities for survival.

Emotional Constraints

The recognition of a threat to oneself (or to one's organization) is accompanied by an acutely uncomfortable feeling of vulnerability. One's very sense of self is vulnerable because one's learned successful behaviors or habits now appear insufficient to protect that sense of self. One way or the other, we each have learned to meet our conscious and unconscious needs through a pattern of habits that define who we are—to ourselves, to those we work with, and to our relations. These needs include security, meaning, and acceptance.

What is more, many persons in this culture, beset by fear, rage, and distrust, are driven by demanding, mostly unconscious wishes for an ordered or disordered world and for power and control. Not everyone is driven by destructive needs, nor do such needs invariably result only in bad consequences. But it is essential to recognize that both private and public documentation reveal such unconscious destructive drives are operative in persons in all sorts of authority positions, including those in high-minded public and private organizations. It is unlikely that the various contenders in issues about ecological management, with its implications for who has power and control, are devoid of such drives.

It follows that traditional, short-term, locally focused beliefs, norms, and forms of actions are likely to be held tightly because they tend to reinforce traditional sociocultural ways that provide the sustaining context for one's learned positive self-image and for meeting personal security needs, including, most important, whom one can trust. Moreover, learning longer-term and more systemic perspectives will almost surely expose unsettling questions about the integrity of the self and organization that have been ignored, denied, or simply not recognized.

Although the *results* of new learning may do a better job of meeting either or both constructive or destructive needs for an invulnerable sense of self than what one has already learned, the *act* of learning may compound one's feelings of vulnerability. Will, in fact, the new learning meet those needs? During unlearning and relearning there are certain to be indeterminate periods of disruption, confusion, and changing circumstances. And in each mind, consciously and unconsciously, questions arise. What must I forgo, while I learn, that rewards and gives meaning to my life now? If I learn these new ways and the beliefs that accompany them, what will become of friends, security, power, status, opportunities for advancement? Or of the deep, comfortable convenience of familiar habits of mind and action? For persons rewarded by their current setting, the threats of unfamiliarity, of being vulnerable, and therefore of deep personal risk are very great, whether they are politicians, executives, administrators, grassroots organizers, true believers in the market and/or environmental sustainability or in one or another social or personal "right."

A typical response to psychological threats such as those elicited by the requirements for learning is to deny that learning is necessary because the opposition will not see reason. It's *their* fault. They want to change the boundaries or they do not want to change the boundaries (whether the boundaries are market opportunities, jobs, environmental protection, tax levels, time frames, or contested "inalienable" rights). Of course, these circumstances may be real, but they can also deflect attention from the profound personal risks involved in being a learner of other ways to perceive and act (e.g., on the time–space relationships between humans and the environment). If the focus is only on ideological and operational matters, then no attention will be given to providing the psychological safety essential for coping with the personal psychological risks that arise when participants are confronted with the need to incorporate learning into the methods for dealing with the challenges of ecological management.

Of course, some may sense in a threat to their organization opportunities to better fulfill their psychological needs by relearning. And, as I wrote earlier, personal, community, or organizational myths may change if people sense an extraordinary opportunity, especially if it is combined with an extraordinary threat to survival. Moreover, the learning opportunity may seem compatible with learned successful behavior already in place. In such circumstances attempting new learning may not of itself seem threatening and may seem worth the risk. It is later, in the midst of the relearning, that the threat to self and organization may become apparent, as it almost always does.

In short, given our snarled societal circumstances, and the beleaguered souls already caught up in them, it will take careful preparation and attention to process as well as purpose if real learning is to be realized. But there is yet another source of constraints that pervades those situations where we must learn how to manage the interaction between humans and the natural environment.

Cognitive Constraints

The cognitive aspects of learning have to do with understanding, using, and evaluating formal information. These cognitive aspects include methodologies for collecting and analyzing data, envisioning simultaneously acting cause–effect loops, evaluating multiple interpretations of the meaning of the data, and specification of policy alternatives.

According to conventional mythology, more information is expected to make learning new answers easier and decision making more decisive. Ironically, just the opposite usually happens. Although not so often the case in the natural sciences, more information about the human condition generally leads to more uncertainty. (Consider that more information about the deeply dangerous condition of nuclear reactors in the former Soviet zone makes current decisions pertaining to the future of nuclear power more uncertain. Given varying dependencies on nuclear power, how will people and governments react and interact if another accident occurs?)

Uncertainty increases when a systems approach is applied because the demonstrable or suspected relationships proliferate. In the ecological realm, as Holling's work reveals, it may be that not everything is connected to everything. But in the human realm everything either is, or can be made to appear, connected to everything else. As a result, usually more information tells us that we need still more information to interpret what information we do have (whether it regards toxic substances, ecological protection, economic projections, welfare policy, social impacts from the greenhouse effect, or the consequences of changes in procedures for public or private decision making).

Another type of information, increasingly pervasive, serves directly or indirectly to encourage doubts about the reliability of sources of information. Consider, for example, the recent documentation that much of what is popularly believed regarding waste disposal is, in fact, misinformation and self-serving oversimplification (Rathje and Murphy

1992). Most non–natural science "data bases" are not reliably mappable into one another. Therefore the consequences flowing from the application of feedback from one or another data source are not reliably predictable. For example, different economists frequently offer multiple, contending predictions that derive from economic information selected to fit their preferred theory. This leads to looking to other than data-based criteria for choosing and acting, which further increases uncertainty.

Persons and organizations view information from their personal and peer-shared myths and boundaries. More information provides an ever-larger pool out of which interested parties can fish differing positions on the history of what has led to current circumstances, on what is now happening, on what needs to be done, and on what the consequences will be. And more information often stimulates the creation of more options, resulting in the creation of still more information.

Arguably, the least commitment people bring to problem solving or problem defining is a commitment to let the information "speak for itself." This is not to say that people do not believe there is an objective truth, but rather that each interest group lays claim to that truth. By its emphasis on rationality as submission to the "facts" or "data," our conventional mythology tends to avoid acknowledging that "objectivity" unavoidably reflects the interests of the perceivers, shaped by their sociocultural, emotional, and cognitive constraints.

Learning to perceive and to evaluate the "facts" differently, including experiencing them from the "rationality" of other interests, and then learning to act differently with regard to them confronts participants with all the threats and vulnerabilities deflected by their conventional habits of perceiving and acting. And, if they can face them, they are also confronted with the very uncomfortable questions, "Where, then, is the truth? Where can I stand with certainty?" As I described early in this paper, a growing body of social interpretation asserts there is no such certainty in the realm of human events (Anderson 1990).

Indeed, in our current world situation, opening oneself or one's group to a larger "data base" reveals the terrifying prospect that the world is now so complex that no one really understands its dynamics and that even rational efforts tend to be washed out or misdirected by processes not understood and consequences not anticipated. Of course, as suggested earlier, those intent on pursuing their interests seldom can risk sociocultural ostracism by acknowledging this to others, and usually not even to themselves. Hence there us a need for a learning approach that acknowledges this situation by being both humble and compassionate (Michael 1989, 1991b; Dror 1994).

In a way we know some of this when we anticipate that politics or other interests will degrade or compromise efforts to establish another viewpoint on which to base policy and actions relating humans and their environment. These surely operate consciously, but I want to emphasize here that there are also powerful unconscious forces at work. By emphasizing these unconscious contributions, I hope to make clear why relearning so often faces steep barriers, even when all the participants believe their intentions are benign. This is all the more so, given the social contexts that reinforce the very constraints or barriers that need to be overcome.

Implications for a Learning Approach to Ecological Management

First, I will briefly summarize the key points already made. The very circumstances that make a learning stance necessary also make it difficult to practice. The turbulence and unpredictability created by an increasingly dense information society result in a rapidly changing context and greater uncertainty because they provide the basis for more creativity, both beneficial and destructive. The consequent requirement for continuous learning implies further threats of vulnerability and challenges to a sense of competence cherished by persons and organizations.

These same circumstances also reduce the slack in the social system. Persons, threatened by increasingly permeable boundaries of all kinds, fight to preserve them, either to keep persons and events out or to sustain those inside. These same persons are confronted by formidable social and natural realities that cannot be denied or successfully displaced onto others or into an indefinite and vague future. As a result, the power of our conventional mythology weakens, and challenges to the boundaries that sustained the mythology grow. These challenges further weaken the mythology. In turn, this results in more opportunities for learning to perceive the world differently and to act differently, to engender new myths and boundaries and thereby to create new processes for ecological management.

Simultaneously, then, there are more opportunities to learn and more difficulties in doing so. In light of this plight, I now turn to some suggestions for advancing the opportunities for individual and organizational learning in the service of ecological management.

Use the Metaphoric Power of Language

Earlier, I mentioned the impact of constructivist ideas on our understanding of the way we create our social realities, including those surrounding problem-posing, problem-solving, and decision-making groups. We construct our realities for ourselves and for each other chiefly through language (Shepard 1978). It is according to these constructed realities that we discover or avoid occasions for learning.

Language derives much of its influence from its metaphoric content. Understanding the nature and power of metaphors illuminates what I have written so far and is essential for appreciating what follows about how to enhance the opportunities for learning to improve ecological management in an uncertain and turbulent world. Metaphors and analogies are different. Analogies treat one situation *as if* it were the same as another, whereas metaphors treat one situation as the *same* as another. Compare "the ship moved as a plow through the waves" to "the ship plows the waves." Analogically, the intent is to say that a ship moves through waves as if it were a plow moving through soil. Metaphorically, the intent is to say that the ship and waves have become the same as a plow and soil. The analogical images are expected to match only in a limited way, but the metaphorical images imply total transformation/identity. Each figure of speach is a powerful way of reperceiving and relearning. But since metaphors imply no limit they can enter the psyche more deeply and unquestioningly.

> The essence of metaphors is understanding and experiencing one kind of thing in terms of another. . . . What we experience with a metaphor is a kind of reverberation down through the network of entailments of other metaphors that awakens and connects our memories of our past . . . experiences and serves as a possible guide for future ones. . . .
>
> Because so many of the concepts that are important to us are either abstract or not clearly delineated in our experience (emotions, ideas, time, etc), we need to grasp them by means of other concepts that we understand more clearly (spatial orientations, objects, etc.). This leads to metaphorical definitions in our con-

> ceptual system. . . . What is real for an individual as a member of a culture is a product both of his social reality and the way that shapes his experience of the physical world. Since much of our social reality is understood in metaphorical terms, and since our conception of the physical world is partly metaphorical, metaphors play a very significant role in determining what is real for us [Lakoff and Johnson 1980].

Metaphors are therefore powerful means for defining boundaries (barriers) and spanning them (bridges). Metaphors reinforce entrenched views of what is real, true, important, or trivial. And, if felicitously chosen, they can ease reframing of issues and actions. They can change the atmosphere in which issues are debated and actions selected and implemented. Metaphors can change expectancies and evaluations.

In the dominant Western construction of reality, war has been a pervasive metaphor for describing peacetime activities and goals, as seen in such phrases as "target audience," "victorious political allies," "defeating the opponents proposal," "attack and destroy the causes of the problem." (Note that sports, the other popular source of metaphors for describing civil society, depends on war metaphors.) These metaphors tacitly emphasize we/they, before/after, winner/loser, beginning/ending, fixed boundaries in time and space, and relationships that map poorly onto the amorphous information world I have been describing and onto the fluid ecological environment. These metaphors encourage the maintenance of boundaries that contribute to our inadequate responses to the problems and opportunities of ecological management. But it is usually by these metaphors (data never stand alone) that activists and policy makers present their proposals.

Sooner or later those involved turn to the media to bolster their case or to undermine others. They do so partially because their spokespersons have also been educated in this war/sports terminology and therefore think and voice their thoughts in its terms. It is very important to realize that they wish to speak the language of the media, whose members also think in these terms and who know that their audiences do too. Therefore much of what the media convey maintains beliefs that are based on boundaries metaphorically reinforced by a now inadequate construction of social reality. A mythology that emphasizes learning requires metaphors that are compatible with the purposes and experiences of learning in a fluid, amorphous world.

Here are examples of categories of metaphors that might serve well in exploring, arguing, and structuring ecological management. These metaphors construct an amorphous, problematic, information-rich world of multiple myths described by such words as reciprocal, resilient, circular, emergent, development, ebb and flow, cultivate, seed, harvest, potential, fittingness, both/and.

There are metaphoric potentials to be derived from the domain of biological growth and development that are also applicable to social/environmental policy. Among these concepts are *things take time, nurturing, maturing, growth, fulfillment, decline,* and *death.* The unpredictable and the unique always accompany the expectable and reliable. Other metaphors reveal patterns of transformation: emergence from seed to bud to bloom to fruit to decay to seed.

There are the metaphoric potentials from the realm of ecology with its concepts of interdependence (you can't do just one thing; diversity, resilience, competition and collaboration, carrying capacity, vulnerability, cyclicity, continuity, and, again, time for development).

There are metaphoric potentials in music: not contradiction but counterpoint, dissonance, harmony, mixed voices and instruments, themes and variations.

Embracing all of these are the learning-related metaphors: discovery, exploration, adventure, questing, knowledge, insight, new experience, risk, vulnerability, error, accomplishment.

Coercion aside, we influence each other through the stories we tell. These stories can be in the form of technical reports rich in theory and data, that is, metaphors that, in this culture, imply authority and objectivity. But the stories that apparently have the strongest influence are those rich in gut-level metaphors that elicit feelings of fear, hope, security, threat. (Of course, for those who live in the culture of science and technology, stories in the form of technical reports can also elicit such feelings.)

At all stages of learning the deliberate or inadvertent metaphoric content of the language used to talk about ecological management issues will shape the story told and the story that unfolds, thereby affecting subsequent learning activities.

Use Myth Reinforcement to Encourage Learning

Those who are products of the Enlightenment tradition have access to mythic fragments that might be amplified to make learning the natural and right condition of human beings. Here are some examples. Learning the right answer is important, but more important is learning to learn, to change, and thereby to grow as a unique human being. Knowledge is supposed to make humans free. but it must be discovered, absorbed, and used to do so. Knowledge is power but it is *new* knowledge or newly organized knowledge that offers the greatest potential for shifting or maintaining power. Science provides much good in our lives, and scientists are members of a community of learners. We admire explorers, and explorers are brave learners about different and often dangerous circumstances. In that subculture of the Enlightenment known as the United States, we pride ourselves on being pragmatists, problem solvers, which means looking for better ways to do things and implementing them—learning. Great artists and athletes practice every day; they actively learn how to perform better.

There is another mythic construction that merits development and it can reach beyond the Western Enlightenment myth in the service of ecological management learning. Here three stories can be told with the same basic "plot": the story of ecological processes, that of formal systems theory, and that of the spiritual oneness of all. It is the myth of connectedness, of interdependence, of mutuality. Persons and groups inclined toward one of these stories could, through the use of appropriate metaphors, come to understand the other stories and thereby learn, perhaps, how to construct a shared perspective appropriate for new ways of ecological management.

Acknowledge Uncertainty and Embrace Errors

Ecological management is concerned with the future, the comparatively short-term future in which policy and practice are created and the long-term future in which they are implemented. The future is inherently uncertain; essentially everything is subject to unpredictable change, especially because we understand so little about the present. Uncertainty, then, is the setting for every extrapolation, assertion, policy, plan, and interpretation of outcomes. Moreover, as described earlier, all social data are uncertain.

Nevertheless, our myths and their associated norms that define what it is to be competent and powerful require that we convey an image of certainty. Conventionally, we hide our uncertainties in public intercourse and we hide the errors that naturally arise in a complex, unpredictable world. To persist in acting out this myth guarantees that no learning can occur. There is nothing left to learn from.

Learning requires that as we construct data, policies, plans, and expected outcomes, we recognize and acknowledge the associated uncertainties. When uncertainties in the outcomes of proposed policy and action are acknowledged, perceived risks and vulnerabilities increase. However, options and the opportunities for resilience also increase (Michael 1973).

Taking advantage of the opportunities to learn from uncertainties requires constructing procedures that *embrace error*. That is, specific uncertainties can be used to design error-detecting capabilities that reach out to embrace the human and natural environment as it emerges over time. In this way a person or organization can see how the uncertainties are unfolding and can adjust the system accordingly. Error embracing is a means for alerting the learner to opportunities to adjust proactively, to be resilient. Of course, learning what kind of error-embracing system works is in itself a learning task.[2]

Minimize the Learner's Sense of Vulnerability

All who participate in change via learning have to deal with their feelings of vulnerability, that is, with fears that in the process of learning new perspectives and actions they will lose their sense of self and status and that what they subsequently undertake may fail. Experience leads me to believe that a major source of failure of efforts to construct learning groups and organizations arises from the distrust created by not acknowledging that such feelings of vulnerability affect *all* the participants, regardless of their role or status and by not explicitly trying to ameliorate them.

The first step in reducing fear and distrust is to create an atmosphere that acknowledges that such vulnerability is part of the experience of being learners. (Once it was acknowledged that fighting men suffered fear in war, the incidence of emotion-based incapacities dropped.) It is counterproductive to pretend that concerns about power, turf protection, personal status, and role security do not matter in ecological management. Sometimes, once these concerns are acknowledged as natural, they become overt issues in the discussion and are explicitly dealt with. If they remain taboo topics, then the distrust engendered, because all involved know these concerns have a major influence, and really can weaken and quite possibly destroy efforts to build responsive learning environments. And without trust learning seldom happens (Michael 1973; Fairtlough 1994).

People, whether they are involved as private citizens or as members of organizations, never operate only in terms of rational evaluations of the "facts." It is important not to pretend that the learning process involves only one kind of rational consideration. Factors such as protecting organizational turf or political expediency also have an effect, as do those unconscious needs described earlier. To conduct problem-solving or problem-posing tasks as if only objective considerations are operating or valid is to create distrust about the whole enterprise—everybody knows it isn't so—thereby undermining the openness required for learning that acknowledges uncertainties and embraces error.

Use Facilitators Rather Than Chairpersons

Events such as public meetings or any organization-based learning situation should benefit from the skilled guidance of a facilitator rather than a conventional chairperson. Complex group processes are always present regardless of the topic or the sophistication of the participants. Conscious and unconscious motives ineluctably play their part. A mere chair, unfamiliar with the power of group dynamics, generally adds to the frustrations and obscurations of the group's task, usually without either the chair or the members recognizing that it could be otherwise. (One of the signs of egregious incompetence at most international meetings is traditional resort to a chair, with the resulting traditional frustrations. Of course, many such meetings are not intended to move matters very much, much less be occasions designed for learning!)

There are well-developed and tested techniques for facilitating a group task, and persons can be trained in these skills. To overlook this requirement for effective learning is to ask for trouble. Some organizations have so benefited from the advantages of group process facilitation that they retain such skilled people on their staffs, shunting them around as needed.

Introduce Training of Group Process Skills

Learning to learn depends on learning skills that enhance task group behavior. Ineptness and obfuscating habits can be replaced with learned constructive skills (Michael 1973; Schein 1988; Gardner 1991; Fairtlough 1994). Use of a facilitator is among these skills—a group must want to be facilitated for a facilitator to succeed—but the effectiveness of that role and of the group is vastly enhanced if other skills are present as well. Ecological management topics and tasks are rich in passions, beliefs, speculations, and conflicts all ensnarled with technical information. Learning is a difficult task at best, and learning interpersonal skills is usually a necessary prelude to other types of learning in which persons must work together. Many persons and organizations have learned such skills.

In Western culture we have difficulty engaging feelings when we are involved in so-called rational tasks. We tend to suppress them in ourselves and sidestep them in others. A common way of doing this is to limit the task discussions to input claimed to be "the facts." But, as described earlier, facts are always the product of selective perceptions, beliefs, and interests, and these are always undergirded by feelings. If the pretense persists that we are dealing only with the facts, the group activity becomes inauthentic and commitment wanes because, covertly, the participants know that their feelings are crucial if their interests are to be adequately engaged in the task.

One way suppressed feelings and, accordingly, suppressed self-image emerge is by interrupting a speaker. Without appropriate skill training most persons are dismayingly, even destructively, poor listeners, especially when the listener also has something to say. As a result, the speaker's thoughts remain unresponded to, the speaker feels not understood, and the conversation meanders. Men especially tend to interrupt (more so if the speaker is a woman), thereby sacrificing the uncompleted thoughts. What is more, the speaker's anger about being interrupted is usually suppressed at the time, but it will emerge subsequently, usually by the first speaker's interrupting another speaker or ignoring another's words.

We lack skills at giving constructive feedback to participants who are muddling a task for the group. Threats to self-image and favored beliefs and habits and the requirement to absorb other perspectives often elicit behaviors that interfere with the group task that is supposed to result in new learning. Interrupting is one such counterproductive response. Others include withdrawal from active participation, resistance to every suggestion, long-windedness, putting down other participants, and scapegoating. Groups tend to put up with such behavior because they know of no ways to confront the person without precipitating even more negative responses. This may indeed be the result, but if group members have learned the skills of giving and receiving feedback, the outcome may well be felicitous.

Because the challenge of learning is also a challenge to one's sense of person—"Who am I?"—it is imperative that learners be conscious of their need for a support group. This special support group is not for reinforcement of a learner's substantive position. It is intended to provide a way of sharing fears, worries, dilemmas, and hopes about the anticipated or experienced changes in oneself with others and to facilitate contact with caring individuals on whom one can depend for empathy. It can be comprised of one person or more, nearby or distant. In my experience, women seem to find it easier to seek this kind of support, whereas most men find it a strange and somewhat threatening interpersonal skill to be learned. This is all the more reason to emphasize its necessity! As more than one learner has observed, "It's cold out there!"

Provide Short-Term Reinforcements/Rewards

Ecological management inherently extends over a long time period, whether to implement or revise policy and program. Continuous learning and relearning under changing circumstances are burdensome. Given a long time frame, the rewards of acknowledgment, appreciation, and accomplishment tend to be indefinitely deferred. Yet to keep the will to learn alive and well there must be rewards, or habit and stasis will reemerge supported by *their* inbuilt, comforting rewards.

Learning for ecological management involves two types of organizations: those usually called bureaucracies (no invidious meaning intended) and task groups set up outside bureaucracies (sometimes at the initiation of the bureaucracies) to affect their policies or conduct. These latter organizations may be short-lived or long-lived. If they are long-lived, their members, as well as those in the bureaucracies, need reinforcement and rewards to maintain their learning mode. As with learning persons, learning organizations benefit from support groups that are familiar with the difficulties that accompany this way of operating and that appreciate the effort.

Thus rituals must be invented that regularly provide recognition and reinforcement for learning. These rituals should involve other stakeholders outside the aforementioned organizations. There is a need for community acknowledgment of the difficulties and desirability of ecological management and the legitimacy of the contending interests in this management. And criteria and rituals are needed that recognize and acknowledge when the situation is moving toward meeting the challenges that make the efforts at learning worthwhile. Simultaneously, it is necessary to acknowledge the risks taken to make the personal and organizational changes that accompany learning. Authentic rituals can be deeply meaningful unifying and vivifying events, and thus require careful attention to their design to avoid their becoming mere cliches.[3] A learning stance becomes them also.

Because ecological management is inherently a long-term activity, organizations must make it an explicit priority to learn how to *remain* learners. Organizations must invent roles and functions that aid this effort. Current examples include ombudspersons and resident critics chosen by insiders plus outside stakeholders and changed every year or two. These are error-embracing activities designed to detect deterioration in a culture of learning (Fairtlough 1994, chapter 9).

Reinforce the Learning Mode by Becoming Educators

Good teachers learn partially through their efforts to learn how to teach better. In part, good teachers learn by listening to (and being educated by) their students. If task force participants are educating members of a bureaucracy or vice versa or educating other stakeholders or learning from them, they too, as educators, must be good listeners, ever ready to learn. By consciously incorporating the role of educator all involved can be encouraged to remain in a learning mode.

Especially important for encouraging others to learn is the opportunity, indeed the necessity, for educators to *model* the kind of learning we have been exploring here; they should practice what they preach.

It is also especially important for educators to choose more appropriate metaphors. By introducing new metaphors educators can affect the language of stakeholders and, by so affecting it, change the overall context within which the social reality is constructed that defines the ecological management challenges.

Use Disasters and Crises as Learning Occasions

Human-made and natural disruptions are here to stay. Habit-governed organizations tend to avoid thinking about such possibilities or to limit thoughts to what damage would be done and what resources would be needed to cope. Seldom are potential crises or disasters *anticipated* as an occasion for learning and for initiating learning-governed organizational behavior. Organizations should be asking questions such as: How would we do things differently if we had the chance? If disaster A or crisis B happens, what could we have educated about beforehand that might open new possibilities for constructive organizational and stakeholder learning *after* the fact?

This approach provides for the seeding of ideas (and metaphors) that might grow in the event of the crisis or disaster. Such occasions disrupt the myth-sustaining social context, thereby opening opportunities for change.

Since the time frame for ecological management is long and the variety of disasters that might occur is large and plausible, attention to such opportunities should not be neglected. Scenario construction and gaming simulation are powerful means for enhancing such thinking and, in the process, encouraging an ongoing learning stance (Schwartz 1991). (And, of course, it is not unknown for those intent on change to manufacture a crisis or the threat of one.)

In Closing

In a short story the Canadian writer Stephen Leacock described his hero as leaping onto his horse and "galloping off in all directions." And so it might seem I have done in this chapter. But what was humor in his story is a serious necessity in mine. We are challenged to cope with the imperatives and difficulties involved with becoming learners intent upon creating and implementing ecological management in a world that *is* galloping off in all directions!

There are two kinds of learning: one for a stable world and one for a world of uncertainty and change. Learning appropriate for the former world has to do with learning the right answers and learning how to adapt and settle into another mode of being and doing. Learning appropriate for our world has to do with learning what are the useful questions to ask and learning how to keep on learning since the questions keep changing.

There will be no place to settle down, no time to stop asking. It is this kind of learning that befits ecological management, the kind that befits our human ecology as it struggles to sustain the natural one.

Acknowledgment

I am grateful to my friend and colleague Edgar Schein for his insights and suggestions pertaining to an earlier draft. I am of course responsible for how I have used them.

NOTES

1. I have extended the implications of three types of constraints from the initial delineation of Schein (1993).

2. Fairtlough (1994) establishes the necessity for and rewards of small, bounded work communities that depend on the creative use of large information flows. Such compartments require internal openness, which, in turn, engenders trust—a felicitous circle. This circle then stimulates reciprocal empowerment and commitment.

3. In our current culture, serious (i.e., uncontaminated) rituals play a comparatively minor part in giving meaning and motive to our lives. Yet their role in validating the meaning of words can be profound (Rappaport 1971). There is, however, a ubiquitous activity in this society, surely relevant to ecological management, that, although intended otherwise, too often becomes empty ritual: legal pronouncements. Consent decrees and legislation often do not result in effective implementation. Ongoing evaluation and learning-based change too often lose out to incentives to maintain the status quo. So what might seem a reward for persistence in efforts at ecological management may be merely a contaminated ritual (DiMento and Hesterman 1993; DiMento 1986).

Part 4

Synthesis

12

Barriers Broken and Bridges Built: A Synthesis

Lance H. Gunderson, C. S. Holling, and Stephen S. Light

The regional examples presented in previous chapters provide a remarkably diverse set of experiences as the author/practitioners of those examples discuss patterns exhibited in the dynamics of coupled human and natural systems. Individually and collectively, the cases expose barriers faced by scientists, resource managers, policy makers, and citizens of the world as they grapple with the complexities of social and ecological systems as well as bridges developed to overcome some of these barriers. The case studies reveal patterns of convergence and consistency, followed by quantum change and fundamental reorientations, as existing strategies, myths, and understanding become overwhelmed by endogenous and exogenous events and trends (Leifer 1989; Tushman and Romanelli 1985). The patterns in general seem to follow the four phases described in the introduction to this volume: exploitation, conservation, crisis, and reconfiguration. The patterns are common, regardless of how big or complex the system, or whether the basic components of the resources are aquatic, terrestrial, or somewhere in between. In this chapter we attempt to compare and contrast the six case histories to examine, first, linkages between observed patterns and theory in a search for an "inner logic" (Morgan 1987) and, second, barriers and bridges that have impeded or promoted restoration or maintenance of the key resources of a region. But first we will indicate how this book intersects with concepts of adaptive management that over a decade ago set the stage and motivated many of the case studies presented here.

The foundations for this book can be found in a sequence of works that have outlined, tested, and expanded the theory and practice of adaptive environmental assessment and management over the past two decades (Holling 1978; Walters 1986; Lee 1993). The first exercise in adaptive management was the Gulf Island Recreational Land Simulation (GIRLS) study in 1968, where the participants attempted to explore ways to bridge gaps among scientific disciplines, technical experts, and policy designers (Holling and Chambers 1973; Holling 1981). Holling, Walters, and their colleagues (Holling 1978; Walters and Hillborn 1978) introduced the concepts of adaptive management and gave the results of various early attempts at implementation. Walters (1986) presented theory and methods for dealing with the uncertainties of managing renewable resources in what is becoming a classic book on adaptive resource management. Most recently, Lee (1993), in lucid and engrossing language, related the experience of using adaptive management concepts to guide decision making in a social and political arena, in a marriage of social learning.

The basic tenets of adaptive management all deal with the unpredictable interactions between people and ecosystems as they evolve together. Ecosystems change, as do the people that attempt to understand and manage the system. Three groups of people commonly interact in the case studies presented here: the resource managers who must manage and make decisions within a framework of existing policies and partial knowledge, scientists who attempt to understand and communicate the systems dynamic, and the citizenry who benefit from or must endure the policies and results of management. The adaptive approach uses a variety of tools to share and communicate understanding of resource issues, to expose key uncertainties, embrace alternatives, develop robust policies, and use their consequences to modify and adapt policies and actions further.

The primary expectation of adaptive management is the unexpected. That is, systems are unpredictable. Natural systems evolve and coevolve with management and change. Surprises are inevitable; hence polices must always be adaptive. Some surprises come from outside. The hurricanes that impact the Everglades, for example, arise in the Atlantic off the coast of Africa, and they are modified in turn by El Niño fluctuations in the Pacific. Similarly, the sugar price supports that foster sugarcane production in Florida are part of a foreign policy response to Fidel Castro in Cuba. But other surprises are generated inside the system as vulnerabilities grow because both the natural system and people evolve as a consequence of myopic development. The eutrophication problems associated with agriculture in the Everglades and associated lawsuits are such an example of an endogenous surprise. Surprises turn into major crises when extreme external events intersect with internally generated vulnerability.

Adaptive management is an inductive approach, relying on comparative studies that blend ecological theories with observation and with the design of planned interventions in nature and with an understanding of human response processes. This book is a continuation of this approach.

In the score of years since the principles of adaptive management were first conceived, it has been applied by a variety of practitioners in a wide range of regional resource and environmental management situations in North America, Europe, Asia, and South America. The examples chosen for this volume cover a range of histories of development regarding the adoption and practice of adaptive management: from the minimal (Chesapeake, Baltic Sea), to the developing (Everglades, Great Lakes) and the established (Columbia River and New Brunswick forests). The prime reason for their choice was to represent a range of complexity in both the ecosystems and institutions in a progression from New Brunswick forests, Everglades, Chesapeake Bay, Columbia River, Great Lakes through the Baltic Sea.

One purpose of this book is to use a comparative approach to reap the benefits from multiple decades of management experience in some of these systems. In those systems with a limited history of applications, the utility of adaptive management is just being discovered and will probably continue to play an important role in the future. Sanderson's essay defines limits as to the extrapolations from these case histories to those in developing countries, especially areas entrained within global economies and incapable of controlling overwhelming exogenous forces. To appreciate how these case histories chronicle the interaction between ecosystem changes and human responses, we now examine the common properties in the dynamics of these adaptive systems.

Dynamics of Adaptive Systems

Linking Ecosystem and Institutional Changes

In the introductory chapter Holling proposed a set of postulates that describe phases or stages in the time course shown by complex systems comprised of nature, people, and institutions. Since the case studies presented here are limited in number and biased in choice, it is inevitable that we would find agreement with the postulates. That is, in no sense is this book a rigorous effort to disprove hypotheses in the traditions of science. The issues are too complex and the territory of inquiry too new to expect that. Rather, in finding agreement with the postulates, we can also detect surprises, unexpected observations that are consistent with but not predicted by theory, observations that could deepen understanding in order to generate the kinds of puzzles from which hypotheses can be suggested and later tested.

First, then, we will use the case studies as empirical demonstrations of the postulates. That will set the stage for the more important task of identifying revealing exceptions to our original expectations. The first postulate proposed that crises inevitably occur as systems (ecological, economic, or social) mature, producing an increasing brittleness or susceptibility to a crisis that signals, or that ought to signal, the need for restructuring the organization of relationships. The second postulate states that in institutions following the crises, informal collaboration develops among groups of people outside the institutions. The purpose is to develop an integrated and shared understanding from which a variety of options for solving the crisis can emerge. Finally, the third postulate states that specific but transient processes result in a reorganization of the management institutions and society, with new goals, policies, and organizational processes in place.

The postulates are a simplified version of the four-phase adaptive cycle as described by Holling (1986, 1992, and chapter 1). The first postulate describes a transition out of a phase of brittle conservatism through some sort of creative destruction (phases 2–3, or K to Ω). The second postulate focuses on the rapid movement of the system toward some reorganization options (phases 3–4, or Ω to α), and the third postulate describes the activities that occur to complete the shift from reorganization options to the appropriate conditions for another exploitative phase (phases 4–1, or α to r). These postulates are qualitatively met in each of the case studies, as shown in the following examples.

New Brunswick

Since the early 1950s, one major crisis and several minor ones have occurred. During this period new technologies (airplanes and pesticides) were introduced and their use matured. The maturation was along parallel tracks of refinements in pesticides used to control budworm, tree harvesting, pulp mill construction, and understanding of forest dynamics. Minor crises occurred when human health impacts were linked to the pesticides. Key pieces of integrated understanding were achieved by the teams of Morris (1963) and the modelers of the 1970s (Clark et al. 1979). The brittleness (defined by a loss of resilience, *sensu* Holling 1973, together with an increase in institutional efforts to control information and action) reflected the complacency of agency staffs that budworm damage was controlled in an efficient and cost-effective manner and that there was plenty of wood available for harvest. In reality, the costs of using pesticides were rapidly increasing because of increases in oil prices and because of adaptation of pesticide application to public pressure. In addition, available stocks of harvestable trees were decreasing because of past harvests and because more and more mature stands over larger areas were gradually deteriorating from the pressure of moderate but persistent budworm defoliation. The major crises occurred during the late 1970s when a forest inventory report finally indicated that there would not be sufficient stock to support the current mills, thereby confirming an earlier prediction of the models. This led to a new law that restructured the licensing and forest management policies, and freed the innovative capacity of local industries within a regional set of goals and constraints. Adaptive responses among the actors began to develop regional forest policy in a way that engaged industrial, environmental, and recreational goals.

Everglades

Major crises occurred when unsuspected variation in the climate (the hurricanes of 1920s, flood of 1947, drought of 1971) was coupled with a system that had little ability to deal with the variation. Other crises have occurred as latent problems resulted from conflicting land uses with the expansion of development and increased pressure on ecosystems. The greater brittleness of the whole system induced by the human constructs that allowed population growth (earthen levee, lack of large-scale water control structures, and increasing water demands of populations) contributed to each of these crises. During and after these crises, a few key people contributed to the understanding that formed the basis for the "new" plan. Work by R. V. Allison laid the foundation of understanding from which the Central and Southern Flood Control Project emerged in 1947. In 1971 it was the results from an informal symposium that provided the vision for the new institutions developed as the water management districts.

Chesapeake Bay

By the mid-1960s scientific and management agencies noticed significant declines in key indicators of the status of aquatic organisms in the bay. This crisis was perceived and amplified by a large cross-section of the people in the area, who had developed a parental attitude toward the resource and a sense of place that embraced the whole region. This pressure resulted in a series of laws and governmental commitment to scientific analysis of the problems. As the impacts became known (nutrient runoff from a variety of sources, toxins), various intergovernmental groups were created to implement plans for recovery.

Columbia River

Three crises were identified that contributed to a transformation of the rules for management in the system. These crises are set around the complex of electrical generating plants developed along the river. One involved the assertion of Native Americans' legal rights to harvest fish, the second involved the rise in the prices of electrical generation (associated with heavy investment in nuclear power), and the third resulted from greater environmental awareness on the part of citizens in the region. These three crises resulted in federal legislation, the Northwest Power Act, which created a public arena to work out these crises and conflicts, and an explicit prescription to use adaptive management principles and practice.

Great Lakes

A number of issues relating to resource use and environmental degradation were all triggered by crises. Eutrophication from phosphorus and the occurrence of toxic contaminants in the Great Lakes were discovered over a period of some 10–20 years prior to the negotiation of the binational Great Lakes Water Quality Agreement (GLWQA) in 1972 and 1978. The implementation of the GLWQA is overseen by the International Joint Commission (IJC), a body established by the 1909 Boundary Waters Treaty. The spread of sea lampreys into the upper Great Lakes, beginning in the 1920s, led to the establishment of the Great Lakes Fishery Commission in 1955, mainly to undertake sea lamprey control measures. The development of some water control structures for Lake Superior (1921) and Lake Ontario (1958) led to recurring studies of further controls over the levels of all the lakes almost every time there are periods of relatively high or low water levels.

Baltic Sea

Starting with the 1950s, crises were recognized in the Baltic similar to those that have occurred in the Chesapeake and Great Lakes systems. These include high concentrations of mercury, other heavy metals and chlorinated pesticides, oil spills, eutrophication from agricultural runoff, and algae blooms. Collapses in the fisheries stocks were also observed and related to both changes in the ecosystem and overharvesting. Scientists in Sweden were critical players in dealing with these crises by developing understanding of causation and linkages between human activities and these crises. The participation of these consortia of scientists helped to trigger international political alliances and regional collaborations.

Even though all the case studies provide evidence of agreement with the postulates, a number of insights appeared during our analysis. One insight was the extreme nature of the recalcitrance or inertia of institutions, and the almost pathological inability to renew or restructure. Chapter 11 indicates that this is not a surprise, because humans create habits and structure for a number of reasons. This institutional rigidity has typically led to failure and resulted in a shift in discussion and actions from agency-controlled arenas to political ones where new institutions were created. Once this transition is made (e.g., new regulations in New Brunswick, creation of the water management districts in the Everglades), the story continues with little resolution.

Another insight was the identification of three critical roles by individuals: a visionary activist (such as Art Marshall in the Everglades); a wise, patient, respected integrator (Gordon Baskerville in New Brunswick, Steve Davis in the Everglades, and Bengt Owe Jansson in the Baltic); and rebel bureaucrats (such as Steve Light in the Everglades or

Jack Vallentyne in the Great Lakes), who play roles described as "novelty detectors" or "loyal heretics" (Trist 1983). The latter two groups at key moments speak "truth to power" (Wildavsky 1979), ignoring personal penalties.

The final insight was in the inability of public panels to function with the same efficacy as counterparts in the private sector. Perhaps public agencies have no clear ways to measure performance, or perhaps board members are chosen to represent a balance of interests that result in competing extremes that politicize and intensify gridlock. Often a board member's agenda is not determined by defined responsibilities for the good of the agency or the region, so that responses become dominated by a personal agenda or one driven by a specific constituency. This last insight mirrors sentiments of Luna Leopold (1990), who stated, "The resource establishment . . . is stuck on the shoals of special interests, lacks a long-term perspective, and has a shortage of public-minded leadership." One antidote for this malaise is offered by Lee (1993) in his erudite work linking adaptive management with bounded conflict resolution within a democratic process.

In spite of these uncertainties, some confident conclusions can be made using the four-phase heuristic: the role and time for key individuals, the persistence of command and control efforts, the development and use of a shadow network. But there is clearly a missing piece that exists in a democratic processs or structure whose discrete form we do not know. Citizen science and the goal for a scientifically literate citizenry (literate in the scientific substance of an issue, its uncertainties, and the ability to understand the process of science) is another part. Citizen juries may well be another. But clearly something is failing and the resolution lies in the society as a whole, a challenge we leave with the reader before we turn to a more detailed description of the lessons learned about the dynamics of institutions and people.

Institutional Dynamics

The four-phase model provides a useful insight into the dynamics of resource management institutions in these examples. Most institutions are established to carry out some set of policies, or mission, and then spend most of their time and energy becoming more efficient in the implementation of these policies. That narrowing of attention causes strategic analysis to wither. The resulting myopia inevitably leads to crises in resource management that seem to occur when expectations are not met or shift with changes in underlying myths (myths as described in chapter 11).

Institutions form the entity from which collective action is taken for a variety of resource management activities (control water levels, harvest trees, spray pests, generate electricity, mitigate health hazards, etc.) to achieve a social or economic goal. Although the definition of an institution can be confused (Gregg et al. 1991), we define an institution as including the sets of rules or conventions that govern the process of decision making, the people that make and execute these decisions, and the edifices created to carry out results. Policy is defined as the principle or plan of action—including laws, treaties, agreements, and other instruments—to achieve goals. At some moment, the policy is no longer viable, either because the original goals are not achieved or because other social goals are compromised or compromise the policy. Political pressure then builds to form new policies. As shown by the case studies and Westley's essay, these changes rarely occur within an agency or specific management group. Indeed, in the successful cases, such as the Chesapeake Bay, institutional changes involve a variety of groups that develop connections both inside and outside the system. Examples from all the case studies support the idea that adaptive, viable institutions exhibit periods of innovation, conservation, creative destruction and renewal. Those that do not adapt become locked into a command-and-control syndrome that can be remarkably persistent.

Activities between the phases of innovation and conservation (phases 1–2, r to K) generally focus on implementing some given policy or sets of policies and increasing the efficiency of their implementation. In New Brunswick, the focus of the forestry sector for almost 40 years since World War II was on increased efficiency of spraying for budworm. In the Everglades the Corps of Engineers and the South Florida Water Management District have implemented the policies of water management as defined by a large-scale federal/state water control project. Prior to the Northwest Power Act, the focus in the Columbia River region was to increase efficiency of electrical power generation and use. In the Great Lakes, environmental regulatory agencies focused on the reduction of phosphorus and a few selected contaminants from effluent discharges going directly into the Lakes, and the Great Lakes Fishery Commission focused most of its efforts on sea lamprey controls.

In many of the cases, a hierarchical bureaucracy is the type of group established to implement the policy. Along with the implementation of policy comes an increase in the monitoring and appraisal of key indicators in the ecosystem. In the Everglades and Great Lakes, an indicator is water quality and levels; in New Brunswick, the state of the insect and harvest; in the Columbia River system, power generation and fish stocks. The monitoring becomes coupled to the implementation; indeed, it becomes the feedback mechanism for determining the efficacy of policy implementation. Great Lakes fish managers become "trapped" into providing fish preferred by client fishermen. Sometimes adaptive behaviors cease after policies are legitimated. Schon (1971) refers to this phenomenon as "dynamic conservatism," or the tendency for institutions to fight to remain the same. And even when policy implementation and evaluation are done within a given bureaucratic framework, most of the time is consumed trying to ignore, bury, forget, avoid, or veer away from recognizing error (Michael 1973). Such resistance to evaluation is part of self-preservation and perpetuation, at least until the next crisis.

Crises in management institutions occur when policy fails. At least two critical factors contribute to the failure: (1) an expectation of one or more target resources in the system is not met and (2) in the myth that provides context for the expectation changes (figure 12.1). Expectations are established by the original policy and refined during the policy implementation phase. Examples of these expectations include bounding the fluctuation of water levels in the Everglades and Great Lakes; maintaining sufficient fishery stocks in the Great Lakes, Baltic, Columbia River, and Chesapeake Bay; and managing sufficient forest inventory in New Brunswick. The expectations are generated at a local space and time scale, and are tempered by experience and study. During policy implementation, the original expectations become a fixed given, not subject to reevaluation. The result is the loss of flexibility and adaptability of the management institution and the loss of resilience of the resource. For example, the strict enforcement of regulation schedules in the Everglades and the minimum water delivery plans created a brittleness in the institutions for dealing with variation outside these standards.

Crises in which myths are transformed are much more subtle and difficult to discern. Examples include the adoption of the idea of the Chesapeake Bay being "everyone's backyard" and the multiple-use concept of the Columbia, including maintaining the fishing rights of Native Americans while also allowing newer uses of the river. Perhaps the one common myth involves the reliance on "technologic solutions" as a panacea for societal problems that has prevailed since World War II.

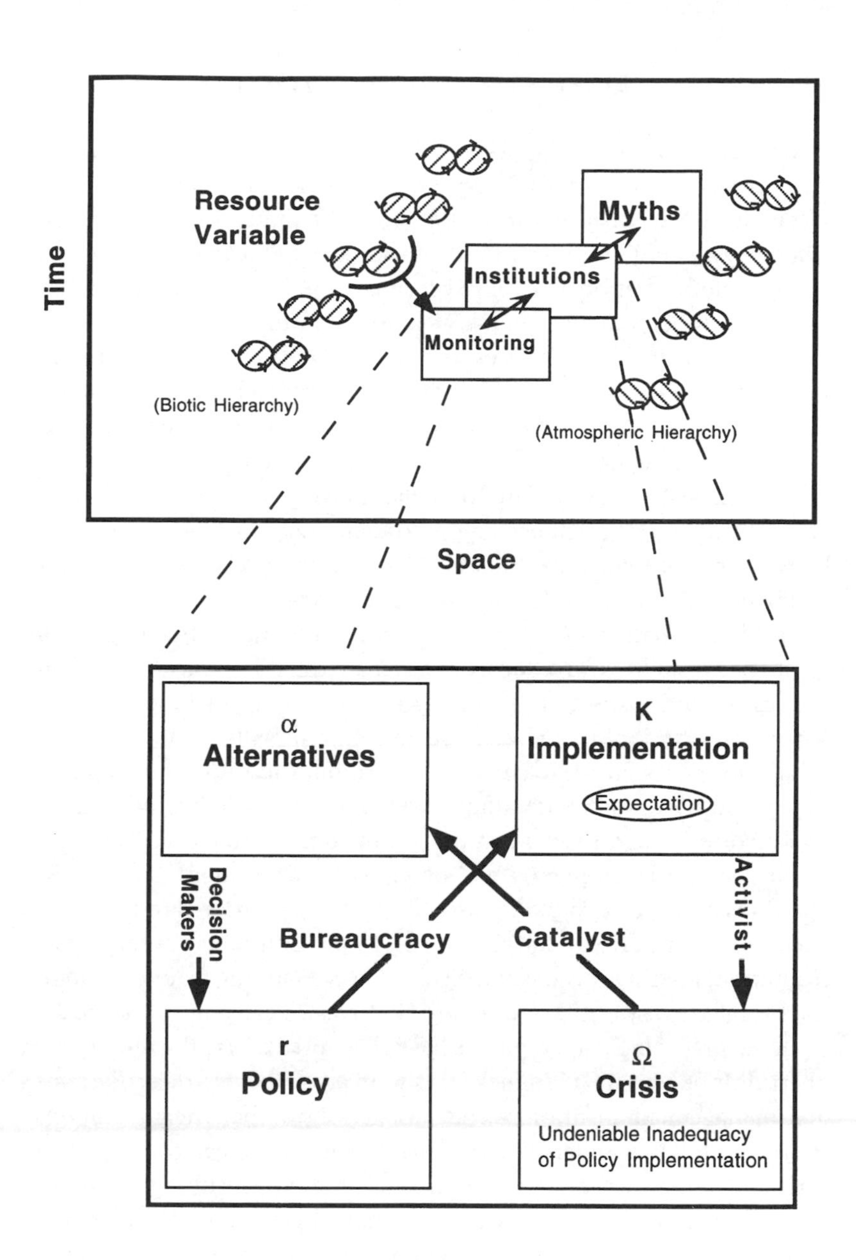

FIGURE 12.1
Nested, interacting adaptive cycles that typify both natural resources (top diagram) and the human institutions established to manage those resources (bottom diagram). The top diagram depicts the relationships and dynamics of key resource variables in a biotic hierarchy (left set of circles), atmospheric hierarchy of structures (right set of circles), and management institutional hierarchy (middle boxes). The bottom diagram is a specific rendition of the four-phase heuristic for institutions.

The aerial spraying in New Brunswick; water control in the Everglades, Great Lakes, and Columbia River; and increased industrialization in the Great Lakes, Chesapeake, and Baltic, can all be linked to the myth that technology would not only increase resource use efficiency, but control all unwanted variation as well! Expectations are linked to other myths, such as views of system stability, variation, and the nature of relationships. As shown by Michael's essay, the relationships between expectation and myth determines, many of the barriers and bridges to sustainability, as will be discussed further later. Now we turn to the next phase of institutional dynamics, the period of renewal and reconfiguration.

The renewal and reconfiguration phase involves the development of alternative plans, and choice among the alternatives. There appear to be two types of reformulation: ones that deal with relatively minor reconfigurations and ones that deal with major reconfigurations. One difference between the two is that the major reconfiguration involves much greater accessibility to resources at space or time domains larger than the region and institutions of the region. Examples include the large federal/state project in the Everglades and Columbia River Basin, where local resources were supplemented by national resources in order to implement new policy with a fundamentally new rearrangement of infrastructure (physical and institutional). In addition to physical changes in the system (dams, pumps, and other constructs), changes in the institutions occur. In two of the cases, dramatic institutional reformation occurred at least once in the recent past. In the Columbia River the new institutional setting was the Northwest Power Planning Council, and in the Everglades it was the South Florida Water Management District.

In the preceding paragraphs we have attempted to use the four-phase heuristic not only to illuminate the coupled dynamics of ecosystems and management institutions, but also to show that the institutions appear to follow similar patterns. Institutions are created or used to carry out policy, crises occur in resource systems, resulting in subsequent institutional transformations. We now turn to a discussion of the groups of people that emerge in these institutional dynamics.

Players in the Institutional Scene

Particular groups or types of people emerge and appear to dominate in the transition among these four phases of institutional change. These groups may always be present or may be ephemeral. Four types or groups appear: bureaucrats, activists, catalysts, and formal decision makers (table 12.1), and each group is discussed as it pertains to the four-phase dynamics.

Bureaucrats carry out activities from the exploitative to conservation phase in the process of implementing given policies. The focus during the implementation phase is on increasing efficiency of the execution of their tasks. For example, bureaucrats focus on self-serving issues and ask such questions as, "Are we doing things right?" rather than, "Are we doing the right things?" Often public institutions become detached from the people they serve; in our own studies, for example, in response to a suggestion for a program of "citizen science," we have heard one statement from a senior manager that captures some senior staffs' basic feeling admirably: "Don't you realize that the public creates these problems?" The insular nature of their operations may contribute to their surprise in the face of the inevitable future crisis and the public's response to it.

Activists are critical to the "creation" of the crises that open the possibility of a shift from the *K* to the omega phase of the four-phase adaptive cycle (table 12.1). This has been one of the major themes of the cultural anthropologists Douglas (1978) and Thompson (1983). Activists identify single issues, real or imagined, that they perceive are inadequately addressed by the institutions. In the cases chosen here, those issues are ones of pollution, threats to human health, or ecosystem deterioration. The activists tenacious focus on narrowly defined goals and their need to identify enemies challenges whatever current goals are, in practice, guiding the institutions. These latter goals typically have not been environmental but social and economic ones, targets that are associated with serving specific influential constituents in industry, agriculture, forestry, fisheries, or cities. The single-issue focus of the activists can be ignored or controlled by the bureaucracy, but if the public can be aroused and the institutional vulnerabilities exposed, a period of crisis can occur that undeniably exposes the inadequacy of existing policy and management.

At that point, the omega phase, a new set of actors begins to emerge. They provide the opportunity to develop one necessary but not sufficient set of activities that can give the cycle an adaptive nature. Crises create the demand for new approaches and ideas. The initial foundation for effective adaptation is formed by developing an integrated understanding of the system, by evolving alternative policies, and by defining

Table 12.1

Attributes of Groups Dominant at Different Phases of Adaptive, Four Phase Cycle

	Phase of Adaptive Cycle				
Attribute	***r–K*** **1–2**	***K–Ω*** **2–3**	***Ω–α*** **3–4**	***α–r*** **4–1**	**A–? 4–?**
Group type	Bureaucracy	Activists	Catalysts	Decision makers strategist	Evolutionary
Activity focus	Self-serving	Insurgence	Unlearning	New learning cooperation	Deep transformation cooperation
Strategy	“Do as before but more”	“Weathering the storm”	“Unlearning yesterday”	“Inventing tomorrow”	
Response to changes	No change	Conflict	Shedding old behaviors	Reframing strategies	Invention
Time horizon	Time of office (linear time)	Present (discontinuous)	Time out (multiple scales)	Near future (multiple scales)	Distant future
Space horizon	Building and holding bounds	Destruction of old bounds	Suspension of bounds	Creating new bounds	
Nature of truth and reality	Constructed	Competing explanations	Discovering what works	Reconfiguring myths	New myths (visionary)

possible futures. Typically, the emergence of a strategic focus has been lost in the tactical and short-term activities of the agencies. In most of the case studies, these alternatives were created by a temporary group of technical people who functioned informally outside traditional institutions but drew upon contacts within them. They make up the transient "shadow network" (Schon 1971) or "epistemic communities" (Haas 1990), where constraints on creativity are removed because of the lack of institutional formality and the self-organized nature of the activity.

The second set of activities that can then actually launch an adaptive lurch into a new regime requires alpha groups that are more formally empowered than the self-organized shadow network. Though often transient, many examples emerge from the case studies of formally established and politically endorsed commissions of various kinds who set upon a task of envisioning alternative futures. The Florida Governor's Conference held in the early 1970s dealing with the Everglades is one such commission. The Great Lakes Science Advisory Board to the International Joint Commission has, from time to time, served as an alpha group. An informal Canadian/U.S. interuniversity seminar has been convened four times since the early 1970s to discuss emerging Great Lakes issues by academics, bureaucrats, and NGOs. Often the alpha groups are empowered to make decisions regarding selection of policy options, and to be effective they need to engage a wide range of the critical people who influence public opinion. Some of these alpha groups generate new views of policy that are effective, whereas others do not. Formally established technical panels such as some blue ribbon commissions that are chosen to provide a balance of interests do not work well because their work focuses on maintaining existing power or policy arrangements. A key ingredient to the success of these alpha groups is in the ability to create credible futures while resolving issues of the past. Therefore these groups truly span the past and future. Successful alpha groups such as the Royal Commissions of Canada (Berger 1978) and Australia (Kakadu Resource Assessment Commission 1991) are able to break the gridlock of the past by creating new policy arrangements. In cases where dramatic changes are proposed, the populace has to be massively engaged, informed, and educated rather than treated as part of the problem!

Examples of all four groups appear in the case studies and are discussed by Westley. The bureaucracies include state, provincial, and federal agencies, such as the SFWMD, U.S. Army Corps of Engineers, and others. The activists tend to be single individuals, such as Tim Wapato in the Columbia case, or small groups like the environmentalists of New Brunswick. The people who facilitate the transition from crisis to reformulation can be either a visionary (see chapter 11) or "shadow groups." The transition groups are in many cases academics or scientists, such as Art Marshall in the Everglades, H. T. Odum in the Baltic Sea, or the Adaptive Environmental Assessment team in New Brunswick. Many of the authors of the case studies (Baskerville, Light, Regier, Francis, Lee, and Jansson) fall into this category. Indeed, much of the activity of adaptive environmental assessment and resource management described by Holling (1978) and Walters (1986), such as the modeling workshops, focus on this transition. The alpha groups (phase 4) come in a variety of configurations. They range from compacts among states or nations (HELCOM in the Baltic, Great Lakes Water Quality Agreement) to elected groups (U.S. Congress, state legislatures) to appointed commissions or boards (Northwest planning council, Water Management Governing Board).

Each of these groups specializes in different sets of activities, strategies, goals and ways of defining and solving problems (table 12.1). Bureaucratic activity is mainly self-serving, involved with completing a mission. Their strategies are to increase efficiency of tasks. The bureaucracies in the regional examples presented earlier are resistant to change during these phases, and define issues and problems in the context of time of office and as the need to maintain and hold boundaries (see chapter 11). The bureaucratic inertia can also be associated with a limited range of professional expertise and a struggle by one group to retain control over problem definition and solution. Activists invoke conflict and insurgence in order to alter these boundaries, and they often view problems in the immediate present (i.e., "Something must be done now!"). The temporary omega groups are geared toward "unlearning" (Nystrom and Starbuck 1984), or forgetting about the past, suspending boundaries, shedding old behaviors, and discovering new possibilities. We believe that the omega and alpha groups develop new learning, transform strategies, and establish new goals. The omega groups that function as shadow networks are clearly seen in the cases, but the form and features of the alpha groups give us the greatest uncertainty. We do see some critical elements. First, there are three types of individuals who play key roles (we'll call them a visionary, respected integrator and loyal heretic). But are there other critical elements that involve engagement of citizenry and broader groups of people. We do not know the details of this process, and refer to Lee (1993) for a more detailed description of "citizen science," but we do see the need for these alpha groups.

We recognize these groups or entities to describe activities, roles, and interactions of these key players in the institutional scene. Each group dominates the scene at different times. These groups interact erratically, emerging at distinct times (e.g., activists during crises), persist for various lengths of time, and then disappear. A theatrical analog would be that the play is much more like an opera than a series of soliloquies. Another facet that emerged from the case studies is that individuals can play multiple roles. The visionaries (see chapter 11) appear to span multiple group activities, perhaps by the judicious use of metaphors for communication. For example, Art Marshall in the Everglades acted in various capacities—as an activist, by injecting conflict; as an omega man, by unlearning the past; and as an alpha group member who reframed new strategies.

We have identified three types of individuals who perform key roles during the omega and alpha phases (the visionary, respected integrator, and loyal heretic). They perform roles that have overlapping features and are mutually supporting. The outside visionary is capable of transforming myths among a wide group of people, spanning a variety of communities—technical, institutional, and political. The wise integrator is respected by players both inside and outside the system and is able to utilize traits of honesty while connecting knowledge to power in spite of countervailing political winds. Examples of these people in the case studies are Gordon Baskerville in New Brunswick, Bengt Owe Jansson in the Baltic, and Steve Davis in the Everglades. The loyal heretic or rebel bureaucrat is critically important in preparing bureaucracies and agencies for change by maintaining strong personal contacts both inside and outside the organization.

These groupings may be helpful for pragmatic managers who would like to translate these concepts into practice. An adaptive organization would likely have all these functional roles, filled by a variety of players. As Westley discusses, the traditional planning-led change process is rarely successful, probably because it is embedded in bureaucracies whose participants, for all intents and purposes, ignore change. Another reason traditional planning is not successful is that it is not perceived as a way to learn (Michael 1973). Organizations need naysayers, rebel bureaucrats who ask not, "Are we doing things right?" but, "Are we doing the right things?" These actors provide the flexibility and responsiveness for an organization and help guide it through omega and alpha phases. Similar functional groups are discussed by Westley in her description of the replacement of traditional hierarchical organizations with more flexible, less structured ones.

The model of adaptive cycles appears to be useful in explaining the dynamics of the resource system under management, the management institutions themselves, and the coupling between the two systems. The coupling is evident in two key subjects: policy and learning. The relationship among policy development, implementation, and failure is discussed earlier, as relating to resource variation and changing myths. The shape, form, and methods of linkages form the conditions for learning opportunities.

Learning Opportunities

We see agreement between various models of learning and the four phases of the adaptive cycle. In chapter 10 Parson and Clark describe the many types of learning from individuals to organizations and derive a parallel between characteristics of biological evolution and social learning, especially in the acceptance of emergent or new ideas. Lee (1993) describes the marriage of adaptive management and the process of political change as social learning. In a simplified abstraction, learning can be categorized as single and double loop (Argyris and Schon 1978). Single-loop learning occurs in resource management during the first two phases (r and K), as existing paradigms are updated with monitoring data and operational experience. Repositories of learning are people (researchers, managers, operators, etc.) and symbolic material. During the crisis and reformation phases, a deeper, or double-loop, type of learning can occur where fundamental assumptions of policy are challenged. This learning is emergent, in that it is rarely predictable before the event (as is most learning; for one may know that one is going to learn, but one usually does not know what one is learning until after the fact).

Learning during the reformation phases appears to have occurred by at least three different modes. Learning can be translocated from knowledge gained in other systems and applied to the system in crisis, in the way that Odum transformed understanding to the Baltic. Another method utilizes the sudden release in local understanding that had accumulated in a separate context, as shown by the work of R. V. Allison in the Everglades or of the rebel bureaucrats in the New Brunswick case. A third mode of learning is more of an "emergent puzzle solution"; that is, learning occurs during the crisis by putting various pieces together. Examples might be the current restoration efforts in the Everglades or the Columbia River, where there are efforts to develop composite solutions that attempt to reconcile ecologic and economic objectives on a regional scale. Another example of this emergent puzzle solving occurred in the Great Lakes, with the portrayal of toxic contamination, especially in piecing together impacts on biota and threats to human health. All these learning modes involve a transformation across space and time scales, which will be discussed later under the barriers and bridges section.

The multiple roles filled by scientists in the case studies change with the phase of system. During the first two phases from *r* to *K*, scientific research reinforces ideas and concepts. Researchers are knowledge brokers, and are often search for the "golden number." As the institution and system mature to the *K* phase, researchers also may be critical to the declaration of a crisis, since many are involved in long-term studies of the resource and have a historical context from which to evaluate change. This is similar to part of the observation by Ludwig and colleagues (1993) that scientists are good at recognizing crises but not at providing solutions. Nevertheless, it is clear that certain types of scientists in the case studies have made dramatic contributions to the reformation process. Some of this argument can be clarified by the type of scientist and by willingness to embrace the uncertainties (an "old" tenet of adaptive management). Deductive rationalists, who practice the science of analyzing pieces, certainly fall in the category of Ludwig and colleagues (1993). Learning during the reformation phase is much more facilitated by the type of scientist who is inductive, integrates human and biophysical elements, and recognizes the inherent uncertainties.

In this section we have used a model of adaptive cycles to illuminate some of the complexity of the coupled dynamics of ecosystems and management institutions. The model is useful for identifying groups involved in the process, and the roles they play, or do not play, in each phase of system dynamics. Also, the model appears to be useful in identifying the types of opportunities and actions for learning that exist in each of the phases. The final section will explore the generalities of the model as they pertain to complex adaptive systems.

Complex Adaptive Systems

The coupled dynamics of nature, society, and resource institutions appear to correspond to the four-phase model. But are the interpretations of the case studies done in a way that force agreement? Perhaps other events were important and were not included in the histories because of the focus on the four-phase model. As Michael points out, since we filter events and thereby construct reality, a model can create a false reality, and almost certainly one that is partial in its explanatory power. The four-phase model provides a heuristic framework for organizing thoughts, especially around the notion of nonlinearities and the discontinuous and adaptive nature of complex systems. It is a model that emphasizes generality and wholeness as a trade-off with precision. But also (and more important) it may be possible that the model captures the ever-changing stages that are exhibited by complex adaptive systems, of which ecosystems and management institutions are two examples. That is the topic we turn to now.

Comparing Ecological and Social Systems

One of the puzzles mentioned in the introductory chapter was the apparently inevitable emergence of a pathology of regional development, at least in its early phases, as the consequence of the initially successful exploitation and management of resources and the environment of a region. In the case studies we examined in the early 1970s, the ecosystem and environment became more vulnerable to surprise and crisis as resilience decreased, the management institutions became more rigid and less responsive both to the resource dynamic and to the public, and the citizenry became more dependent and with fewer options for self-reliance. We saw the form of this pathology emerging in the early stages of testing and developing theories, methods, and case study examples of adaptive environmental assessment and management. They and the diagnosis were summarized in Holling (1986).

Those cases involved a number of different examples of forest development, of fisheries exploitation, of semiarid grazing systems, and of disease management in crops and people. We have greatly expanded and deepened the case studies and tests since then, adding examples that are chapters in this book. But two of the original examples continued to provide insights and are featured here in their present, more expanded forms as the New Brunswick regional forest case (Baskerville, chapter 2) and the Columbia River regional development case (Lee, chapter 5). It was those studies that began to clarify the pathology and expose the way out of it.

In those two examples the initial diagnoses of the pathology up to the early 1970s were as follows:

- Successful suppression of spruce budworm populations during the 1950s and 1960s in eastern Canada, using insecticide, certainly preserved the pulp and paper industry in the short term by partially reducing defoliation by the insect so that tree mortality was delayed. This encouraged expansion of pulp mills but left the forest, and hence the economy, more vulnerable to an outbreak that would cause more intense and more extensive tree mortality than had ever been experienced. That is, the short-term success of spraying led to moderate levels of infestation that spread and persisted over larger areas, demanding ever more vigilance and control.
- Effective protection and enhancement of salmon spawning through use of fish hatcheries on the west coast of North America quickly led to more predictable and larger catches by both sport and commercial fishermen. That triggered increasing fishing and investment pressure in both sectors, pressure that caused more and more of the less productive natural stocks to become locally extinct. That left the fishing industry precariously dependent on a few artificially enhanced stocks whose productivity began declining in a system where larger-scale physical oceanic changes began to contribute to unexpected impacts on distribution and abundance of fish.

In both those cases, however, by the 1980s we began to realize that the phase of a growing pathology was transient and could be broken by a spasmodic readjustment, an adaptive lurch of learning that created new opportunity. It is that creation of something fundamentally novel that gives an evolutionary character to development of a region that might make sustainable development an achievable reality rather than an oxymoron. The motive for this book thus became an effort to improve the definition of the conditions for such learning—the bridges that could facilitate it, the barriers that could impede it.

Once the pathology was defined and it was understood that it might be possible to break out of it for a new phase of development, the four-phase adaptive cycle seemed to emerge as a logical thought model to organize inquiry. But its original form, described in the introductory chapter, was strictly a representation of an ecosystem cycle. That original ecological application has been expanded extensively and its predictions about structure and function of ecosystems have been tested enough to show its power and relevance, at least in that field of inquiry (Holling 1992). But until now we have avoided the temptation to transfer it from ecological systems to social ones. There have been too many examples in history of inappropriate transfer of insight from one field to another. In such cases the effort can be an embarrassment at best; at worst it is a rationale for social evil (witness the excesses of social Darwinism prior to World War II).

Now, however, we are willing to take the chance for two reasons. First, the notion of a complex adaptive system as an entity in itself has emerged from a number of different fields and those independent insights have begun successfully to cross-fertilize each other. Those insights, and the people articulating them, come from developmental biology and genetics, evolutionary biology, physics, economics, ecology, and computer science (Waldrop 1992). Since the late 1980s, the Santa Fe Institute has been providing a prime focus for exploring and deepening these insights and for opening opportunities provided by the identification of adaptive complex systems as a specific object of inquiry, whatever form it takes—economic, social, biological, or ecological.

But much more important, there were growing hints that the model's limitations could be usefully explored and perhaps that the theory expanded by applying it to specific examples of development linking people, nature, and regional economies. Rather than forcing an ecological model on social systems, therefore, our hope was to expose its inadequacies and perhaps to expand its generality. That is what brought us here.

At its foundation, the model is essentially a tautology of birth, growth and maturation, death and renewal that must apply to any living system and that might apply to some nonliving ones as well. But viewed as such, the issue becomes one not of inappropriate transfer from one field to another, but of general applicability. Explaining everything explains nothing!

The critical feature of the model that can distinguish among different systems lies in the phase of renewal, the alpha phase of box 4 of figure 1.2 in the introduction. Renewal can simply mean the endless repetition of the same initializing condition for the four-phase cycle. That was what was implied in the initial application of the model to ecosystems. For human systems, however, that would mean humanity was tied to a rack of determinism doomed to repeat the lessons of history with no option for individual will. And certainly much of history of development is just that. How else could Marchetti (1987) so consistently describe the development of various technologies with a simple logistic curve—a box 1 to box 2, an *r* to *K* conceit?

Novelty, and the different futures that novelty creates, however, can unfold as the consequences of the unpredictable opportunities provided by the box 4 alpha phase, where, for a transient period, constraints are released and dissociated accumulated capital can nucleate elements never connected before into unfamiliar and new configurations. The novelty appears in many forms in the case studies—as new institutions (the South Florida Water Management District in the Everglades), as new policies (new rules of forest management in New Brunswick), and as new suites of physical structures (dams and fish ladders along the Columbia River).

Although box 4 (alpha phase) conditions represent the stage most vulnerable to erosion and to the loss of accumulated capital, this is also the stage that has the potential to jump to unexpectedly different and more productive systems. For example, when fire frequency declined along the prairie–forest border in Minnesota during the Little Ice Age, 400 years ago (Grimm 1983), extensive areas changed from oak savanna to maple forest. It was the change in the release function of fire at the omega phase that opened opportunity for rapid transformation of vegetation at the reorganization or alpha phase. Unexpected combinations of previously independent species developed affinities among each other to give them a key role in future structuring of the ecosystem. That is not too different from the emergence of new innovative business, when individuals with previously separate experiences combine their different skills in a novel way to achieve a new goal during periods of turbulence.

To explore those similarities and differences between ecological and social systems further, it is useful now to determine if related concepts and theories, particularly ones with an empirical base, have been developed in the social sciences.

Social Change Theories

We can usefully identify three classes of social theories of change. The first is the life cycle representation so common to many fields and to the logistic formulation that Marchetti (1987) uses to such good purpose. These life cycle/logistic representations imply growth to some sustained plateau, with senescent elements replaced from some unknown pool. In ecology that was the foundation for Clement's model of ecosystem succession described earlier. In organizational theory that is the foundation for representing the time course for products, processes, and organizations (Kimberly and Miles 1980). In economics, however, new expansions of theory expose the abrupt nature of the flip from one mature product to a competing one (Arthur 1990), much as we describe here for the shift from phase 2 to phase 3, from K to Ω. Mature products are seen as capturing a market, and, through increasing returns to scale, for a time freezing out superior, competing innovations.

That addition leads to a second class of social change theories that contrast gradualist life cycle models with revolutionary change ones. Gersick (1991) has reviewed these using another biological theory as a template for describing change in complex systems. This is Eldredge and Gould's view (1972) of biological evolution as proceeding by punctuated equilibria, rather than by gradual incremental change. They claim that the fossil record suggests that species persist for long periods in essentially the same form or equilibrium and that new species arise abruptly in sudden adaptive explosions of rapid change. That representation is consistent with the behavior generated by the four-phase cycle but aggregates the four into two stages—one prolonged period of gradual change and one of rapid transformation. The theory emerged as a description of the fossil record, with explanations for the sudden changes ranging from the consequences of external disturbances (e.g., planetesimal impacts on the earth) to internal senescent/reorganization sequences.

Similar representations have been proposed in the social sciences. For example, in the history of some sciences, Kuhn (1970) distinguished the alternations between long periods of normal science and sudden scientific revolutions leading to a paradigm shift. Abernathy and Utterback (1982) distinguish gradual from radical innovation sequences in industry. The Miller and Friesen (1984) theory of organizational adaptation contrasts periods of momentum with those of revolution, and Levinson (1978) sees individual human development as periods of stability alternating with abrupt rapid transitions.

Such theories identify so-called deep structures that provide the sustaining rules for the gradual incremental changes that occur throughout the "equilibrium" periods. Revolutions are seen as being brief periods when a system's deep structure collapses to become subsequently re-formed around new strategies, power, and alignments.

We earlier described related entities when we apply the four-phase cycle to ecosystems. For example, stands of even-aged trees are the slow, structural variables that for long periods can provide the context for dynamic interactions of fast variables, such as needles, insect defoliators, and their predators. The even-aged tree stands have a speed, or turnover rate, of approximately 100 years and a spatial grain of 10s of meters. The fast variables have a turnover rate of a year and a spatial grain of centimeters—two orders of magnitude faster and smaller. For long periods the budworm is controlled by predators at low densities, allowing trees to grow slowly to maturity. The "revolution" occurs when the control of budworm collapses because growth of the forest—the slow, structural variable—dilutes the effects of predation, and an outbreak of the insect—the fast variable—is generated that kills trees over large areas.

But this, as in the case of the social revolutionary change theories, is more a description of a phenomenon than an explanation of its causes. Recognizing the different variables that control each of the four phases deepens understanding of the dynamics.

There are some detailed differences between these social revolutionary change theories and the four-phase adaptive cycle, but the fundamental difference is that the boom-and-bust dynamic and the opportunities that at times are constrained and at other times opened emerge from the interaction among variables that characterize and control each of the four phases. The behavior emerges from the way control shifts from *r*, to *K*, to Ω, to α variables and processes and back to a new or a repeated *r* set. We will expand that description of shifting controls in a moment, but before doing so, one final set of social theories of change must be described, because they come close to and deepen the insights of the four-phase adaptive cycle.

These deeper theories explicitly recognize the four-phase properties of complex evolving systems and the tensions they generate that produce stages of growth and transformation. For example, the Austrian economist Schumpeter (1950) saw socioeconomic transformations proceeding such that market forces controlled the *r* phase of innovation; institutional hierarchies, monopolism and social rigidity controlled the *K* phase of consolidation; forces of "creative destruction" triggered the release, or omega phase; and technological invention determined the source for a phase transformation at the alpha phase.

Like all such theories of revolutionary change, the most precise insights have been provided by delineating the properties of the release, or omega, phase. In contrast, the renewal, or alpha, phase lies in a fog of mystery by reason of its inherently unpredictable nature. But Schumpeter's designation of capitalism as a "perennial gale of creative destruction" highlights precisely the same paradox in ecosystems at the transition from box 2 of consolidation, or *K*, to box 3 of release, or omega. There is obviously a destructive element to the collapse of a company or to the occurrence of an intensive fire in a mature forest. But there is also a creative element, because previously tightly bound capital is released: money, skills, and knowledge in a business sector; organized carbon and nutrients in a forest.

An even more specific typology comes from cultural anthropology in the works of Douglas (1978) and Thompson (1983). Four explicit types of individuals or institutions are identified, and these are organized within two axes very similar to the ones in figure 1.2 of the introduction. The *r* phase is designated as the entrepreneur, the *K* phase is the caste or bureaucracy, the omega phase represents the sects, and the alpha phase stands for the ineffectual individual (Douglas 1978; Thompson 1983). The insights provided by their descriptions of sects resonate with attributes of the release processes that we describe for ecosystems. The sects are described as being small and tightly organized, often around a charismatic leader with a strong, singular ideological purpose. Their power emerges only occasionally, when their tenacious allegiance to internal rules and purpose intersects with the vulnerability of a mature and rigid bureaucracy. This captures their role in triggering release and, for us, has been particularly helpful in illuminating our understanding of the role of the more extreme types of environmental activists in the earlier analyses of the case studies.

Their description of the critical box 4, alpha phase, however, only partly echoes our description of the ecological analog. They do see the dissociated nature of the elements of box 4, describing them as the atomized individuals with no control over their own destiny, caught by whatever winds of change are generated by the other players. But in ecological systems the role for that phase is as a repository of the capital that has accumulated during earlier phases of growth and maturation—*r* to *K*. Its dissociated, weakly connected state is the very attribute that makes unexpected combinations of associations possible and the influence of the individual most likely. It is the flywheel of the whole system, whose properties determine whether there is a repetition of past cycles, collapse of those cycles, or the emergence of a new system distinguished by its novelty.

A recent example from social responses to crisis helps make the point. After the devastation caused by Hurricane Andrew in south Florida in 1992, the neighborhoods that renewed most rapidly and effectively were ones that drew upon the accumulated experience and contacts of the individuals themselves. The role of elderly people was particularly revealing and critical (Guillette 1993). It was their personal contacts, developed over their working lives, that allowed them to break through bureaucratic blockages to access aid and support available from federal, state, and municipal agencies and from insurance companies. Contacts and friendships are as much accumulated capital as are money and skills in the alpha phase of renewal.

This fourth, or alpha, phase is the one least understood, by us as well, because of its inherently unpredictable nature. The only rigorous theory we have encountered that gives it some specificity is from one facet of chaos theory (Stewart 1989). One of the key points of chaos theory is that slight changes in initial conditions can generate a great complexity of behavior and unpredictable outcomes. One of the favorite examples comes from a simple model of the atmosphere developed by Lorenz (Stewart 1989), which showed that slight departures from initial conditions of weather lead to widely divergent futures. The resulting behavior looks random, although within a bounded domain, but is completely deterministic. There is an inherent unpredictability to outcomes. Lorenz named this the butterfly effect, dramatizing the phenomenon with an analogy that a butterfly flapping its wings in Beijing now can change storm patterns in New York next month.

There have been many examples of chaotic behavior identified or proposed in physical, biological, and social systems. As in any new theory that gives fresh insight, chaos theory has generated an exuberant search for other examples, driven by the yearning for universality. Is healthy brain function chaotic and unhealthy function stable? Does heart function have chaotic patterns? Planetary orbits? But for ecosystems, at least, the question should be not, "Are they chaotic?" but, "When are they chaotic?"

For long periods, ecosystems develop growing connectivity and predictability as they progress through the r to K sequence. During this period the conditions that generate chaotic behavior are unlikely. But those same conditions produce a brittleness that sets the condition for the release, or creative destruction phase. That does, then, lead to the conditions for chaotic behavior during the brief period when the cycle achieves the weakly connected state of the box 4 phase. It is this organization that allows systems to exploit, in Kauffman's terms (1993), the edge of chaos, where adaptive opportunity lies.

To summarize, at times behavior is determined by the *r*-strategists, by the pioneers and opportunists of box 1. They set the conditions for control to shift to the *K*-strategists—to the effective competitors and consolidators of position and power. As a result, resilience is reduced, controls are intensified, and the system becomes an accident waiting to happen. As the shift to the omega phase occurs, the slow, extensive variables lose their control of behavior; rather, fast variables assume control and release the capital that was stored in tightly organized form. This capital then becomes dissociated in the alpha phase, where a new set of variables, processes, and random events slows the leakage of capital out of the system, mobilizes it in accessible forms, and precipitates unexpected associations between previously independent variables.

We have explored in this book possible similarities and differences between actual ecological and social systems and between theories of change developed in each field. It has led us to an expansion of the original concepts we had by formalizing two features that distinguish our arguments. One has to do with the adaptive character of the opportunity that is opened by the back loop of the four-phase cycle—from omega to alpha. The other has to do with the nested nature of the elements that comprise a complex ecological or social system. We will deal with each in turn in the next section, where we attempt a theoretical synthesis.

A Theoretical Synthesis

The Adaptive Cycle

In contrast to existing social change theories, the four-phase adaptive cycle emphasizes a loop from hierarchical consolidation in box 2 to two phases of destruction and reorganization, where innovation and chance assume a dominant role (figure 1.2). That reorganization phase occurs when a rare and unexpected intervention or event can shape new futures as an act of creating opportunity. The tight organization and hierarchical control of box 2, or *K*, that precludes alternatives is broken because of maturing brittleness that often intersects with external events that provide the proximate trigger for the change. The resulting loss of control leads to the release of the accumulated capital (nutrients and organized carbon in ecosystems; money, skills, contacts, and experience in organizations) and to its decay or dissociation into constituent elements in the alpha phase, box 4. At this stage the system becomes ill defined and loosely coupled. It contains the paradox of being in a state most likely to collapse but also one most likely to be transformed by innovation. In human systems it is the stage where the individual, for good or ill, has the greatest potential for influencing the future. The disassociated nature of the alpha phase is the very condition that makes both a possible.

Note that as the system cycles through all of its four phases, although control shifts from one set of variables, processes, and events to another, all variables and processes other than the ones controlling at the moment are present and functioning in either a maintenance or a "holding" pattern. For example, pioneer species or entrepreneurs are present during the consolidation phases; some trees and bureaucrats (or at least the seeds and saplings of each) persist through the release and reorganization sequence; soil processes function throughout all phases. That functional diversity keeps critical actors in the wings or in a supporting role, while the lead shifts for a period to others.

Although we see fundamental similarity between adaptive ecological and adaptive human systems, we suspect that the human ones have much greater powers for both rigidity and novelty. The ability of the bureaucracy of a government agency to control information and resist change seems to show a level of individual and group ingenuity and persistence that reflects conscious control by dedicated and intelligent individuals as well as the unconscious part of the organization and culture of bureaucracies. Collaterally, the culture of organizations imposes unconscious constraints. And certainly there are some empires and some institutions that have long endured in their same basic form. But that observation might simply reflect the frustration the authors have experienced in dealing with inflexible bureaucracies. Alternatively, the possibility exists that the locus and speed of the adaptive cycle can be changed by conscious design so that renewal of norms, values, and other traits occurs internally while maintaining much of the overall organizational structure.

Similarly, we guess that the very ingenuity that allows survival of a bureaucracy also provides uniquely human opportunities for change and novelty. Human innovations create new environments independent of, or in concert with, external changes. The development of sedentary agriculture might have been conditioned by climate changes or the industrial revolution by steam technology, but it was human innovations that literally created new worlds. At the least, these cultural transformations are of a magnitude similar to historical transformations of biological evolution that created land-dwelling animals or warm-blooded ones. But again, that view might be more a reflection of our human arrogance or hope than of reality.

Hierarchies and Panarchies

The second feature that distinguishes the scheme presented here concerns the manner in which elements of complex adaptive systems nest inside one another. Simon (1974) was one of the first to argue the adaptive significance of such structures. He called them hierarchies, but not in the sense of a top–down sequence of authoritative control. Rather, semiautonomous levels are formed from the interactions among a set of variables that share similar speeds, such as information flow rates (and, we would add, spatial geometries). Each level communicates a small set of information or quantity of material to the next higher, and slower (coarse in geometric terms), level. An example for a forested landscape was presented in the introduction as figure 1.3.

As long as the transfer of information, material, or energy from one level to the other is maintained, the interactions within the levels themselves can be transformed, or the variables changed without the whole system losing its integrity. As a consequence, the wide latitude that this structure allows for "experimentation" within levels greatly increases the speed of evolution and resiliency of the system.

Ecologists were inspired by this seminal article of Simon's to transfer the term *hierarchy* to ecological systems (Allen and Starr 1982; O'Neill et al. 1986). But subsequently, the structural, top–down aspect has tended to dominate, reinforced by the proper, everyday definition of *hierarchy* as involving vertical authority and control. The dynamic, adaptive nature of such nested structures has tended to be lost. It certainly is true that slower and larger levels set the context within which faster and slower ones function. Thus a forest stand moderates the climate within the stand to narrow the range of temperature variation that the species experience.

But missing in this representation is the dynamic of each level that is organized in the four-phase cycle of birth, growth and maturation, death, and renewal. That cycle is the engine that periodically generates the variability and novelty upon which experimentation depends. As a consequence of the periodic but transient phase of destruction and reorganization, as the cycle progresses from the omega phase to the alpha, the variables can become reshuffled, perhaps to establish different relationships, perhaps to be open to foreign and entirely novel entrants. This explicitly introduces mutations and rearrangements as a periodic process within each hierarchical level in a way that partially insulates the resulting experiments so as not to destroy the integrity of the whole structure. Hence species, physical factors, and biophysical processes in ecosystems and roles, ideas, and people in institutions can be reshuffled periodically and reinvented to explore the consequences of novel associations during the alpha phase.

The organization and functions we now see embracing both ecological and human systems are therefore ones that contain a nested set of the four-box cycles, in which periodic reshuffling within levels maintains adaptive opportunity and the interactions across levels maintain integrity.

Since the word *hierarchy* is so burdened by the rigid, top–down nature of its common meaning, we prefer to invent another term that captures the adaptive and punctuated evolutionary nature of adaptive cycles that are nested one within the other across space and time scales. We call them *panarchies,* drawing on the image of the Greek god Pan—the universal god of nature. This hoofed, horned, hairy, and horny deity (Hughes 1986) represents the all-pervasive spiritual power of nature and has a personality and role that is described in sections of the Orphic Hymns. Thus Pan is described as

> Goat-legged, enthusiastic, lover of ecstasy, dancing among stars,
> Weaving the harmony of the cosmos into playful song
>
> TRANSLATED BY HUGHES (1986)

In addition to this creative role, he has a destabilizing role that is captured in the word *panic,* directly derived from one facet of his prickly personality. His attributes are described in ways that resonate with the attributes of the four-phase adaptive cycle—as the creative and motive power of universal nature, the controller and arranger of the four elements—earth, water, air, and fire (or perhaps, of K, Ω, r, and α). He therefore represents the inherent features of the synthesis that has emerged in this comparison of ecological and social systems.

Two examples of panarchies, one ecological and one social, are presented as figures 12.2 and 12.3. The ecological panarchy is from the Everglades of Florida and indicates the interactions among vegetation structures (individual plants through plant communities and landscape units) and atmospheric phenomena, from small-scale thunderstorms to El Niño or global fluctuations. The vegetation and atmospheric panarchies provide the stage or bounds within which animals (wading birds or humans) make decisions and operate. The structural elements each go through a four-phase dynamic—for example, grass plants cycle nutrients and biomass on an annual cycle, while plant communities turn over on decadal intervals. The linkages among these units depend upon

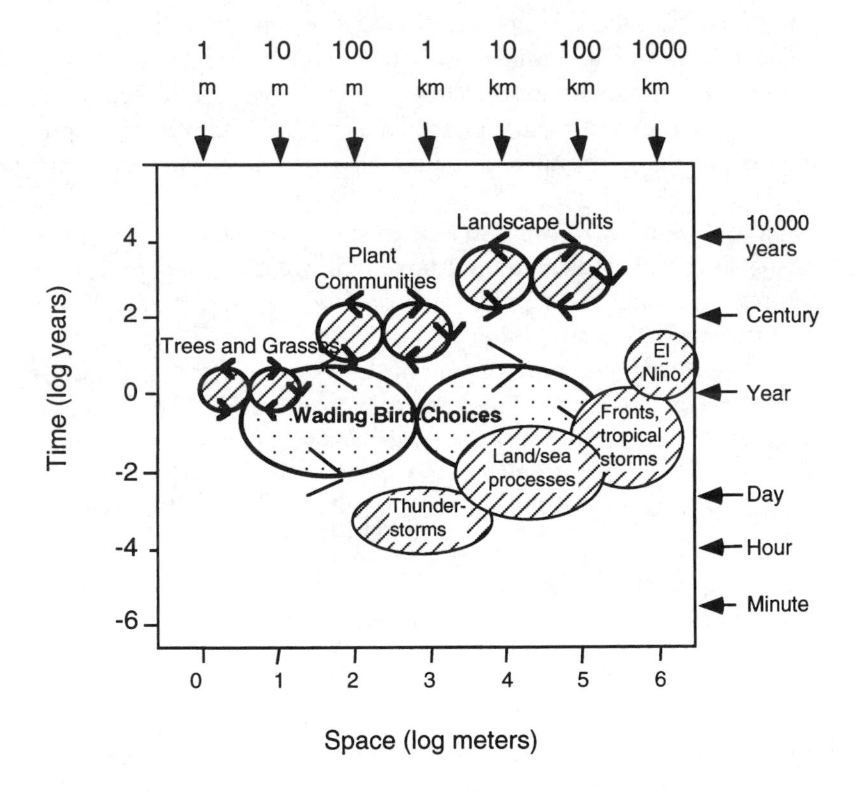

FIGURE 12.2

Ecological panarchy indicating the four-phase adaptive cycles of key vegetation structures (trees, plant communities, and landscape units) and wading bird choices with the interaction of key atmospheric phenomena across space and time scales. The centroid of each cycle represents an approximate time to complete the four phases of the adaptive cycle.

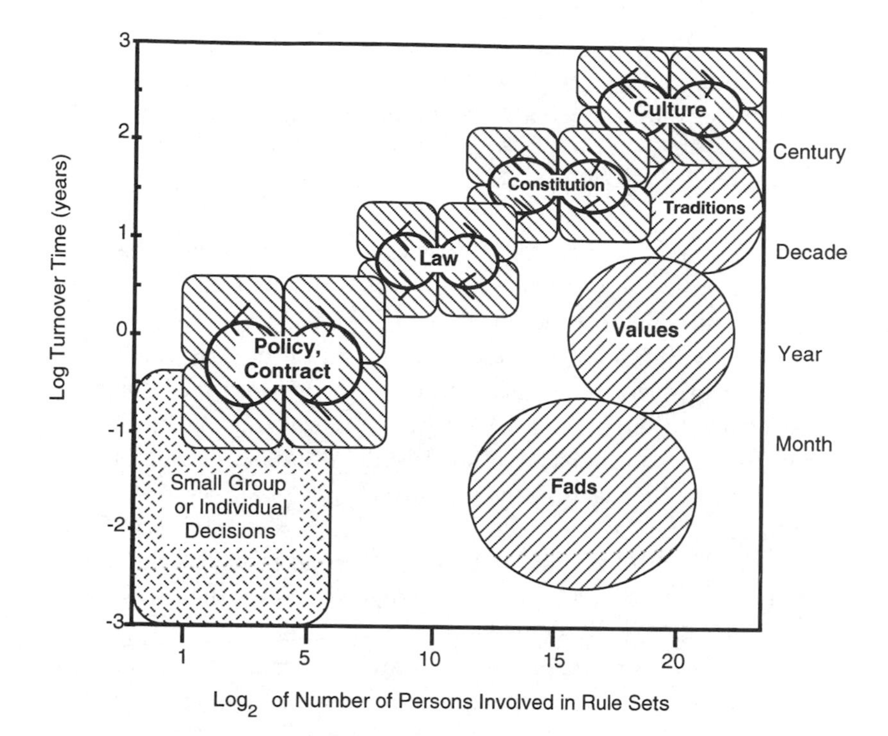

FIGURE 12.4
Map of institutional panarchy, indicating key structures of human institutions. The centroid of each cycle represents an approximate time to complete the four phases of the adaptive cycle. (Kai Lee created the general framework for this diagram.)

the internal phase and synchrony of phases of the different units across scales.

Similar relationships are found in human institutions (figure 12.3), where each level or unit in the panarchy has an intrinsic four-phase dynamic but is coupled with others across space and time scales. In this relationship the spatial surrogate is number of people involved in the decision process. We see the formal rule sets going through four phases, whereas more transient phenomena, such as fads and values, may not.

Institutional Panarchies

The preceding synthesis offers some insight and opportunity to speculate about the structure and function of management institutions, perhaps one of the most complicated aspects of resource management (Walters and Holling 1990; Ingram et al. 1984). The key issues of management institutions deal with recognition of scale domain of rule sets, the nested or embedded nature of rule sets, interactions and revisions of rule sets, and how each scale of rules deals with the issue of uncertainty.

The concept of hierarchical rule sets and their interactions has long been both a conceptual underpinning and a source of controversy among decision theorists. A "hierarchy of decision levels" or the "nesting of rules within rules" is at the root of some of the most challenging conceptual problems of decision theory and provides the source for considerable debate (Ciriacy-Wantrup and Parsons 1967; Ostrom 1986). How these various levels are linked and influence each other is important to understanding institutional change processes. All too frequently attention has been focused on cataloging the manifestations of changes and too little consideration has been given to the patterns of stability and change and ultimately the logic of change processes themselves (Jantsch 1976; Morgan 1987).

Another reason the nested nature of institutional systems has been frequently ignored in theory and praxis relates to complexities of interaction among the levels. In the past there has been a strong tendency among management theorists to reduce forms of knowledge to a "closed system framework" to rid themselves of all ultimate uncertainty (Thompson 1967). This tendency has resulted in theorists and practitioners alike becoming captives of one decision level, often the operational level, where efficiency is the preeminent criterion. Such attempts to eliminate uncertainty have proven counterproductive. Furthermore, this fascination with the operational level has come at the expense and neglect of understanding deeper levels of structure, function, and change. In recent decades resource managers and theorists have begun to take the problems of complexity and uncertainty more seriously (Michael 1973; Holling 1978; Waldrop 1992). Institutional systems are increasingly recognized as open and evolutionary, with uncertainty as a central conceptual issue of design.

The concept of panarchies in institutions, like ecosystems, allows for a description of simultaneously changing structures and functions, at multiple stages or levels in order to cope with dynamism and complexity in the world. The panarchical view puts forth a description of institutions as emergent sets of loosely connected and nested decision assemblages much like Simon's (1962) description, rather than a rigid, hierarchical view of the organization as a top–down, command and control outfit. Although closely linked with external influences, during certain phases of change, institutional systems are capable of exhibiting considerable rigidity, staying power, and resistance. Each level in an institutional panarchy operates at distinctive time and space scale and pursues dissimilar goals.

Institutional levels and their assemblages manifest themselves and operate within two distinct types of rule sets, one deliberate and formal and the other emergent and informal. The former rules take on a pattern graded hierarchically and display themselves as constitutional, collective-choice, and operational rules (Ostrom 1990). The more emergent and tacit rules reveal themselves as myths, strategies, and operating premises (Mintzberg 1978; 1990)

At least three distinct functional scales (the macro, meso, and micro) are recognized within institutional systems (Ciriacy-Wantrup and Parsons 1967; Hedberg and Jonsson 1977; Ostrom 1990). Assemblages of rule sets or stability domains that appear at distinct levels are in a constant state of flux analogous to the four-phase dynamic Holling discovered in ecological systems. These structures track sets of variables that operate at similar speeds. The actual number of assemblies or subassemblies within or among the various system levels is thought to depend upon the character of the task environment (i.e., social and natural), the type and/or source of risk encountered, and transaction costs. If the task environment is placid, the system appropriately bounded, and transaction costs low, few assemblages among levels may suffice for the functions. On the other hand, if the task environment is disturbed or turbulent, transaction costs are high, and knowledge about the system is inadequate, a greater number of assemblages embedded in the larger system may be required to accomplish needed functions.

Constitutional choice rules and myths dominate the macro level and have slow turnover rates, on the order of 50 years to centuries (figure 12.3). They operate at the cultural or societal level and are commonly considered beyond political reach. The learning incorporated at this macro level includes our most fundamental understandings of how the world works, our basic assumptions, and our postulates. The broadest rule sets deal with the most uncertain elements of understanding and generally define the methods by which rules of the other levels are made. The assumptions and rules at this deepest level of an institutional system cannot be tested directly. Rather, constitutional rules and myths are tested indirectly, through the influence they manifest on the other meso- and microlevels.

The mesolevel is dominated by policy-making and management processes that function under considerable political influence. Collective-choice rules and strategies are prevalent at this level. The turnover rate at the mesolevel is measured in decades. This level acts as a bridge between our most deeply held convictions and what is perceived as reality (Hedberg and Jonsson 1977). It is a crucial level, where the integrity and long-term viability of institutional systems are tested. Policy decisions are made and strategy provides direction for action that is either consistent or inconsistent with past practices or assumptions. Strategies become "hypotheses-in-action" that test through feedback the viability of assumptions inherent in myths (i.e., double-loop learning). It is at the mesolevel that there has been a propensity for resource managers and scientists to design natural system variability away artificially. The resulting oversimplification of the system is dysfunctional and inevitably contains latent surprises. At this level the inevitability of uncertainty must be recognized and selective risk taking must be pursued. Processes for searching and learning are required to detect and correct error and connect new knowledge with power (Thompson 1967; Wildavsky 1979; Michael 1973).

The microlevel is where routine, programmed actions are taken. To the extent possible, uncertainty is squeezed out or reduced to statistical probabilities. The goal of such actions is to provide consistency and efficiency that build competency. Formal rules or standard operating procedures dominate the scene. The emergent aspects of this level manifest themselves as operating premises that appear immutable. Fundamental assumptions are not tested, and learning is confined to the single-loop type.

How these embedded levels or assemblages interact is not well understood. We hypothesize that the three levels of rule assemblages and variables operate somewhat autonomously within certain phases of the adaptive cycle, going through periods of stability and instability. At a given level in the panarchy during the transition from *r* to *K*, actions that reduce, ignore, or avoid uncertainty shrink the stability domain and make the system more vulnerable to creative destruction. The omega phase links smaller to larger scales, when seemingly minor activities or broad-scale movements are the agents of change. During the alpha reorganization linkages appear to establish across levels in the panarchy; the extent of each linkage may depend upon the magnitude or complexity of the crisis.

The implications of the foregoing discussion are manifold. First, for would-be agents of change, it is paramount to understand in what phase of the adaptive cycle institutional systems and their subassemblies are functioning. Certain levels in the system may be ripe for change, whereas others may seem invincible. Second, within the four-phase dynamic the window of opportunity to inject novelty into the system and its components is limited but detectable. Third, profound change and learning operate at scales that are intergenerational. Functions at micro and mesolevels are entrained by slow-moving variables for long periods, but at critical moments, key cross-scale linkages become important. At these moments microprocesses and events transform larger-scale ones.

The preceding theoretical synthesis explored the use of the four-phase heuristic for explaining the dynamics of complex, adaptive systems, and the interaction of these dynamics across different space and time scales in panarchical relationships. In the next section, we move to a discussion of specific barriers and bridges that were recognized in the case studies.

Barriers Broken and Bridges Built

In this final section we shall discuss the barriers and bridges to renewal and sustainability that have emerged in the case studies. The barriers and bridges appear to fall into three categories: (1) the ways in which humans interpret and understand nature and resources, (2) the design and practice of human institutions, and (3) the interaction between people and ecosystems. To contrast what works from what does not, both barriers and bridges will be presented in each of these categories, starting with how humans view and understand nature.

Understanding Nature and the Nature of Understanding

Barriers and bridges emerged from the ways in which nature and human nature are conceived and understood. The paradigms of relationships among processes and structure in nature can be either a barrier or a bridge. The mode in which science is practiced can also be either hindrance or conveyance.

The paradigms of how nature operates are critical to management of natural resources. The key bridges are provided by views that include a systems perspective, interdisciplinary, nonlinearity, and cross-scale views. These traits are included in the emerging view of complex, adaptive systems, as mentioned earlier. A systems perspective was mentioned by all the case study authors as a critical bridge.

Inductive and deductive modes of inquiry are identified in the introduction. Both are necessary and useful, even though at times contradictory. The deductive mode can be a barrier. When the system is in a crisis or transition phase, deductive modes can be used to create "pseudo-facts" that perpetuate or deny a crisis. One example was in the Everglades, when the denial by the farmers that their nutrient additions were causing changes in the vegetation succession was compounded by scientists hired by the farmers. Mission-oriented research, whose goal is to validate existing theory (rather than invalidate hypotheses, as in science), generally hinders the development of alternative views and leads the system into an intellectual brittleness. This is characterized by a comment made to one of us by the superintendent of Everglades National Park, who stated that the "goal of science should be to provide data precise enough to be defensible in court." Traditional funding pathways tend to support existing dogma rather than to support contrasting explanations.

The inductive mode, including the use of computer models, can act as a bridge. The many examples of the application of this mode form the basis of the AEAM approach (Holling,1978; Walters 1986; Lee 1993). Integrative, multidisciplinary teams that utilized the wealth of human experience were crucial in unlocking the intellectual gridlock in New Brunswick (Baskerville, chapter 2), the Everglades (Light et al., chapter 3), the Columbia River Basin (Lee, chapter 5). In many of these cases the inductive mode led to a reexamination of the appropriate spatial and temporal scale of the issue at hand.

Design and Practice of Institutions

People create institutions in a variety of ways, from top–down to bottom–up. In this sense institutions are the instruments developed by people to create and enact policy. The movement in the Chesapeake during the 1960s and 1970s was a grassroots or bottom–up organization that created dramatic changes. During the same period, dramatic changes occurred in the Baltic via the HELCOM group, where the international treaty precipitated policy down to the local level. Both methods proved to be bridges in terms of engaging the population, and enacting a suite of laws and policy. Perhaps the essence of successful institutions is in an ability to span adaptively across hierarchical or panarchical levels in space and time. Sanderson, in his essay, came to similar conclusions: systems of greater size and complexity have more management and institutional flexibility.

Throughout the case studies the emergence of nongovernmental groups (NGOs) or epistemic communities (Haas 1990) has formed critical bridges, which appear to fill gaps in existing institutional hierarchies and serve as conduits or media for information flow. This is especially true in a period of growing dissatisfaction and mistrust in existing institutions. Examples of these bridging organizations are the Great Lakes Union (a binational NGO in the Great Lakes), the Everglades Coalition, and others. These groups tend to focus on collaboration and therefore create and enhance connections that are otherwise lacking in the traditional management institutions.

Another bridge has emerged in the movement away from traditional bureaucratic entities toward more "adaptive" institutions with an explicit strategic mission. One example is the Northwest Power Council, where adaptive concepts are formally internalized and part of the operational principles. As Westley points out in her essay, the adaptive groups focus on innovating, on reviewing assumptions, and on building skills of reflection. These adaptive groups are especially flexible in a rapidly changing world. These are the groups that not only tolerate but nurture "rebellious" bureaucrats who are willing to declare the "emperor has no clothes."

Humans and Nature

One of the most pervasive problems in the interaction between humans and their environment deals with defining scope in space and duration of time for natural resource issues. Other key interactions deal with the development (or lack) of a sense of place and the use of metaphors in valuation of the resources.

The way in which resource issues are defined will determine much of the subsequent actions. The expansion of the space and time scale of an issue, without losing contact with the local and fast dimensions, acted as a bridge in most of the case studies.

Barriers arose from the mismatch of scales and from a failure to acknowledge processes occurring at broader scopes, a tendency to orient to past values, and an inability to translate measures or values across scales. One key to understanding the hydrodynamics of the Baltic (Jansson and Velner, chapter 7) was the correlation between inflows and atmospheric pressure in the North Atlantic. Many problems remain because of orientation to past assumptions. Agriculture in the Everglades continues to thrive because of foreign policy decisions made over 30 years ago. In New Brunswick, forest management is based on old formulas, and despite dire predictions, people have been comfortable to continue with the status quo. A critical barrier still remains in the translation of values across space and time scales. Many of the water quality and fisheries issues (Chesapeake, Great Lakes, Columbia, and the Baltic) involve translating not only measures but values across scales. This is described in part as the " intergenerational" problem by economists.

Recognition of processes at the appropriate space and time scales is reported by all the case studies as a clear bridge to improved management. As Baskerville states about the forest problem, "It is not *what* is the problem, but *where* is the problem." In the Everglades, the work of Art Marshall was aimed at generating and evaluating management and policy at the level of the entire system. In the Chesapeake, Great Lakes, and Baltic computer models and evaluation of the entire water shed were critical to development of robust pollution policies.

One of the best methods for experiencing a broader range of space and time scales is the use of computer models. The models accelerate time and compress space in order to connect the system dynamics with personal perceptions. The models also surface "hidden" assumptions, which is critical to their use as devices to improve communication.

One of the challenges is to develop understanding and methodology that spans across space and time scales. Most science works well across large ranges of scales only when the processes are obeying the same scaling rules across the whole range. Bounding the problem and issues to a narrow, fixed domain in space and time (i.e., as is typically done when models are developed) is useful for many solutions, but suffers when changes occur in processes at scale ranges that have been ignored.

This is especially true in ecology, the "science of middle-number complexity." There are positive signs, with the transformation of hierarchy theory (Allen and Starr 1982; O'Neill et al. 1986) and the development of analyses of cross-scale interactions (Wiens 1990; Holling 1992), that new methods can be developed that span scales.

Interaction Between People and Their Environment

Clear bridges emerged from the use of symbols to communicate understanding and develop a sharing of values. The clearest case was the Chesapeake, where the notion of everyone's backyard mobilized tremendous citizen and political action. The use of these symbols is indicative of the use of metaphors in communicating, as Michael's essay points out. The shared values for some systems, such as the Everglades, are international ones, where the international reputation of the Everglades was in large part due to the poignant writings of Marjory Stoneman Douglas (1978). In chapter 5 Lee discusses the use of native American metaphors in establishing the goal of shared cultural uses of the Columbia River's resources. In the Great Lakes system the seemingly paradoxical use of the phrase *ecosystem health and integrity* is a much more powerful metaphor for a sustainable system than a more rigorous definition that involves one stability domain.

One key bridge emerged from the development of policy that was consistent with system dynamics. The Everglades study was key in revealing this relationship; the rapidly accumulating human population combined with more rapid ecosystem dynamics (hydrologic variation occurs yearly and decadally) leads to the rapid cycling of the adaptive cycles described earlier. That is, the system must learn and reformulate policy almost continually because the ecosystem dynamics are so rapid! The case of the New Brunswick forests shows just the opposite: a resource system that changes on a time scale of multiple generations. As Baskerville states, "to be effective, policy must enjoy the support of the public for as long as it takes to implement the policy and for the natural system to respond to the policy actions." Ignoring the dynamics of the natural system or attempting to control variation leads to inevitable brittleness and crisis.

The final key bridge has been in the development of integrated policies. Singular, quick-fix solutions tend not to work over the long term because they generate a rapid return of crises and erosion of trust by those who put faith in the public institutions. Examples from the case studies include the Columbia River, where the production of electricity was paramount to other resources, such as fisheries and other uses.

In the review of the barriers and bridges across the six regional ecosystems, a few surprises emerged that can be described as new classes of problems and new ways of understanding and dealing with those problems. The new problems involve interactions in unforeseen ways across space and time scales. The problems are not fixed to specific areas or to a linear or discrete time pattern. This is true from the Everglades, where problems and issues vary from a few meters (around an alligator hole) to the world sugar market and its implications for agriculture. Other examples have similar properties; that is, processes and structures interact across scales, yet most of tools for understanding (experimental science, modeling) and management (determination of property rights, for example) are for the most part fixed in scope. New scale-spanning methods and theories of understanding are creating a bridge to deal with the cross-scale dimensions of resource issues. New techniques are needed to measure processes and patterns (Francis and Regier, chapter 6) as they vary from a spawning patch on a tributary to the entire Great Lakes Basin. New technologies, such as improved imaging techniques and bigger and faster computers, are contributing in this field (Costanza and Maxwell 1990; Costanza, Wainger, Folke, and Maler, 1993). But this all points to conceptual reconfigurations as well, because as Michael points out, we need new myths, new words and meaning in our language, and quickly.

Many aspects of our world are changing rapidly. In the case studies the surprises appear to be coming more frequently and appear to have broader implications. This is true in other resource problems, such as ozone depletion, AIDS, and other global issues. But how do resource managers cope with this increasing complexity? We will attempt some suggestions to answers for this question in the next and last section as we look to the past for a summary of key conclusions and implications for the future.

Lessons from the Past, Glimpses of the Future

Among the key lessons learned from this review are ones that deal with issues of connections across scales between problems and solutions, the critical roles by management, the need for strategic approaches to management policies, and finally the need for public engagement. All these lessons are the consequences of the dynamics of complex adaptive systems and of the interactions across space and time scales. We will deal with each lesson in turn.

1. *Spanning scales.* Recognition and understanding of key interactions across space and time scales provide robust solutions to crises and pathways out of gridlock. This is not simply a matter of scale matching (i.e., determining the appropriate scale for the problem or issue at hand) but of recognition of how processes translate across scales. One paradox that emerged was how institutions with a seemingly appropriate spatial venue (South Florida Water Management District in the Everglades, Provincial Government in New Brunswick Forest, and Northwest Planning Council in the Columbia River Basin) were unable to adapt to a series of temporal crises. Certainly, one way in which scales are spanned is through the application of overarching rules or policy. Two examples are the grassroots movement to save Chesapeake Bay (Costanza and Greer, chapter 4), and the HELCOM treaty in the Baltic (Jansson and Velner, chapter 7). But this is not the only mechanism, because at critical times, individual acts (Art Marshall in the Everglades, chapter 3) can create entirely new visions and policies.
2. *Key roles in management and institutions.* We have seen the need for at least four different roles by players in the management scene: bureaucrats who implement policy, activists who declare crises, catalysts who frame alternatives, and decision makers who decide upon new policies. These roles are seemingly filled as needed by formal structures (agencies, for example) and as informal emergent entities such as epistemic communities, shadow networks, and other nongovernment agencies. Institutions that desire to be more flexible or adaptive should contain all these roles in order to continually learn, generate system understanding, and resolve the challenge of both simultaneously implementing and revising policy.
3. *Strategic policies.* We see the need for more strategic, long-term, broad-scale, and integrative policies. These are policies that successfully create credible futures and at the same time resolve

a pathologic past. Often the management institutions and policies become focused on fast events and lose track of the slower, critical processes that are the basis of the next surprise. Corollary to the development of strategic policies is the generation of integrative understanding and recognition of policies as plausible, alternative hypotheses for understanding. It is that foundation of civic science within a democratic political process that Lee (1993) develops so perceptively in his recent book.

4. *Public education and involvement.* Finally, we see the involvement and education of the people who are part of the system as crucial to building resilient solutions and removing gridlock. The people who live in these areas or who are affected by the polices often become ignored or detached from the institutions established to serve them. They provide the "pool" for creative and adaptive solutions.

We close with a paragraph of hope: Hope that the lessons from our past provide some understanding for coping with an uncertain future, hope that sustainability is part of that future, and hope that this work has provided some insight about how to break down barriers and build new bridges for a sustainable future of these complex, adaptive systems of humans and nature.

References

Abelson, Robert P. 1976. Script processing in attitude formation and decision making. In John S. Carroll and John W. Payne, eds., *Cognition and Social Behavior*, pp. 33–45. Hillsdale, NJ: Lawrence Erlbaum Associates.

Abernathy, W. and J. Utterback. 1982. Patterns in industrial innovation. In M. Tushman and W. Moore, eds., *Readings in the Management of Innovation*, pp. 97–108. Boston: Pitman.

Ackefors, H. 1971. Mercury pollution in Sweden with special reference to conditions in the water habitat. *Proceedings of the Royal Society of London.* 177:365–387.

Adler, E. and Beverly Crawford. 1991. *Progress in Postwar International Relations.* New York: Columbia University Press.

Adler, Emanuel. 1991. Cognitive evolution: A dynamic approach for the study of international relations and their progress. In E. Adler and B. Crawford, eds., *Progress in Postwar International Relations*, pp. 43–88. New York: Columbia University Press.

Adler, Emanuel. 1992. The emergence of cooperation: National epistemic communities and the international revolution of the idea of nuclear arms control. *International Organization* 46(1): 101–145.

Aguilar, J. 1967. *Scanning the Business Environment.* New York: Macmillan.
Albaum, G. 1964. Horizontal information flow: An exploratory study. *Academy of Management* (March): 21–33.
Allee, D. J. and L. B. Dworsky. 1990. Breaking the incremental trap: Achieving unified management of the Great Lakes ecosystem. Paper to the joint meeting of the American and Canadian Water Resources Association. Toronto, April 1990.
Allen, T. F. H. and T. B. Starr. 1982. *Hierarchy: Perspectives for Ecological Complexity.* Chicago: University of Chicago Press.
Allen, T. F. H., B. L. Bandurski, and A. W. King. 1991. The ecosystem approach: Initial report of a multi-year project of the Ecological Committee. Review draft to International Joint Commission/Science Advisory Board, Windsor, Ontario.
Allison, Graham T. 1971. *Essence of Decision.* Boston: Little, Brown.
Almkvist, L. 1982. Baltic marine mammals—a status report. Copenhagen: International Council for the Exploration of the Sea (ICES), 1982. 16 pp.
Amy, Douglas J. 1987. *The Politics of Environmental Mediation.* New York: Columbia University Press.
Anderson, W. 1990. *Reality Isn't What It Used to Be.* New York: Harper & Row.
Andersson, Ạ. 1991. Nights and days in Kotka. *Framtider International* 1:3. Stockholm: Roos Tryckerier A. B.
Anonymous. 1974. Research programs for investigations of the Baltic as a natural resource with special reference to pollution problems. International Council for the Exploration of the Sea Cooperative Research Report 42, Copenhagen.
Anonymous. 1987a. First Baltic Sea pollution compilation. *Baltic Sea Environment Proceedings* 20. Helsinki Commission, Helsinki.
Anonymous. 1987b. Swedish action program against marine pollution. Swedish Environmental Protection Agency, Solna.
Anonymous. 1990a. Ecological plankton research of the Baltic Sea. Project Pelag. Final report 1987–1989. Helsinki: Pelag Press. 194 pp.
Anonymous. 1990b. Marine pollution '90. Action program. Swedish Environmental Protection Agency, Solna. 165 pp.
Anonymous. 1993a. Complex analysis of the hazard related to the captured German chemical weapon dumped in the Baltic Sea. National Report of the Russian Federation, Moscow.
Anonymous. 1993b. Second Baltic Sea pollution load compilation. *Baltic Sea Environment Proceedings* 45. Helsinki Commission, Helsinki. 161 pp.
Argote, L., S. L. Beckman, and D. Epple. 1990. The persistence and transfer of learning in industrial settings. *Management Science* 36(2):140–154.
Argyris, C. and D. A. Schon. 1978. *Organizational Learning: A Theory of Action Perspective.* Reading, MA: Addison-Wesley.

Argyris, C. 1977. Double loop learning in organizations. *Harvard Business Review* 55 (September–October):115–125.

Aronfreed, J. 1976. Moral development from the standpoint of a general psychological theory. In T. Likona, ed., *Moral Development and Behavior.* New York: Holt, Rinehart and Winston.

Arthur, B. 1990. Positive feedback in the economy. *Scientific American* 262:92–99.

Arthur, B. W. 1984. Competing technologies and economic prediction. *Options, International Institute for Applied Systems Analysis* 2:10–13.

Arthur, B. W. 1989. Competing technologies, increasing returns, and lock-in by historical events. *Economic Journal* 99:116–131.

Ascher, W. and R. Healy. 1990. *Natural Resource Policymaking in Developing Countries.* Durham, NC: Duke University Press.

Ashworth, W. 1986. *The Late Great Lakes: An Environmental History.* Toronto: Collins Publishers.

Astley, G. 1984. Toward an appreciation of collective strategy. *Academy of Management Review* 9:526–535.

Astley, W. G. 1985. The two ecologies: Population and community perspectives on organizational evolution. *Administrative Science Quarterly* 30:224–241.

Aubert, E. J. 1984. Advisory mechanisms supporting the mission of the IJC. In *Review of the Great Lakes Water Quality Agreement, Working Papers and Discussion,* pp. 68–86. Washington, DC: National Academy Press.

Axelrod, R. 1984. *The Evolution of Cooperation.* New York: Basic Books.

Balch, R. E. 1946. The spruce budworm and forest management in the Maritime Provinces. Canadian Department of Agriculture, Entomology Division, Proceedings Publication No. 60.

Balch, R. E. 1952. The spruce budworm and aerial forest spraying. *Canadian Geographical Journal* 45:200–209.

Baldwin, N. S., R. W. Saalfeld, M. A. Ross, and H. J. Buettner. 1979. Commercial fish production in the Great Lakes, 1867–1977. Technical Report 3 (superseding 1962 edition), Great Lakes Fishery Commission, Ann Arbor. 187 pp.

Bancroft, G. T. 1989. Status and conservation of wading birds in the Everglades. *American Birds* 43:1258–1265.

Bandura, A. 1973. *Aggression: A Social Learning Analysis.* Englewood Cliffs, NJ: Prentice-Hall.

Bandura, A. 1977. *Social Learning Theory.* Englewood Cliffs, NJ: Prentice-Hall.

Barkow, J., R. Cosmides, and L. Tooby. 1992. *The Adapted Mind: Evolutionary Psychology and the Generation of Culture.* New York: Oxford University Press.

Baskerville, G. L. 1955. Effects of the spruce budworm outbreak on softwood reproduction, Green River. Remeasurement Report Project M-436, Forestry Canada, Fredericton, N.B. 14 pp.

Baskerville, G. L. 1960. Mortality in immature fir following severe budworm defoliation. *Forestry Chronicle* 36:342–345.

Baskerville, G. L. 1965. Deterioration and replacement in two over-mature forest stands. Canada Department of Forestry, Publication 1125. 16 pp.

Baskerville, G. L. 1975a. Spruce budworm—super silviculturist. *Forestry Chronicle* 51:1–6.

Baskerville, G. L. 1975b. Spruce budworm—the answer is forest management—or is it? *Forestry Chronicle* 51:23–26.

Baskerville, G. L. 1977a. Inconsistencies between allowable cut formulae and forest dynamics. Unpublished paper at SYLVICON '77, Annual Meeting Maritime Section, Canadian Institute of Forestry, Fredericton, N.B.

Baskerville, G. L. 1977b. New Brunswick's budworm future. *Pulp and Paper Canada.* June. 7 pp.

Baskerville, G. L. 1978. Experts, fools and why the budworm problem seems so complex. *Forestry Chronicle* 54:11–13.

Baskerville, G. L. 1979. Implementation of adaptive approaches in provincial and federal agencies. International Institute for Applied Systems Analysis, CP-79–9. 45 pp.

Baskerville, G. L. 1982. The spruce–fir wood supply in New Brunswick. New Brunswick Department of Natural Resources and Energy, Fredericton. 15 pp.

Baskerville, G. L. 1983. Good forest management: A commitment to action. New Brunswick Department of Natural Resources and Energy, Fredericton. 12 pp.

Baskerville, G. L. 1986. Understanding forest management. *Forestry Chronicle* 62:339–347.

Baskerville, G. L. 1987. Implementation of the Crown Lands & Forests Act: Observations and comments on the process. New Brunswick Department of Natural Resources and Energy, Fredericton. 57 pp.

Baskerville, G. L. 1988a. Redevelopment of a degrading forest system. *Ambio* 17:314–322.

Baskerville, G. L. 1988b. Management of publicly owned forests. *Forestry Chronicle* 64:193–198.

Baskerville, G. L. 1990. Forestry and sustainable development: the new and sustainable development, Heritage Resources Center, University of Waterloo, Ontario. pp 40–42.

Baskerville, G. L. and K. L. Brown. 1985. The different worlds of scientists and reporters. *Journal of Forestry* 83:490–493.

Baskerville, G. L. and S. Kleinschmidt. 1981. A dynamic model of growth in defoliated fir stands. *Canadian Journal of Forestry Research* 11:206–214.

Baskerville, G. L. and D. MacFarlane. 1975. Cutting out the budworm. Unpublished paper at SYLVICON '75, Annual Meeting Maritime Section, Canadian Institute of Forestry, Fredericton, N.B. 6 pp.

Baskerville, G. L. and D. A. MacLean. 1979. Budworm-caused mortality and 20-year recovery in immature balsam fir stands. Canadian Forestry Service, Information Report M-X-102. 23 pp.

Bates, J. S. 1957. *Report of the New Brunswick Forest Development Commission.* Fredericton, New Brunswick: Queens Printer. 153 pp.

Bateson, G. 1942a. Morale and national character. In G. Watson, ed., *Civilian Morale.* Society for the Psychological Study of Social Issues. Boston: Houghton Mifflin.

Bateson, G. 1942b. Social planning and the concept of deutero-learning. In *Science, Philosophy and Religion: Second Symposium.* New York: Harper & Row.

Bateson, G. 1967. Cybernetic explanation. *American Behavioral Scientist* 10(8):29–32.

Bateson, G. 1972. *Steps to an Ecology of Mind.* New York: Ballantine.

Beer, S. 1972. *Brain of the Firm.* New York: Herder and Herder.

Bengtsson, B. E., Ą. Bengtsson, and M. Himberg. 1985. Fish deformities and metal pollution. *Ambio* 14:32–35.

Benjamin, W. 1969. Theses on the philosophy of history. In H. Arendt, ed., *Illuminations,* pp. 253–264. New York: Schocken.

Bennett, A. O. 1990. Theories of individual, organizational and governmental learning and the rise and fall of Soviet military interventionism, 1973–1983. Doctoral dissertation in public policy. Harvard University, Cambridge.

Berger, T. R. 1978. Northern frontier, northern homeland: The report of the Mackenzie Valley Pipeline Inquiry. Minister of Supply and Services, Ottawa.

Bergman, A. and M. Olsson. 1985. Pathology of Baltic grey seal and ringed seal females with special reference to adrenocortical hyperplasia: Is environmental pollution the cause of a widely distributed disease syndrome? Proceedings from the Symposium on the Seals in the Baltic and Eurasian Lakes. *Finnish Game Research* 44:47–62.

Bernstein, B. 1971. *Class, Codes and Control.* London: Routledge and Kegan Paul.

Berst, A. H. and G. R. Spangler. 1973. Lake Huron: The ecology of the fish community and man's effect on it. Technical Report No. 21. Ann Arbor: Great Lakes Fisheries Commission. 41 pp.

Bertram, G. and R. Thorp. 1978. *Peru, 1890–1977: Growth and Policy in an Open Economy*. New York: Columbia University Press.

Bets, D. A., M. Gilbertson, and H. Hudson. 1990. Proceedings of the expert consultation meeting on bald eagles. Report to International Joint Commission/Science Advisory Board, Windsor, Ontario.

Bettenhausen, K. and J. K. Murnighan. 1985. The emergence of norms in competitive decision-making groups. *Administrative Science Quarterly* 30:350–372.

Bingham, G. 1986. *Resolving Environmental Disputes: A Decade of Experience.* Washington, DC: Conservation Foundation.

Blaikie, P. and H. Brookfield. 1987. *Land Degradation and Society*. London: Methuen.

Blake, N. M. 1980. *Land into Water—Water into Land: A History of Water Management in Florida (Everglades)*. Gainesville: University Presses of Florida.

Blomqvist, S., A. Frank, and L. R. Petersson. 1987. Metals in liver and kidney tissues of autumn-migrating dunlin *Calidris alpina* and curlew sandpiper *Calidris ferruginea* staging at the Baltic Sea. *Marine Ecological Program Series* 35:1–13.

Blumm, M. C. and A. Simrin. 1991. The unraveling of the parity promise: Hydropower, salmon and endangered species in the Columbia Basin. *Environmental Law* 21:657–744.

Boczek, B. A. 1989. The Baltic region in its historical context. In A. H. Westing, ed., *Comprehensive Security for the Baltic: An Environmental Approach,* pp. 23–34. International Peace Research Institute, United Nations Environment Program. London: Sage Publications.

Bohm, P. 1981. *Deposit–Refund Systems: Resources for the Future, Inc.* Baltimore: Johns Hopkins University Press.

Bonneville Power Administration. 1988. *Programs in Perspective.* Portland, OR: Bonneville Power Administration.

Borg, K., H. Wanntorp, K. Erne, and E. Hanko. 1966. Mercury poisoning in Swedish wildlife. *Journal of Applied Ecology* 3:171–172.

Bormann, F. H. and G. E. Likens. 1981. *Patterns and Process in a Forested Ecosystem.* New York: Springer-Verlag.

Botkin, D. B. 1990. *Discordant Harmonies.* New York: Oxford University Press.

Bourassa, R. 1985. *Power from the North.* Englewood Cliffs, NJ: Prentice-Hall.

Bowman, E. H. 1990. Strategy changes: Possible worlds and actual minds. In J. W. Fredrickson, ed., *Perspectives on Strategic Management,* pp. 9–37. New York: Harper & Row.

Boyce, F. M., M. N. Charlton, D. Rathke, C. H. Mortimer, and J. R. Bennett. 1988. Lake Erie bi-national study, 1979–1980. *Journal of Great Lakes Research* 13(4):406–840.

Boyd, R. and P. J. Richerson. 1985. *Culture and the Evolutionary Process.* Chicago: University of Chicago Press.

Boyle, R. H. and R. M. Mechem. 1981. There's trouble in paradise. *Sports Illustrated* 54(6):82–96.

Boynton, W., J. Garber, R. Summers, and W. M. Kemp. 1994. Inputs, transformations, and transport of nitrogen and phosphorus in Chesapeakbe Bay and selected tributaries. *Estuaries* (in press).

Braat, L. C. and I. Steetskamp. 1991. Ecological economic analysis for regional sustainable development. In R. Costanza, ed., *Ecological Economics: The Science and Management of Sustainability,* pp. 269–288. New York: Columbia University Press.

Braybrooke, D. and C. E. Lindblom. 1970. *A Strategy of Decision: Policy Evaluation as Social Process.* New York: The Free Press.

Braybrooke, D. 1987. *Philosophy of Social Science.* Englewood Cliffs, NJ: Prentice-Hall.

Breslauer, G. 1987. Ideology and learning in Soviet third world policy. *World Politics* 39:3–432.

Breslauer, G. and P. E. Tetlock. 1991. *Learning in U.S. and Soviet Foreign Policy.* Boulder, CO: Westview Press.

Brockner, J. and J. Z. Rubin. 1985. *Entrapment in Escalating Conflicts: A Social Psychological Analysis.* New York: Springer-Verlag.

Broman, D. and B. Ganning. 1985. Bivalve molluscs (*Mytilus edulis* and *Macoma baltica*) for monitoring diffuse oil pollution in a northern Baltic archipelago. *Ambio* 14:23–28.

Broman, D., A. Colmsjö, and C. Näf. 1987. Characterization of the PAC profile in settling particulates from the urban waters of Stockholm. *Bulletin Environmental Contamination and Toxicology* 38:1020–1028.

Bromley, D., W. D. Feeny, M. A. McKean, P. Peters, J. L. Gilles, R. J. Oakerson, C. F. Runge, and J. T. Thomson. 1992. *Making the Commons Work: Theory, Practice.* San Francisco: ICS Press.

Brookman, D. H., C. Copra, D. J. Ecobichon, C. T. Kang, L. Ritter, and J. Thurson. 1984. Assessment of the potential of insecticides, emulsifiers and solvent mixtures to enhance viral infection in cultured mammalian cells. *Applied and Environmental Microbiology* (January): 84:80–83.

Brooks, H. 1986. The typology of surprises in technology, institutions and development. In W. C. Clark and R. E. Munn, eds., *Sustainable Development of the Biosphere,* pp. 325–347. International Institute for Applied Systems Analysis, Laxenburg, Austria.

Brose, D. S. 1976. *The Late Prehistory of the Lake Erie Drainage Basin.* Cleveland Museum of Natural History. 355 pp.

Brown, A. H. 1948. Haunting heart of the Everglades. *National Geographic* 93:145–174.

Brown, J. S. and P. Duguid. 1991. Organizational learning and communities of practice. *Organization Science* 2(1):40–52.

Brown, K. B. 1950. The development of the forest and forest industries of New Brunswick. New Brunswick Resources Development Board, Fredericton. 100 pp.

Bruner, J. 1979. *On Knowing: Essays for the Left Hand,* 2nd Edition. Cambridge, MA: Harvard University Press.

Bruner, J. 1986. *Actual Minds, Possible Worlds.* Cambridge, MA: Harvard University Press.

Brunsson, N. 1982. The irrationality of action and action rationality: Decisions, ideologies and organizational actions. *Journal of Management Studies* 19:1.

Brunsson, N. 1985. *The Irrational Organization.* New York: John Wiley.

Bulkley, J. W. and A. P. Mathews. 1973. Water quality relationships in the Great Lakes: Analyses of a survey questionnaire. *Proceedings of the 16th Conference on Great Lakes Research.* Huron, MI: International Association for Great Lakes Research, pp. 872–879.

Bulkley, J. W., S. J. Wright, and D. Wright. 1984. Preliminary study of the diversion of 283 $m^3 s^{-1}$ (10,000 cfs) from Lake Superior to the Missouri River Basin. *Journal of Hydrology* 68:461–472.

Bureau of the Census. *County and City Data Book 1952.* Washington, DC: U.S. Government Printing Office.

Bureau of the Census. *County and City Data Book 1972.* Washington, DC: U.S. Government Printing Office.

Bureau of the Census. *County and City Data Book 1988.* Washington, DC: U.S. Government Printing Office.

Burgelman, R. 1983a. A model of interaction, strategic behavior, corporate context and the concept of strategy. *Academy of Management Review* 8(1):61–90.

Burgelman, R. 1983b. A process model of internal corporate venturing in a diversified major firm. *Administrative Science Quarterly* 28:223–244.

Burgelman, R. 1983c. Corporate entrepreneurship and strategic management: Insights from a process study. *Management Science* 29(3):1349–1364.

Burke, K. 1950. *A Rhetoric of Motives.* Englewood Cliffs, NJ: Prentice-Hall.

Burns, N. M. 1985. *Erie, the Lake that Survived.* Totowa, NJ: Rowman and Allenheld.

Burt, W. H. 1972. *Mammals of the Great Lakes Region,* 2nd Edition. Ann Arbor: University of Michigan Press.

Cairncoss, F. 1992. *Costing the Earth.* Boston: Harvard Business School Press.

Caldwell, L. K. 1988. *Perspectives on Ecosystem Management for the Great Lakes.* Albany: State University of New York Press.

Caldwell, L. K. 1991. *Emerging Boundary Environmental Challenges and Institutional Issues: Canada and the United States.* Paper to the Trilateral Conference on the United States and Its Boundary Commissions, U.S.–Mexico and U.S.–Canada. Boca Grande, FL, April 1991.

Campbell, D. T. 1960. Blind variation and selective retention in creative thought as in other knowledge processes. *Psychological Review* 67:380–400.

Campbell, D. T. 1974. Evolutionary epistemology. In P. A. Schilpp, ed., *The Philosophy of Karl Popper.* La Salle, IL: Open Court.

Canada–Ontario. 1990. Technical summary. Second report of Canada under the 1987 protocol to the 1978 Great Lakes Water Quality Agreement, prepared under the 1986 Canada–Ontario Agreement respecting Great Lakes water quality.

Carr, A. 1967. Alligators—dragons in distress. *National Geographic* 131:133–148.

Carr, R. S. and J. S. Beriault. 1984. Prehistoric man in southern Florida. In P. J. Gleason, ed., *Environments of South Florida, Present and Past, Memoir II,* pp. 1–14. Coral Gables: Miami Geological Society.

Carroll, G. R. 1984. Organizational ecology. *Annual Review of Sociology* 10:71–93.

Carrow, R. 1985. Improving the process. Proceedings of Public Affairs and Forest Management, Pesticides in Forestry Seminar. Canadian Pulp and Paper Association, Toronto. 12 pp.

Carter, L. J. 1974. *The Florida Experience: Land and Water Policy in a Growth State.* Publication for Resources Future, Inc. Baltimore: Johns Hopkins University Press.

Carter, N. E. 1991. Efficacy of *Bt* in New Brunswick 1988–90, pp E113–E116. *Proceedings of Canadian Pulp and Paper Association,* Annual Meeting. Montreal, Quebec, March 1990.

Casti, J. L. 1979. *Connectivity, Complexity and Catastrophe in Large-Scale Systems.* New York: John Wiley.

Cavalli-Sforza, L. L. and M. W. Feldman. 1981. *Cultural Transmission and Evolution: A Quantitative Approach.* Princeton, NJ: Princeton University Press.

Caverhill. P. Z. 1917. Forest policy in New Brunswick. Master of Science Thesis, University of New Brunswick, Fredericton. 72 pp.

Cederwall, H. and R. Elmgren. 1980. Biomass increase of macrobenthic fauna demonstrates eutrophication of the Baltic Sea. *Ophelia* 1(Suppl): 287–304.

Central and Southern Flood Control District. 1950. A report on water resources of Everglades National Park. Central and Southern Flood Control District, West Palm Beach.

Chase, A. 1987. *Playing God in Yellowstone.* San Diego: Harvest/HBJ.

Chomsky, N. 1959. A review of B. F. Skinner's *Verbal Behavior. Language* 35:26–58.

Chomsky, Noam. 1989. *Necessary Illusions: Thought Control in Democratic Societies.* Boston: South End Press.

Christiansen, E. 1980. *Northern Crusades: The Baltic and the Catholic Frontier, 1100–1525.* Minneapolis: University of Minnesota Press.

Christie, W. J. 1973. A review of the changes in the fish species composition of Lake Ontario. Technical Report No. 23. Great Lakes Fishery Commission, Ann Arbor. 65 pp.

Christie, W. J. 1991. Thoughts about past events and future prospects for the biotic system of Lake Ontario. Presentation to the International Association for Great Lakes Research, Buffalo. June.

Christie, W. J., M. Becker, J. W. Cowden, and J. R. Vallentyne. 1986. Managing the Great Lakes Basin as home. *Journal of Great Lakes Research* 12(1):2–17.

Ciriacy-Wantrup, S. V. 1976. Water policy and economic optimizing: Some conceptual problems in water research. *Journal of the American Economic Association* Spring:179–189.

Ciriacy-Wantrup, S. V. and J. J. Parsons. 1967. *Natural Resources: Quality and Quantity.* Berkeley: University of California Press. 217 pp.

Clark, W. C. 1985. Scales of climate impacts. *Climate Change* 7:5–27.

Clark, W. C. and G. Majone. 1985. The critical appraisal of scientific inquiries with policy implications. *Science, Technology and Human Values* 10(3): 6–19.

Clark, W. C. and R. E. Munn. 1986. *Sustainable Development of the Biosphere.* Cambridge: Cambridge University Press.

Clark, W. C., D. D. Jones, and C. S. Holling. 1979. Lessons for ecological policy design: A case study of cosystem management. *Ecological Modelling* 7:1–53.

Clark, W C. 1986. Sustainable development of the biosphere: Themes for a research program. In W. C. Clark and R. E. Munn, eds., *Sustainable Development of the Biosphere,* pp 5–48. IIASA, Laxenburg, Austria. Cambridge: Cambridge University Press,

Clark, W. C. 1989. The human ecology of global environmental change. *International Social Science Journal* 41:315–345.

Clarke, M. 1977. An economic and environmental assessment of the Florida Everglades sugarcane industry. Master of Science Thesis, Johns Hopkins University, Baltimore. 140 pp.

Clements, F. E. 1916. Plant succession: An analysis of the development of vegetation. *Publication of the Carnegie Institute of Washington* 242:1–512.

Clifford, W. L. T. 1981. Wood supply analysis by simulation. Master of Science Thesis, University of New Brunswick, Fredericton. 98 pp.

Cohen, S. 1986. Impacts of CO_2-induced climatic change on water resources in the Great Lakes Basin. *Climate Change* 8:135–153.

Colborn, T. E., A. Davidson, S. N. Green, R. A. Hodge, C. I. Jackson, and R. A. Liroff. 1990. *Great Lakes, Great Legacy?* The Conservation Foundation, Washington, DC and The Institute for Research on Public Policy, Ottawa.

Colby, P. J. 1977. Proceedings of the 1976 Percid International Symposium, *Journal of the Fisheries Research Board of Canada* 34(10): 1445–1999.

Cole, M. 1985. The zone of proximal development: Where culture and cognition create each other. In J. V. Wertsch, ed., *Culture, Communication and Cognition: Vygotskian Perspectives.* Cambridge: Cambridge University Press.

Cole, M. and S. Scribner. 1974. *Culture and Thought.* New York: John Wiley.

Collins, H. M. and T. J. Pinch. 1978. The construction of the paranormal: nothing unscientific is happening, In R. Wallis, ed., *Rejected Knowledge: Sociological Review Monograph.* Keele: University of Keele, Northwest Territories, Canada.

Collins, R. 1981. On the microfoundations of macrosociology. *American Journal of Sociology* 86(5):984–1013.

Conger, J. 1989. *The Charismatic Leader.* San Francisco: Jossey-Bass.

Cooper, R. N. 1989. International cooperation in public health as a prologue to macroeconomic cooperation. In *Can Nations Agree? Issues in International Economic Cooperation.* The Brookings Institution, Washington, DC.

Cornwell, G. W., R. L. Downing, A. R. Marshall, and J. N. Layne. 1970. Everglades water and its ecological implications. Report of the Special Team to Study the Everglades, Florida Wildlife Federation, Miami.

Costanza, R. 1987. Social traps and environmental policy. *BioScience* 37:407–412.

Costanza, R. 1991. *Ecological Economics: The Science and Management of Sustainability.* New York: Columbia University Press.

Costanza, R. and L. Cornwell. 1991. An experimental analysis of the effectiveness of an environmental assurance bonding system on player behavior in a simulated firm. Final report for U.S. Environmental Protection Agency, contract #CR-815393-01-0, Office of Policy, Planning and Evaluation, Washington, DC

Costanza, R. and L. Cornwell. 1992. The 4P approach to dealing with scientific uncertainty. *Environment* 34:12–20, 42.

Costanza, R. and T. Maxwell. 1990. Spatial ecosystem modelling using parallel processors. *Ecological Modelling* 58:159–183.

Costanza, R. and C. Perrings. 1990. A flexible assurance bonding system for improved environmental management. *Ecological Economics* 2:57–76.

Costanza, R., M. Kemp, and W. Boynton. 1993. Predictability, scale and biodiversity in coastal and estuarine ecosystems: Implications for management. *Ambio* 22:88–96.

Costanza, R., F. H. Sklar and M. L. White. 1990. Modeling coastal landscape dynamics. *BioScience* 40:91–107.

Costanza, R., L. Wainger, C. Folke, and K.-G. Maler. 1993. Modeling complex ecological economic systems. *BioScience* 43:545–555.

Council of Great Lakes Industries. 1991. Environmental workgroup biannual activities report and 1992 program. Council of Great Lakes Industries, Detroit. 4 pp.

Crispin, S. 1991. Using natural heritage data to monitor Great Lakes ecosystem health. *Endangered Species Update* 8(5,6): 1–4.

Croley, T. 1990. Laurentian Great Lakes double-CO_2 climate change hydrological impacts. *Climate Change* 17:27–47.

Cross, J. G. and M. J. Guyer. 1980. *Social Traps.* Ann Arbor: University of Michigan Press.

Crutzen, P. J. and F. Arnold. 1986. Nitric acid cloud formation in the cold Antarctic stratosphere: A major cause for the springtime ozone hole. *Nature* 324:651–655.

Curtis, H. and N. S. Barnes. 1989. *Biology.* 5th Edition. New York: Worth.

Cyert, R. and J. March. 1963. *A Behavioral Theory of the Firm.* Englewood Cliffs, NJ: Prentice-Hall.

Daft, R. and V. Buenger. 1990. Hitching a ride on a fast train to nowhere: The past and future of strategic management research. pp. 81–103. In J. W. Fredrickson, ed., *Perspectives on Strategic Management.* New York: Harper & Row.

Daft, R. and R. Lengel. 1984. Information richness. In B. Straw, ed., *Research in Organizational Behavior,* vol. 6: pp. 191–233. Greenwich, CT: JAI Press.

Dau, F. W. 1934. *Florida Old and New.* New York: G. P. Putnam and Sons.

Davidson, M. 1983. *Uncommon Sense: The Life and Thought of Ludwig von Bertalannfy (1901–1972), Father of General Systems Theory.* Los Angeles: Tarcher.

Davis, J. H., Jr. 1943. The natural features of southern Florida, especially the vegetation and the Everglades. Bulletin 25. Florida Geological Survey, Tallahassee, FL.

Davis, K. P. 1954. *American Forest Management.* New York: McGraw-Hill.

Davis, M. B. 1986. Climatic instability, time lags and community disequilibrium. In J. Diamond and T. Case, eds., *Community Ecology,* pp. 269–284. New York: Harper & Row.

Davis, S. M. 1989. Sawgrass and cattail production in relation to nutrient supply in the Everglades. In R. R. Sharitz and J. W. Gibbons, eds., *Fresh Water Wetlands and Wildlife,* pp. 325–341. Oak Ridge, TN: Office of Scientific and Technical Information, U.S. Department of Energy.

Davis, S. M. 1990. Growth, decomposition and nutrient retention of sawgrass and cattail in the Everglades. Technical Publication 90–03. South Florida Water Management District, West Palm Beach.

Davis, S. M. 1991. Phosphorus inputs and vegetation sensitivity in an oligotrophic Everglades ecosystem. South Florida Water Management District, West Palm Beach. 37 pp.

Davis, S. M. and J. C. Ogden. 1994. Towards ecosystem restoration. In S. M. Davis and J. C. Ogden, eds., *Everglades: The Ecosystem and Its Restoration.* Delray Beach, FL: St. Lucie Press.

Davis, S. M., L. H. Gunderson, W. Park, J. Richardson, and J. Mattson. 1994. Landscape dimension, composition and function in a changing Everglades ecosystem. In S. M. Davis and J. C. Ogden, eds., *Everglades: The Ecosystem and Its Restoration.* Delray Beach, FL: St. Lucie Press.

Dawkins, Richard. 1976. *The Selfish Gene.* New York: Oxford University Press.

DeBellevue, E. and R. Costanza. 1991. Unified generic ecosystem and land use model (UGEALUM): a modeling approach for simulation and evaluation of landscapes with an application of the Patuxent River Watershed. In J. A. Mihursky and A. Chaney, eds., *New Perspectives in the Chesapeake System: A Research and Management Partnership,* pp. 265–275. Publication No. 137. Solomons, MD: Chesapeake Research Consortium.

DeGrove, J. M. 1958. Approaches to water resource development in central and southern Florida, 1845–1947. Public Administration Clearing Service, University of Florida, Gainesville.

DeGrove, J. M. 1973. *The Central and Southern Florida Flood Control Project: A Study in Intergovernmental Cooperation and Public Administration.* Ann Arbor, MI: University Microfilms, A Xerox Company.

DeGrove, J. M. 1978. Administrative systems for water management in Florida. Southeast Conference on Legal and Administrative Systems for

Water Allocations, Virginia Polytechnic Institute and State University, Blacksburg.

Delcourt, H. R., P. A. Delcourt, and T. I. Webb. 1983. Dynamic plant ecology: The spectrum of vegetational change in space and time. *Quaternary Science Reviews* 1:153–175.

Delcourt, P. A. and H. R. Delcourt. 1987a. *Long-Term Forest Dynamics of the Temperate Zone: A Case Study of Late-Quaternary Forests in Eastern North America.* New York: Springer-Verlag.

Delcourt, P. A. and H. R. Delcourt. 1987b. Late-quaternary dynamics of temperate forests: Applications of paleoecology to issues of global environmental change. *Quaternary Science Reviews* 6(2):129–146.

Delfino, J. J. 1979. Toxic substances in the Great Lakes. *Environmental Science and Technology* 13(12):1460–1468.

Denis, L. G. 1916. *Water Works and Sewerage Systems of Canada.* Ottawa Commission of Conservation.

Denton, G. H. and T. J. Hughes. 1981. *The Last Great Ice Sheet.* New York: Wiley-Interscience.

Department of Environmental Regulation. 1988. Lake Okeechobee Technical Advisory Council, Interim Report to the Florida Legislature. Florida Department of Environmental Regulation, Tallahassee.

Department of Environmental Regulation. 1990. Lake Okeechobee Technical Advisory Committee. Final Report to the Florida Legislature. Florida Department of Environmental Regulation. Tallahassee, Florida.

Department of Natural Resources and Energy. 1981. New Brunswick forest inventory 1978. New Brunswick Department of Natural Resources and Energy, Fredericton. 141 pp.

Department of Natural Resources and Energy. 1987a. On the performance of Crown timber licensees. New Brunswick Department of Natural Resources and Energy, Fredericton. 71 pp.

Department of Natural Resources and Energy. 1987b. New Brunswick forest development survey: Field manual. New Brunswick Department of Natural Resources and Energy, Fredericton. 46 pp.

Department of Natural Resources and Energy. 1987c. New Brunswick Crown Lands and Forests Act—the first five years. New Brunswick Department of Natural Resources and Energy, Fredericton. 37 pp.

Department of Natural Resources and Energy. 1989. New Brunswick forest inventory 1986. New Brunswick Department of Natural Resources and Energy, Fredericton. 302 pp.

Department of Natural Resources and Energy. 1991a. Forest management manual for Crown lands. New Brunswick Department of Natural Resources and Energy, Fredericton. 89 pp.

Department of Natural Resources and Energy. 1991b. Forest land habitat management program. New Brunswick Department of Natural Resources and Energy, Fredericton. 57 pp.

Desowitz, J. W. 1991. *The Malaria Capers.* New York: W. W. Norton.

Department of Natural Resources and Energy. 1958. New Brunswick forest inventory. New Brunswick Department of Natural Resources and Energy, Fredericton. 110 pp.

Dewey, J. 1927. *The Public and Its Problems: An Essay in Political Inquiry.* Chicago: Gate Books.

Dewey, J. 1938. *Logic: The Theory of Enquiry.* New York: Holt.

Dewey, J. 1980 (original 1929). *The Quest for Certainty: A Study of the Relation of Knowledge to Action.* New York: Perigee Books.

Diamond, C. 1990. An analysis of public subsidies and externalities affecting water use in south Florida. Report of Joint Center for Environmental and Urban Problems, Florida Atlantic University/Florida International University, Ft. Lauderdale.

Dickson, R. R. 1973. The prediction of major Baltic inflows. *Deutsche Hydrografische Zeitschrift* 26:97.

Dietz, Klaus. 1967. Epidemics and rumors: A survey. *Journal of the Royal Statistical Society* Series A 130:505–528.

DiMento, J. 1986. *Environmental Law and American Business: Dilemmas of Compliance.* New York: Plenum.

DiMento, J. and D. Hesterman. 1993. Ordering the elephants to dance: consent decrees and organizational behavior. *Journal of Urban and Contemporary Law* 43:301–339.

Dohrenwend, R. E. 1977. Evapotranspiration patterns in Florida. *Florida Scientist* 40:184–192.

Donovan, A., L. Laudan, and R. Laudan. 1988. *Scrutinizing Science: Empirical Studies of Scientific Change.* Dordrecht: Kluwer.

Dore, R. P. 1961. Function and cause. *American Sociological Review* 26:843–853.

Dortch, M. S., C. F. Cerco, and D. L. Robey. 1988. Work plan for three-dimensional time-varying, hydrodynamic and water quality model of Chesapeake Bay. Miscellaneous Paper EL-88–9. Department of the Army, Waterways Experiment Station, Corps of Engineers, Vicksburg, MS.

Douglas, M. 1978. Cultural bias. Occasional paper for the Royal Anthropological Institute No. 35. Royal Anthropological Institute, London.

Douglas, M. 1986. *How Institutions Think.* Syracuse: Syracuse University Press.

Douglas, M. S. 1947 (reprinted 1986). *The Everglades: River of Grass.* Miami: Banyan Books.

Dovell, J. E. 1942. Brief history of the Florida Everglades. *Proceedings of Soil & Crop Science Society of Florida* 4:132–161.

Dovell, J. E. 1947. A history of the Everglades of Florida. Ph.D. Thesis. University of North Carolina, Chapel Hill.

Dror, Y. 1994. Governance for the 21st century. New York: Club of Rome.

Druehl, L. D., M. Cackette, and J. M. D'Auria. 1988. Geographical and temporal distribution of iodine-131 in the brown seaweed *Fucus* subsequent to the Chernobyl incident. *Marine Biology* 98:125–129.

Dryzek, J. S. 1987. *Rational Ecology: Environment and Political Economy.* Oxford: Basil Blackwood.

Dubinskas, F. 1988. *Making Time: Ethnographies of High-Technology Organizations.* Philadelphia: Temple University Press.

Dunn, W. 1980. The two-communities metaphor and models of knowledge use. *Knowledge* 1(June): 515–536.

Durkheim, E. 1938. *The Rules of Sociological Method,* 8th Edition, translated by S. A. Solovay and J. H. Mueller, edited by G. E. G. Catlin. New York: The Free Press.

Dworsky, L. B. 1986. Changes in management: International Great Lakes. *Water Forum '86, World Water Issues in Evolution,* vol. 2. pp. 2185–2172. New York: American Society of Civil Engineers.

Dworsky, L. B. 1988. The Great Lakes: 1955–1985. In L. K. Caldwell, ed., *Perspectives on Ecosystem Management for the Great Lakes: A Reader.* Albany: State University of New York Press.

Dworsky, L. B., G. R. Francis, and C. F. Swezey. 1974. Management of the international Great Lakes. *Natural Resources Journal* 14(1): 103–138.

Dybern, B. I. and H. P. Hansen. 1989. Baltic Sea patchiness experiment. PEX '86, Part 1: General Report. International Council for the Exploration of the Sea Cooperative Research Report 163:1–100.

Ecosystem Objectives Work Group. 1992. Interim report of the ecosystem objectives work group on ecosystem objectives and their environmental indicators for Lake Ontario. U.S. Environmental Protection Agency, Region II, New York, and Environment Canada, Great Lakes Office, Toronto.

Eder, Klaus. 1987. Learning and the evolution of social systems: An epigenetic perspective. In M. Schmid and F. M. Wuketits, eds., *Evolutionary Theory in Social Science.* Dordrecht: D. Reidel.

Edmundston Applied Technology Group 1991. Growth and yield project report. New Brunswick Department of Natural Resources and Energy, Fredericton. 15 pp.

Edwards, C. J. and H. A. Regier. 1990. An ecosystem approach to the integrity of the Great Lakes in turbulent times. Great Lakes Fishery Commission, Special Publication 90-4, Ann Arbor, MI.

Ehrlich, P. R. and A. H. Ehrlich. 1990. *The Population Explosion.* New York: Simon and Schuster. 320 pp.

Eichenlaub, V. 1979. *Weather and Climate of the Great Lakes Region.* Notre Dame: University of Notre Dame Press.

Ela, J. and R. A. King. 1977. *The Faces of the Great Lakes.* San Francisco: Sierra Club Books.

Eldredge, N. and S. Gould. 1972. Punctuated equilibria: An alternative to phyletic gradualism. In T. J. Schopf, ed., *Models in Paleobiology,* pp. 82–115. San Francisco: Freeman, Cooper and Company.

Eldredge, N. 1985. *Unfinished Synthesis: Biological Hierarchies and Modern Evolutionary Thought.* New York: Oxford University Press.

Elgin, D. S. and R. A. Bushnell. 1977. The limits to complexity: Are bureaucracies becoming unmanageable? *The Futurist* December:337–349.

Elliott, J. E. 1980. Marx and Schumpeter on capitalism's creative destruction: A comparative restatement. *Quarterly Journal of Economics* 95:46–58.

Elmgren R. 1984. Trophic dynamics in the enclosed, brackish Baltic Sea. *Rapp. P.-v. Réun. Cons. Int. Explor. Mer* 183:149–166.

Elmgren R. and U. Larsson. 1987. Analys av ett kustomrades eutrofiering med hjälp av förändringar i näringsämnesbelastningen: Exemplet Himmerfjärden. Nordforsk. *Miljövardsserien Publikation* 1:297–319. (In Swedish.)

Elmgren, R., R. Rosenberg, A. B. Andersin, S. Evans, P. Kangas, J. Lassig, E. Leppäkoski, and R. Varmo. 1984. Benthic macro- and meiofauna in the Gulf of Bothnia (northern Baltic). *Finnish Marine Research* 250:3–18.

Elster, J. 1979. *Ulysses and the Sirens: Studies in Rationality and Irrationality.* Cambridge: Cambridge University Press.

Elster, J. 1986. *Rational Choice.* New York: New York University Press.

Elster, J. 1989. *The Cement of Society: A Study of Social Order.* Cambridge: Cambridge University Press.

Elster, Jon 1983. *Explaining Technical Change: A Case Study in the Philosophy of Science.* Cambridge: Cambridge University Press.

Environment Canada. 1982. Review and evaluation of adaptive environmental assessment and management. Vancouver.

Environment Canada. 1991. Toxic chemicals in the Great Lakes and associated effects. Synopsis; Volume I, Contaminant levels and trends; Volume II, Effects. Departments of Fisheries and Oceans, and Health and Welfare Canada. Ministry of Supply and Services, Ottawa.

Erdle, T. A. 1990. Concept and practice of integrated harvest and protection design in management of eastern spruce-fir forests. PhD Thesis, University of New Brunswick, Fredericton.

Erdle, T. A. and C. Frame. 1991. Timber management in New Brunswick. Canada's timber resources: Proceedings of Victoria Conference, pp 87–95. Forestry Canada Information Report PI-X-101.

Erdle, T. A., and G. L. Baskerville. 1986. Optimizing timber yields in New Brunswick forests. In *Ecological Knowledge and Environmental Problem Solving,* pp. 275–300. Washington, DC: National Academy Press.

Etheredge, L. S. 1985. *Can Governments Learn? American Foreign Policy and Central America.* Elmsford, NY: Pergamon Press.

Evans, D. J. and L. H. Hemmingway. 1984. Northwest power planning: Origins and strategies. *Northwest Environmental Journal* 1:1–22.

Fairtlough, G. 1994. *Creative Compartments.* London: Adamantine.

Farb, P. 1965. Disaster threatens the Everglades. *Audubon Magazine,* September 1965.

Federal Reserve Bank of Chicago and the Great Lakes Commission. 1991. The Great Lakes economy: Looking north and south. Chicago.

Feldman, D. C. 1984. The development and enforcement of group norms. *Academy of Management Review* 9:47–53.

Fellows, E. S. 1988. New Brunswick's natural resources: 150 years of stewardship. New Brunswick Department of Natural Resources and Energy, Fredericton. 194 pp.

Festinger, L. 1957. *A Theory of Cognitive Dissonance.* Stanford: Stanford University Press.

Fishlow, A. 1990. The Latin American state. *Journal of Economic Perspectives* 4(3):61–74.

Fleck, L. 1979. *The Genesis and Development of a Scientific Fact,* translation by F. Bradley and T. J. Trenn of 1935 German edition. Chicago: University of Chicago Press.

Flieger, B. W. 1940. The post-budworm forest in the area between Dungarvon River and the northwest branch of the Miramichi River. Report, New Brunswick Department of Natural Resources and Energy, Fredericton. 38 pp.

Fogleman, V. M. 1987. Worst case analysis: A continued requirement under the National Environmental Policy Act. *Columbia Journal of Environmental Law* 13:53.

Folke, C. and N. Kautsky. 1989. The role of ecosystems for a sustainable development of aquaculture. *Ambio* 18:234–243.

Fonselius, S. 1969. Hydrography of the Baltic Deep Basins III. Fishery Board Swedish Series Hydrography 23. Lund. 97 pp.

Fowle, D. 1988. Using information to cope with risks in spruce budworm control programs in the Maritime Provinces. In *Information Needs in Risk Management,* pp 157–175. Toronto: Institute of Environmental Studies, University of Toronto.

Fox, A. B. 1959. *The Power of Small States: Diplomacy in World War II.* Chicago: University of Chicago Press.

Francis, G. R. 1990. University-based initiatives in favor of ecosystem rehabilitation for the North American Great Lakes. *Higher Education in Europe* 15(4):76–85.

Francis, G. R., A. P. Grima, H. A. Regier, and T. H. Whillans. 1985. A prospectus for the management of the Long Point ecosystem. Technical Report No. 43, Great Lakes Fishery Commission, Ann Arbor, MI.

Francis, G. R., J. J. Magnuson, H. A. Regier, and D. R. Talhelm. 1979. Rehabilitating Great Lakes ecosystems. Technical Report No. 37, Great Lakes Fishery Commission, Ann Arbor, MI. 99 pp.

Frankel, E., 1976. Corpuscular optics and the wave theory of light: The science and politics of a revolution in physics. *Social Studies of Science* 6:141–184.

Frederick, P. C. and M. W. Collopy. 1989. Nesting success of five *Ciconiiform* species in relation to water conditions in the Florida Everglades. *The Auk* 106:625–634.

Frederico, A. 1981. Lake Okeechobee water quality studies and eutrophication assessment. Technical Report, South Florida Water Management District, West Palm Beach.

Friedmann, John. 1987. *Planning in the Public Domain: From Knowledge to Action.* Princeton, NJ: Princeton University Press.

Fulton, R. J., P. F. Karrow, P. La Salle, and D. R. Grant. 1986. Summary of quaternary stratigraphy and history, eastern Canada. In V. Sibrava, D. Q. Bowan, and G. M. Richmond. Quaternary glaciation in the Northern Hemisphere. *Quaternary Science Review* 5:211–228.

Funtowicz, S. O. and J. R. Ravetz. 1991. A new scientific methodology for global environmental issues. In R. Costanza, ed., *Ecological Economics: The Science and Management of Sustainability*, pp 137–152. New York: Columbia University Press.

Gadgil, M., F. Berkes and C. Folke. 1993. Indigenous knowledge for biodiversity conservation. *Ambio* 22:151–156.

Ganning, B. and U. Billing. 1974. Effects on community metabolism of oil and chemically dispersed oil on Baltic bladder wrack, *Fucus vesiculosus.* In L. R. Beynon and E. B. Cowell, eds., *Ecological Aspects of the Toxicity of Oils and Dispersants,* pp. 53–61. New York: John Wiley.

Ganning, B. and F. Wulff. 1969. Seasonal and diurnal variations of some chemical and physical factors in brackish water rock pools, polluted by birds. *Limnologica* 7:85. (Abstract).

Gannon, P. T. 1978. Influence of earth surface and cloud properties on south Florida sea breeze. National Oceanic and Atmospheric Administration Technical Report ERL 402-NHELM2, Department of Commerce, Washington, DC.

Gardner, H. 1987. *The Mind's New Science: A History of the Cognitive Revolution.* New York: Basic Books.

Gardner, J. W. 1991. *Building Community.* Washington, DC: The Independent Sector.

Geertz, C. 1976. Funeral rites in Java. In W. A. Lessa and E. Z. Vogt, eds., *Reader in Comparative Religions.* New York: Harper & Row.

Gentry, R. C. 1984. Hurricanes in south Florida. In P. J. Gleason, ed., *Environments of South Florida: Present and Past, Memoir II,* pp. 510–517. Coral Gables, FL: Miami Geological Society.

Gerlach, L. P. and V. Hines. 1970. *People, Power, Change.* New York: Bobbs-Merrill.

Gerlach, S. A. 1985. Wurde der 1981 in der Deutschen Bucht beobachtete Sauerstoffmangel durch anthropogene Nährstoff-Frachten begünstigt? Wasser Berlin '85. Kongressvorträge. Hersusgeber AMK Berlin. Wissenschaftsverlag V. Spiess.

Gersick, C. J. G. 1991. Revolutionary change theories: A multilevel exploration of the punctuated equilibrium paradigm. *Academy of Management Review* 16:10–36.

Geuss, Raymond. 1981. *The Idea of a Critical Theory: Habermas and the Frankfurt School.* Cambridge: Cambridge University Press.

Giddens, A. 1984. *The Construction of Society: Outline of the Theory of Structuration.* Oxford: Policy Press.

Gilbert, G. N. and M. Mulkay. 1984. *Opening Pandora's Box: A Sociological Analysis of Scientists' Discourse.* Cambridge: Cambridge University Press.

Gilbertson, M., T. Kubiak, J. Ludwig and G. Fox. 1991. Great Lakes embryo mortality, edema and deformities syndrome (GLEMEDS) in colonial fish-eating birds: Similarity to chick-edema disease. *Journal of Toxicology and Environmental Health* 33:455–520.

Gimpl, M. L. and S. R. Dakin. 1984. Management and magic. *California Management Review* Fall: 125–136

Gleason, P. J., A. D. Cohen, P. Stone, W. G. Smith, H. K. Brooks, R. Goodrick, and W. Spackman, Jr. 1984. The environmental significance of Holocene sediments from the Everglades and saline tidal plains. In P. J. Gleason, ed., *Environments of South Florida: Present and Past, Memoir II,* pp. 287–341. Miami Geological Society, Coral Gables.

Gootenberg, Paul. 1989. *Between Silver and Guano: Commercial Policy and the State in Postindependence Peru.* Princeton, NJ: Princeton University Press.

Gouldner, A. 1976. *Dialectic of Ideology and Technology.* London: Macmillan.

Graham, F. 1990. Kite vs. stork. *Audubon* May:104–112.

Granovetter, M. 1973. The strengths of weak ties. *American Journal of Sociology* 78(6): 1360–1380.

Granovetter, Mark. 1985. Economic action and social structure: The problem of embeddedness. *American Journal of Sociology* 91(3):481–510.

Gray, B. 1985. Conditions facilitating interorganizational collaboration. *Human Relations* 38:911–936.

Gray, B. 1989. *Collaborating: Finding Common Ground for Multiparty Problems.* San Francisco: Jossey-Bass.

Gray, B. and T. Hay. 1986. Political limits to interorganizational consensus and change. *Journal of Applied Behavioral Science* 22:95–112.

Great Lakes Commission. 1991. The Indiana Declaration. *Advisor* 3(4): 4, 8–9. Ann Arbor.

Great Lakes Fishery Commission. 1980. Proceedings of the Sea Lamprey International Symposium, sponsored by the Great Lakes Fishery Commission. *Canadian Journal of Fisheries and Aquatic Sciences* 37(11):1585–2215.

Great Lakes Fishery Commission. 1992. Strategy vision of the Great Lakes Fishery Commission for the decade of the 1990's. Ann Arbor.

Great Lakes United. 1987. Unfulfilled promises: A citizens' review of the International Great Lakes Water Quality Agreement. Buffalo.

Great Lakes United. 1991. Broken agreement: The failure of the United States and Canada to implement the Great Lakes Water Quality Agreement. Buffalo.

Greenbank, D. O. 1956. The role of climate and dispersal in the initiation of outbreaks of the spruce budworm in New Brunswick: The role of climate. *Canadian Journal of Zoology* 34:453–476.

Greenbank, D. O. 1957. The role of climate and dispersal in the initiation of outbreaks of the spruce budworm in New Brunswick: The role of dispersal. *Canadian Journal of Zoology* 5:385–403.

Greenbank, D. O., G. W. Shaefer, and R. C. Rainey. 1980. Spruce budworm moth light and dispersal: New understanding from canopy observations, radar and aircraft. *Memoirs of the Canadian Entomological Society* No. 110, 49 pp.

Gregg, F., S. M. Born, W. B. Lord, and M. Waterstone. 1991. Institutional response to a changing water policy environment. Water Resources Research Center, University of Arizona, Tucson. 141 pp.

Grimm, E. C. 1983. Chronology and dynamics of vegetation change in the prairie–woodland region of southern Minnesota. *New Phytologist* 93:311–335.

Guillette, E. 1993. Role of the aged in community recovery following hurricane Andrew. Quick Response Report QR56. Natural Hazards Research and Applications Center, Denver.

Guire, K. E. and E. G. Voss. 1963. Distributions of distinctive shoreline plants in the Great Lakes region. *The Michigan Botanist* 2(4): 99–114.

Gunderson, L. H. 1992. Spatial and temporal hierarchies in the Everglades ecosystem with implications for water management. Ph.D. Dissertation, University of Florida, Gainesville.

Gunderson, L. H. 1994. Vegetation of the Everglades: Composition and determinants. In S. M. Davis and J. C. Ogden, eds., *The Everglades: The Ecosystem and Its Restoration.* Del Ray Beach, FL: St. Lucie Press.

Gunderson, L. H. and W. F. Loftus. 1993. The Everglades. In W. H. Martin, S. G. Boyce, and A. C. Echternacht, eds., *Biodiversity of the Southeastern United States.* New York: John Wiley.

Haas, E. 1990. *When Knowledge Is Power: Three Models of Change in International Organizations.* Berkeley: University of California Press.

Haas, P. 1990. *Saving the Mediterranean: The Politics of International Environmental Cooperation.* New York: Columbia University Press.

Habermas, J. 1979. *Communication and the Evolution of Society.* Boston: Beacon Press.

Häfele, W. 1974. Hypotheticality and the new challenges: The pathfinder role of nuclear energy. *Minerva* 10:303–323.

Hall, P. A. 1989. *The Political Power of Economic Ideas: Keynesianism Across Nations.* Princeton, NJ: Princeton University Press.

Hall, P. A. 1990. Policy paradigms, social learning and the state: The case of economic policy-making in Britain. Mimeographed.

Hall, T. H. 1978. Toward a framework for forest management decision making in New Brunswick. New Brunswick Department of Natural Resources and Energy, Fredericton. 83 pp.

Hall, T. H. 1981. Forest management decision-making: Art or science. *Forestry Chronicle* 57:233–238.

Hallberg, R. O. 1973. The microbiological C-N-S cycles in sediments and their effect on the ecology of the sediment–water interface. *Oikos Supplement* 15:51–62.

Hamel, G. and C. K. Prahalad. 1989. Strategic intent. *Harvard Business Review* 67:63–76.

Handel, W. 1979. Normative expectations and the emergence of meaning as solutions to problems: Convergence of structuralist and interactionist views. *American Journal of Sociology* 84:855–881.

Hannan, M. T. and J. Freeman. 1977. The population ecology of organizations. *American Journal of Sociology* 82:929–964.

Hannan, M. T. and J. Freeman. 1984. Structural inertia and organizational change. *American Sociological Review* 49(April): 149–164.

Hansson, S. 1984. Oljeutsläpp i Östersjön: djuren drabbas, växterna skonas. *Forskning och Framsteg* 5:20–24. (In Swedish.)

Hardin, G. 1968. The tragedy of the commons. *Science* 162:1243–1248.

Harper, R. M. 1927. Natural resources of southern Florida. 18th Annual Report, Florida Geological Survey, Tallahassee.

Harris, H. J., D. R. Talhelm, J. J. Magnuson, and A. M. Forbes. 1982. Green Bay in the future—a rehabilitative prospectus. Technical Report No. 38, Great Lakes Fishery Commission, Ann Arbor.

Harris, L. D. 1993. Some spatial aspects of biodiversity conservation. In M. A. Fenger, E. H. Miller, J. A. Johnson, and E. J. R. Williams, eds., *Our Living Legacy: Proceedings of a Symposium on Biodiversity in British Columbia,* pp. 97–108. Royal British Columbia Museum, Victoria.

Harrison, R. P. 1992. *Forests: Shadow of Civilization.* Chicago: Chicago University Press.

Harsanyi, J. C. 1986. Advances in understanding rational behavior. In J. Elster, ed., *Rational Choice.* New York: New York University Press.

Harshberger, J. W. 1914. The vegetation of south Florida, south of 27 degrees 30' north, exclusive of the Florida Keys. *Transaction of Wagner Free Institute of Science,* Philadelphia. 7:49–189.

Hartmann, H. 1990. Climate change impacts on Laurentian Great Lakes. *Climate Change* 17:49–67.

Hartwell, J. H., H. Klein, and B. F. Joyner. 1964. Preliminary evaluation of hydrologic situation in Everglades National Park, Florida. U.S. Geological Survey, Miami. Open-File Report FL-63001. 25 pp.

Hatcher, J. D. and F. M. M. White 1985. Final report. Task force on chemicals in the environment and human reproductive problems in New Brunswick. New Brunswick Department of Health, Fredericton. 269 pp.

Havighurst, W. 1975. *The Long Ships Passing: The Story of the Great Lakes,* Revised Edition. New York: Macmillan. 360 pp.

Hayes, R. H. 1985. Strategic planning: Forward in reverse? *Harvard Business Review* 63(Nov/Dec):111–119.

Hays, Samuel P. 1959. Conservation and the gospel of efficiency: the progressive conservation movement, 1890–1920. Cambridge: Harvard University Press.

Hays, S. P. and B. D. Hays. 1987. *Beauty, Health and Permanence: Environmental Politics in the United States, 1955–1985.* Cambridge: Cambridge University Press.

Hebb, D. 1954. The social significance of animal studies. In G. Lindzey, ed., *Handbook of Social Psychology,* Vol. II. Reading, MA: Addison-Wesley.

Hebert, P. N. C., C. C. Wilson, M. H. Murdoch, and R. Lazar. 1991. Demography and ecological impacts of the invading mollusc, *Dreissena polymorpha. Canadian Journal of Zoology* 69(2):405–409.

Heclo, H. 1974. *Modern Social Politics in Britain and Sweden.* New Haven: Yale University Press.

Hedberg, B. and S. Jönsson. 1977. Strategy formulation as a discontinuous process. *International Studies of Management and Organization.* Summer:88–109.

Hedberg, B. 1981. How organizations learn and unlearn. In P. C. Nystrom and W. H. Starbuck, eds., *Handbook of Organizational Design,* Vol. 1, *Adapting Organizations to Their Environments,* pp. 3–27. New York: Oxford University Press.

Heider, F. 1958. *The Psychology of Interpersonal Relations.* New York: John Wiley.

Heilpren, A. 1887. Explorations of the west coast of Florida and in the Okeechobee Wilderness. *Transaction of Wagner Free Institute of Science,* Philadelphia. 134 pp.

Heiner, R. A. 1983. The origin of predictable behavior. *American Economic Review* 73(4):560–595.

Heino, A. 1972. The Palva oil tanker disaster in the Finnish southwestern archipelago. *Aqua Fenn* 1972:116–121.

Heintz, H. T. and H. C. Jenkins-Smith. 1988. Advocacy coalitions and the practice of policy analysis. *Policy Sciences* 21:263–277.

Hela, I. 1952. Remarks on the climate of south Florida. *Bulletin of Marine Science* 2:438–447.

Helle, E., M. Olsson and S. Jensen. 1976. DDT and PCB levels and reproduction in the ringed seal from the Bothnian Bay. *Ambio* 5:188–189.

Hempel, G. 1975. An interdisciplinary marine project at the University of Kiel, Sonderforschungsbereich 95. *Meerentutkimuslait/Havs-forskningsinst.* Skr No 239:162–166.

Hengeveld, H. 1991. Understanding atmospheric change: A survey of the background science and implications of climatic change and ozone depletion. Environment Canada. State of the Environment Report No. 91–2. 67 pp.

Hilborn, R. 1987. Living with uncertainty in resource management. *North American Journal of Fisheries Management* 7:1–5.

Hiltgartner, S. and C. L. Bosk. 1988. The rise and fall of social problems: A public arenas model. *American Journal of Sociology* 94:53–78.

Hirschman, A. O. 1970. The search for paradigms as a hindrance to understanding. *World Politics* 22:329–343.

Hirschman, A. O. 1981. The rise and decline of development economics. In *Essays in Trespassing: Economics to Politics and Beyond,* pp. 1–24. Cambridge: Cambridge University Press.

Hirschman, A. O. 1989. How the Keynesian revolution was exported from the United States and other comments. In P. A. Hall, ed., *The Political Power of Economic Ideas,* pp. 347–359. Princeton, NJ: Princeton University Press.

Hochberg, J. 1964 (1978, 2nd Edition). *Perception.* Englewood Cliffs, NJ: Prentice-Hall.

Hochschild, A. R. 1983. *The Managed Heart.* Berkeley: University of California Press.

Hogarth, R. M. and M. W. Reder. 1986. *Rational Choice: The Contrast Between Economics and Psychology.* Chicago: University of Chicago Press.

Höglund, H. 1972. On the Bohuslän herring during the great herring fishery period in the eighteenth century. Institute of Marine Research, Lysekil Ser. Biol., Report No. 20. 86 pp.

Holland, J. H., K. J. Holyoak, R. E. Nisbett, and P. R.Thagard. 1986. *Induction: Processes of Inference, Learning, and Discovery.* Cambridge, MA: MIT Press.

Holling, C. S. 1973. Resilience and stability of ecological systems. *Annual Review of Ecology and Systematics* 4:1–23.

Holling, C. S. 1976. Resilience and stability of ecosystems. In E. Jantsch and C. H. Waddington, eds., *Evolution and Consciousness: Human Systems in Transition,* pp. 73–92. Reading, MA: Addison-Wesley.

Holling, C. S. 1978. *Adaptive Environmental Assessment and Management.* London: John Wiley.

Holling, C. S. 1980. Forest insects, forest fires and resilience. In H. Mooney, J. M. Bonnicksen, N. L. Christensen, J. E. Lotan, and W. A. Reiners, eds., Fire Regimes and Ecosystem Properties. USDA Forest Service General Technical Report WO-26.

Holling, C. S. 1981. Highlights of adaptive environmental assessment and management. Institute of Resource Ecology, Vancouver.

Holling, C. S. 1986. Resilience of ecosystems; local surprise and global change. In W. C. Clark and R. E. Munn, eds., *Sustainable Development of the Biosphere,* pp. 292–317. Cambridge: Cambridge University Press.

Holling, C. S. 1987. Simplifying the complex: The paradigms of ecological function and structure. *European Journal of Operational Research* 30:139–146.

Holling, C. S. 1988. Temperate forest insect outbreaks, tropical deforestation and migratory birds. *Memoirs of the Entomological Society of Canada* 146:21–32.

Holling, C. S. 1991. Role of insects in structuring the boreal landscape. In H. H. Shugart, R. Leemans, and G. B. Bonan, eds., *A Systems Analysis of the Global Boreal Forest,* pp. 170–191. Cambridge: Cambridge University Press.

Holling, C. S. 1992. Cross-scale morphology, geometry and dynamics of ecosystems. *Ecological Monographs* 62(4):447–502.

Holling, C. S. and A. D. Chambers. 1973. Resource science: The nurture of an infant. *Bioscience* 23(1):13–20.

Holling, C. S., G. B. Dantzig, W. C. Clark, D. D. Jones, G. L. Baskerville, and R. M. Peterman. 1977. Quantitative evaluation of pest management options: The spruce budworm case study. Symposium on Concepts and Practice of Integrated Pest Management in Forestry, pp. 82–102. 15th International Congress of Entomology, Washington, DC, August 1976.

Holling, C. S., G. B. Dantzig and C. Winkler. 1986. Determining optimal policies for ecosystems. In M. Kallio, A. E. Andersson, R. Seppala, and A. Morgan, eds., *System Analysis in Forestry and Forest Industries,* pp. 453–473. Amsterdam: Elsevier Science Publishers.

Holling, C. S., D. D. Jones, and W. C. Clark. 1976. Ecological policy design: Lessons from a study of forest/pest management. Institute of Resource Ecology, University of British Columbia, Vancouver. 89 pp.

Holling, C. S., L. H. Gunderson, and C. J. Walters. 1994. The structure and dynamics of the Everglades system: Guidelines for ecosystem restoration. In S. Davis and J. Ogden, eds., *Everglades: The Ecosystem and Its Restoration,* pp. 741–756. Del Ray Beach, FL: St. Lucie Press.

Holling, C. S., D. W. Schindler, B. Walker, and J. Roughgarden. 1994. Biodiversity in the functioning of ecosystems: An ecological primer and synthesis. In C. Perrings, C. S. Holling, B. O. Jansson, and K. G. Mäler, eds., *Biodiversity: Ecological and Economic Foundations.* Cambridge: Cambridge University Press.

Homans, G. C. 1964. Bringing men back in. *American Sociological Review* 29(5): 809–818.

Hook, O. and A. G. Johnels. 1972. The breeding and distribution of the grey seal, *Halichoerus grypus* Fab, in the Baltic Sea, with observations on other seals of the area. *Proceedings of the Royal Society of London* 182:37–58.

Horton, T. 1989. *Bay Country.* New York: Tickner and Fields.

Horton, T. and W. M. Eichbaum. 1991. *Turning the Tide: Saving the Chesapeake Bay.* Washington, DC: Island Press.

Hough, J. L. 1962. Geologic framework. In H. J. Pincus, ed., *Great Lakes Basin.* AAAS Publication 71. Washington, DC, pp. 3–27.

Hubbs, C. L. and K. F. Lagler. 1949. Fishes of the Great Lakes region. Cranbrook Institute of Science, Bulletin No. 26.

Hughes, J. D. 1986. Pan: Environmental ethics in classical polytheism. In E. C. Hargrove, ed., *Religion and Environmental Crisis.* Athens: University of Georgia Press.

Hunt, Earl. 1989. Cognitive science: Definition, status and questions. *Annual Review of Psychology* 40:603–630.

Huser, Tom. 1989. Into the fifth decade. History of the South Florida Water Management District 1949–1989. South Florida Water Management District, West Palm Beach.

Hutchinson, B. 1981. Running on empty. *New Florida,* October.

Hwang, C. C. 1990. Coupling estuarine hydrodynamic and water quality models. Ph.D. Dissertation, School of Engineering and Applied Sciences, University of Virginia, Charlottesville.

HydroQual, Inc. 1987. A steady state coupled hydrodynamic/water quality model of the eutrophication and anoxia process in Chesapeake Bay. Report for U.S. Environmental Protection Agency Contract No. 68–03–3319 to Battelle Ocean Sciences, for U.S. Environmental Protection Agency Chesapeake Bay Program, Annapolis.

HydroQual, Inc. 1989. Development and calibration of a coupled hydrodynamic/water quality/sediment flux model of Chesapeake Bay. Final Report, Task Order Two. U.S. Army Corps of Engineers, Waterways Experiment Station. Vicksburg, MS. Contract No. DACW39–88-D0035.

Ingram, H. M., D. E. Mann, G. D. Weatherford and H. J. Cortner. 1984. Guidelines for improving institutional analysis in water resources planning. *Water Resources Research* 20(3):323–334.

Insoff, H. I. 1984. *Implanting Strategic Management.* Englewood Cliffs, NJ: Prentice-Hall. 510 pp.

International Joint Commission. 1951. Report of the International Joint Commission, United States and Canada, on the pollution of boundary waters. Washington, DC, and Ottawa.

International Joint Commission. 1976. Further regulation of the Great Lakes. An IJC report to the governments of Canada and the United States. Washington, DC, and Ottawa.

International Joint Commission. 1978. Environmental management strategy for the Great Lakes system. Final report from the International Reference Group on Great Lakes Pollution from Land Use Activities (PLUARG). Washington, DC, and Ottawa.

International Joint Commission. 1981. Lake Erie water level study. Main report, International Lake Erie Regulation Study Board. Washington, DC, and Ottawa.

International Joint Commission. 1985. Great Lakes diversions and consumptive uses. A report to the governments of the United States and Canada under the 1977 Reference. Washington, DC, and Ottawa.

International Joint Commission/Water Quality Board. 1985. 1985 Report on Great Lakes water quality. Report to the International Joint Commission. Great Lakes Office, Windsor.

International Joint Commission. 1988. Revised Great Lake Water Quality Agreement of 1978 as amended by protocol signed November 18, 1987. Washington, DC, and Ottawa.

International Joint Commission. 1990. Fifth biennial report on Great Lakes water quality. 2 volumes. Washington, DC, and Ottawa.

International Joint Commission. 1993. Methods of alleviating the adverse consequences of fluctuating water levels in the Great Lakes–St. Lawrence River basin. A report to the governments of Canada and the United States. International Joint Commission, Ottawa and Washington, DC.

International Joint Commission/Council of Great Lakes Research Managers. 1991. A proposed framework for developing indicators of ecosystem health for the Great Lakes region. Report to the IJC from the Council of Great Lakes Research Managers. Great Lakes Office, Windsor.

International Joint Commission/Great Lakes Fishery Commission. 1990. Exotic species and the shipping industry: The Great Lakes–St. Lawrence at risk. Special report to the governments of the United States and Canada. Washington, DC, and Ottawa.

International Joint Commission/Project Management Team. 1989. Living with the Lakes: Challenges and opportunities. A progress report to the International Joint Commission. Washington, DC, and Ottawa.

International Joint Commission/Science Advisory Board. 1989. Great Lakes Science Advisory Board Report to the International Joint Commission. International Joint Commission, Ottawa and Washington DC.

International Joint Commission/Water Quality Board. 1989. 1989 Report on Great Lakes water quality. Report to the International Joint Commission. Great Lakes Office, Windsor.

International Joint Commission/Science Advisory Board. 1979. Anticipatory planning for the Great Lakes. Volume 1, Summary; Volume 2, Reports from working groups. Great Lakes Office, Windsor.

International Joint Commission/Science Advisory Board. 1991. Great Lakes Science Advisory Board report to the International Joint Commission. Ottawa and Washington, DC

International Joint Commission/Task Force. 1991. Toward a state of the Great Lakes Basin ecosystem report. Discussion draft, report of a Task Force. Great Lakes Office, Windsor.

International Joint Commission/Water Quality Board. 1975. Radioactivity subcommittee report. Appendix D, Great Lakes Water Quality 1975. Great Lakes Office, Windsor.

International Joint Commission/Water Quality Board. 1991. Review and evaluation of the Great Lakes RAP Remedial Action Plan Program, 1991. Washington, DC, and Ottawa.

Irland, L. C. 1980. Pulpwood, pesticides and people. Controlling spruce budworm in northeastern North America. *Environmental Management* 4:381–389.

Irving, H. J. and F. E. Webb. 1984a. State of the art in forest insect control in Canada as reflected by protection against spruce budworm. Annual Meeting. Entomological Society of Canada. St. Andrews, NB. 28 pp.

Irving, H. J. and F. E. Webb. 1984b. Questions of safety in aerial application of chemicals against the spruce budworm in New Brunswick. Woodlands Section, Canadian Pulp and Paper Association, WSI No. 2934. 6 pp.

Izreali, D. N. 1975. The middle manager and the tactics of power expansion. *Sloan Management Review* 17(Winter): 57–70.

Izuno, Forrest T., A. B. Bottcher, and W. A. Davis. 1992. Agricultural water quality sampling strategies. Cooperative Extension Service, Institute of Food and Agricultural Sciences, University of Florida, Gainesville.

Jameson, A. 1971. The lesson of the Askö Lab. *New Scientist* (December) 9:78–80.

Jansson, A. M. 1966. Diatoms and microfauna—producers and consumers in the *Cladophora* belt. *Veröff. Inst. Meeresforsch. Bremerhaven, Sonderb.* 2:281–288.

Jansson, A. M. 1976. Integration of economy and ecology. An outlook for the eighties. Proceedings of the Wallenberg Symposia. Stockholm. 240 pp.

Jansson, A. M. 1991. On the significance of open boundaries for an ecologically sustainable development of human societies. In R. Costanza, ed., *Ecological Economics: The Science and Management of Sustainability,* pp 449–458. New York: Columbia University Press.

Jansson, A. M. and J. Zuchetto. 1978. Energy, economic and ecological relationships for Gotland, Sweden. A regional system study. *Ecological Bulletin* 28, Stockholm. 154 pp.

Jansson, B. O. 1972. Ecosystem approach to the Baltic problem. Ecological Research Commission. Bulletin 16. 82 pp.

Jansson, B. O. 1980. Natural systems of the Baltic Sea. *Ambio* 9:128–136.

Jansson, B. O. 1988. Organization of institutes of marine sciences. *UHÄ* (Swedish National Board of Universities and Colleges) *Report* 1988:11. Stockholm.

Jansson, B. O. 1990. The Baltic pilot project. In CRE (European Rector's Conference), Project COPERNICUS. Geneva.

Jansson, B. O., W. Wilmot, and F. Wulff. 1984. Coupling the subsystems—the Baltic Sea as a case study. In M. J. R. Fasham, ed., *Flows of Energy and Materials in Marine Ecosystems,* pp. 5494–5595. New York: Plenum.

Jantsch, E. 1976. *Evolution and Consciousness: Human Systems in Transition.* Reading, MA: Addison-Wesley.

Jensen, S. 1966. Report of a new chemical hazard. *New Scientist* 32:612.

Jensen, S., A. G. Johnels, M. Olsson, and G. Otterlind. 1969. DDT and PCB in marine animals from Swedish waters. *Nature* 224:247–250.

Jernclöw, A. and R. Rosenberg. 1976. Stress tolerance of ecosystems. *Environmental Conservation* 3:43–46.

Jervis, R. 1976. *Perception and Misperception in International Politics.* Princeton, NJ: Princeton University Press.

Johnels, A. G., T. Westermark, W. Berg, G. Persson, and B. Sjöstrand. 1967. Pike (*Esox lucius* L.) and some other aquatic organisms in Sweden as indicators of mercury contamination in the environment. *Oikos* 18:323–333.

Johnson, L. 1958. A survey of water resources of Everglades National Park, Florida. Report to the Superintendent, Everglades National Park and National Park Service. 36 pp. (Mimeographed.)

Johnson, L. 1974. *Beyond the Fourth Generation.* Gainesville: University Presses of Florida.

Johsson, P. 1992. Östersjöns miljöhistoria avspeglas i bottnarna. In B. Kullinger AB, ed., Östersjön—ett hav i förändring. Naturvetenskapliga forskningrạdets ạrsbok, pp. 41–51. Uppsala. (In Swedish.)

Jonsson, S. A. and R. A. Lundin. 1977. Myths and wishful thinking as management tools. In P. C. Nystrom and W. H. Starbuck, eds., *Prescriptive Models of Organizations.* Amsterdam: North Holland.

Judd, A. 1991. Troubled waters: A five part series. *Lakeland Ledger,* March 3–7. Lakeland, Florida.

Kaasik, T. O. 1989. The geography of the Baltic region. In A. H. Westing, ed., *Comprehensive Security for the Baltic: An Environmental Approach,* pp. 15–22. International Peace Research Institute, United Nations Environment Program. London: Sage.

Kahneman, D. and A. Tversky. 1979. Prospect theory: An analysis of decision under risk. *Econometrica* 47:263–291.

Kahneman, D. and A. Tversky. 1981. The framing of decisions and the psychology of choice. *Science* 211:453–458.

Kahrl, W. 1978. *California Water Atlas.* Los Altos: William Kauffmann.

Kakadu Conservation Zone Resource Assessment Commission. 1991. Draft Report, Volumes 1 and 2, Commonwealth of Australia, Canberra.

Kanfer, F. H. 1971. The maintenance of behavior by self-generated stimuli and reinforcement. In A. Jacobs and L. Sachs, eds., *The Psychology of Private Events.* New York: Academic Press.

Kanter, R. 1983. *The Change Masters.* New York: Simon and Schuster.

Kanter, R. M. 1989. Becoming PALs: Pooling, allying and linking across companies. *Academy of Management Executive* 33:183–193.

Karrow, P. F. and P. E. Calkin. 1985. Quaternary evolution of the Great Lakes. Geological Association of Canada, Special Paper 30. St. Johns, Newfoundland.

Katzenstein, P. J. 1985. *Small States in World Markets: Industrial Policy in Europe.* Ithaca: Cornell University Press.

Kauffman, S. A. 1993. *Origins of Order: Self-organization and Selection in Evolution.* Oxford: Oxford University Press.

Kaufman, L. 1992. Catastrophic change in species-rich freshwater ecosystems: The lessons of Lake Victoria. *BioScience* 42:846–858.

Kautsky, N., H. Kautsky, U. Kautsky, and M. Waern. 1986. Decreased depth penetration of *Fucus vesiculosus* (L.) since the 1940s indicates eutrophication of the Baltic Sea. *Marine Ecological Progress Series* 28:1–8.

Kay, N. M. 1979. *The Innovating Firm: A Behavioral Theory of Corporate R & D.* New York: St. Martin's Press.

Keddy, C. J. and M. J. Sharp. 1989. Atlantic coastal plain flora conservation in Ontario. Report for the Natural Heritage League and World Wildlife Fund. World Wildlife Fund, Toronto. 35 pp.

Kent State University. 1975. *Current Institutional Arrangements, Lake Erie Basin.* Center for Urban Regionalism and Environmental Systems, Kent, Ohio. 26 pp.

Keohane, R. O. and J. Nye. 1987. Power and interdependence revisited. *International Organization* 41(4):725–753.

Kettela, E. G. 1975. Aerial spraying for protection of forests infested with spruce budworm. *Forestry Chronicle* 51:7–8.

Kettela, E. G. 1982. A cartographic history of spruce budworm defoliation, 1963–1981, in eastern North America. Forestry Canada Information Report DFC-X-14. Ottawa.

Kihlström, J. E., M. Olsson, S. Jensen, Å. Johansson, J. Ahlbom, and Å. Bergman. 1992. Effects of PCB and different fractions of PCB on the reproduction of the mink (*Mustela vision*). *Ambio* 21:563–569.

Kimberley, J. R. 1981. Managerial Innovation. In N. C. Nystrom and W. H. Starbuck, eds., *Handbook of Organizational Design*, pp. 84–101. Oxford: Oxford University Press.

Kimberly, J. R., and J. R. Miles. 1980. *The Organizational Life Cycle.* San Francisco: Jossey-Bass.

Kingdon, J. W. 1984. *Agendas, Alternatives and Public Policies.* Boston: Little, Brown.

Kitschelt, H. 1986. Four theories of public policy making and fast breeder reactor development. *International Organization* 40(1): 65–104.

Koshland, D. E., Jr. 1990. Priority one: Rescue the environment. *Science* 247:777.

Krushelnicki, B. and L. Botts. 1987. *The Great Lakes: An Environmental Atlas and Resource Book.* Environment Canada and U.S. Environmental Protection Agency, Toronto and Chicago.

Kubiak, T. J., H. J. Harris, L. M. Smith, T. R. Schwartz, D. L. Stalling, J. A. Trick, L. Sileo, D. E. Docharty, and T. C. Erdman. 1989. Microcontaminants and reproductive impairment of the Forster's tern on Green Bay, Lake Michigan—1983. *Archives of Environmental Contamination and Toxicology* 18:706–727.

Kuhn, T. S. 1970. *The Structure of Scientific Revolutions.* 2nd Edition. Chicago: University of Chicago Press.

Kuresoo, R. and E. Leibak. 1992. Estonia—increasing pressure on valuable natural heritage. *World Wildlife Federation Baltic Bulletin* 3/4:5–11. Uppsala.

Kushlan, J. A. 1987. External threats and internal management: The hydrologic regulation of the Everglades, Florida, USA. *Environmental Management* 11:109–119.

Lakatos, I. 1970. Falsification and the methodology of scientific research programs, In I. Lakatos and E. Musgrave, eds., *Criticism and the Growth of Knowledge.* Cambridge: Cambridge University Press.

Lakoff, G. and M. Johnson. 1980. *Metaphors We Live By.* Chicago: University of Chicago Press.

Landner, L. and Ạ. Hagström. 1975. Oil spill protection in the Baltic Sea. *Journal of the Water Pollution Control Federation* 47:796–809.

Larsen, C. E. 1987. The Great Lakes: Troubled waters. *National Geographic* 172(1): 2–31.

Larson, D. W. 1985. *Origins of Containment: A Psychological Explanation.* Princeton, NJ: Princeton University Press.

Larsson, Ạ., T. Andersson, L. Förlin, and J. Härdig. 1988. Physiological disturbances in fish exposed to breached kraft mill effluents. *Water Science and Technology* 20:67–76.

Larsson, U. and Ạ. Hagström. 1982. Fractionated phytoplankton primary production, exudate release and bacterial production in a Baltic eutrophication gradient. *Marine Biology* 67:57–70.

Larsson, U., R. Elmgren, and F. Wulff. 1985. Eutrophication and the Baltic Sea: Causes and consequences. *Ambio* 14:9–14.

Latour, B. and S. Woolgar. 1986. *Laboratory Life: The Construction of Scientific Facts.* Princeton, NJ: Princeton University Press.

Laudan, L. 1977. *Progress and Its Problems: Toward a Theory of Scientific Growth.* Berkeley: University of California Press.

Leach, E. 1964. Anthropological aspects of language: Animal categories and verbal abuse. In E. H. Lenneberg, ed., *New Directions in the Study of Language*. pp. 23–63. Boston: M.I.T. Press.

Lee, Kai N. 1991a. Rebuilding confidence: Salmon, science and law in the Columbia Basin. *Environmental Law* 21:745–805.

Lee, Kai N. 1991b. Unconventional power: Energy efficiency and environmental rehabilitation under the Northwest Power Act. *Annual Review of Energy and the Environment* 16:337–364.

Lee, Kai N. 1993. *Compass and Gyroscope: Integrating Science and Politics for the Environment.* Washington, DC: Island Press.

Lehtinen, K. J., M. Notini, J. Mattsson, and L. Landner. 1988. Disappearance of bladder/wrack (*Fucus vesiculosus* L.) in the Baltic Sea: Relation to pulp-mill chlorate. *Ambio* 17:387–393.

Leifer, R. 1989. Understanding organizational transformation using a dissipative structure model. *Human Relations* 42(10):899–916.

Leman, A. B. and I. A. Leman. 1970. Great Lakes megalopolis: From civilization to ecumenization. Ministry of State for Urban Affairs, Ottawa.

Leopold, L. 1969. Environmental impact of the Big Cypress Swamp Jetport. U.S. Department of the Interior, Washington, DC.

Leopold, L. B. 1990. Ethos, equity and the water resource. *Environment* 32(2): 16–20, 37–41.

Levi-Strauss, C. 1955. The structural study of myth. *Journal of American Folklore* 67:428–444.

Levieil, D. P. and B. Orlove. 1990. Local control of aquatic resources: Community and ecology in Lake Titicaca, Peru. *American Anthropologist* 92:362–382.

Levin, J. V. 1960. *The Export Economies: Their Pattern of Development in Historical Perspective.* Cambridge, MA: Harvard University Press.

Levin, S. A. 1992. The problem of pattern and scale in ecology. *Ecology* 73:1943–1967.

Levinson, D. J. 1978. *The Seasons of a Man's Life.* New York: Knopf.

Levitt, B. and J. G. March. 1990. Chester I. Barnard and the intelligence of learning. In O. E. Williamson, ed., *Organization Theory: From Chester Barnard to the Present and Beyond.* pp 11–37. Oxford: Oxford University Press.

Levitt, B. and J. G. March. 1988. Organizational learning. *Annual Review of Sociology* 14:319–340.

Lewin, A. Y. and C. Wolf. 1975. The theory of organizational slack: A critical review. In E. Shifler, ed., *Proceedings of the Twentieth International Meeting, the Institute of Management Science,* pp. 648–654. Vol. 2, Jerusalem: Jerusalem Academic Press.

Light, S. S. 1983. Anatomy of Surprise: A study of resiliency in water supply management institutions during drought. Ph.D. dissertation, University of Michigan, Ann Arbor.

Light, S. S., J. R. Wodraska and J. Sabrina. 1989. The southern Everglades—evolution of water management. *National Forum* 69(1): 11–14.

Lindblom, C. E. 1959. The "science" of muddling through. *Public Administration Review* 19:79–88.

Lindblom, C. E. and D. K. Cohen. 1979. *Usable Knowledge: Social Science and Social Problem-Solving.* New Haven: Yale University Press.

Lindén, O. 1974. Effects of oil spill dispersants on the early development of Baltic herring. *Annales Zoologica Fennica* 11:141–148.

Lindén, O., R. Elmgren, and P. Boehm. 1979. The Tsesis oil spill: Its impact on the coastal ecosystem of the Baltic Sea. *Ambio* 8:244–253.

Lindholm, T., and T. Virtanen. 1992. A bloom of *Prymnesium parvum* Carter in a small coastal inlet in Dragsfjärd, southwestern Finalnd. *Environmental Toxicology and Water Quality: An International Journal* 7:165–170.

Lindwall, B. and A. Alm. 1983. Status of the bladder-wrack community in the Svartö–Ödängla archipelago and in 16 reference localities along the coast of the county of Kalmar. *Contributions from the University of Kalmar* 5. (In Swedish.)

Lisk, J. 1967. *Struggle for Supremacy in the Baltic, 1600–1725.* New York: Funk and Wagnalls.

Loftus, K. H. and H. A. Regier. 1972. Proceedings of the 1971 symposium on salmonid communities in oligotrophic lakes. *Journal of the Fisheries Research Board of Canada* 29(6):611–986.

Lorenz, J. 1963. Deterministic non-periodic flow. *Journal of the Atmospheric Sciences* 20:130.

Loucks, O. L. 1962. A forest classification for the Maritime provinces. *Proceedings of the Nova Scotia Institute* 25:85–167.

Lovelock, J. 1988. *The Ages of Gaia.* New York: W. W. Norton.

Lowe-McConnell, R. H. 1993. Fish faunas of the African great lakes: Origins, diversity and vulnerability. *Conservation Biology* 7:634–643.

Luce, R. D. and H. Raiffa. 1957. *Games and Decisions.* New York: John Wiley.

Ludwig, D., R. Hilborn, and C. J. Walters. 1993. Uncertainty, resource exploitation and conservation: Lessons from history. *Science* 260:17, 36.

Lundberg, C. and J. E. Kihlström. 1973. DDT and the frequency of implanted ova in the mouse. *Bulletin of Environmental Contamination and Toxicology* 9:267–270.

Lundgren, L. 1974. Vattenförorening: Debatten i Sverige 1890–1921. Lund: Berlingska. (In Swedish.)

Luther, H., G. Hällfors, A. Lappalainen, and P. Kangas. 1975. Littoral benthos of the Northern Baltic Sea. I. Introduction. *Internationale Revue ges der gesamten Hydrobiologie* 60: 289–286.

Lynne, G. D. and J. Burkhardt. 1990. The evolution of water institutions in Florida: A neoinstitutional perspective. *Journal of Economic Issues* 24:1059–1077.

MacLean D. M. and T. E. Erdle. 1984. A method to determine the effects of spruce budworm on stand yield and wood supply projections in New Brunswick. *Forest Chronicle* 60:167–173.

MacVicar, T. K. 1987. Rescuing the Everglades. *Civil Engineering* 57(8):40–42.

MacVicar, T. K. and S. S. T. Lin. 1984. Historical rainfall activity in central and southern Florida: Average, return period estimates and selected extremes. In P. J. Gleason, ed., *Environments of South Florida: Present and Past, Memoir II.* Coral Gables, FL: Miami Geological Society.

Majone, G. 1980. Policies as theories. *Omega* 8:151–162.

Majone, G. 1989. *Evidence, Argument and Persuasion in the Policy Process.* New Haven: Yale University Press.

Majone, G. 1991. Cross-national sources of regulatory policymaking in Europe and the United States. *Journal of Public Policy* 11(1): 79–106.

Malling, E. 1991. Recipes for sacred cow. School of Journalism and Communications, University of Regina, Regina. 26 pp.

Maloney, F. E., R. C. Ausness, and J. Scott Morris. 1972. *A Model Water Code with Commentary.* Gainesville: University of Florida Press.

Mansfield, E. 1968. *The Economics of Technical Change.* New York: W. W. Norton.

March, J. G. and J. P. Olson. 1976. *Ambiguity and Choice in Organizations.* Bergen: Universitetsforlaget.

Marchetti, C. 1987. Infrastructures for movement. *Technological Forecasting and Social Change* 32:373–393.

Marks, S. R. 1977. Multiple roles and role strain: Some notes on human energy, time and commitment. *American Sociological Review* 42:921–926.

Marquis, R. J. and E. G. Voss. 1981. Distributions of some western North American plants disjunct in the Great Lakes region. *The Michigan Botanist* 20:53–82.

Marsh, G. P. 1864. *Man and Nature.* Cambridge, MA: Harvard University Press.

Marshall, A. R. 1972. The Kissimmee–Okeechobee Basin: A report to the Florida Cabinet. Division of Applied Ecology, University of Miami, Coral Gables, FL.

Marshall, K. B. 1975. The spruce budworm and the dollar in New Brunswick. *Forestry Chronicle* 51:9–12.

Marston, A. H., S. H. McCrory, and G. B. Hills. 1927. Report of Everglades Engineering Board of Review to Board of Commissioners of Everglades Drainage District. Everglades Engineering Board of Review, Tallahassee.

Marx, K. 1972. The eighteenth brumaire of Louis Bonaparte. In R. C. Tucker, ed., *The Marx–Engels Reader,* pp. 436–525. New York: W. W. Norton.

May, R. M. 1973. *Stability and Complexity in Model Ecosystems.* Princeton, NJ: Princeton University Press. 235 pp.

May, R. M. 1977. Thresholds and breakpoints in ecosystems with a multiplicity of stable states. *Nature* 269:471–477.

McArthur, R. H. 1955. Fluctuations in animal populations and a measure of community stability. *Ecology* 36:533–536.

McCaffery, P. 1981. Politics of the Kissimmee River Survey Review. Kissimmee River–Lake Okeechobee–Everglades Coordinating Council, Tallahassee, FL.

McCloskey, D. N. 1985 *The Rhetoric of Economics.* Madison: University of Wisconsin Press.

McCluney, W. R. 1971. *What You Can Do to Stop the Environmental Destruction of South Florida.* Coral Gables, FL: University of Miami Press.

McGuire, W. J. 1966. Attitudes and opinions. *Annual Review of Psychology* 17:36–59.

McNeill, W. 1982. The care and repair of public myth. *Foreign Affairs* 6(1): 1982.

McNeill, W. H. 1979. *The Human Condition: An Ecological and Historical View.* Princeton, NJ: Princeton University Press.

Melvasaalo, T., J. Pawlak, J. Grasshof, L. Thorell, and A. Tsiban. 1981. Assessment of the effects of pollution on the natural resources of the Baltic Sea, 1980. *Baltic Sea Environment Proceedings* 5B, Helsinki. 426 pp.

Michael, D. N. 1973. *On Learning to Plan—and Planning to Learn.* San Francisco: Jossey-Bass Publishers. 341 pp.

Michael, D. N. 1984. Reason's shadow: Notes on the psychodynamics of obstruction. *Technological Forecasting and Social Change* 26(2): 149–153.

Michael, D. N. 1989. Planning and forecasting in an incoherent context. *Technological Forecasting and Social Change* 36(1/2):79–87.

Michael, D. N. 1991a. Governing by learning in an information society. *ICIS Forum* 21:31–37.

Michael, D. N. 1991b. Leadership's shadow: the dilemma of denial. *Futures* 23(1):69–79.

Milbrath, L. W. 1988. A governance structure designed to learn would better protect the Great Lakes ecosystem. In L. K. Caldwell, ed., *Per-*

spectives on Ecosystem Management for the Great Lakes, pp. 141–167. Albany: State University of New York Press.

Miller, D. 1992. The Icarus paradox: how exceptional companies bring their own downfall. *Business Horizons* January/February:24–35.

Miller, D. and C. P. H. Friesen. 1984. *Organizations: A Quantum View.* Englewood Cliffs, NJ: Prentice-Hall.

Miller, G. A. 1979. Images and models, similes and metaphors. In A. Ortony, ed., *Metaphor and Thought.* Cambridge: Cambridge University Press.

Miller, R. B. 1913. Report on New Brunswick reconnaissance work. New Brunswick Department of Natural Resources and Energy. Fredericton. 120 pp.

Minsky, M. 1975. A framework for representing knowledge. In P. H. Winston, ed., *The Psychology of Computer Vision.* New York: McGraw-Hill.

Mintzberg, H. 1978. Patterns in strategy formation. *Management Science* 24(9):934–948.

Mintzberg, H. and J. Waters. 1982. Tracking strategy in an entrepreneurial firm. *Academy of Management Journal* 25:465–499.

Mintzberg, H. and A. McHugh. 1985. Strategy formation in adhocracy. *Administrative Science Quarterly* 30(2):160–197.

Mintzberg, H. and J. A. Waters. 1985. Of strategies, deliberate and emergent. *Strategic Management Journal* 6(3):257–272.

Mintzberg, H. 1990. Strategy formation: Schools of thought. In J. W. Frederickson, ed., *Perspectives on Strategic Management,* pp. 105–235. New York: Harper & Row.

Mintzberg, H. and F. Westley. 1992. Cycles of organizational change. *Strategic Management Journal* 13:39–59.

Mintzberg, H. and J. Rose. Strategic management upside down: A study of McGill University from 1829 to 1980. McGill University, Montreal. In preparation.

Mischel, W. 1968. *Personality and Assessment.* New York: John Wiley.

Morgan, A. V. 1987. Late Wisconsin and early Holocene paleoenvironments of east-central North America based on assemblages of fossil Coleoptera. In W. F. Ruddiman and H. E. J. Wright, eds., *North America and Adjacent Oceans During the Last Deglaciation,* pp. 353–370. Boulder: Geological Society of America.

Morgan, G. 1986. *Images of Organization.* Beverly Hills, CA: Sage.

Morris, R. F. 1963. The dynamics of epidemic spruce budworm populations. *Memoirs of the Entomological Society of Canada* 21:332.

Morse, S. S. 1993. *Emerging Viruses.* Oxford: Oxford University Press.

Mulkay, M. 1979. *Science and the Sociology of Knowledge.* London: George Allen and Unwin.

Musgrove, M. 1991. Editorial. *Miami Herald,* January 4, 1991.

Myers, R. 1983. Site susceptibility to invasion by the exotic tree *Melaleuca quinquenervia* in south Florida. *Journal of Applied Ecology* 20:645–658.

National Oceanic and Atmosphere Administration. 1989. *Lake Erie Estuarine Systems: Issues, Resources, Status and Management.* U.S. Department of Commerce, Washington, DC.

National Research Council. 1989. *Great Lakes Water Levels: Shoreline Dilemmas.* Report of a colloquium sponsored by the Water Science and Technology Board, No. 4. Washington, DC: National Academy Press. 160 pp.

National Research Council. 1993. *Sustainable Agriculture and the Environment in the Humid Tropics.* Committee on Sustainable Agriculture and the Environment in the Humid Tropics. Washington, DC: National Academy Press.

National Research Council and the Royal Society of Canada. 1985. The Great Lakes Water Quality Agreement: An evolving instrument for ecosystem management. Washington, DC: National Academy Press.

National Wildlife Federation and the Canadian Institute for Environmental Law and Policy. 1991. A prescription for healthy Great Lakes. Ann Arbor and Toronto.

Naughton, L. and S. Sanderson. 1993. Property, politics and wildlife conservation. Department of Political Science, University of Florida, Gainesville.

Nehlsen, W., J. E. Williams, and J. Lichatowich. 1991. Pacific salmon at the crossroads: Stocks at risk from California, Oregon, Idaho and Washington. *Fisheries* 16(2): 4ff.

Nehring, D., H. P. Hansen, L. A. Jörgensen, A. Trzosinska, F. Wulff, and A. Yurkovskis. 1990. Nutrients. *Baltic Sea Environment Proceedings* 35B:109–152. Helsinki.

Nelson, R. R. and S. G. Winter. 1982. *An Evolutionary Theory of Economic Change.* Cambridge: Harvard University Press.

New Brunswick Cabinet Secretariat. 1976. Report of the Task Force for Evaluation of Budworm Control Alternatives. Fredericton. 210 pp.

New Directions. 1991. Reducing and eliminating toxic substances emissions: An action plan for Canada. Ottawa.

Newell, A. and H. A. Simon. 1972. *Human Problem-Solving.* Englewood Cliffs, NJ: Prentice-Hall.

Newell, R. I. E. 1988. Ecological changes in Chesapeake Bay: Are they the result of overharvesting the American oyster, *Crassostrea virginica? Understanding the Estuary: Advances in Chesapeake Bay Research* 129:536–546.

Nicolis, G. and I. Prigogine. 1977. *Self-organization in Non-equilibrium Systems.* New York: Wiley-Interscience.

Niemi, Ạ.1979. Blue-green algal blooms and the N:P ratio in the Baltic Sea. *Acta Botanica Fennica* 110:57–61.

Nisbett, R. and L. Ross. 1980. *Human Inference: Strategies and Shortcomings of Social Judgment.* Englewood Cliffs, NJ: Prentice-Hall.

Nissling, A. and L. Westin. 1991. Egg mortality and hatching rate of Baltic cod (*Gadus morhua*) in different salinities. *Marine Biology* 111:29–32.

Nonaka, I. 1988. Creating organizational order out of chaos. *California Management Review* 30(3):57–73.

Norgaard, R. B. and R. B. Howarth. 1991. Sustainability and discounting the future. In R. Costanza, ed., *Ecological Economics: The Science and Management of Sustainability,* pp. 88–101. New York: Columbia University Press.

North, D. C. 1990. *Institutions, Institutional Change and Economic Performance.* New York: Cambridge University Press.

Northwest Power Planning Council. 1987. Columbia River Basin Fish and Wildlife Program. Portland.

Northwest Power Planning Council. 1991. Northwest Conservation and Electric Power Plan. Portland.

Notini, M. 1978. Long-term effects of an oil spill on *Fucus* macrofauna in a small bay. *Journal Fisheries Research Board of Canada* 35:745–753.

Nuclear Awareness Project. 1990. *Great Lakes Nuclear Hot Spots.* Interpretive Map. Ontario.

Nye, J. S. 1987. Nuclear learning and U.S.–Soviet security regimes. *International Organization* 41(3):371–402.

Nystrom, N. C. and W. H. Starbuck. 1981. *Handbook of Organizational Design.* Oxford: Oxford University Press.

Nystrom, P. C. and W. H. Starbuck. 1984. To avoid organizational crises, unlearn. *Organizational Dynamics* 12(Spring):53–65.

Odén, B. 1980. Human systems in the Baltic area. *Ambio* 9:116–127.

Odsjö, T. and C. Edelstam. 1972. Effect of toxic chemicals on bird life. *Bulletin International Council for Bird Preservation* 12:114–120.

Odsjö, T. and M. Olsson. 1987. Övervakning av miljögifter i levande organismer. Statens naturvạrdsverk, Solna, Sweden. (In Swedish.) 49 pp.

Odum, E. P. 1969. *Ecology.* Philadelphia: Saunders.

Odum, H. T. 1971. *Environment, Power and Society.* New York: Wiley-Interscience.

Odum, H. T. 1977. The south Florida study: Seeking a balance of man and nature. Report from Center for Wetlands, University of Florida, Gainesville.

Odum, H. T. and C. M. Hoskin. 1958. Comparative studies of the metabolism of marine waters. *Publication of the Institute of Marine Science, University of Texas.* 5:16–46.

Ogden, J. C. 1978. Recent population trends of colonial wading birds on the Atlantic and Gulf coastal plains. In A. Sprunt, J. Ogden, and R. Winckler, eds., *Wading Birds,* pp. 137–153. Research Report No. 7. New York: National Audubon Society.

Oliver, C. 1990. Determinants of interorganizational relationships: Integration and future directions. *Academy of Management Review* 15(2):241–265.

Olivieri, I., D. Couvet, and P. H. Gouyon. 1990. The genetics of transient populations: Research at the metapopulation level. *Tree* 5(7):207–210.

Olsson, M., B. Karlsson, and E. Ahnland. 1992. Seals and seal protection: A presentation of a Swedish research project. *Ambio* 21:494–496.

Olsson, M. and L. Reutergård. 1986. DDT and PCB pollution trends in the Swedish aquatic environment. *Ambio* 15:103–109.

O'Neill, R. V., D. L. DeAngelis, J. B. Waide, and T. F. H. Allen. 1986. *A Hierarchical Concept of Ecosystems.* Princeton, NJ: Princeton University Press.

Opp, K. D. 1982. The evolutionary emergence of norms. *British Journal of Sociology* 21:139–149.

Orians, G. H. 1975. Diversity, stability and maturity in natural ecosystems. In W. H. van Dobben and R. H. Lowe-McConnell, eds., *Unifying Concepts in Ecology,* pp. 139–150. The Hague: Junk.

Ostrom, E. 1986. An agenda for the study of institutions. *Public Choice* 48(3):3–25.

Ostrom, E. 1990. *Governing the Commons: The Evolution of Institutions for Collective Action.* Cambridge: Cambridge University Press.

Otterlind, G. 1986. Swedish fishery research and marine eutrophication. *Medd. Havsfiskelab. Lysekil* 316. 16 pp. (In Swedish.)

Pacyna, J. M., S. Larssen, and A. Semb. 1991. European survey for NO_x emissions with emphasis on eastern Europe. *Atmospheric Environment* 25A:425–440.

Paehlke, R. C. 1989. *Environmentalism and the Future of Progressive Politics.* New Haven: Yale University Press.

Parker, G. G. 1984. Hydrology of the pre-drainage system of the Everglades in southern Florida. In P. J. Gleason, ed., *Environments of South Florida: Present and Past, Memoir II.* Coral Gables, FL: Miami Geological Society.

Parker, G. G., G. E. Ferguson, and S. K. Love. 1955. Water resources of southeastern Florida with special reference to geology and groundwater of the Miami area. Water Supply Paper 1255. U.S. Geological Survey, Washington, DC.

Pascale, R. 1984. Perspectives on strategy: The real story behind Honda's success. *California Management Review* 26(3):47–72.

Pascale, R. 1987. The problem of strategy. Paper presented at the Seventh Annual Conference of the Strategic Management Society, Boston.

Pearse, P. H. 1987. Property rights in the development of natural resource policies in Canada. E. B. Eddy Lecture, Faculty of Forestry, University of Toronto, Toronto. 16 pp.

Pelkonen, K. and P. Tulkki. 1972. The Palva oil tanker disaster in the Finnish southwest archipelago III. The littoral fauna of the oil polluted area. *Aqua Fennica* 1972:129–141.

Penn and Schoen Associates, Inc. 1988. Report on the public opinion research project for the South Florida Water Management District of Elites and the General Public. New York. 183 pp.

Perrings, C. 1991. Reserved rationality and the precautionary principle: Technological change, time and uncertainty in environmental decision making. In R. Costanza, ed., *Ecological Economics: The Science and Management of Sustainability*, pp. 153–166. New York: Columbia University Press.

Peters, T. J. and R. H. Waterman. 1982. *In Search of Excellence.* New York: Harper & Row.

Phillips, D. W. and J. A. W. McCulloch. 1972. The climate of the Great Lakes Basin. Climatological Studies No. 20, Atmospheric Environment Service. Environment Canada, Toronto.

Platt, J. 1973. Social traps. *American Psychologist* 28:642–651.

Polanyi, M. 1958. *Personal Knowledge: Toward a Post-critical Philosophy.* Chicago: University of Chicago Press.

Polanyi, M. 1966. *The Tacit Dimension.* Garden City, NY: Doubleday.

Popper, K. R. 1959. *The Logic of Scientific Discovery.* London: Hutchinson.

Popper, K. R. 1963. *Conjectures and Refutations: The Growth of Scientific Knowledge.* London: Routledge and Kegan Paul.

Popper, K. R. 1972. *Objective Knowledge: An Evolutionary Approach.* Oxford: Oxford University Press.

PRC Environmental Management. 1986. Performance bonding. A final report prepared for the Office of Waste Programs and Enforcement, U.S. Environmental Protection Agency, Washington, DC.

Premiers Roundtable. 1992. Towards sustainable development in New Brunswick: a plan for action. Queen's Printer, Province of New Brunswick. 27 pp.

Prendergast, J. R., R. M. Quinn, J. H. Lawton, B. C. Eversham, and J. W. Gibbons. 1993. Rare species, the coincidence of diversity hotspots and conservation strategies. *Nature* 365:335–337.

Prigogine, I. 1980. *From Being to Becoming: Time and Complexity in the Physical Sciences.* New York: W. H. Freeman.

Prince, G. H. 1921. Forest survey report 1921. New Brunswick Department of Natural Resources and Energy, Fredericton. 350 pp.

Quinn, J. B. 1985. Managing innovation: Controlled chaos. *Harvard Business Review* 63(March/June): 73–84.

Ragotzkie, R. A. 1990. Great Lakes research: editorial. *Journal of Great Lakes Research* 16(1): 1–2.

Ransom, S., S. Hinings, and R. Greenwood. 1980. The structure of organizational structures. *Administrative Science Quarterly* 25:117.

Raphael, C. N. 1987. Prehistoric and historic wetland heritage of the Upper Great Lakes. *Michigan Academician* 19(3): 331–365.

Rapoport, A. and A. M. Chammah. 1965. *Prisoner's Dilemma: A Study in Conflict and Cooperation.* Ann Arbor: University of Michigan Press.

Rappaport, R. 1971. The sacred in human evolution. *Annual Review of Ecology and Systematics* 2:23–44.

Rapport, D. J. and H. A. Regier. 1992. Disturbance and stress effects on ecological systems. In B. C. Patten, ed., *Complex Ecology: The Part–Whole Relation in Ecosystems.* Englewood Cliffs, NJ: Prentice-Hall.

Raschke, E. 1992. Baltic Sea Experiment (Baltex), Geesthacht, Germany.

Rasmussen, E. M. 1985. El Niño and variations in climate. *American Scientist* 73:168–177.

Rathje, W. and C. Murphy. 1992. *Rubbish.* New York: Harper-Collins.

Rathke, D. E. and C. J. Edwards. 1985. A review of trends in Lake Erie water quality with emphasis on the 1978–1979 intensive survey. Report to the Surveillance Committee, International Joint Commission/Water Quality Board, Windsor, Ontario.

Rawson Academy of Aquatic Science. 1989. Towards an ecosystem charter for the Great Lakes–St. Lawrence. Occasional Paper No. 1, the Rawson Academy of Aquatic Science, Ottawa.

Redford, K. H. and S. E. Sanderson. 1992. The brief barren marriage of biodiversity and sustainability. *Bulletin of the Ecological Society of America* 73:36–39.

Reed, F. L. C. and G. L. Baskerville. 1991. A contemporary perspective on silviculture investments. *Journal of Business Administration* 19:161–185.

Regier, H. A. 1990. Indicators of ecosystem integrity. In D. MacKenzie, ed., *Proceedings of the International Symposium on Ecosystem Indicators,* pp. 183–200. Amsterdam: Elsevier Science Publications.

Regier, H. A. 1992. Ecosystem integrity in the Great Lakes Basin: A historical sketch of ideas and actions. *Journal of Aquatic Ecosystem Health* 1(1): 25–37.

Regier, H. A. and G. L. Baskerville. 1986. Sustainable redevelopment of regional ecosystems degraded by exploitive development. In W. C. Clark and R. Munn, eds., *Sustainable Development in the Biosphere,* pp. 75–103. Cambridge: Cambridge University Press.

Regier, H. A. and E. Bronson. 1992. New perspectives on sustainable development and barriers to relevant information. *Environmental Monitoring and Management* 20:111–120.

Regier, H. A. and R. L. France. 1990. Perspectives on the meaning of ecosystem integrity in 1975. In C. J. Edwards and H. A. Regier, eds., *An Ecosystem Approach to the Integrity of the Great Lakes in Turbulent Times,* pp. 1–15. Great Lakes Fishery Commission, Special Publication 90–4, Ann Arbor.

Regier, H. A. and J. R. Goodier. 1992. Irruption of sea lamprey in the upper Great Lakes: Analogous events to those that may follow climate warming. In M. H. Glantz, ed., *Climate Variability, Climate Change and Fisheries,* pp. 185–211. Cambridge: Cambridge University Press.

Richards, J. 1990. Land transformation. In B. L. Turner II, W. C. Clark, R. W. Kates, J. F. Richards, J. T. Mathews, and W. B. Meyer, eds., *The Earth as Transformed by Human Action: Global and Regional Changes in the Biosphere Over the Past 300 Years,* pp. 163–178. Cambridge: Cambridge University Press.

Robbin, D. M. 1984. A new Holocene sea level curve for the upper Florida Keys and Florida reef tract. In P. J. Gleason, ed., *Environments of South Florida: Present and Past, Memoir II,* pp. 437–458. Coral Gables, FL: Miami Geological Society.

Robertson, W. B., Jr., G. Sprugel, and L. Sumner. 1966. Everglades National Park Natural Sciences Research Plan. U.S. Department of the Interior, National Park Service, Washington, DC.

Robinson, J. 1993. The limits of caring: Sustainable living and the loss of biodiversity. *Conservation Biology* 7:20–28.

Robinson, M. and L. Tiger. 1991. *Man and Beast Revisited.* Washington, DC: Smithsonian Institute.

Rogers, E. M. and F. F. Shoemaker, 1971. *Communication of Innovations.* New York: The Free Press.

Rogoff, B. 1990. *Apprenticeship in Thinking: Cognitive Development in Social Context.* New York: Oxford University Press.

Rönnberg, O. 1981. Traffic effects on rocky-shore algae in the Archipelago Sea, Southwest Finland. *Acta Academiae Aboensis,* Series B 41(3): 86.

Ropelewski, C. F. and M. S. Halpert. 1987. Global and regional scale precipitation patterns associated with the El Niño/Southern Oscillation. *Monthly Weather Review* 115:1606–1626.

Rosa, F. and N. M. Burns. 1987. Lake Erie central basin oxygen depletion changes from 1929–1980. *Journal of Great Lakes Research* 13(4):684–696.

Rose, R. 1991. What is lesson-drawing? *Journal of Public Policy* 11(1): 3–30.

Rosenberg, R., H. Blanck, and O. Lindahl. 1988. Silent spring in the sea. *Ambio* 17:289–290.

Rosenberg, R., R. Elmgren, S. Fleischer, P. Jonsson, G. Persson, and H. Dahlin. 1990. Marine eutrophication case studies in Sweden. *Ambio* 19:102–108.

Rouse, J. 1987. *Knowledge and Power: Toward a Political Philosophy of Science.* Ithaca: Cornell University Press.

Rowe, J. S. 1959. Forest regions of Canada. Bulletin No. 123, Forestry Branch, Department of Northern Affairs and National Resources, Ottawa. 71 pp.

Rubenstein, D. and R. W. Woodman. 1984. Spiderman and the Burma Raiders: Collateral organizational theory in action. *Journal of Applied Behavioral Science* 20(1): 1–21.

Rudestam, L. G., S. Hansson, S. Johansson, and U. Larsson. 1992. Dynamics of planktivory in a coastal area of the northern Baltic Sea. *Marine Ecology Progress Series* 80:159–173.

Rumelhart, D. 1980. Schemata: The building blocks of cognition. In R. Spiro, ed., *Theoretical Issues in Reading Comprehension.* Hillsdale, NJ: Lawrence Erlbaum.

Runyon, K. 1984. Review of analytical economic studies of forest insect protection programs. Maritime Center, Forestry Canada, Fredericton. 86 pp.

Rushton, J. P. 1982. Altruism and society: A social learning perspective. *Ethics* 92:425–446.

Sabatier, P. A. 1988. An advocacy coalition framework of policy change and the role of policy-oriented learning therein. *Policy Sciences* 21:129–168.

Sabatier, P. A. and H. C. Jenkins-Smith. 1988. Policy change and policy-oriented learning: Exploring an advocacy coalition framework. *Policy Sciences* 21:123–127.

Salter, L. 1988. *Mandated Science: Science and Scientists in the Making of Standards.* Dordrecht: Kluwer.

Sandel, M. 1982. *Liberalism and the Limits of Justice.* New York: Cambridge University Press.

Sanderson, S. E. 1992. *The Politics of Trade in Latin American Development.* Stanford: Stanford University Press.

Sarason, S. 1972. *The Creation of Settings and the Future Societies.* San Francisco: Jossey-Bass.

Saulesleya, A. 1986. *Great Lakes Climatological Atlas.* Atmospheric Environment Service, Environment Canada, Ottawa.

Schalk, R. 1986. Estimating salmon and steelhead usage in the Columbia Basin before 1850: The anthropological perspective. *Northwest Environmental Journal* 2:1–30.

Schank, R. and R. P. Abelson. 1977. *Scripts, Plans, Goals and Understanding: An Enquiry into Human Knowledge Structure.* Hillsdale, NJ: Lawrence Erlbaum.

Schein, E. 1985. (2nd Edition 1992.) *Organizational Culture and Leadership: A Dynamic View.* San Francisco: Jossey-Bass.

Schein, E. 1988. *Process Consultation,* Vols. 1 and 2. Reading, MA: Addison-Wesley.

Schein, E. 1992. *Organization Culture and Leadership.* San Francisco: Jossey-Bass.

Schein, E. 1993. How can organizations learn faster: The challenge of entering the green room. *Sloan Management Review* 34:85–92.

Schmid, M. 1987. Collective action and the selection of rules. Some notes on the evolutionary paradigm in social theory. In M. Schmid and F. M. Wuketits, eds., *Evolutionary Theory in Social Science.* Dordrecht: D. Reidel.

Schmid, M. and F. M. Wuketits. 1987. *Evolutionary Theory in Social Science.* Dordrecht: D. Reidel.

Schneider, D. J., A. H. Hastorf, and P. C. Ellsworth. 1970. *Person Perception,* 2nd Edition. Reading, MA: Addison-Wesley.

Scholl, D. W. and M. Stuiver. 1984. Florida submergence curve revised: Its relation to coastal sedimentation rates. In P. J. Gleason, ed., *Environments of South Florida: Present and Past, Memoir II.* Coral Gables, FL: Miami Geological Socicty.

Schon, D. A. 1971. *Beyond the Stable State.* New York: W. W. Norton.

Schon, D. A. 1979. Generative metaphor: A perspective on problem-setting in social policy. In A. Ortony, ed., *Metaphor and Thought.* Cambridge: Cambridge University Press.

Schubel, J. R. 1981. *The Living Chesapeake.* Baltimore: Johns Hopkins University Press.

Schubel, J. R. 1986. *The Life and Death of the Chesapeake Bay.* College Park: Maryland Sea Grant.

Schultz, S. 1970. Der Lebensraum Ostsee. Ökologische Probleme in einem geschichteten Brackwassermeer. *Biologische Rundschau* 8:208–218.

Schumacher, E. F. 1973. *Small Is Beautiful: Economics as If People Mattered.* New York: Harper & Row.

Schumacher, E. F. 1977. *A Guide to the Perplexed.* London: Harper & Row.

Schumpeter, J. A. 1934. *The Theory of Economic Development.* Cambridge, MA: Harvard University Press.

Schumpeter, J. A. 1950. *Capitalism, Socialism and Democracy.* New York: Harper & Row.

Schwartz, P. 1991. *The Art of the Long View.* New York: Doubleday.

Sen Gupta, R. 1972. Photosynthetic production and its regulating factors in the Baltic. *Sea Marine Biology* 18:82–92.

Shaffer, G. 1979. On the phosphorus and oxygen dynamics of the Baltic Sea. *Contributions of the Askö Laboratory, University of Stockholm,* Stockholm 26:1–90.

Shaffer, G. and U. Rönner. 1984. Denitrification of the Baltic proper deep water. *Deep-Sea Research* 31:197–220.

Shaffer, M. L. 1985. The metapopulation and species conservation: The special case of the northern spotted owl. In R. J. Gutierrez and A. B. Carey, eds., *Ecology and Management of the Spotted Owl.* General Technical Report, USDA Forest Service, Pacific Northwest Forest and Range Equipment Station, Portland.

Shepard, P. 1978. *Thinking Animals.* New York: Viking.

Shklar, J. N. 1964. *Legalism.* Cambridge, MA: Harvard University Press.

Siegel, S. 1957. Level of aspiration and decision-making. *Psychological Review* 64:253–262.

Simon, H. A. 1955. A behavioral model of rational choice. *Quarterly Journal of Economics* 69:99–118.

Simon, H. A. 1959. Theories of decision-making in economics. *American Economic Review* 49:253–283.

Simon, H. A. 1962. The architecture of complexity. *Proceedings American Philosophical Society* 106:467–482.

Simon, H. A. 1974. The organization of complex systems. In H. H. Pattee, ed., *Hierarchy Theory: The Challenge of Complex Systems.* New York: George Braziller.

Simon, H. A. 1983. *Reason in Human Affairs.* Stanford: Stanford University Press.

Simon, J. L. and H. Kahn. 1984. *The Resourceful Earth: A Response to Global 2000.* Oxford: Basil Blackwell.

Simonov, A. 1990. Katasrofa tankera "Globe Assismi" v portu Klaipeda i ee ekologicheskie posledstviaj (rezultat issledovanij po mezhuvedomstvennoi programme). Moskva: Moskovskoje otdlenie Gigdrometeoizdata. 231 pp. (In Russian.)

Simpson, C. T. 1920. *In Lower Florida Wilds.* New York: Putnam.

Singh, J. V. and C. J. Lumsden. 1990. Theory and research in organizational ecology. *Annual Review of Sociology* 16:161–195.

Skinner, B. F. 1957. *Verbal Behavior.* New York: Appleton-Century-Crofts.

Skinner, B. F. 1971. *Beyond Freedom and Dignity.* New York: Knopf.

Slocombe, D. S. 1990. Assessing transformation and sustainability in the Great Lakes Basin. *GeoJournal* 21(3): 251–272.

Small, J. K. 1929. *From Eden to Sahara, Florida's Tragedy.* Lancaster, PA: Science Press. 140 pp.

Smith, B. 1848. Report on reconnaissance of the Everglades. Report to the Secretary of the Treasury, Washington, DC.

Smith, P. G. R. 1987. Towards the protection of Great Lakes Natural Heritage Areas. Technical Paper No. 2, Heritage Resources Centre, University of Waterloo, Ontario.

Smith, T. J., III, H. Hudson, G. V. N. Powell, M. B. Robblee, and P. J. Isdale. 1989. Freshwater flow from the Everglades to Florida Bay: A historical reconstruction based on fluorescent banding in the coral *Solenastrea bournoni. Bulletin of Marine Science* 44:12–24.

Södergren, A., B. E. Bengtsson, P. Jonsson, S. Lagergren, Ạ. Larsson, M. Olsson, and L. Renberg. 1988. Summary of results from the Swedish project of Environment and Cellulose. *Water Science and Technology* 20:49–60.

Söderqvist, T. 1986. *The Ecologists, from Merry Naturalists to Saviors of the Nation.* Stockholm: Almqvist and Wiksell.

South Florida Water Management District. 1978. Water Use and Supply Development Plan. West Palm Beach.

South Florida Water Management District. 1981. Water quality management strategy for Lake Okeechobee. West Palm Beach.

Spencer, R., J. Kirton, and K. R. Nossel. 1981. *The International Joint Commission Seventy Years On.* Toronto: Centre for International Studies, University of Toronto.

Spender, J. C. 1989. *Industry Recipes: The Nature and Sources of Managerial Judgement.* Oxford: Basil Blackwell.

Spitzer, W. O. 1982. Report of the New Brunswick Task Force on the Environment and Reyes Syndrome. *Clinical and Investigative Medicine* 5(2/3): 201–214.

Spooner, B. 1987. Insiders and outsiders in Baluchistan: Western and indigenous perspectives on ecology and development. In P. D. Little and M. M. Horowitz, eds., *Lands at Risk in the Third World,* pp. 58–68. Boulder: Westview Press.

Steinbruner, J. 1974. *The Cybernetic Theory of Decision.* Princeton, NJ: Princeton University Press.

Steinmo, S., K. Thelen, and F. Longstreth. 1992. *Structuring Politics: Historical Institutionalism in Comparative Analysis.* Cambridge: Cambridge University Press.

Stephens, G. W. 1930. *The St. Lawrence Waterway Project.* Toronto: Louis Carrier. 460 pp.

Stephens, J. C. 1984. Subsidence of organic soils in the Florida Everglades—a review and update. In P. J. Gleason, ed., *Environments of South Florida:*

Present and Past, Memoir II, pp. 375–384. Coral Gables, FL: Miami Geological Society.

Stephens, J. C. and L. Johnson. 1951. Subsidence of organic soils in the upper Everglades region of Florida. *Proceedings of Soil and Crop Science Society of Florida* 11:191–237.

Stevens, R. J. J. 1988. A review of Lake Ontario water quality with emphasis on the 1981–1982 intensive years. Report to the Surveillance Subcommittee, International Joint Commission Water Quality Board. International Joint Commission, Great Lakes Office, Windsor, Ontario.

Stewart, I. 1989. *Does God Play Dice? The Mathematics of Chaos.* Oxford: Basil Blackwell.

Stigebrandt, A. 1983. A model for the exchange of water and salt between the Baltic and the Skagerrak. *Journal of Physical Oceanography* 13:411–427.

Stigebrandt, A. and F. Wulff. 1987. A model for the dynamics of nutrients and oxygen in the Baltic proper. *Journal of Marine Research* 45:729–759.

Stuart, R. B. 1989. Social learning theory: A vanishing or expanding presence? *Psychology: A Journal of Human Behavior* 26(1): 35–50.

Summers, R. 1990. Description of the Patuxent Watershed nonpoint source water quality monitoring and modeling program data base and data management system. Technical Report No. 56. Maryland Department of Health and Mental Hygiene, Office of Environmental Programs, Water Management Administration, Division of Modeling and Analysis, Baltimore.

Swain, J. M. and F. C. Craighead. 1924. Studies on the spruce budworm. Canada Department of Agriculture Technical Bulletin 37, Ottawa.

Tabb, D. C., T. R. Alexander, and E. J. Heald. 1972. Comment on environmental constraints and description of major plant community description for the aerojet properties in southern Dade County, Florida. Rosenstiel School of Marine and Atmospheric Science, University of Miami, Coral Gables, FL.

Talhelm, D. R. 1988a. The International Great Lakes Sport Fishery of 1980. Great Lakes Fishery Commission, Special Publication 88-4, Ann Arbor, MI.

Talhelm, D. R. 1988b. Economics of Great Lakes fisheries: A 1985 assessment. Great Lakes Fishery Commission, Technical Report No. 54, Ann Arbor, MI.

Tamashiro, H. 1984. Algorithms, heuristics, and the artificial intelligence modelling of strategic statecraft. In D. A. Sylvan and S. Chan, eds., *Foreign Policy Decision Making.* New York: Praeger.

Tanner, H. H. 1987. *Atlas of Great Lakes Indian History.* Norman: University of Oklahoma Press.

Taylor, M. 1987. *The Possibility of Cooperation.* Cambridge: Cambridge University Press.

Teger, A. I. 1980. *Too Much Invested to Quit.* New York: Pergamon.

Thielen, B. 1963. Florida rides a space-age boom. *National Geographic* 124:858–903.

Thomas, C. Y. 1988. *The Poor and the Powerless: Economic Policy and Change in the Caribbean.* New York: Monthly Review Press.

Thomas, T. M. 1970. A detailed analysis of climatological and hydrological records of south Florida with reference to man's influence upon ecosystem evolution. Technical Report 70-2. University of Miami, Coral Gables, FL.

Thompson, J. D. 1967. *Organization in Action.* New York: McGraw-Hill.

Thompson, M. 1983. Postscript: a cultural basis for comparison. In H. C. Kunreuther and J. Linnerooth, eds., *Risk Analysis and Decision Processes,* pp. 232–262. Berlin: Springer-Verlag.

Throgmorton, J. A. 1991. The rhetorics of policy analysis. *Policy Sciences* 24(2):153–179.

Thurow, F. 1989. Fishery resources of the Baltic region. In A. H. Westing, ed., *Comprehensive Security for the Baltic: An Environmental Approach,* pp. 54–61. International Peace Research Institute, United Nations Environment Program. London: Sage.

Thurston, H. 1982. The enemy above. *Harrowsmith* 6:36–55.

Tietenberg, T. H. 1992. *Environmental and Natural Resource Economics.* 3rd Edition. New York: Harper Collins.

Tippie, V. K. 1984. An environmental characterization of Chesapeake Bay and a framework for action. In V. S. Kennedy, ed., *The Estuary as a Filter,* pp. 467–488. Orlando, FL: Academic Press.

Toth, L. A. 1993. Principles and guidelines for restoration of river/floodplain ecosystems—Kissimmee River, Florida. In J. Cairns, ed., *Rehabilitating Damaged Ecosystems.* Boca Raton, FL: CRC Press.

Toth, L. A. and N. G. Aumen. 1993. Integration of multiple issues in environmental restoration and resource enhancement projects in central south Florida. In J. Cairns, ed., *Implementing Environmental Management.* Chelsea, MI: Lewis Publishers.

Tothill, J. D. 1922a. Notes on the outbreaks of spruce budworm, forest tent caterpillar and larch sawfly in New Brunswick. *Proceedings Acadian Entomological Society* 8:172–182.

Tothill, J. D. 1922b. An estimate of the damage done in New Brunswick by spruce budworm. Entomology Laboratory, Department of Agriculture Canada, Fredericton. 4 pp.

Toulmin, S. 1972. *Human Understanding.* Princeton, NJ: Princeton University Press.

Treaty between the Government of the United States of America and the Government of Canada concerning Pacific salmon, Treaty Document No. 99–2 (entered in force, March 18, 1985).

Trist, E. 1983. Referent organizations and the development of interorganizational domains. *Human Relations* 36:269–284.

Tuchman, B. W. 1978. *A Distant Mirror.* New York: Ballantine Books.

Tushman, M. L. and E. Romanelli. 1985. Organizational evolution: A metamorphosis model of convergence and reorientation. *Research in Organizational Behavior* 7:171–122.

Tversky, A. and D. Kahneman. 1974. Judgment under uncertainty: Heuristics and biases. *Science* 185:1124–1131.

Tweeddale, R. E. 1974. Report of the Forest Resources Study. Government of New Brunswick, Fredericton. 362 pp.

Udehn, L. 1987. *Methodological Individualism: A Critical Appraisal.* Uppsala: Uppsala Universiteta.

United Nations Environmental Program. 1989. Basel convention on the control of transboundary movements of hazardous wastes and their disposal. United Nations, New York. 91 pp.

United States Army Corps of Engineers. 1990. Draft general design memorandum for modified water deliveries to Everglades National Park. District Office, Jacksonville.

United States Congress. 1911. Everglades of Florida: acts, reports and other papers, state and national, relating to the Everglades of the state of Florida and their reclamation. Senate Document 89, 62nd United States Congress, 1st session. Government Printing Office, Washington, DC

United States Congress. 1980. Public Law 96–501, 16 USC 839a et seq., 94 Stat. 2697–2736. Pacific Northwest Electric Power Planning and Conservation Act. Cited in the text as "Northwest Power Act." U.S. Government Printing Office, Washington, DC.

United States Department of Agriculture. 1954. Census of Agriculture. U.S. Government Printing Office, Washington, DC.

United States Department of Agriculture. 1974. Census of Agriculture. U.S. Government Printing Office, Washington, DC.

United States Department of Agriculture. 1987. Census of Agriculture. U.S. Government Printing Office, Washington, DC.

United States House of Representatives. 1948. House Document 643. U.S. Government Printing Office, Washington, DC.

Vallentyne, J. R., and A. Beeton. 1988. The ecosystem approach to managing human uses and abuses of natural resources in the Great Lakes Basin. *Environmental Conservation* 15(1):58–62.

VanDyne, D. and C. Gilbertson. 1978. *Estimating U.S. Livestock and Poultry*

Manure and Nutrient Production. U.S. Department of Agriculture, Washington, DC.

Van Voris, P., R. V. O'Neill, W. R. Emanuel, and H. H., Shugart. 1980. Functional complexity and ecosystem stability. *Ecology* 61:1352–1360.

Varty, I. W. 1975. Forest spraying and environmental integrity. *Forestry Chronicle* 51:12–15.

Vigiland, L., K. Hawkes, and A. C. Wilson. 1991. African populations and the evolution of human mitochondrial DNA. *Science* 253:1503–1507.

Vignoles, C. B. 1823. *Observations upon Floridas.* New York: Bliss and White.

Vitousek, P. M. and P. A. Matson. 1984. Mechanisms of nitrogen retention in forest ecosystems: A field experiment. *Science* 225:51–52.

Voigt, K. 1983. Baltic Sea: Pollution problems and natural environmental changes. *Impact of Science on Society* 33:413–420.

Voipio, A. 1981. *The Baltic Sea.* Amsterdam: Elsevier Science Publishing Company. 418 pp.

Volkman, J. M. 1992. Making room in the ark: The endangered species act and the Columbia River Basin. *Environment* 34(4):18–20, 37–43.

Vredenburg, H. and F. Westley. 1993. The creation and implementation of sustainable development strategies in Canadian ecosystems: A participative action research, paper presented at the Administrative Science Association of Canada Conference, Lake Louise, Alberta.

Vygotsky, L. S. 1978. *Mind in Society: The Development of Higher Psychological Processes.* Cambridge: Harvard University Press.

Waern, M. 1973. Nutrients and their influence on the algae in the Stockholm Archipelago during 1970. *Oikos* 15:153–154.

Waern, M. and S. Pekkari. 1973. Outflow studies. Nutrients and their influence on the algae in the Stockholm Archipelago during 1970. *Oikos* 15:155–170.

Wagner, J. I. and P. C. Rosendahl. 1987. History and development of water delivery schedules for Everglades National Park through 1982. Unpublished technical report. U.S. Department of Interior, National Park Service, South Florida Research Center, Homestead.

Waldrop, M. M. 1992. *Complexity.* New York: Simon and Schuster.

Walin, G. 1977. A theoretical framework for the description of estuaries. *Tellus* 29:128–136.

Walker, B. H., D. Ludwig, C. S. Holling, and R. M. Peterman. 1969. Stability of semi-arid savanna grazing systems. *Ecology* 69:473–498.

Walker, B. H. 1981. Is succession a viable concept in African savanna ecosystems? In D. C. West, H. H. Shugart, and D. B. Botkin, eds., *Forest Succession: Concepts and Application,* pp. 431–447. New York: Springer-Verlag.

Walker, B. H. and A. R. E. Sinclair. 1990. Problems of development aid. *Nature* 343:587.

Wallace, A. 1961. *Culture and Personality.* New York: Random House.

Wallace, A. 1966. *Religion: An Anthropological View.* New York: Random House.

Walters, C. J. 1986. *Adaptive Management of Renewable Resources.* New York: Macmillan.

Walters, C. J. and C. S. Holling. 1990. Large scale management experiments and learning by doing. *Ecology* 71(6):2060–2068.

Walters, C. J. and L. H. Gunderson. 1994. Screening water policy alternatives for ecological restoration in the Everglades. In S. M. Davis and J.C. Ogden, eds., *Everglades, the Ecosystem and Its Restoration.* Delray Beach, FL: St. Lucie Press.

Walters, C. J., L. H. Gunderson, and C. S. Holling. 1992. Experimental policies for water management in the Everglades. *Ecological Applications* 2(2):189–202.

Walter, C. J. and R. Hilborn. 1978. Ecological optimization and adaptive management. *Annual Review of Ecology and Systematics* 9:157–188.

Wang, E. C., T. Erdle, and T. Rousell. 1987. FORMAN wood supply model users manual. Vanguard Forest Management Services, Fredericton.

Ward, F. 1967. The imperiled Everglades. *National Geographic* 141:1–27.

Warner, W. 1976. *Beautiful Swimmers.* Boston: Little, Brown.

Water Resources Management Committee. 1990. Annual report of the Water Resources Management Committee to the governors and premiers of the Great Lakes states and provinces prepared at the request of the Council of Great Lakes Governors. Water Resources Management Committee, Chicago. 32 pp.

Waters, T. F. 1987. *The Superior North Shore: A Natural History of Lake Superior's Northern Lands and Waters.* Minneapolis: University of Minnesota Press.

Watson, R. S. 1984. New Brunswick's forest policy—facing the future. *Forestry Chronicle* 60:71–74.

Webb, F. E. and H. J. Irving. 1983. My fir lady—the New Brunswick production with its acts and fancies. *Forestry Chronicle* 59:118–122.

Webber, D. 1983. Obstacles to the utilization of systematic policy analysis. *Knowledge* 4:534–560.

Weber, M. 1922. *Economy and Society.* Berkeley: University of California Press.

Weick, K., 1991, The nontraditional quality of organizational learning. *Organization Science* 2(1):116–124.

Weinberg, A. M. 1972. Science and trans-science. *Minerva* 10:209–222.

Weiner, R., D. Selejeski, E. Quintero, S. Coon, and M. Walsch. 1993. Parasitic bacteria cue oyster larvae to set on fertile benthic biofilms. In R. Guerrero and C. Pedros-Alio, eds., *Trends in Microbial Ecology.* Barcelona: Spanish Society for Microbiology.

Weiss, C. H. 1977. Research for policy's sake: The enlightenment function of social research. *Policy Analysis* 3:531–545.

Weiss, C. H. 1982. Policy research in the context of diffuse decision-making. *Policy Studies Review Annual* 6:19–36.

Weller, P. 1989. *Natural Heritage Areas and Programs in the U.S. Great Lakes States.* Unpublished report to the International Joint Commission, Science Advisory Board. Great Lakes Office, International Joint Commission, Windsor, Ontario.

Weller, P. 1990. *Fresh Water Seas: Saving the Great Lakes.* Toronto: Between the Lines Press.

Wells, L. and A. L. McLain. 1973. Lake Michigan: Man's effects on native fish stocks and other biota. Technical Report No. 20. Great Lakes Fishery Commission, Ann Arbor, MI. 55 pp.

Wertsch, J. 1985a. *Vygotsky and the Social Formation of Mind.* Cambridge: Harvard University Press.

Wertsch, J. 1985b. *Culture, Communication and Cognition: Vygotskian Perspectives.* New York: Cambridge University Press.

Wertsch, J. 1991. *Voices of the Mind: a Sociocultural Approach to Mediated Action.* Cambridge: Harvard University Press.

West, D. C., H. H. Shugart, and D. B. Botkin. 1981. *Forest Succession: Concepts and Application.* New York: Springer-Verlag.

Westley, F. 1977. Searching for surrender. *American Behavioral Science* 20(6):925–940.

Westley, F. 1990. Middle managers and strategy: Microdynamics of inclusion. *Strategic Management Journal* 11:337–352.

Westley, F. 1991. Bob Geldof and Live Aid: The effective side of global social innovation. *Human Relations* 44(10):1011–1036.

Westley, F. 1992. Vision worlds: Strategic vision as social interaction. *Advances in Strategic Management* 8:271–305.

Westley, F. and H. Mintzberg. 1989. Strategic management and visionary leadership. *Strategic Management Journal* 10:17–32.

Westley, F. and H. Vredenburg. 1991a. Strategic bridging: The alliances of business and environmentalists. *Journal of Applied Behavioral Science* 27(1):65–90.

Westley, F. and H. Vredenburg. 1991b. Three models of interorganizing. Paper presented at the Society for Strategic Management. Toronto, October.

Westley, F. and H. Vredenburg. 1992. Managing the Ark: Interorganizational collaboration and the preservation of biodiversity. McGill Working Paper No. 92-11-16, McGill University, Montreal.

Wiens, J. A. 1990. On the use of "grain size" in ecology. *Functional Ecology* 4:720.

Wildavsky, A. B. 1979. *Speaking Truth to Power: The Art and Craft of Policy Analysis.* Boston: Little, Brown. 439 pp.

Wildavsky, A. B. and M. Douglas. 1982. *Risk in Culture: An Essay on the Selection of Technical and Environmental Dangers.* Berkeley: University of California.

Wilensky, H. 1967. *Organizational Intelligence.* New York: Basic Books.

Wilkinson, C. F. and D. K. Conner. 1983. The law of the Pacific salmon fishery. *University of Kansas Law Review* 32:17–109.

Williamson, O. E. 1975. *Markets and Hierarchies: Analysis and Antitrust Implications.* New York: The Free Press.

Williamson, O. E. 1985. *The Economic Institutions of Capitalism.* New York: The Free Press.

Willoughby, H. L. 1898. *Across the Everglades.* Philadelphia: J. B. Lippincott. 192 pp.

Willoughby, W. R. 1961. *The St. Lawrence Waterway: A Study in Politics and Diplomacy.* Madison: University of Wisconsin Press.

Wilson, E. O. 1978. *On Human Nature.* Cambridge: Harvard University Press.

Wolin, Sheldon. 1960. *Politics and Vision: Continuity and Innovation in Western Political Thought.* Boston: Little, Brown.

World Bank. 1992. *World Development Report: Development and the Environment.* Oxford: Oxford University Press.

World Commission on Environment and Development. 1987. *Our Common Future.* Oxford: Oxford University Press.

World Conservation Monitoring Centre. 1992. *Global Biodiversity: Status of the Earth's Living Resources.* London: Chapman and Hall.

Worster, D. 1985. *Rivers of Empire: Water, Aridity and the Growth of the American West.* New York: Pantheon.

Worth, D. 1987. Environmental responses of water conservation area 2A to reduction in regulation schedule and marsh drawdown. Technical publication 88-2, South Florida Water Management District, West Palm Beach.

Wren, T. E. 1982. Social learning theory, self-regulation and morality. *Ethics* 92:409–424.

Wright, J. O. 1909. Extract from a report on the drainage of the Everglades of Florida. U.S. Department of Agriculture, Office of Experiment Station. Capital Publications, Tallahassee.

Wright, J. V. 1972. *Ontario Prehistory: An Eleven-Thousand-Year Archaeological Outline.* National Museum of Man, Ottawa.

Wulff, F. 1990. Large-scale environmental effects and ecological processes in the Baltic Sea. Research program for the period 1990–1995 and background documents. Swedish Environmental Protection Agency Report 3849, Stockholm. 225 pp.

Wulff, F. and Ạ. Niemi. 1992. Priorities for the restoration of the Baltic Sea—a scientific perspective. *Ambio* 21:193–195.

Wulff, F., L. Rahm, P. Jonsson, L. Brydsten, T. Ahl, and Ạ. Granmo. 1993. A mass-balance model of chlorinated organic matter for the Baltic Sea—a challenge for ecotoxicology. *Ambio* 22:27–31.

Wulff, F., A. Stigebrandt, and L. Rahm. 1990. Nutrient dynamics of the Baltic Sea. *Ambio* 19:126–133.

Wynne, B. 1976. C. G. Barkla and the J Phenomenon: A case study of the treatment of deviance in physics. *Social Studies of Science* 6:307–347.

Yates, S. 1983. Florida's broken rain machine. *The Amicus Journal* Fall:48–55.

Yee, P., R. Edgett, and A. Eberhardt. 1990. Great Lakes–St. Lawrence river regulation: What it means and how it works. Environment Canada, Ontario Region, Burlington, and U.S. Army Corps of Engineers, North Central Division, Chicago.

Young, R. C. 1988. Is population ecology a useful paradigm for the study of organizations? *American Journal of Sociology* 94(1):1–24.

Zald, M. and J. D. McCarthy. 1987. *Social Movements in an Organizational Society.* New Brunswick, NJ: Transaction Books.

Zaleski, J. and C. Wojewòdka. 1972. *Europa Baltyka* (Baltic Europe). Warszawa: Ossolineum. (In Polish, summary in English, Russian and German.)

Index